Traffic and Granular Flow '11

AF181683

Valery V. Kozlov • Alexander P. Buslaev
Alexander S. Bugaev • Marina V. Yashina
Andreas Schadschneider • Michael Schreckenberg
Editors

Traffic and Granular Flow '11

Springer

Editors
Valery V. Kozlov
Steklov Mathematical Institute
Russian Academy of Sciences
Moscow, Russia

Alexander P. Buslaev
Moscow State Automobile
Road Technical University
Moscow, Russia

Alexander S. Bugaev
Electronics Institute and
Kotel'nikov Radioengineeering
Russian Academy of Sciences
Moscow, Russia

Marina V. Yashina
Mathematical Cybernetics Department
Moscow Technical University
of Communications and Informatics
Moscow, Russia

Andreas Schadschneider
Institut für Theoretische Physik
Universität zu Köln
Köln, Germany

Michael Schreckenberg
Physik von Transport und Verkehr
Universität Duisburg-Essen
Duisburg, Germany

ISBN 978-3-662-52172-4 ISBN 978-3-642-39669-4 (eBook)
DOI 10.1007/978-3-642-39669-4
Springer Heidelberg New York Dordrecht London

Mathematics Subject Classification (2010): 00B20, 60G99, 68U20, 70K99, 82C22, 82C31, 82D99, 91F99

© Springer-Verlag Berlin Heidelberg 2013
Softcover reprint of the hardcover 1st edition 2013
This work is subject to copyright. All rights are reserved by the Publisher, whether the whole or part of the material is concerned, specifically the rights of translation, reprinting, reuse of illustrations, recitation, broadcasting, reproduction on microfilms or in any other physical way, and transmission or information storage and retrieval, electronic adaptation, computer software, or by similar or dissimilar methodology now known or hereafter developed. Exempted from this legal reservation are brief excerpts in connection with reviews or scholarly analysis or material supplied specifically for the purpose of being entered and executed on a computer system, for exclusive use by the purchaser of the work. Duplication of this publication or parts thereof is permitted only under the provisions of the Copyright Law of the Publisher's location, in its current version, and permission for use must always be obtained from Springer. Permissions for use may be obtained through RightsLink at the Copyright Clearance Center. Violations are liable to prosecution under the respective Copyright Law.
The use of general descriptive names, registered names, trademarks, service marks, etc. in this publication does not imply, even in the absence of a specific statement, that such names are exempt from the relevant protective laws and regulations and therefore free for general use.
While the advice and information in this book are believed to be true and accurate at the date of publication, neither the authors nor the editors nor the publisher can accept any legal responsibility for any errors or omissions that may be made. The publisher makes no warranty, express or implied, with respect to the material contained herein.

Printed on acid-free paper

Springer is part of Springer Science+Business Media (www.springer.com)

Preface

The conference series *Traffic and Granular Flow* has been established in 1995 when the first conference was held in Jülich, Germany. It brought together experts interested in complex transport processes which appear to be rather different at first sight. Granular flow is fully determined by the laws of physics whereas in vehicular traffic the behavior of human drivers plays a decisive role which makes non-physical aspects like psychology relevant.

The first meeting was a big success, as were the following meetings which were held bi-annually in Duisburg (1997), Stuttgart (1999), Nagoya (2001), Delft (2003), Berlin (2005), Paris (2007) and Shanghai (2009). In this time new focus areas have emerged, like pedestrian and evacuation dynamics or transport in biological systems and computer networks. Experts from all over the world with rather different backgrounds ranging from engineering to physics and computer science presented their newest results. Besides this strong interdisciplinarity the full range of scientific tools has always been represented with contributions of experimental, empirical, theoretical and mathematical nature.

The conference *Traffic and Granular Flow '11 (TGF'11)* was held from September 28 to October 1, 2011 in Moscow at the Moscow Technical University of Communications and Informatics (MTUCI). It was attended by around 150 people and 28 research groups from Russia and 28 foreign groups have presented their works. Despite some difficulties with communications, primarily with countries in Asia and America, the vast majority of registered participants took part in the TGF'11. In these proceedings we also present the papers and works of those who could not attend, e.g. due visa problems.

The conference was characterized by the increased use of software and hardware means of monitoring and simulating transport processes, from urban and highway traffic and pedestrian crowds to the flow of granular media. Because of this, today it is possible to address many problems of calibration and validation of model approaches. This is most systematically developed in the works of scientists from

Germany. Another feature of many presentations was the importance of probabilistic approaches for the modeling of socio-technical systems (including flows).

However, most of the presented works used a different strategy: the formulation of the model, the choice of calculation methods and the illustration of the proposed approach by investigation of some transport processes scenarios. In many presentations at the conference the dynamics of processes was illustrated by computer animation. A significant problem, the solution of which at the moment is vague at best, is the transition from local modeling and simulation for simple structures (i.e. rings or segments) to the modeling and management of e.g. urban road networks. In this case the level of detail describing a particular part of the network is bound to be reduced while including the key links. In some cases, at the moment by schematic examples, general trends in the development of the network load with the driver awareness taken into account were established.

Simulation of pedestrian flows, which began much later, shows a rapid development and increasing exchange of ideas with the modeling of traffic flows. The processing of empirical data and experimental study of the phenomenon makes a significant contribution to it. In a number of works of the pedestrian community, self-organization phenomena and their origin are intensively studied. This experience should be transferred to the vehicular traffic community.

The TGF'11 conference has also successfully integrated to its ranks researchers from countries with developing economies, for which the traffic problems have dramatically increased over a short time. It is certain that complex research problems of socio-technical systems, such as traffic and other systems with the presence of the human factor, must be solved together.

The conference was held with financial support of the Russian Foundation for Basic Research (RFBR) under the direct supervision of the Russian Academy of Sciences (RAS). The organizing committee would like to thank Academician Boris N. Chetverushkin, member of Russian Academy of Sciences, Professor, Doctor of Physical and Mathematical Sciences, director of Keldysh Institute of Applied Mathematics of the Russian Academy of Sciences and Prof. Artem S. Adzhemov, Doctor of Technical Sciences, rector of the Moscow Technical University of Communications and Informatics for their great support and invaluable contribution to organization of the conference. Also we would like to thank Leonid M. Lipsitz, Head of State Traffic Policy Department, Ministry of Transport of Russian Federation, for active participation and the methodological support of the conference.

We hope that the next conference *Traffic and Granular Flow '13* will continue to facilitate the scientific exchange in the field of complex transport problems. It will not only help to increase our theoretical understanding of traffic-like systems, but also provide solutions to problems of everyday life.

For the anniversary of its 10th edition the conference will return to the location of the inaugural meeting, the Research Center in Jülich, Germany, from September 25–27, 2013.

Moscow, Köln, Duisburg
May 2013

Valery V. Kozlov
Alexander P. Buslaev
Alexander S. Bugaev
Marina V. Yashina
Andreas Schadschneider
Michael Schreckenberg

Contents

Contributors

Larisa Afanasyeva Lomonosov Moscow State University, Moscow, Russia

Madani Ammi Institut de Physique de Rennes, UMR U.Rennes1-CNRS 6251, Université de Rennes 1, Rennes CEDEX, France

Antonia Pérez Arias Intelligent Sensor-Actuator-Systems Laboratory (ISAS), Karlsruhe Institute of Technology (KIT), Karlsruhe, Germany

Jens Boberski Department of Theoretical Physics, University of Duisburg-Essen, Duisburg, Germany

Martins Brics Institute of Physics, Rostock University, Rostock, Germany

Ekaterina Bulinskaya Lomonosov Moscow State University, Moscow, Russia

Mario Campanella Department of Transport & Planning, Delft University of Technology, Delft, The Netherlands

Armando Carrillo Mathematics Department, Universidad de Sonora, Hermosillo, Sonora, México

Hilmi Berk Celikoglu Civil Engineering Department, Istanbul Technical University, Istanbul, Turkey

Partha Chakroborty Department of Civil Engineering, Indian Institute of Technology Kanpur, Kanpur, India

Ujjal Chattaraj Department of Civil Engineering, National Institute of Technology Rourkela, Rourkela, India

Yury V. Chekhovich Computing Center of the Russian Academy of Science, Moscow, Russia

Bokui Chen Department of Modern Physics, University of Science and Technology of China, Hefei, China

Gregory Cherniak JSC COS&HT, Dolgoprudny, Moscow Region, Russia

Mohcine Chraibi Jülich Supercomputing Centre, Forschungszentrum Jülich GmbH, Jülich, Germany

Computer Simulation for Fire Safety and Pedestrian Traffic, Bergische Universität Wuppertal, Wuppertal, Germany

Natalia Churbanova Keldysh Institute of Applied Mathematics RAS, Moscow, Russia

Winnie Daamen Department of Transport & Planning, Delft University of Technology, Delft, The Netherlands

Maria Davidich Siemens AG, München, Germany

F. Decoupigny Laboratoire Espace, UMR CNRS 6012 and University of Nice, Nice Cedex 3, France

Joaquín Delgado Departamento de Matemáticas, Universidad Autónoma Metropolitana Iztapalapa, México, D.F., México

Onur Deniz Civil Engineering Department, Istanbul Technical University, Istanbul, Turkey

Christian Eilhardt Institut für Theoretische Physik, Universität zu Köln, Köln, Germany

Gunnar Flötteröd KTH Royal Institute of Technology, Stockholm, Sweden

Yun-Pang Flötteröd DLR German Aerospace Center, Cologne, Germany

Martina Freialdenhoven Jülich Supercomputing Centre, Forschungszentrum Jülich GmbH, Jülich, Germany

Simon Friedberger Intelligent Sensor-Actuator-Systems Laboratory (ISAS), Karlsruhe Institute of Technology (KIT), Karlsruhe, Germany

Ilya Furmanov Keldysh Institute of Applied Mathematics RAS, Moscow, Russia

Cyril Furtlehner INRIA Saclay – LRI, Bat. 490, Université Paris-Sud, Orsay cedex, France

Paula A. Gago Instituto de Física de Líquidos y Sistemas Biológicos (CONICET La Plata, UNLP), La Plata, Argentina

E. V. Gasnikova Faculty of Applied Mathematics and Control, Moscow Institute of Physics and Technology (State University), Moskovskaya oblast, Russia

Konstantin K. Glukharev Institute of Machines Science named after A.A. Blagonravov of RAS, Moscow, Russia

Moscow Institute of Physics and Technology (State University), Dolgoprudny, Moscow Region, Russia

Jorge D. González Departamento de Matemáticas, Universidad Autónoma Metropolitana Iztapalapa, México, D.F., México

Sh. E. Guseynov Institute of Mathematical Sciences and Information Technologies, University of Liepaja, Liepaja, Latvia

S. M. V. Gwynne Hughes Associates, UK

Fire Safety Engineering Group, University of Greenwich, Greenwich, UK

Habib Haj-Salem IFSTTAR, GRETTIA, Noisy le Grand Cedex, France

Uwe D. Hanebeck Intelligent Sensor-Actuator-Systems Laboratory (ISAS), Karlsruhe Institute of Technology (KIT), Karlsruhe, Germany

Stefan Hengst PTV Planung Transport Verkehr AG, Karlsruhe, Germany

Serge P. Hoogendoorn Department of Transport & Planning, Delft University of Technology, Delft, The Netherlands

Takashi Imamura Research Center for Advanced Science and Technology, The University of Tokyo, Komaba, Meguro-ku, Tokyo 153-8904, Japan

Nikita P. Ivkin Moscow Institute of Physics and Technology (State University), Dolgoprudny, Moscow Region, Russia

Mayuko Iwamoto Graduate School of Advanced Mathematical Sciences, Meiji University, Nakano-ku, Tokyo, Japan

Ivan N. Kalinin Institute of Machines Science named after A.A. Blagonravov of RAS, Moscow, Russia

Moscow Institute of Physics and Technology (State University), Dolgoprudny, Moscow Region, Russia

J. Kaupužs Institute of Mathematical Sciences and Information Technologies, University of Liepaja, Liepaja, Latvia

Armel Ulrich Kemloh Wagoum Jülich Supercomputing Centre, Forschungszentrum Jülich GmbH, Jülich, Germany

W. Klingsch Institute for Building Material Technology and Fire Safety Science, Wuppertal University, Wuppertal, Germany

Victor L. Knoop Department of Transport & Planning, Delft University of Technology, Delft, The Netherlands

Gerta Köster University of Applied Sciences, München, Germany

Tobias Kretz PTV Planung Transport Verkehr AG, Karlsruhe, Germany

Reinhart Kühne German Aerospace Center, Institute for Transportation Research, Berlin, Germany

Jean-Marc Lasgouttes INRIA Paris–Rocquencourt, Le Chesnay cedex, France

Sylvain Lassarre IFSTTAR – GRETTIA, Noisy le Grand, France

Jean-Patrick Lebacque IFSTTAR, GRETTIA, Noisy le Grand Cedex, France

Reinhard Mahnke Institute of Physics, Rostock University, Rostock, Germany

G. Maignant Laboratoire Espace, UMR CNRS 6012 and University of Nice, Nice Cedex 3, France

Maniya Maleki Department of Physics, Institute for Advanced Studies in Basic Sciences (IASBS), Zanjan, Iran

Vadim Malyshev Faculty of Mechanics and Mathematics, Lomonosov Moscow State University, Moscow, Russia

Anatoly Manita Faculty of Mechanics and Mathematics, Lomonosov Moscow State University, Moscow, Russia

Tsubasa Masui Graduate School of Science and Technology, Meiji University, Tama-ku, Kawasaki, Kanagawa, Japan

Robabeh Moosavi Department of Physics, Institute for Advanced Studies in Basic Sciences (IASBS), Zanjan, Iran

T. A. Nagapetyan Faculty of Applied Mathematics and Control, Moscow Institute of Physics and Technology (State University), Moskovskaya oblast, Russia

Katsuhiro Nishinari Research Center for Advanced Science and Technology, The University of Tokyo, Komaba, Meguro-ku, Tokyo, Japan

B. Nkonga Laboratoire Dieudonné, UMR CNRS 6621 and University of Nice, Nice Cedex 2, France

Stefan Nowak Institut für Theoretische Physik, Universität zu Köln, Köln, Germany

Luc Oger Institut de Physique de Rennes, UMR U.Rennes1-CNRS 6251, Université de Rennes 1, Rennes CEDEX, France

Daniel R. Parisi CONICET and Instituto Tecnológico de Buenos Aires, C. A. de Buenos Aires, Argentina

Luis A. Pugnaloni Instituto de Física de Líquidos y Sistemas Biológicos (CONICET La Plata, UNLP), La Plata, Argentina

Michel Rascle Laboratoire Dieudonné, UMR CNRS 6621 and University of Nice, Nice Cedex 2, France

J. Rimšāns Institute of Mathematical Sciences and Information Technologies, University of Liepaja, Liepaja, Latvia

Christian Rogsch Fire Protection Engineering, Bonhoefferstr. 16, D-67435 Neustadt an der Weinstraße, Germany

Michel Roussignol Paris-Est University – LAMA, Marne-la-Vallée, France

Heather J. Ruskin Dublin City University, Dublin 9, Ireland

Patricia Saavedra Departamento de Matemáticas, Universidad Autónoma Metropolitana Iztapalapa, México, D.F., México

Maxim Samsonov INRIA Saclay – LRI, Bat. 490, Université Paris-Sud, Orsay cedex, France

M. Luisa Sandoval Departamento de Matemáticas, Universidad Autónoma Metropolitana Iztapalapa, México, D.F., México

Tibye Saumtally Université Paris Est, GRETTIA, Noisy le Grand Cedex, France

Andreas Schadschneider Institut für Theoretische Physik, Universität zu Köln, Köln, Germany

M. Reza Shaebani Department of Theoretical Physics, University of Duisburg-Essen, Duisburg, Germany

Armin Seyfried Jülich Supercomputing Centre, Forschungszentrum Jülich GmbH, Jülich, Germany

Computer Simulation for Fire Safety and Pedestrian Traffic, Bergische Universität Wuppertal, Wuppertal, Germany

A. A. Siddiqui Fire Safety Engineering Group, University of Greenwich, Greenwich, UK

Akiyasu Tomoeda Meiji Institute for Advanced Study of Mathematical Sciences, Meiji University, Nakano-ku, Tokyo, Japan

CREST, Japan Science and Technology Agency, Nakano-ku, Tokyo, Japan

Wei Tong Department of Modern Physics, University of Science and Technology of China, Hefei, China

Antoine Tordeux Paris-Est University – LVMT, Marne-la-Vallée, France

János Török Department of Theoretical Physics, University of Duisburg-Essen, Duisburg, Germany

Marina Trapeznikova Keldysh Institute of Applied Mathematics RAS, Moscow, Russia

Daishin Ueyama Meiji Institute for Advanced Study of Mathematical Sciences, Meiji University, Nakano-ku, Tokyo, Japan

Graduate School of Advanced Mathematical Sciences, Meiji University, Nakano-ku, Tokyo, Japan

Nikolay M. Ulyukov Institute of Machines Science named after A.A. Blagonravov of RAS, Moscow, Russia

Moscow Institute of Physics and Technology (State University), Dolgoprudny, Moscow Region, Russia

Alexandre Valance Institut de Physique de Rennes, UMR U.Rennes1-CNRS 6251, Université de Rennes 1, Rennes CEDEX, France

Andrey M. Valuev Institute of Machines Science named after A.A. Blagonravov of RAS, Moscow, Russia

Moscow Institute of Physics and Technology (State University), Dolgoprudny, Moscow Region, Russia

Hans van Lint Department of Transport & Planning, Delft University of Technology, Delft, The Netherlands

Jelena Vasic Dublin City University, Dublin 9, Ireland

Rosa Maria Velasco Physics Department, UAM–Iztapalapa, México, D.F., México

Fernando Verduzco Mathematics Department, Universidad de Sonora, Hermosillo, Sonora, México

A. A. Vinogradov Moscow Technical University of Communications and Informatics (MTUCI), Moscow, Russia

Peter Wagner DLR German Aerospace Center, Cologne, Germany

Binghong Wang Department of Modern Physics, University of Science and Technology of China, Hefei, China

The Research Center for Complex System Science, University of Shanghai for Science and Technology and Shanghai Academy of System Science, Shanghai, China

Hans Weber Department of Physics, Luleå University of Technology, Luleå, Sweden

Dietrich E. Wolf Department of Theoretical Physics, University of Duisburg-Essen, Duisburg, Germany

Alexander A. Yakukhnov Moscow Institute of Physics and Technology (State University), Dolgoprudny, Moscow Region, Russia

Daichi Yanagisawa College of Science, Ibaraki University, Bunkyo, Mito, Ibaraki, Japan

M. V. Yashina Moscow Technical University of Communications and Informatics (MTUCI), Moscow, Russia

Andrei Zamyatin Faculty of Mechanics and Mathematics, Lomonosov Moscow State University, Moscow, Russia

Jun Zhang Institute for Building Material Technology and Fire Safety Science, Wuppertal University, Wuppertal, Germany

Wenyao Zhang Department of Modern Physics, University of Science and Technology of China, Hefei, China

Part I
Highway and Urban Vehicular Traffic

The Fundamental Diagram on the Ring Geometry for Particle Processes with Acceleration/Braking Asymmetry

Cyril Furtlehner, Jean-Marc Lasgouttes, and Maxim Samsonov

Abstract The slow-to-start mechanism is known to play an important role in the particular shape of the fundamental diagram of traffic and to be associated to hysteresis effects of traffic flow. We study this question in the context of stochastic processes, namely exclusion and queueing processes, by including explicitly an asymmetry between deceleration and acceleration in their formulation. Spatial condensation phenomena and metastability are observed, depending on the level of the aforementioned asymmetry. The relationship between these two families of models is analyzed on the ring geometry, to yield a large deviations formulation of the fundamental diagram (FD).

1 Introduction

In the microscopic theory of traffic [1, 2], an asymmetry between acceleration and deceleration, observed for example in the headway distribution, is empirically known to be responsible for the way spontaneous congestion occurs, as can be seen experimentally on a ring geometry for example [3]. This is often referred as the slow-to-start mechanism, not present in the original cellular automaton of Nagel-Schreckenberg [4], but added in refined versions like the "velocity dependent randomized" one (VDR) [5], which exhibits a first order phase transition between the fluid and the congested phases and hysteresis phenomena [6] associated to metastable states. This is partly reflected in the shape of the FD. In this work, we analyze this property in the context of stochastic processes. We propose a

C. Furtlehner (✉) · M. Samsonov
INRIA Saclay – LRI, Bat. 490, Université Paris-Sud – 91405, Orsay cedex, France
e-mail: cyril.furtlehner@inria.fr

J.-M. Lasgouttes
INRIA Paris–Rocquencourt – Domaine de Voluceau B.P. 105 – 78153, Le Chesnay cedex, France
e-mail: jean-marc.lasgouttes@inria.fr

V.V. Kozlov et al. (eds.), *Traffic and Granular Flow '11*,
DOI 10.1007/978-3-642-39669-4_1, © Springer-Verlag Berlin Heidelberg 2013

minimal extension of TASEP to include that mechanism, and a way to compute the corresponding FD on the ring geometry, where

- Some equivalence between particle systems and queueing processes holds, either exactly or with good approximation, and can be exploited to compute the FD;
- The relation between the density of vehicles and the flux is a well defined quantity in the thermodynamic limit on this geometry, since the number of cars is conserved (canonical ensemble).

The paper is organized as follows: in Sect. 2, we define a family of exclusion processes relevant to traffic and we discuss how these can be reformulated in terms of a tandem queue process with a dynamically coupled stochastic service rate. The invariant measure of such a process at steady-state is determined in Sect. 3. Finally, in Sect. 4, we discuss the computation of the FD in the canonical ensemble (with a fixed number of cars), when a product form of the equivalent queueing process holds.

2 Multi-speed Exclusion Processes

2.1 Model Definition

In the totally asymmetric version of the exclusion process (TASEP), particles move randomly on a 1-d lattice, always in the same direction, hopping from one site to the next at times following a Poisson process and under condition that the next site is vacant. In the Nagel-Schreckenberg cellular automaton, the dynamics is parallel and the vehicles' speeds are encoded in the number of slots that they can cover in one jump. This speed can adapt stochastically, depending on the available space in front of the particle. We propose to combine the braking and accelerating features of the Nagel-Schreckenberg models, with the locality of the simple ASEP model, in which only two consecutive sites do interact at a given time. For this, we allow each particle to change stochastically its hopping rate, depending on the state of the next site. For a 2-speed model, let A (resp. B) denote a site occupied by a fast (resp. slow) vehicle, and let O denote empty sites. The model is defined by the following set of reactions, involving pairs of neighboring sites:

$$AO \xrightarrow{\mu_a} OA \qquad \text{simple move of fast vehicle,} \tag{1}$$

$$BO \xrightarrow{\mu_b} OB \qquad \text{simple move of slow vehicle,} \tag{2}$$

$$BO \xrightarrow{\gamma} AO \qquad \text{slow vehicle spontaneously accelerates,} \tag{3}$$

$$A\bar{O} \xrightarrow{\delta} B\bar{O} \qquad \text{fast vehicle brakes behind another one } (\bar{O} = A \text{ or } B). \tag{4}$$

Here μ_a, μ_b, γ and δ denote the transition rates. The first couple of transitions corresponds to normal movements of vehicles. The other two encode the fact that a slow vehicle tends to accelerate when there is space ahead (3), while in the opposite case (4), it tends slow down. The asymmetry between braking and acceleration is explicitly present in the model with γ different from δ. Our model can be viewed as the sequential version of the Appert-Santen model [7], in which there is one single speed, but particles have two states (at rest and moving). In the remainder of the paper, we will consider this model on the ring geometry.

Let us first notice that this model contains and generalizes several sub-models which are known to be integrable with particular rates. The hopping part (1) and (2) is just the totally asymmetric exclusion process [8, 9] (TASEP) when $\mu_a = \mu_b$, which is known to be integrable. The acceleration/deceleration dynamics is equivalent to the coagulation/decoagulation models, which are known to be solvable by the empty interval method and by free fermions for particular sets of rates [10], but the whole process is presumably not integrable.

2.2 *Relation to Tandem Queues*

In some cases, the model can be exactly reformulated in terms of generalized queueing processes (or zero range processes in the statistical physics parlance), where the service rate of each server follows itself a stochastic dynamics [11]. In this preceding work, however, we considered exclusion processes involving three consecutive sites interactions. In fact, in the present case, a large family of sub-models can be mapped onto such queueing processes. The mapping works only on the ring geometry, by identifying servers either with

(i) Cars: clients are the empty sites;
(ii) Empty sites: clients are the vehicles.

In our case, the mapping of type (i) is exact. In the corresponding queueing process, servers are associated either with fast or slow cars, having then service rates μ_a or μ_b. Slow queues become fast at rate γ, conditionally to having at least one client, while empty fast queues become slow at rate δ.

The mapping of type (ii) is more informative with respect to jam distribution [11], but is not possible with transitions limited to 2-consecutive sites interactions, because in that case homogeneity is not maintained in the clusters, and information beyond the number of cars and the rate of the car leaving the queue is needed to know the service rate of the queue. Nevertheless, to be able to get meaningful information on the jam structure, i.e. on the long range correlations of the model associated to cluster formations, this is the mapping that we will consider later in Sect. 3.

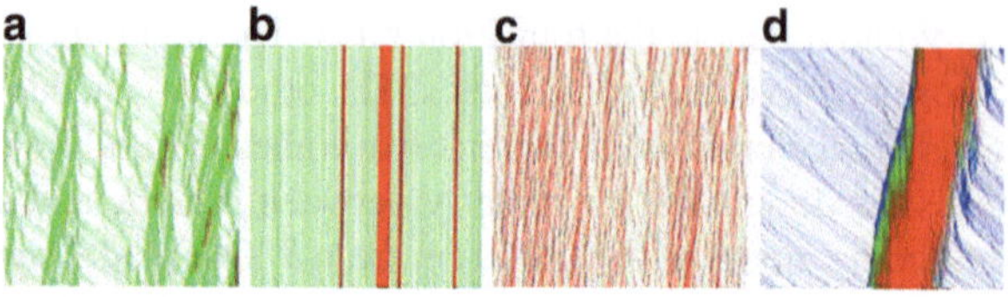

Fig. 1 Space-time diagram for multi-speed exclusion process with two speed levels (**a**)–(**c**) and three in (**d**). Time is going downward and particles to the *right*. *Red*, *green* and *blue* represent different speeds in increasing order. The size of the system is 3,000 except for (**b**) where it is 100,000. Setting are $\mu_a = 100$, $\mu_b = 10$, $\gamma = 100$, $\delta = 2$ for (**a**) and (**b**) and $\delta = 10$ for (**c**), all with density $d = 0.2$

2.3 Numerical Observations

Some numerical observations are illustrated on Fig. 1. When no asymmetry between braking and accelerating is present ($\gamma = \delta$), as in TASEP on a ring, no spontaneous jam formation occurs. As the density d of cars increases, one observes a smooth transition between a TASEP of fast particles for small d to a TASEP of slow particles around $d \simeq 1$. When the ratio γ/δ is reduced, there is a proliferation of small jams associated with slow vehicles as seen on Fig. 1c. Instead when it increases slow car are less present as well as small jams as shown on Fig. 1a. Above some threshold of this ratio γ/δ, which depends on the density, a condensation phenomenon occurs as seen on Fig. 1b; one or more large jams contain a finite fraction of the total number of cars. It is therefore tempting to understand this phase separation with the ZRP interpretation [12] of a spatial condensation phenomena in the canonical ensemble [13]. In this viewpoint, we expect a condensation mechanism to occur if the apparition of slow vehicles is a sufficiently rare event [11]. In our context, the probability that a vehicle joining a cluster of particles is still of type A when it reaches the front is decaying exponentially

$$p(\mu = \mu_a) = e^{-\delta t(s)} = p_s,$$

with $t(s)$ the waiting time in a queue of size s. Assuming stationarity and the possible acceleration of the front car, we obtain self-consistently

$$t(s) = s\Big[\frac{1}{\mu_a}p_s + \frac{\mu_a + \gamma}{\mu_a(\mu_b + \gamma)}(1 - p_s)\Big].$$

This observation will be made more precise in Sect. 3. It implies qualitatively that starting with a distribution of jams with different effective service rates, as time evolves, long-lived jams with effective service rates slightly above $\mu_b + \gamma(\mu_a - \mu_b)/(\mu_a + \gamma)$ are able to survive. If the situation with a single jam is not stable, which is observed below some threshold density, then no large jam may develop at all, and only small fluctuations are to be observed. Typically this occurs when

$\delta \ll \mu_b$ and $\delta < \gamma$ (Fig. 1b). In contrary when $\delta > \gamma$ one likely observes the kind of jams in Fig. 1c.

We have also simulated a model with three speed levels, see Fig. 1d. In that case, small jams may have different speeds, depending on which type of slow car is leading. Then a cascade mechanism takes place, slow speed regions generate even slower speed clusters of cars and so on, and some kind of synchronized flow is observed.

3 Queues with Two-State Service Rates

The mapping of type (ii) defined in Sect. 2.2, while not exact here, may help us to capture the correlations associated to cluster formations. To this end, consider first the dynamics and steady state regime of a single cluster of vehicles of size n_t and assume that vehicles of type A [resp. B] join the queue with rate λ_a [resp. λ_b], while they leave the queue with rate μ_a [resp. μ_b]. $\lambda \stackrel{\text{def}}{=} \lambda_a + \lambda_b$ represents the intensity of the incoming process, and particles decelerate with rate δ in the bulk of the queue. At position n in the queue, counting from the back end position, the probability for a particle to be still of type A is then

$$p_n \stackrel{\text{def}}{=} \frac{\lambda_a}{\lambda} r^n \qquad \text{with} \qquad r \stackrel{\text{def}}{=} \frac{\lambda}{\lambda + \delta}.$$

Since the front end interface of the cluster has no causal effect on the rest of the queue, except on the front vehicle which may accelerate with rate γ, we can consider the dynamics of the sequence independently of the motion of the front interface. Under the additional assumption of independence between speed labels in the bulk of the queue, we consider the joint probability $P_t(n, \tau) = \mathbb{P}(n_t = n, \mu_t = \mu_a \tau + \mu_b \bar{\tau})$, i.e. the joint probability that the queue has n clients and its front car is of type A ($\tau = 1$) or B ($\bar{\tau} \stackrel{\text{def}}{=} 1 - \tau = 1$). With $p_n(\tau) \stackrel{\text{def}}{=} p_n \tau + (1 - p_n)\bar{\tau}$, the master equation then reads

$$\begin{aligned}
\frac{dP_t(n,\tau)}{dt} =& \lambda\big(P_t(n-1,\tau) - P_t(n,\tau)\big) + \big(\mu_a P_t(n+1,1) + \mu_b P_t(n+1,0)\big) p_n(\tau) \\
&- (\mu_a \tau + \mu_b \bar{\tau}) P_t(n,\tau) + \gamma(\tau - \bar{\tau}) P_t(n,0), \qquad n \geq 2 \\
\frac{dP_t(1,\tau)}{dt} =& (\lambda_a \tau + \lambda_b \bar{\tau}) P_t(0) - \lambda P_t(1,\tau) + \big(\mu_a P_t(2,1) + \mu_b P_t(2,0)\big) p_1(\tau) \\
&- (\mu_a \tau + \mu_b \bar{\tau}) P_t(1,\tau) + \gamma(\tau - \bar{\tau}) P_t(1,0), \\
\frac{dP_t(0)}{dt} =& -\lambda P_t(0) + \mu_a P_t(1,1) + \mu_b P_t(1,0).
\end{aligned}$$

It is a special case of a queueing process with a 2-level dynamically coupled stochastic service rate generalizing queueing processes with stochastic service [10, 11, 14]. In the stationary regime, we denote

$$\pi_n^a \stackrel{\text{def}}{=} P(n, \mu = \mu_a), \qquad \pi_n^b \stackrel{\text{def}}{=} P(n, \mu = \mu_b),$$

and $\pi_0 = P(n = 0)$. Consider the generating functions

$$g_{a,b}(z) \stackrel{\text{def}}{=} \sum_{n=1}^{\infty} \pi_n^{a,b} z^n \qquad \text{and} \qquad g(z) \stackrel{\text{def}}{=} \pi_0 + g_a(z) + g_b(z),$$

and let

$$\lambda^2 u \stackrel{\text{def}}{=} \lambda_a(\mu_a - \mu_b) \qquad \lambda^2 w \stackrel{\text{def}}{=} (\mu_a\mu_b + \lambda_a\mu_a + \lambda_b\mu_b + \gamma\mu_a)\pi_0$$

$$\lambda^2 v \stackrel{\text{def}}{=} (\lambda_a\mu_b + \lambda_b\mu_a)\pi_0 \qquad \Delta \stackrel{\text{def}}{=} (\lambda - \gamma + \mu_a - \mu_b)^2 + 4\lambda\gamma.$$

The invariant measure is obtained as follows [15]:

Theorem 1. *(i) $g(z)$ satisfies the functional equation of the type*

$$ug(rz) = vz + w - (z - z^+)(z - z^-)g(z),$$

where $z^{\pm}$ are given by

$$z^{\pm} = \frac{1}{2\lambda}\big(\mu_a + \mu_b + \lambda + \gamma \pm \sqrt{\Delta}\big).$$

(ii) The solution reads:

$$g(z) = \sum_{n=0}^{\infty} (-u)^n \frac{vr^n z + w}{\prod_{k=0}^{n+1}(zr^k - z^+)(zr^k - z^-)}.$$

Using the following partial balance relation,

$$\lambda\pi_n = \mu_a\pi_{n+1}^a + \mu_b\pi_{n+1}^b, \qquad \forall n \geq 0,$$

the reduced generating functions $g_{a,b}$ are then obtained as

$$g_a(z) = \frac{\mu_b}{\mu_a - \mu_b}\pi_0 + \frac{\lambda z - \mu_b}{\mu_a - \mu_b}g(z), \qquad g_b(z) = \frac{\mu_a}{\mu_b - \mu_a}\pi_0 + \frac{\lambda z - \mu_a}{\mu_b - \mu_a}g(z). \tag{5}$$

Upon using Cauchy integrals, the $\pi_n^{a,b}$ are then given as sums of geometric laws. From the radius of convergence z^- of g, the limit of ergodicity is obtained for $z^- \geq 1$, i.e. for

$$\lambda \leq \mu_b + \gamma \frac{\mu_a - \mu_b}{\mu_a + \gamma}.$$

4 Computing the Fundamental Diagram of Product Measures

4.1 Fundamental Diagram

In practice, points plotted in experimental FD studies are obtained by averaging data from static loop detectors over a few minutes (see e.g. [2]), but spatial average is much easier to obtain analytically. The equivalence between time and space averaging is not an obvious assumption, but since jams are moving, space and time correlations are combined in some way [16] and we consider this assumption to be quite safe. In what follows, we will therefore compute the FD along with its fluctuations, by considering the conditional probability measure $P_\lambda(\phi|d)$ for a closed system, where

$$\begin{cases} d = \dfrac{N}{N+L}, \\ \phi = \dfrac{\Phi}{N+L}, \end{cases} \quad \text{with} \quad \begin{cases} L & \text{number of queues} \\ N = \displaystyle\sum_{i=1}^{L} n_i & \text{number of vehicles} \\ \Phi = \displaystyle\sum_{i=1}^{L} \mu_i \mathbb{1}_{\{n_i>0\}} & \text{integrated flow} \end{cases}$$

are spatial averaged quantities and represent respectively the density and the traffic flow. N and L are fixed, which means that we are working with the canonical ensemble. If we assume that we are in the conditions of having a product form [17] (see [15] for details in this context) for the stationary distribution with individual probabilities $\pi^\lambda(n,\mu)$ associated to each queue taken in isolation, then, taking into account the constraints yields the following form of the joint probability measure:

$$P(\{n_i,\mu_i\}) = \frac{\delta\big(N - \sum_{i=1}^{L} n_i\big)}{Z_L[N]} \prod_{i=1}^{L} \pi^\lambda(n_i,\mu_i),$$

with the canonical partition function

$$Z_L[N] \stackrel{\text{def}}{=} \sum_{\{n_i,\mu_i\}} \delta\big(N - \textstyle\sum_{i=1}^{L} n_i\big) \prod_{i=1}^{L} \pi^\lambda(n_i,\mu_i),$$

where δ denotes now the usual Dirac function. When this form is not exact but constitutes a mean-field ansatz, some flow conservation conditions have to be imposed, which for the model under consideration concern λ and λ_a and read

$$\lambda_a = \mu_a \sum_{n=1}^{\infty} \pi_n^a, \qquad \lambda_b = \mu_b \sum_{n=1}^{\infty} \pi_n^b = \lambda - \lambda_a.$$

The density-flow conditional probability distribution takes the form

$$P(\phi|d) = \frac{Z_L[N,\Phi]}{Z_L[N]},$$

with

$$Z_L[N,\Phi] \stackrel{\text{def}}{=} \sum_{\{n_i,\mu_i\}} \delta\big(N - \textstyle\sum_{i=1}^L n_i\big)\delta\big(\Phi - \sum_{i=1}^L \mu_i \mathbb{1}_{\{n_i>0\}}\big) \prod_{i=1}^L \pi^\lambda(n_i,\mu_i).$$

$Z_L[N]$ and $Z_L[N,\Phi]$ represent respectively the probability of having N vehicles and the joint probability for having at the same time N vehicles and a flux Φ, under the unconstrained product form. Under this product form, we expect d and ϕ to satisfy a large deviations principle (see e.g. [18]) i.e. that there exist two rate functions $I(d)$ and $J(d,\phi)$ such that, for large L,

$$Z_L(N) \asymp e^{-LI(d)}, \qquad Z_L[N,\Phi] \asymp e^{-LJ(d,\phi)}.$$

In other words, we expect a large deviations version of the fundamental diagram of the form

$$P(\phi|d) \asymp e^{-LK(\phi|d)}, \qquad \text{with} \qquad K(\phi|d) \stackrel{\text{def}}{=} J(d,\phi) - I(d).$$

When there is one single constraint like for $Z_L(N)$, the large deviations expression can be obtained by saddle point techniques [13, 19]. For more than one constraint it seems easier to work variationally. Let us introduce the cumulant generating function h associated to π^λ,

$$h(s,t) \stackrel{\text{def}}{=} \log\Big[\sum_{n=0,\mu}^{\infty} \pi^\lambda(n,\mu)e^{sn+t\mu}\Big],$$

assuming by convention that the rate μ is zero in absence of client. After some computations (see [15] for details), we obtain I and J as Legendre transforms of the cumulant generating function, namely:

$$J(d,\phi) = \frac{d}{1-d}\lambda_n(d,\phi) + \frac{\phi}{1-d}\lambda_\mu(d,\phi) - h\big(\lambda_n(d,\phi),\lambda_\mu(d,\phi)\big),$$

$$I(d) = \frac{d}{1-d}\lambda_n'(d) - h\big(\lambda_n'(d),0\big),$$

with $\lambda_n(d,\phi)$ and $\lambda_\mu(d,\phi)$ the Lagrange multipliers, conjugate respectively to the density and flux, and solution of the equations

$$\begin{cases} \dfrac{\partial h}{\partial s}(\lambda_n(d,\phi),\lambda_\mu(d,\phi)) = \dfrac{d}{1-d}, \\ \dfrac{\partial h}{\partial t}(\lambda_n(d,\phi),\lambda_\mu(d,\phi)) = \dfrac{\phi}{1-d}, \\ \dfrac{\partial h}{\partial s}(\lambda_n'(d),0) = \dfrac{d}{1-d}. \end{cases}$$

The ordinary FD $\phi(d)$ is the minimizer of $K(\phi|d)$ and actually corresponds to

$$K(\phi(d)|d) = 0.$$

The small i.e. Gaussian fluctuations are then obtained by expanding K at second order in $\phi - \phi(d)$. Denoting the dual Hessian,

$$H^\star(s,t) \stackrel{\text{def}}{=} \begin{bmatrix} h_{ss} \; h_{st} \\ h_{ts} \; h_{tt} \end{bmatrix},$$

with the use of shorthand notations for the derivatives, representing the covariant matrix between the charges of the queues and the flux, we find the following expression for the variance of the FD:

$$\mathrm{Var}(\phi|d) = \frac{(1-d)^2}{L}\left(H^{\star-1}{}_{tt}\right)^{-1}.$$

Under the canonical ensemble constraint, the single queue distribution can be obtained from the partition function as

$$\begin{aligned} p_{CE}(n,\mu) &= p_\lambda(n,\mu)\frac{Z_{L-1}(N-n)}{Z_L(N)} \\ &\simeq p_\lambda(n,\mu)\exp L\Big[h\big(s(d-x),0\big) - h\big(s(d),0\big) \\ &\qquad\qquad + \frac{d-x}{1-d-x}s(d-x) - \frac{d}{1-d}s(d)\Big], \end{aligned}$$

with $x \stackrel{\text{def}}{=} n/(N+L)$ and the density constraint satisfied by $s(d)$:

$$\frac{\partial h}{\partial s}(s(d),0) = \frac{d}{1-d},$$

A direct comparison with the 2-speed TASEP of cluster size distributions is shown on Fig. 2. The correspondance is rather accurate, in particular for the

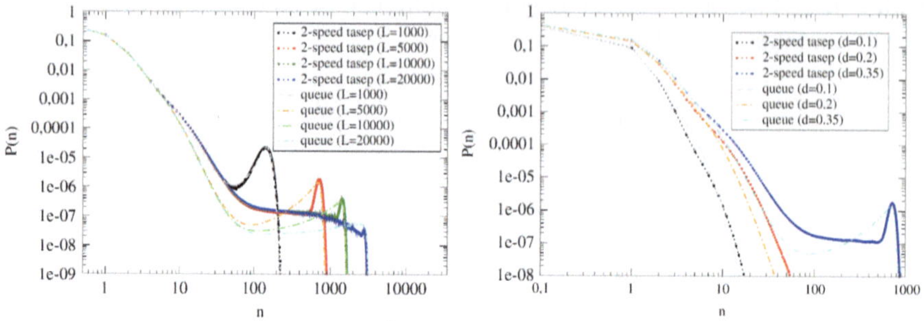

Fig. 2 Comparison of particle cluster vs. queue's size distribution for set of parameters $\mu_a = 10\mu_b = 100$, $\gamma = 10$ and $\delta = 1$ for various sizes L with fixed density $d = 0.35$ (*left*) and various densities with fixed size $L = 5{,}000$ (*right*), the number of queues being $(1 - d)L$

presence and location of the bumps signaling apparition of wide jams. In both cases, condensation is observed as a finite size effect, the bumps being replaced by a plateau at larger system size as seen on Fig. 2a.

4.2 Comparison with 2-Speed TASEP

As a sanity check, let us apply first this formula to the simple $M/M/1$ queue corresponding to one single speed level ($\mu_a = \mu_b = \mu$), i.e. a TASEP process. The rate of arrival λ is set by convenience to $\lambda = d\mu$. The cumulant-generating function then reads

$$h(s,t) = \log(1-d) + \log \frac{1 + de^s(e^{tv} - 1)}{1 - de^s}.$$

The FD rate function for TASEP then reads ($\bar{d} \stackrel{\text{def}}{=} 1 - d$)

$$K(\phi|d) = \frac{d}{\bar{d}} \log \frac{\mu d - \phi}{\mu d^2} + \frac{\phi}{\mu \bar{d}} \log \frac{\phi^2}{(\mu d - \phi)(\mu \bar{d} - \phi)} - \log \frac{\mu \bar{d}^{\,2}}{\mu \bar{d} - \phi},$$

yielding an expression for FD and variance that matches direct computation [15]:

$$\phi(d) = \mathbb{E}(\phi|d) = \mu d \bar{d}, \qquad \mathrm{Var}(\phi|d) = \frac{\mu^2}{N + L}\, d^2 \bar{d}^{\,2}.$$

In the case of the model of Sect. 3, the assumption of independence between servers is not valid anymore, but corresponds to a mean-field approximation [15]. From (5), we get for the cumulant generating function

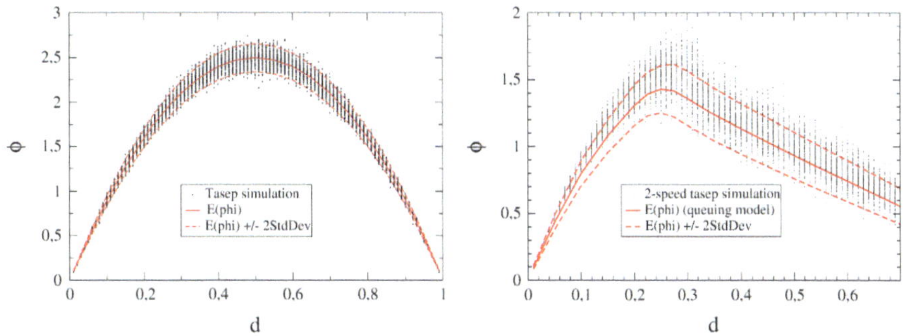

Fig. 3 Comparison between simulation for the simple TASEP FD (*left*) and the 2-speed TASEP with $\mu_a = 10 \times \mu_b = 10 \times \gamma = 100 \times \delta$ (*right*) with small deviations results (two standard deviations) obtained from the queuing processes

$$h(s,t) = \log\Big(\pi_0\big(1 + \frac{\mu_b e^{\mu_a t} - \mu_a e^{\mu_b t}}{\mu_a - \mu_b}\big) + \frac{(\mu_a - \lambda e^s)e^{\mu_b t} + (\lambda e^s - \mu_b)e^{\mu_a t}}{\mu_a - \mu_b}\, g(e^s)\Big),$$

from which the Legendre transform as well as the Hessian $H^*(s^*, 0)$ for the small fluctuations can be obtained, where s^* is the point which satisfies

$$e^{s^*} g'(s^*) = \frac{d}{1-d}.$$

A comparison with the corresponding 2-speed TASEP is shown on Fig. 3. The observed discrepancy can be traced back to various approximations made in the mean-field theory, in particular the assumption of independence between successive servers. This could be possibly cured by a refined mean field approach, replacing the product form ansatz with a Bethe approximation. This requires to solve exactly the 2-servers problem, but the large deviations approach would follow the same lines.

Acknowledgements This work was supported by the French National Research Agency (ANR) grant No ANR-08-SYSC-017.

References

1. M. Schreckenberg, A. Schadschneider, K. Nagel, and N. Ito. Discrete stochastic models for traffic flow. *Phys. Rev.*, **E51**:2339, 1995.
2. B.S. Kerner. *The Physics of Traffic*. Springer Verlag, 2005.
3. Y. Sugiyama et al. Traffic jams without bottlenecks: experimental evidence for the physical mechanism of the formation of a jam. *New Jrl. Phys.*, 10:1–7, 2008.
4. K. Nagel and M. Schreckenberg. A cellular automaton model for freeway traffic. *J. Phys. I*,*2*:2221–2229, 1992.

5. R. Barlović, L. Santen, A. Schadschneider, and M. Schreckenberg. Metastable states in cellular automata for traffic flow. *Eur. Phys. J.*, B5:793, 1998.
6. M. Blank. Hysteresis phenomenon in deterministic traffic flows. *J. Stat. Phys.*, 120:627–658, 2005.
7. C. Appert and L. Santen. Boundary induced phase transitions in driven lattice gases with meta-stable states. *Phys. Rev. Lett.*, 86:2498, 2001.
8. T. M. Liggett. *Interacting Particle Systems*. Springer, Berlin, 2005.
9. F. Spitzer. Interaction of Markov processes. *Adv. Math.*, **5**:246, 1970.
10. M. Şamsonov, C. Furtlehner, and J.-M. Lasgouttes. Exactly solvable stochastic processes for traffic modelling. Technical Report 7278, INRIA, 2010.
11. C. Furtlehner and J.-M. Lasgouttes. A queueing theory approach for a multi-speed exclusion process. In *Traffic and Granular Flow ' 07*, pages 129–138, 2007.
12. J. Kaupužs, R. Mahnke, and R. J. Harris. Zero-range model of traffic flow. *Phys. Rev. E*, 72:056125, 2005.
13. M. R. Evans, S. N. Majumdar, and R. K. P. Zia. Canonical analysis of condensation in factorized steady states. *J. Stat. Phys.*, 123(2):357–390, 2006.
14. C.M. Harris. Queues with state-dependant stochastic service rate. *Operation Research*, 15:117–130, 1967.
15. C. Furtlehner, J.-M. Lasgouttes, and M. Samsonov. One-dimensional particle processes with acceleration/braking asymmetry. *J. Stat. Phys.*, 147:1113:1144, 2012.
16. K. Nagel and M. Paczuski. Emergent traffic jams. *Phys. Rev. E*, 51(4):2909–2918, 1995.
17. F. P. Kelly. *Reversibility and stochastic networks*. John Wiley & Sons Ltd., 1979. Wiley Series in Probability and Mathematical Statistics.
18. H. Touchette. The large deviation approach to statistical mechanics. *Physics Reports*, 478:1–69, 2009.
19. G. Fayolle and J.-M. Lasgouttes. Asymptotics and scalings for large closed product-form networks via the Central Limit Theorem. *Markov Proc. Rel. Fields*, 2(2):317–348, 1996.

A Bogdanov–Takens Bifurcation in Generic Continuous Second Order Traffic Flow Models

Armando Carrillo, Joaquín Delgado, Patricia Saavedra, Rosa Maria Velasco, and Fernando Verduzco

Abstract We consider the continuous model of Kerner–Konhäuser for traffic flow given by a second order PDE for the velocity and density. Assuming conservation of cars, traveling waves solution of the PDE are reduced to a dynamical system in the plane. We describe the bifurcations set of critical points and show that there is a curve in the set of parameters consisting of Bogdanov–Takens bifurcation points. In particular there exists Hopf, homoclinic and saddle node bifurcation curves. For each Hopf point a one parameter family of limit cyles exists. Thus we prove the existence of solitons solutions in the form of one bump traveling waves.

1 Introduction

Macroscopic traffic models are based on the analogy with a continuous one dimensional flow. Conservation of number of cars leads to conservation of mass

$$\frac{\partial \rho}{\partial t} + \frac{\partial \rho V}{\partial x} = 0 \tag{1}$$

A. Carrillo · F. Verduzco
Mathematics Department, Universidad de Sonora, Blvd. Luis Encinas y Rosales S/N, Col. Centro, Hermosillo, Sonora, México
e-mail: carrillo@gauss.mat.uson.mx; verduzco@gauss.mat.uson.mx

J. Delgado (✉) · P. Saavedra
Mathematics Department, UAM–Iztapalapa, Avenida San Rafael Atlixco 186, México, D.F. 09340, México
e-mail: jdf@xanum.uam.mx

R.M. Velasco
Physics Department, UAM–Iztapalapa, Avenida San Rafael Atlixco 186, México, D.F. 09340, México
e-mail: rmvb@xanum.uam.mx

V.V. Kozlov et al. (eds.), *Traffic and Granular Flow '11*,
DOI 10.1007/978-3-642-39669-4_2, © Springer-Verlag Berlin Heidelberg 2013

where $\rho(t,x)$ is density and $V(t,x)$ the average velocity of cars. In the kinetic models the continuity equation is supplied with the law of motion given by the Navier–Stokes equation

$$\rho\left(\frac{\partial V}{\partial t}+V\frac{\partial V}{\partial x}\right)=\frac{\partial}{\partial x}\left(\eta\frac{\partial V}{\partial x}\right)-\frac{\partial p}{\partial x}+X \tag{2}$$

where η is viscosity and p the local pressure, being proportional to the variance ("temperature") of the traffic $\Theta(x,t)$, namely the average of the squared differences of the individual cars and the average velocity. The "external forces" in the Kerner-Konhäuser model are represented by driver's tendency to acquire a safe velocity $V_e(\rho)$ with a relaxation time τ,

$$X=\rho\frac{V_e(\rho)-V}{\tau}.$$

When $\Theta=\Theta_0$ and $\eta=\eta_0$ are supposed constants, the equation of motion (2) simplifies to

$$\rho\left(\frac{\partial V}{\partial t}+V\frac{\partial V}{\partial x}\right)=-\Theta_0\frac{\partial\rho}{\partial x}+\frac{\rho(V_e(\rho)-V)}{\tau}+\eta_0\frac{\partial^2 V}{\partial x^2} \tag{3}$$

We look for travelling wave solutions of (1) and (3), then the change of variables $\xi=x+V_g t$ transform (1) into the quadrature which can be immediately solved:

$$\rho(V+V_g)=Q_g. \tag{4}$$

Here, the arbitrary constant Q_q represents the local flux as measured by an observer moving with the same velocity V_g as the travelling wave.

Following [5] introduce adimensional variables

$$z=\rho_{max}\xi,\quad v=\frac{V}{V_{max}},\quad v_g=\frac{V_g}{V_{max}},\quad q_g=\frac{Q_g}{\rho_{max}V_{max}} \tag{5}$$

where ρ_{max} and V_{max} are some maximum reference values of the density depending on the empirical law $V_e(\rho)$. Also let

$$v_e(\rho)=\frac{V_e(\rho)}{V_{max}},\quad \theta_0=\frac{\Theta_0}{V_{max}^2},\quad \lambda=\frac{V_{max}}{\eta_0},\quad \mu=\frac{1}{\rho_{max}\eta_0\tau}. \tag{6}$$

Here $V_e(V)$ is obtained through $V_e(\rho)$ by means of (4). Substitution of (4) into the equation of motion (3) yields dynamical system

$$\begin{aligned}\frac{dv}{dz}&=y,\\ \frac{dy}{dz}&=\lambda q_g\left[1-\frac{\theta_0}{(v+v_g)^2}\right]y-\mu q_g\left(\frac{v_e(v)-v}{v+v_g}\right).\end{aligned} \tag{7}$$

In adimensional form, (4) becomes

$$\rho/\rho_{max} = \frac{q_g}{v + v_g}. \tag{8}$$

Here and in what follows, we will take the parameter values λ, μ as given for the model, and will analyze the dynamical behavior with respect to the parameters v_g, q_g.

2 Fundamental Diagram

Dynamics of (7) depend on the explicit form of the constitutive relationship defined by the fundamental diagram $V_e(\rho)$. A typical form of the curve $V_e(\rho)$ must meet the following properties that we will state as hypotheses.

Hypothesis I. $V_e(\rho)$ is a monotone decreasing function defined for $\rho \geq 0$ and $V_e(\rho_{jam}) = 0$ for some $\rho_{jam} > 0$.
Hypothesis II. $V_e(\rho)$ is bounded from below.
Hypothesis III. $\lim_{\rho \to 0} V_e(\rho) = V_f > 0$ and $\lim_{\rho \to 0} \rho^2 V_e'(\rho) = 0$.

Hypothesis I is explicitly adopted in [2]; ρ_{jam} can be interpreted as jam density where traffic is stuck. The condition $V_e'(\rho) = o(\rho^2)$ in Hypothesis II allows for a plateau in the graph of the curve $V_e(\rho)$ for small values of density; V_f is then the average velocity in free traffic and can be determined by speed limits.

Historically, diverse fundamental diagrams have been considered. Either fit to empirical data, or deduced from theoretical bases. Some examples are shown in Fig. 1. Explicit formulas are:

1. Greenshields: $V = V_{max}(1 - \rho/\rho_{jam})$.
2. Greenberg: $V = V_{max} \ln(\rho_{jam}/\rho)$.
3. Newell: $V = V_{max} \exp(-\lambda\rho)$.
4. Kerner–Konhäuser: $V = V_{max}\left(\frac{1}{1+\exp(\frac{\rho/\rho_{max}-0.25}{0.06})} - 3.72 \times 10^{-6}\right)$.

It can be verified that all except Newell's fundmental diagram, satisfy Hypothesis I, and only (3) and (4) satisfy hypothesis II. Kerner-Konhäuser's can be considered the most accurate fundamental diagram, since it was fitted from a large set of data using double induction detectors. Kerner and Konhäuser [2] and [3] (Greenberg was fitted from 18 data circa 1959). Although Hypothesis III seems reasonable, neither of the fundamental diagrams (1)–(4) satisfy this condition. We argue that it can be mathematically sustained since experimental data are available only for strictly positive ρ; put it in another way, modern technologies such as GPS could be used to get more accurate data for small densities. Thus one can take Kerner–Konhäuser model as an empirical validated for $\rho > \delta$ where δ is some positive small density, and modify it in such a way that Hypothesis III is true.

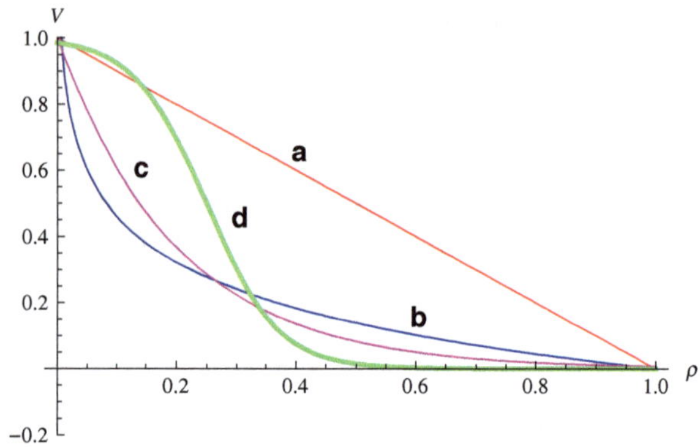

Fig. 1 Fundamental diagrams of mean velocity V vs. density ρ. (**a**) Greenshields; (**b**) Greenberg; (**c**) Newell; (**d**) Kerner–Konhäuser

Hypothesis II implies that $\lim_{\rho\to\infty} V_e(\rho)$ exists. This limit is zero for Newell's diagram and equals $-3.72 \times 10^{-6} V_{max}$ for Kerner–Konhäuser's. Although $\rho \to \infty$ does not make sense physically, it will be useful from a mathematical point of view to allow this limit. This amounts to consider an arbitrary large value of ρ_{jam} (as Newell's diagram). In fact, in (8) $v \to -v_g$ as $\rho \to \infty$ and this corresponds to the limit of an observer moving with the wave seeing it standing.

We denote by $v_e(v)$ the expression $v_e(\rho)$ when ρ is expressed through (8) as a function of v (see 6).

The following properties of $v_e(v)$ are readily obtained.

Lemma 1. *Let $V_e(\rho)$ satisfy Hypothesis I, II and III. Then $v_e(v)$ is a monotone increasing function of v and: (1) $\lim_{v\to -v_g} v_e(v)$ exists; (2) $\lim_{v\to -v_g} v_e'(v) = 0$ and, (3) $\lim_{v\to\infty} v_e(v) = v_f$.*

Proof. Since (4) can be written as

$$v + v_g = \frac{q_g}{\rho/\rho_{max}} \tag{9}$$

it follows that ρ/ρ_{max} is a decreasing function of v, and thus $v_e(v)$ is an increasing function of v, for $v > -v_g$. If $v \to -v_g$ then $\rho \to \infty$ but then $\lim v_e(\rho)$ exists; this proves (1). Using the chain rule

$$v_e'(v) = v_e'(\rho(v))\rho'(v) = v_e'(\rho(v))\left(-\frac{q_g/\rho_{max}}{(v+v_g)^2}\right) = -\frac{v_e'(\rho)\rho^2}{q_g\rho_{max}}$$

the last term tends to zero by Hypothesis III; this proves (2). Finally if $v \to \infty$ then $\rho \to 0$ but then $v_e(\rho) \to 0$. This completes the proof. □

The above lemma shows that given "reasonable" fundamental diagram (namely satisfying Hypothesis I–III) function $v_e(v)$ is sigmoidal: within the interval $[-v_g, \infty]$, $v_e(v)$ approaches its limits as $v \to -v_g$ or ∞, asymptotically flat.

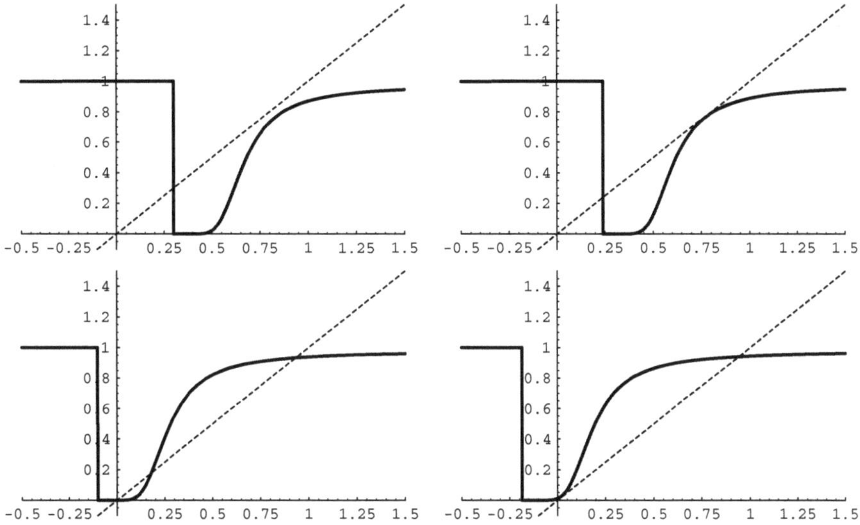

Fig. 2 Possible intersections $v_e(v) = v$ leading to critical points of system (7). Only the graph to the *right* of the *vertical curve* $v = -v_g$ must be considered. From *left* to *right* from *top* to *bottom*, the number of critical points are: zero; one; three and two. A non–generic situation of tangency at the inflection point is not shown

2.1 *Critical Points*

Critical points of system (7) are given by the equations $y = 0$ and $v_e(v) = v$. The following results describes a generical situation.

Proposition 1. *Let $V_e(\rho)$ satisfy Hypothesis I–III. Then there exists either 1, 2 or 3 critical points of system (7).*

Proof. Since the graph of $v_e(v)$ is sigmoidal, and the critical points are given by the intersection of the graphs of $v_e(v)$ and the straight line $v = v$ the result follows. □

Figure 2 shows several possibilities for the intersection of the curve $v_e(v) = v$, with 0, 1, 2 or 3 critical points.

3 The Bogdanov–Takens Bifurcation

In this section we will show that the Kerner-Konhäuser ODE (7) undergoes a Bogdanov-Takens (BT) bifurcation. The BT bifurcation occurs in a two parameter dynamical system in the plane (see also it generalization to n-dimensions [1]) when for some value of the parameter vector, the dynamical system has a critical point with a double non–semisimple eigenvalue and some non-degeneracy conditions are satisfied. We state the conditions and its normal form according to Kuznetzov [4].

Theorem 1 (Bogdanov–Takens). *Suppose that a planar system*

$$\dot{x} = f(x,\alpha), \quad x \in \mathbb{R}^2, \quad \alpha \in \mathbb{R}^2, \tag{10}$$

with smooth f, has at $\alpha = 0$ the equilibrium $x = 0$ with a double zero eigenvalue

$$\lambda_{1,2}(0) = 0.$$

Assume that the following genericity conditions are satisfied

(BT.0) The Jacobian matrix $A(0) = f_x(0,0) \neq 0$;
(BT.1) $a_{20}(0) + b_{11}(0) \neq 0$;
(BT.2) $b_{20}(0) \neq 0$;
(BT.3) The map

$$(x,\alpha) \mapsto \left(f(x,\alpha), tr\left(\frac{\partial f(x,\alpha)}{\partial x}\right), det\left(\frac{\partial f(x,\alpha)}{\partial x}\right)\right)$$

is regular at the point $(x,\alpha) = (0,0)$.
Then there exists smooth invertible transformations smoothly depending on the parameters, a direction–preserving time reparametrization, and smooth invertible parameters changes, which together reduce the system to

$$\dot{\eta}_1 = \eta_2 \tag{11}$$

$$\dot{\eta}_2 = \beta_1 + \beta_2\eta_1 + \eta_1^2 + s\eta_1\eta_2 + O(||\eta||^3) \tag{12}$$

where $s = sign[b_{20}(0)(a_{20}(0) + b_{11}(0))] = \pm 1$.

Remark 1. The coefficients appearing in the hypotheses (BT.1), (BT.2) are defined as follows: Let v_0, v_1 be generalized right eigenvectors of $A_0 = A(0)$ and w_0, w_1 generalized left eigenvectors, such that $A_0 v_0 = 0$, $A_0 v_1 = v_0$ and $A^T w_1 = 0$, $A^T w_0 = w_1$. Then by a linear change of coordinates $x = y_1 v_0 + y_2 v_1$ system (10) can be expanded in a Taylor series as

$$\begin{aligned}
\dot{y}_1 &= y_2 + a_{00}(\alpha) + a_{10}(\alpha)y_1 + a_{01}(\alpha)y_2 \\
&\quad + \frac{1}{2}a_{20}(\alpha)y_1^2 + a_{11}(\alpha)y_1y_2 + \frac{1}{2}a_{02}(\alpha)y_2^2 + P_1(y,\alpha) \\
\dot{y}_2 &= b_{00}(\alpha) + b_{10}(\alpha)y_1 + b_{01}(\alpha)y_2 \\
&\quad + \frac{1}{2}b_{20}(\alpha)y_1^2 + b_{11}(\alpha)y_1y_2 + \frac{1}{2}b_{02}(\alpha)y_2^2 + P_2(y,\alpha)
\end{aligned}$$

where $a_{kl}(\alpha)$, $P_{1,2}(y,\alpha)$ are smooth functions of their arguments and

$$a_{00}(0) = a_{10}(0) = a_{01}(0) = b_{00}(0) = b_{10}(0) = b_{01}(0) = 0.$$

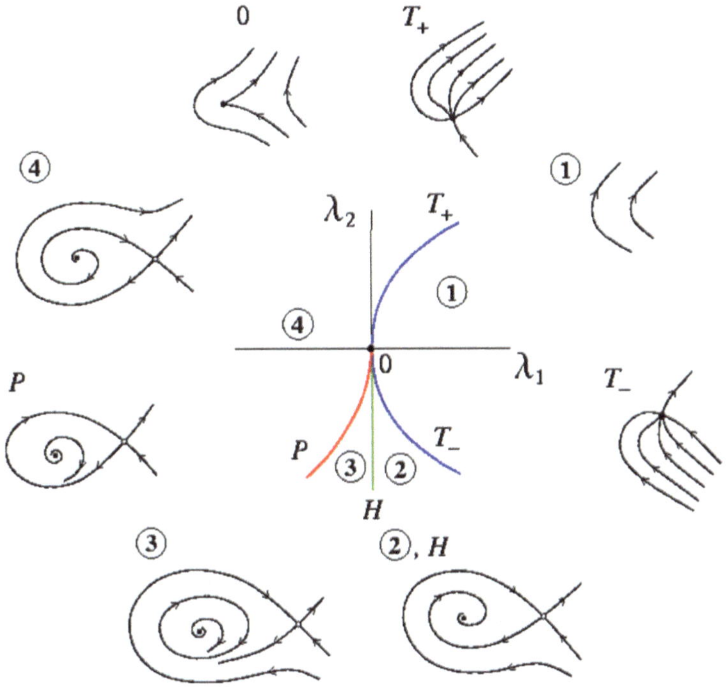

Fig. 3 Bogdanov–Takens bifurcation

Remark 2. The normal form (12) is a *universal unfolding*, that is all nearby dynamical systems satisfying the hypotheses of the theorem is topologically equivalent to the normal form with the $O(||\eta||^3)$ terms deleted. Figure 3 shows the different scenarios depending in the parameter space β_1–β_2.

Lemma 2. *Let $V_e(\rho)$ satisfy Hypothesis I–III. Suppose that for some values of the parameters v_g, q_g the curves $v_e(v)$ and $v = v$ have a tangency at $v = v_c$. Then there exists θ_0 such that system (7) has a critical point with a double zero eigenvalue and its linear part is not semisimple.*

Proof. The linear part at v_c is

$$A_0 = \begin{pmatrix} 0 & 1 \\ -\frac{q_g\mu(v_e'(v_c)-1)}{v+v_g} & q_g\left(1-\frac{\theta_0}{(v_c+v_g)^2}\right)\lambda \end{pmatrix}$$

since the curves $v = v_e(v)$, $v = v$ have a tangency at $v = v_c$ then $v_e'(v_c) = 1$ and A_0 reduces to

$$A_0 = \begin{pmatrix} 0 & 1 \\ 0 & q_g\left(1-\frac{\theta_0}{(v_c+v_g)^2}\right)\lambda \end{pmatrix}$$

choosing $\theta_0 = (v_c + v_g)^2$ reduces to

$$A_0 = \begin{pmatrix} 0 & 1 \\ 0 & 0 \end{pmatrix} \tag{13}$$

□

Theorem 2. *Let $V_e(\rho)$ satisfy Hypothesis I–III. Suppose that for some values of the parameters v_g, q_g the curves $v_e(v)$, $v = v$ have a tangency at $v = v_c$ with $v_e''(v_c) \neq 0$. Then there exists θ_0 such that system (7) has the linear part (13). Moreover, if the following generic condition*

$$\left(\frac{\partial^2 v_e}{\partial v^2} - \frac{\partial^2 v_e}{\partial v_g \partial v}\right)\frac{\partial v_e}{\partial q_g} + \frac{\partial^2 v_e}{\partial q_g \partial v}\frac{\partial v_e}{\partial v_g} \neq 0 \tag{14}$$

is satisfied at the critical value of the point and parameters, then the system undergoes a Bogdanov-Takens bifurcation.

Proof. According to the remark after the Bogdanov–Takens theorem, the generalized right eigenvectors v_0, v_1 and the generalized left eigenvectors w_1, w_0 as the canonical vectors $(1, 0)$, $(0, 1)$ in $\mathbb{R}^2$, respectively, and the coordinates (y_1, y_2) as (v, y) coordinates of the dynamical system (7). Then, denoting as $f(y_1, y_2)$ the vector field defining (7), the coefficients appearing in conditions (BT.1), (BT.2) are

$$a_{20}(0) = \frac{\partial^2}{\partial y_1^2} w_0 \cdot f(y_1, y_2) = \frac{\partial^2}{\partial v^2} y = 0$$

$$\begin{aligned} b_{20}(0) &= \frac{\partial^2}{\partial y_1^2} w_1 \cdot f(y_1, y_2) \\ &= \frac{\partial^2}{\partial v^2}\left(\lambda q_g \left[1 - \frac{\theta_0}{(v + v_g)^2}\right] y - \mu q_g \left(\frac{v_e(v) - v}{v + v_g}\right)\right) = -\mu \frac{q_g v_e''(v_c)}{v_c + v_g} \end{aligned}$$

$$\begin{aligned} b_{11}(0) &= \frac{\partial^2}{\partial y_1 \partial y_2} w_1 \cdot f(y_1, y_2) \\ &= \frac{\partial^2}{\partial y_1 \partial y_2}\left(\lambda q_g \left[1 - \frac{\theta_0}{(v + v_g)^2}\right] y - \mu q_g \left(\frac{v_e(v) - v}{v + v_g}\right)\right) = \frac{2\mu q_g \theta_0}{(v_c + v_g)^3} \end{aligned}$$

we immediately verify that

$$a_{20}(0) + b_{11}(0) = \frac{2\mu q_g \theta_0}{(v_c + v_g)^3} \neq 0$$

$$b_{20} = -\mu \frac{q_g v_e''(v_c)}{v_c + v_g} \neq 0$$

the last condition being assumed by hypothesis.

A straightforward computation shows that the matrix of the map in condition (BT.3) at $v = v_c$, $y = 0, vg, q_g$ is

$$\begin{pmatrix} 0 & 1 & 0 & 0 \\ 0 & 0 & -\frac{\mu q_g {v_e}'_{v_g}}{v_c+v_g} & -\frac{\mu q_g {v_e}'_{q_g}}{v_c+v_g} \\ \frac{2\lambda q_g}{v_c+v_q} & 0 & \frac{2\lambda q_g}{v_c+v_g} & 0 \\ \frac{\mu q_g {v_e}''_{v}}{v_c+v_g} & -\frac{2\lambda q_g}{v_c+v_q} & -\frac{\mu q_g {v_e}'_{v_g}}{(v_c+v_g)^2} + \frac{\mu q_g {v_e}''_{v_g,v}}{v_c+v_g} & -\frac{\mu q_g {v_e}'_{q_g}}{(v_c+v_g)^2} + \frac{\mu q_g {v_e}''_{q_g,v}}{v_c+v_g} \end{pmatrix}$$

where ${v_e}'$, ${v_e}''$ denote first and second order partial derivatives with respect to the subindex variables and are evaluated at (q_g, v_g, v_c). The nonvanishing of the determinant is equivalent to the condition

$$({v_e}''_{v} - {v_e}''_{v_g,v}){v_e}'_{q_g} + {v_e}'_{v_g}{v_e}''_{q_g,v} \neq 0 \tag{15}$$

which is the same as (14). □

4 The Kerner-Konhäuser Model

For the Kerner-Konhäuser model one takes

$$v_e(q_g, v_g, v) = \frac{1}{1 + \exp\left[\frac{\frac{q_g}{v+v_g} - d_2}{d_3}\right]} + d_1 \tag{16}$$

with the empirical values [2]

$$d_1 = -3.72 \times 10^{-6}, \quad d_2 = 0.25, \quad d_3 = 0.06$$

The set of critical points and parameters

$$\{(q_g, v_g, v_c) \mid v_e(v_c) - v_c = 0\}$$

form a three-dimensional surface, given as the zero level set of a function $F(q_g, v_g, v_c) = v_e(q_g, v_g, v_c) - v_c = 0$. It is shown in Fig. 1. There the vertical axis is $x = v_g + v_c$ and we restrict to $x > 0$. The solid curves which will referred as γ are the loci where the projection $(v_g, q_g, v_c) \mapsto (v_g, q_g)$ is not surjective, namely the curve defined by the set of equations

$$F(q_g, v_g, v_c) = v_e(v_c) - v_c = 0,$$
$$F_v(q_g, v_g, v_c) = v_e'(v_c) - 1 = 0.$$

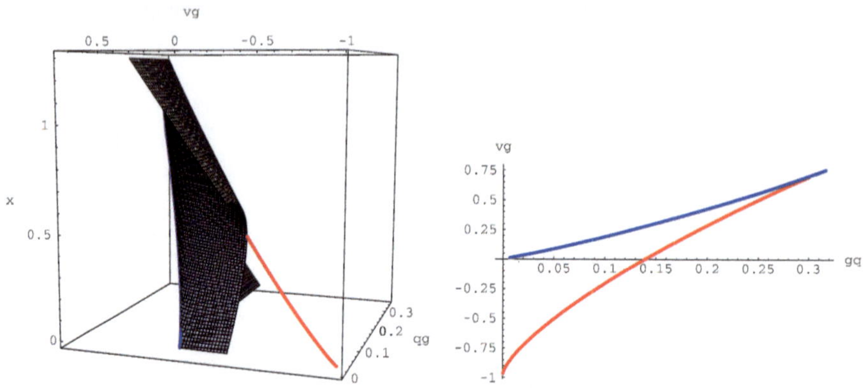

Fig. 4 Surface of critical points and singular locus of the projection. The *solid curve* is γ

These were computed numerically by solving the above equations. Their projections on the q_g–v_g plane is a cusp curve (see Fig. 4) and characterize the values of parameters (q_g, v_g) such that the curve $v_e(v)$ is tangent to the graph of the identity. On the other hand the solid curve in the three dimensional space q_g–v_g–v_c (or q_g–v_g–x) can be lifted to a curve $\tilde{\gamma}$ in four dimensional space θ_0–q_g–v_g–v_c where $\theta_0 = \sqrt{v_c + v_g}$ and thus γ can be viewed as the projection of $\tilde{\gamma}$ and represent values (θ_0, q_g, v_g) such that the linear part at the critical point has the form (13).

Proposition 2. *The conditions of Theorem 2 are satisfied for the Kerner–Konhäuser model (16) whenever d_1 is small an negative, and $d_2, d_3 > 0$.*

Proof. The condition $v_e''(v_c) = 0$ reduces to

$$e^{\frac{q_g}{d_3 x}}(q_g - 2d_3 x) - e^{d_2/d_3}(q_g + 2d_3 x) = 0, \tag{17}$$

and the nondegenericity condition (14) to

$$e^{\frac{q_g}{d_3 x}}(q_g - d_3 x) + e^{d_2/d_3}(q_g + d_3 x) = 0. \tag{18}$$

The change of variable $q_g = zx$ (recall $x = v + v_g$, thus $z = \rho/\rho_{max}$) and reduces each equation to the trivial one $x = 0$ which is discarded since $v \neq -v_g$, or

$$2d_3 e^{d_2/d_3} + (2d_3 - z)e^{z/d_3} + e^{d_2/d_3} z = 0 \tag{19}$$

$$d_3 e^{d_2/d_3} + (d_3 - z)e^{z/d_3} + e^{d_2/d_3} z = 0 \tag{20}$$

Taking the difference of these yields $d_3 e^{z/d_3} = -d_3 e^{d_2/d_3}$. Since all constants d_1, d_2, d_3 are positive, this last equation has no solution for $z > 0$. □

Corollary 1. *The Kerner–Konhäuser model (3) with the fundamental diagram (16) under the assumption (1) has traveling wave solutions in the unbounded domain* $x \in (-\infty, \infty)$ *and in the bounded domain* $[0, L]$ *with periodic boundary conditions.*

Proof. From the BT Theorem 2, it follows that there exists Hopf limit cycles $(v(z), y(z))$, where $z = \rho_{max}\xi$ and $\xi = x + V_g t$ (see (5)). This yields a traveling wave solution of (3) in the form $V(x,t) = V_{max} v(x + V_g t)$. If T is commensurable with L then the traveling wave satisfies also periodic boundary conditions. □

References

1. Carrillo, F.A., Verduzco, F. Delgado, J. (2010) Analysis of the Takens-Bogdanov Bifurcation on m-parameterized Vector Fields. International Journal of Bifurcation and Chaos, Vol. 20, No. 4,
2. Kerner, B.S. and Konhäuser, P. (1993) Cluster effect in initially homogeneous traffic flow. Phys. Rev. E, **48**, No.4, R2335–R2338.
3. Kerner, B.S., Konhäuser, P. and Schilke, P. (1995) Deterministic spontaneous appearence of traffic jams in slighty inhomogeneous traffic flow. Physical Review E, **51**, 6243–6248.
4. Y. Kuznetzov, *Elements of Applied Bifurcation Theory*, Appl. Math. Ser. **112**, 3d. ed., Springer
5. P. Saavedra & Velasco R.M. Phase space analysis for hydrodynamic traffic models (2009) Phys. Rev. E **79**, 0066103.

Model of Vehicular Traffic by Hilliges and Weidlich Revisited

Martins Brics, Reinhard Mahnke, and Reinhart Kühne

Abstract Driving of cars on a highway is a complex process which can be described by different means using continuous or discrete basic equations. It always leads to equations of motion with asymmetric interaction.

In 1994 Martin Hilliges and Wolfgang Weidlich from University of Stuttgart developed a phenomenological modeling for dynamic traffic flow in networks, published in Hilliges and Weidlich (Transpn Res B 29:407–431, 1995). The authors Hilliges and Weidlich introduce the model in its discrete formulation, carry out a continuous approximation and investigate stationary solutions with respect to stability analytically.

Here we consider the Hilliges–Weidlich–Model once again using our optimal velocity function already introduced in previous papers by Mahnke and Pieret (Phys Rev E 56:2666–2671, 1997). We solve the equations of motion given by two coupled partial differential equations numerically and discuss the moving staedy state profiles of density as well as speed. As a first result we present the long-time behaviour. The investitations are still going on and comparisons to related research (Sugiyama Y, Masuoka K, Ishida T (2009) Dynamics of dissipative system with asymmetric interaction and N–body problem for the emergence of moving cluster. In: Appert-Rolland C, Chevoir F, Gondret P, Lassarre S, Lebacque J-P, Schreckenberg M (eds) Traffic and Granular Flow'07. Springer, Berlin, pp 556–563; Wagner, Phil Trans R Soc A 368:4481–4495, 2010) are in preparartion.

M. Brics (✉) · R. Mahnke
Institute of Physics, Rostock University, D–18051 Rostock, Germany
e-mail: martins.brics2@uni-rostock.de; reinhard.mahnke@uni-rostock.de

R. Kühne
German Aerospace Center, Institute for Transportation Research, D–12489 Berlin, Germany
e-mail: reinhart.kuehne@dlr.de

V.V. Kozlov et al. (eds.), *Traffic and Granular Flow '11*,
DOI 10.1007/978-3-642-39669-4_3, © Springer-Verlag Berlin Heidelberg 2013

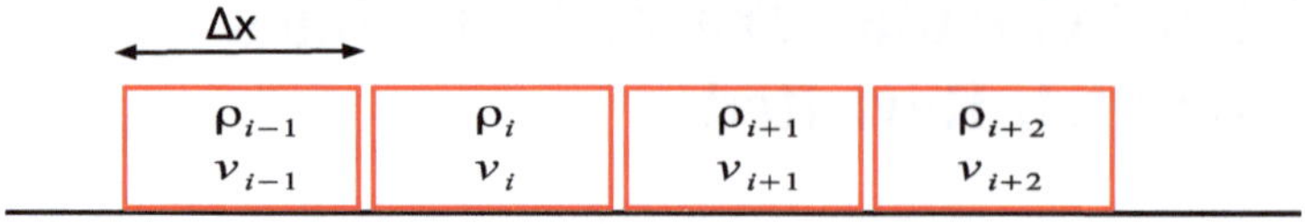

Fig. 1 Road segment with length L divided into cells with length Δx

1 The Discrete Model by Hilliges and Weidlich and its Continuous Formulation

Following [1] we have a road segment with some length L. We divide it into N cells with size Δx such that $L = N\Delta x$. In each cell we assume that density ρ_i and velocity v_i are constant (see Fig. 1).

The density ρ_i in cell i can change only because of inflow from cell $i-1$ and outflow to cell $i+1$, therefore

$$\frac{\partial \rho_i}{\partial t}\Delta x = j_i - j_{i+1} = \rho_{i-1}v_i - \rho_i v_{i+1} \,. \tag{1}$$

Equation (1) underlines that traffic is forward oriented. The inflow flux from cell $i-1$ to cell i is defined as $j_i = \rho_{i-1}v_i$.

If we expand density and velocity in Taylor series up to second order at cell i, then

$$\begin{aligned} \rho_{i-1} &= \rho_i - \Delta x \frac{\partial \rho_i}{\partial x} + \frac{(\Delta x)^2}{2}\frac{\partial^2 \rho_i}{\partial x^2} \\ v_{i+1} &= v_i + \Delta x \frac{\partial v_i}{\partial x} + \frac{(\Delta x)^2}{2}\frac{\partial^2 v_i}{\partial x^2} \end{aligned} \tag{2}$$

and replace v_i by v and ρ_i by ρ we end up with a partial differential equation for the density

$$\frac{\partial \rho}{\partial t} + \frac{\partial}{\partial x}(\rho v) = \frac{\Delta x}{2}\frac{\partial}{\partial x}\left(\frac{\partial \rho}{\partial x}v - \frac{\partial v}{\partial x}\rho\right) . \tag{3}$$

For the velocity we take the well-known relaxation ansatz

$$\frac{Dv}{Dt} = \frac{1}{\tau}\left(V_{opt}(\rho) - v\right) , \tag{4}$$

where $\frac{D}{Dt} = \frac{\partial}{\partial t} + v\frac{\partial}{\partial x}$ is the material derivative and V_{opt} is a given optimal velocity function. It forms the following set of equations

$$\begin{aligned} \frac{\partial \rho}{\partial t} + \frac{\partial}{\partial x}(\rho v) &= \frac{\Delta x}{2}\frac{\partial}{\partial x}\left(\frac{\partial \rho}{\partial x}v - \frac{\partial v}{\partial x}\rho\right), \\ \frac{\partial v}{\partial t} + v\frac{\partial v}{\partial x} &= \frac{1}{\tau}\left(V_{opt}(\rho) - v\right). \end{aligned} \tag{5}$$

To investigate both equations periodic boundary conditions are chosen. As optimal velocity function we use

$$V_{opt}(\rho) = v_{max}\frac{1}{D^2\rho^2 + 1} \tag{6}$$

previously introduced by Mahnke and Pieret [2] to understand car clustering.

By using dimensionless variables $\tilde{x} = \frac{x}{D}$, $\tilde{t} = \frac{t}{\tau}$, $\tilde{\rho} = \rho D$, $\tilde{v} = \frac{v}{v_{max}}$, $\Delta\tilde{x} = \frac{\Delta x}{D}$, $\alpha = \frac{D}{\tau v_{max}}$ the derived set of equations (5) transforms into

$$\begin{aligned} \frac{\partial \tilde{\rho}}{\partial \tilde{t}} &= -\frac{1}{\alpha}\frac{\partial}{\partial \tilde{x}}(\tilde{\rho}\tilde{v}) + \frac{1}{\alpha}\frac{\Delta\tilde{x}}{2}\frac{\partial}{\partial \tilde{x}}\left(\frac{\partial \tilde{\rho}}{\partial \tilde{x}}\tilde{v} - \frac{\partial \tilde{v}}{\partial \tilde{x}}\tilde{\rho}\right), \\ \frac{\partial \tilde{v}}{\partial \tilde{t}} &= -\frac{1}{\alpha}\tilde{v}\frac{\partial \tilde{v}}{\partial \tilde{x}} + \left(\frac{1}{\tilde{\rho}^2 + 1} - \tilde{v}\right). \end{aligned} \tag{7}$$

2 Temporal Development of Car Cluster Solutions

To solve numerically the system of coupled partial differential equations (7) the spatial derivatives are approximated by central finite differences

$$\frac{\partial \tilde{\rho}_i}{\partial \tilde{x}} = \frac{\tilde{\rho}_{i+1} - \tilde{\rho}_{i-1}}{2\tilde{h}} \quad ; \quad \frac{\partial \tilde{v}_i}{\partial \tilde{x}} = \frac{\tilde{v}_{i+1} - \tilde{v}_{i-1}}{2\tilde{h}} \tag{8}$$

$$\frac{\partial^2 \tilde{\rho}_i}{\partial \tilde{x}^2} = \frac{\tilde{\rho}_{i+1} - 2\tilde{\rho}_i + \tilde{\rho}_{i-1}}{\tilde{h}^2} \quad ; \quad \frac{\partial^2 \tilde{v}_i}{\partial \tilde{x}^2} = \frac{\tilde{v}_{i+1} - 2\tilde{v}_i + \tilde{v}_{i-1}}{\tilde{h}^2}, \tag{9}$$

where $\tilde{h} = \Delta\tilde{x}$ is a step in space. Then Eqs. (7) together with periodic boundary conditions and given initial conditions are solved using Runge-Kutta 4th order method. Note that by applying central finite differences to the first equation of (7) we end up back to Eq. (1).

For some parameter values of α and $\Delta\tilde{x}$ we see that the homogeneous flow solution is unstable and clusters are formed (see Fig. 2). However Eq. (7) describes

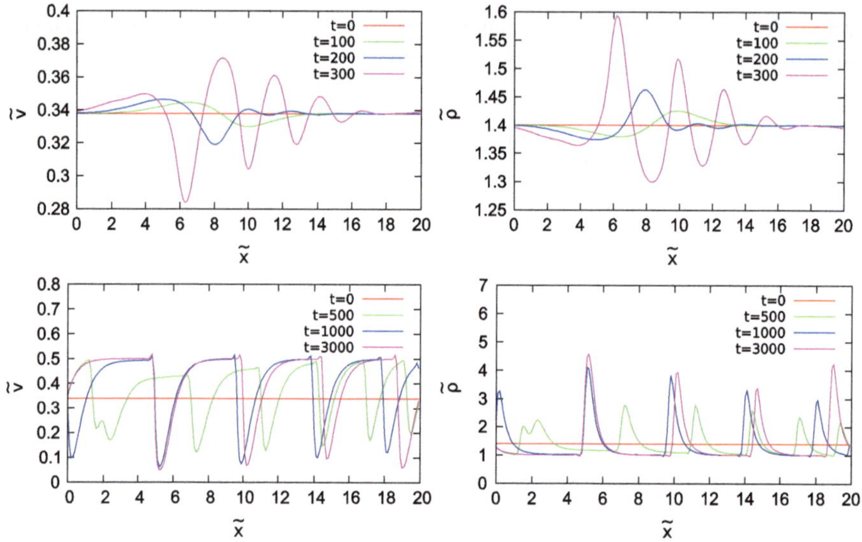

Fig. 2 Solution of Eq. (7) with $\alpha = 4.0$, $\tilde{L} = 20$, $\Delta\tilde{x} = 0.1$ for different time moments starting with the homogeneous solution $\tilde{\rho}(\tilde{t}) = \tilde{\rho}_h = 1.4$, $\tilde{v} = \frac{1}{1+\tilde{\rho}_h^2}$

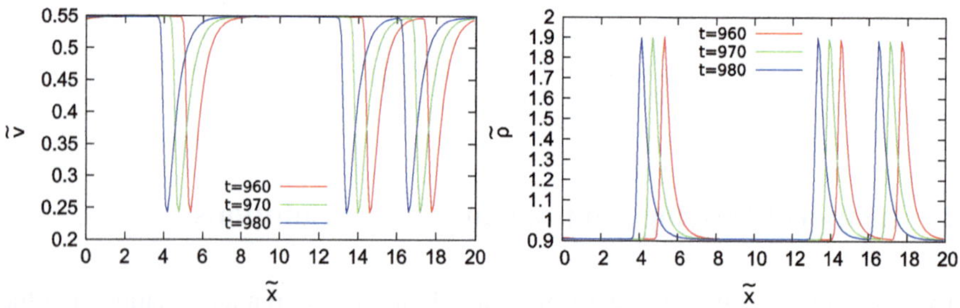

Fig. 3 Clusters moving in the opposite of driving direction of cars (cars drive such that $\tilde{x}$ increases)

two types of clusters, those which are moving in the opposite direction to driving direction of cars (see Fig. 3) and in the same direction (see Fig. 4). We get clusters which are moving in the same direction as cars if we start with lower values of space averaged initial density as initial condition. This type of cluster is usually not observed in microscopic traffic models, however, one is able to see this in real traffic data analysis and other traffic models.

If we tray to simulate Eqs. (7) for longer times, we see, that the number of clusters reduces and after some time there are only few clusters existing. Looking at the density profiles in Fig. 5 it seems that they are going to merge together, but it will take too long computational time. Unfortunately we are still not able to get initial conditions such that a one-cluster-solution can be reached in shorter time.

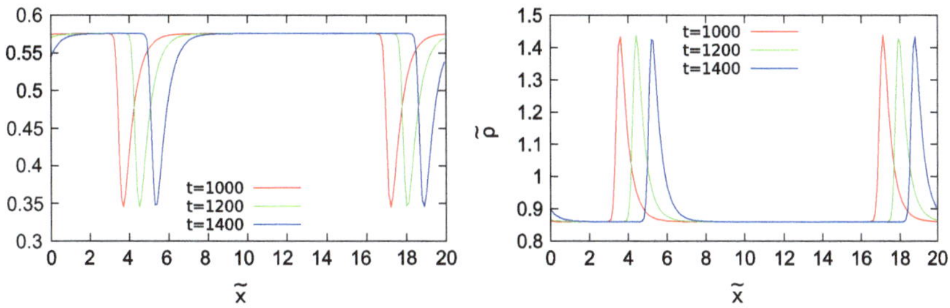

Fig. 4 Clusters moving in the same direction as cars (cars drive such that $\tilde{x}$ increases)

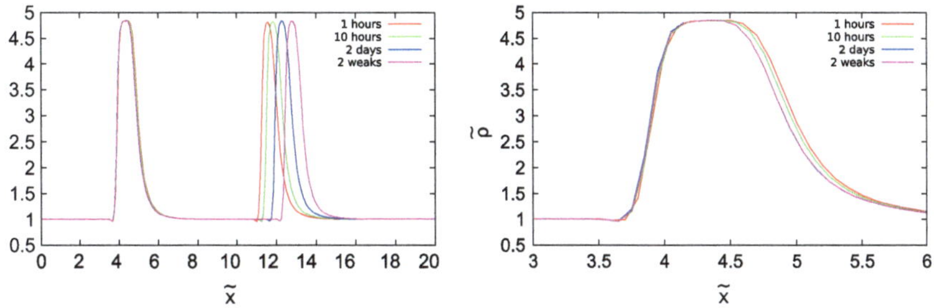

Fig. 5 Long-time solution of Fig. 2 (*right-hand part*)

3 Remarks on Steady States Profiles

If we want to get stationary moving profiles which moves with speed $\tilde{v}_g$, then the transformation to a new coordinate system $\tilde{\xi} = \tilde{x} - \frac{1}{\alpha}\tilde{v}_g\tilde{t}$ is useful.

The equations of motion after transformation into moving frame system can be written as

$$\begin{aligned} \frac{\partial\tilde{\rho}}{\partial\tilde{t}} - \frac{1}{\alpha}\tilde{v}_g\frac{\partial\rho}{\partial\tilde{\xi}} &= -\frac{1}{\alpha}\frac{\partial}{\partial\tilde{\xi}}(\tilde{\rho}\tilde{v}) + \frac{1}{\alpha}\frac{\Delta\tilde{x}}{2}\frac{\partial}{\partial\tilde{\xi}}\left(\frac{\partial\tilde{\rho}}{\partial\tilde{x}}\tilde{v} - \frac{\partial\tilde{v}}{\partial\tilde{x}}\tilde{\rho}\right), \\ \frac{\partial\tilde{v}}{\partial\tilde{t}} - \frac{1}{\alpha}\tilde{v}_g\frac{\partial v}{\partial\tilde{\xi}} &= -\frac{1}{\alpha}\tilde{v}\frac{\partial\tilde{v}}{\partial\tilde{\xi}} + \left(\frac{1}{\tilde{\rho}^2+1} - \tilde{v}\right). \end{aligned} \tag{10}$$

Then for such profiles we have

$$\frac{\partial\tilde{v}(\tilde{\xi},\tilde{t})}{\partial\tilde{t}} = \frac{\partial\tilde{\rho}(\tilde{\xi},\tilde{t})}{\partial\tilde{t}} = 0\,. \tag{11}$$

To calculate such a profile we have to solve the following equations

$$\begin{aligned} \frac{\partial\tilde{\rho}}{\partial\tilde{\xi}} &= \frac{2}{\Delta\tilde{x}}\left(\tilde{\rho}\left(1 - \frac{\tilde{v}_g}{\tilde{v}}\right) - \frac{C_g}{\tilde{v}}\right) + \frac{\alpha\tilde{\rho}}{\tilde{v}(\tilde{v}-\tilde{v}_g)}\left(\frac{1}{\tilde{\rho}^2+1} - \tilde{v}\right) \\ \frac{\partial\tilde{v}}{\partial\tilde{\xi}} &= \frac{\alpha}{\tilde{v}-\tilde{v}_g}\left(\frac{1}{\tilde{\rho}^2+1} - \tilde{v}\right), \end{aligned} \tag{12}$$

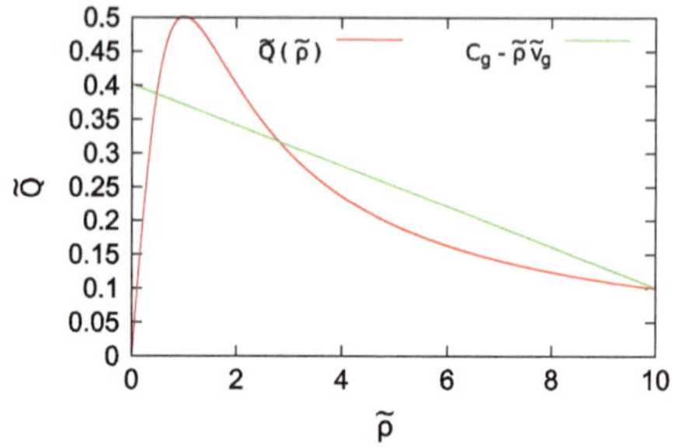

Fig. 6 Graphical solution of Eqs. (15) with $C_g = 0.4$ and $\tilde{v}_g = -0.03$. Solutions are at the intersections of curves

where C_g is an integration constant which physical meaning of flux in this moving reference frame $C_g = \tilde{\rho}_h(\tilde{v}_h - \tilde{v}_g)$ and $0 \leq C_g \leq 0.5 - \tilde{\rho}_h \tilde{v}_g$. Here $(\tilde{\rho}_h, \tilde{v}_h)$ is the homogeneous solution in the stationary reference frame.

3.1 Homogeneous Solutions

The equations for a homogeneous solution are given as

$$\begin{aligned} C_g &= \tilde{\rho}\left(\tilde{v} - \tilde{v}_g\right) \\ 0 &= \alpha\left(\frac{1}{\tilde{\rho}^2 + 1} - \tilde{v}\right), \end{aligned} \tag{13}$$

and if we introduce the dimensionless flux $\tilde{Q} = \tilde{\rho}\tilde{v}$ which is given in this case as

$$\tilde{Q}(\tilde{\rho}) = \frac{\rho}{\rho^2 + 1} \tag{14}$$

then Eqs. (13) can be written as

$$\begin{aligned} C_g + \tilde{\rho}\tilde{v}_g &= \tilde{Q}(\tilde{\rho}), \\ \tilde{v} &= \frac{1}{\tilde{\rho}^2 + 1}. \end{aligned} \tag{15}$$

The Eqs. (15) can be simply solved graphically and an example with $C_g = 0.4$ and $\tilde{v}_g = -0.03$ can be found in Fig. 6. The solutions are the intersection points. Note that for negative $\tilde{v}_g$ the value C_g can be larger then 0.5. In general as we can conclude from Fig. 6 for positive $\tilde{v}_g$ we can get up to two solutions, and for negative $\tilde{v}_g$ up to three solutions. Let us look to a situation with the values $\tilde{\rho}_h = 1.4$ and $\tilde{v}_g = -0.079$. This case corresponds to $C = \tilde{\rho}_h \tilde{v}_h \approx 0.473$ with $C_g \approx 0.584$. Graphical solutions of Eq. 15 can be seen in Fig. 7. As we can see in Fig. 7 we have three solutions $\tilde{\rho} \approx \{\tilde{\rho}_h = 1.4, 1.075, 4.80\}$. We also see that $\tilde{\rho} = \tilde{\rho}_h$ is still a

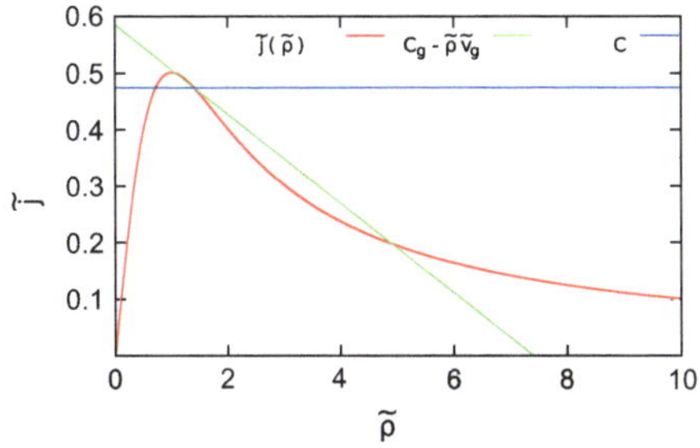

Fig. 7 Graphical solution of Eq. (15) for $C \approx 0.473$ and $\tilde{v}_g = -0.079$. Solutions are at crossings of lines

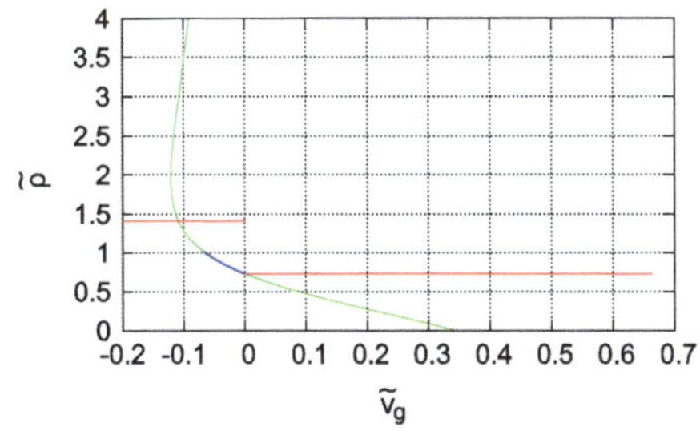

Fig. 8 Valid solution of Eq. (15) as function of $\tilde{v}_g$ for $C \approx 0.473$. *Red line* shows stationary homogeneous solution, *green line* shows solutions for which small and large perturbations propagate both with either positive or negative speeds. *Blue line* shows the solution for which small perturbations propagate with positive speed and large perturbations with negative

solution of the stationary problem in moving reference frame and it can be shown that this is valid for every velocity $\tilde{v}_g$ and every density $\tilde{\rho}_h$. If we now look to Fig. 5 we actually see that the peak density in clusters is $\tilde{\rho}_c \approx 4.80$ and for free flow $\tilde{\rho}_f \approx 1.08$. So these two non-trivial solutions have some meaning, see also Fig. 8. Related research has been published [3, 4].

References

1. M. Hilliges, W. Weidlich: A phenomenological modeling for dynamic traffic flow in networks, Transpn. Res.-B **29**, 407–431 (1995)
2. R. Mahnke, N. Pieret: Stochastic master-equation approach to aggregation in freeway traffic, Phys. Rev. E **56**, 2666–2671 (1997)
3. Y. Sugiyama, K. Masuoka, T. Ishida: Dynamics of dissipative system with asymmetric interaction and N–body problem for the emergence of moving cluster, In: Traffic and Granular Flow '07 (Eds.: C. Appert-Rolland, F. Chevoir, Ph. Gondret, S. Lassarre, J.-P. Lebacque, M. Schreckenberg), pp. 556–563, Springer, Berlin, 2009
4. P. Wagner: Fluid-dynamical and microscopic description of traffic flow: a data-driven comparison, Phil. Trans. R. Soc. A **368**, 4481–4495 (2010)

Stationary State Properties of a Microscopic Traffic Flow Model Mixing Stochastic Transport and Car-Following

Sylvain Lassarre, Michel Roussignol, Andreas Schadschneider, and Antoine Tordeux

Abstract We study the stationary properties of a microscopic traffic flow model related to a continuous time mass transport process. It is a stochastic collision-free mapping of a classical deterministic first order car-following model calibrated by the targeted speed function and the driver reaction time. The stationary states of the model are analytically treated for vanishing reaction time. Some approximations are calculated, assuming a product form of the invariant measure. When the reaction time is strictly positive, the process is studied by simulation. A relation between the parameters and the propagation of kinematic *stop-and-go* waves is identified as identical to the well-known stability condition of the car-following model. The results underline a negative impact of the driver reaction time parameter on the homogeneity of the flow in stationary state.

S. Lassarre (✉)
IFSTTAR – GRETTIA, 2 rue de la Butte Verte, 93166 Noisy le Grand, France
e-mail: sylvain.lassarre@ifstarr.fr

M. Roussignol
Paris-Est University – LAMA, bv. Descartes, 77454 Marne-la-Vallée, France
e-mail: michel.roussignol@univ-mlv.fr

A. Schadschneider
Institut für Theoretische Physik, Universität zu Köln, 50937 Köln, Germany
e-mail: as@thp.uni-koeln.de

A. Tordeux
Paris-Est University – LVMT, 19 rue Alfred Nobel, 77455 Marne-la-Vallée, France
e-mail: antoine.tordeux@enpc.fr

V.V. Kozlov et al. (eds.), *Traffic and Granular Flow '11*,
DOI 10.1007/978-3-642-39669-4_4, © Springer-Verlag Berlin Heidelberg 2013

1 Introduction

Stochastic transport processes, extracted from the systems of interacting particles, can be used to model traffic flow [1]. The approach represents an alternative to the modelling of traffic flow by systems of differential equations and is an extension in continuous time of cellular automata models. Markovian jump processes such as *exclusion* [2, 3], *zero-range* [2, 4, 5], *random average* [6, 7] or *misanthrope* [8] processes can describe microscopic traffic flows in the totally asymmetric case and for specific interpretations of the model parameters.

The totally asymmetric simple exclusion process (TASEP) is a basic model that has been studied extensively and is often used as a theoretical tool. The TASEP with nearest-vehicle interaction (which can be mapped to a zero-range process [5, 9, 10]) as well as with Arrhenius interaction [11] are employed to model traffic flows. The zero-range process is used to model the evolution of platoons in [12]. Recently, the misanthrope process is applied to describe microscopic multi-lane traffic flow [13] or mesoscopic ones [14]. These approaches, based on particle systems, are defined on a discrete space. The totally asymmetric random average process (TARAP), studied in [15], allows the microscopic modelling of traffic flows in continuous space.

On the other hand, microscopic traffic models based on ordinary or delayed differential systems are developed since the 1950s. The approach assumes interactions of the vehicles with their predecessors. Fundamental parameters, initially estimated from statistical observations, of the targeted speed as a function of the distance gap and the driver reaction time are sufficient to reproduce reasonably the driver's behavior. Models such as the first order one by Newell [16] or the second order *Optimal Velocity* model [17] are well understood. Notably, conditions of linear stability of homogeneous configurations are known. The first one imposes the derivative of the targeted speed function to be strictly positive, the second one a derivative strictly positive and strictly less than the inverse of two times the reaction time [17].

A new model combining the two approaches is proposed. The model is a stochastic mass transport process mapping a discretisation scheme of the Newell car-following model. The stochastic process we use is close to the TARAP. It models the evolution of vehicles in continuous time on a continuous space.

The paper is organised as follows. In a first part, we study the basic model without reaction time. The model behaviour in the stationary state is described analytically with the help of results taken from the literature. In a second part, we define and study the mass transport model including a strictly positive driver reaction time parameter. The stationary state of the process is investigated by simulations. Its stability properties are underlined according to the values of the parameters. We conclude by comparing the properties of the model to empirical observations and classical traffic theories.

2 The Basic Model with No Reaction Time

Let us start by mapping the TARAP to the Newell first order car-following model with no delay.

2.1 Model Definition

The model represents the evolution in continuous time of the distance gaps of vehicles on a one-way road. Since the vehicle order remains the same, this representation is an exact mapping of a line of vehicles [9, 15]. When a vehicle moves, a part of its distance gap is transmitted to the following vehicle. This representation allows the use of zero-range processes in discrete space case or random average processes in the continuous space case that is studied here.

$\eta = (\Delta_i)_i \in E = \left(\mathbb{R}^+\right)^{\mathbb{Z}}$ denotes the process of the vehicles distance gaps on an infinite lane. The jump size and jump rate of the process are defined by a mapping of the Newell first order car-following model: $\frac{\mathrm{d}x_i}{\mathrm{d}t}(t) = V\big(\Delta_i(t)\big)$ where $x_i(t)$ is the position of the vehicle i at the instant t. $\Delta_i = x_{i+1} - x_i - \ell$, with the vehicle length ℓ, is the distance gap of the vehicle i that depends on the position of its predecessor $i+1$. $V : \Delta \mapsto \min\{V_{\max}, \Delta/T\}$ is the positive non-decreasing and piecewise linear targeted speed function. $V_{\max} > 0$ is the maximal desired speed in the free traffic state and $T > 0$ is the targeted time gap for a vehicle in the pursuit case. For a vehicle i the explicit eulerian discretisation scheme of its motion with time step $\delta t > 0$ is

$$x_i(t+\delta t) = x_i(t) + \delta t \cdot V\big(\Delta_i(t)\big)$$

with the position x and the distance gap Δ to the predecessor. In this discrete time model the jump size of vehicle i is $\delta t V(\Delta_i)$ and the jump time is δt. One may consider a stochastic model with a jump rate equal to $1/\delta t$ (and the mean jump time δt) with jump size $\delta t V(\Delta_i)$. For this model, the generator is

$$\mathscr{L} f(\eta) = \sum_{i \in I_f} \frac{1}{\delta t}[f(\eta^i) - f(\eta)]\mathbb{1}_{\{\Delta_i > \delta t\, V(\Delta_i)\}} \tag{1}$$

with $\eta^i = (\Delta^i_j)_j$ such that $\Delta^i_j = \Delta_j$ if $j \neq i$ and $j \neq i-1$, $\Delta^i_i = \Delta_i - \delta t V(\Delta_i)$ and $\Delta^i_{i-1} = \Delta_{i-1} + \delta t V(\Delta_i)$. The generator is an operator of a function f depending on the finite set of coordinates I_f.

The Chapman-Kolmogorov equation allows to describe the marginal first order momentum of the process. As expected, one obtains the Newell car-following form applied to the distance gap ($(.)'$ denotes the time derivative):

$$\big(\mathbb{E}(\Delta_i(t))\big)' = \mathscr{L}\,\mathbb{E}(\Delta_i(t)) = \mathbb{E}\big(V(\Delta_{i+1}(t)) - V(\Delta_i(t))\big).$$

Vehicles trajectories are independent for the free case where $V(\Delta) = V_{\max}$ is constant. The model is a totally asymmetric random average process [7] when $V(\Delta) = \Delta/T$ is a non-decreasing linear function. The generator of a TARAP is

$$\mathscr{L} f(\eta) = \sum_{i \in I_f} \lambda \int p(\mathrm{d}u) \, [f(\eta^i(u)) - f(\eta)] \, \mathbb{1}_{\{\Delta_i > 0\}} \tag{2}$$

with $\eta^i = (\Delta^i_j)_j$ such that $\Delta^i_j = \Delta_j$ if $j \neq i$ and $j \neq i-1$, $\Delta^i_i(u) = u\Delta_i$ and $\Delta^i_{i-1}(u) = \Delta_{i-1} + (1-u)\Delta_i$. $p(\mathrm{d}u)$ is the distribution on $[0, 1]$ of the distance gap fraction jumping, concentrated on $1 - \delta t/T$ in the traffic model, and $\lambda > 0$ is the constant jump rate corresponding to the inverse of the time step δt.

The vehicles jump successively and the jump times of each vehicle are continuous. More precisely, the jump time of a vehicle is an homogeneous poissonian process. The jump size does not exceeded the distance gap.

The quantity $\delta t > 0$ is not a physical parameter but a tool of the modelling that should be close to 0 (at least $\delta t \leq T$). In the following, one focuses on the limit case where $\delta t \to 0$ and targeted speed function $V(\Delta) = \Delta/T$.

2.2 *Stationary State Description*

When the jump size remains constant (free case), vehicle trajectories are independent. The stationary distribution of this process is a product form and the invariant marginal distributions are independent and identical distributed and exponential. This is not the case for the TARAP. In contrast to the zero-range process whose invariant distribution is a product form [2, 4, 5], the invariant distribution of the TARAP is not known for any p distribution. However, explicit formulas for the first and second moments of the marginal invariant distribution have been derived [7].

First and Second Moment of the Marginal Invariant Distribution

If $\alpha \in E$ is the space-homogeneous initial distribution of the process, i.e. $\mathbb{P}(\eta_0 = \alpha) = 1$ with the initial system state η_0, such that $\mathbb{E}\alpha_i = D$ for all i and $\sum_i |\mathbb{E}\alpha_0\alpha_i - D^2| < \infty$, the first and second order moment of the marginal distribution satisfy [7]:

$$\mathbb{E}\Delta_i = D \qquad \forall i \forall t \qquad \text{and} \qquad \begin{aligned} \lim_{t\to\infty} \mathbb{E}\Delta_i\Delta_j &= D^2 & \forall i \neq j \\ \lim_{t\to\infty} \mathbb{E}\Delta_i^2 &= \frac{r}{s} D^2 & \forall i \end{aligned}$$

with $r = \int (1-x)p(x)\mathrm{d}x$ and $s = \int x(1-x)p(x)\mathrm{d}x$. The first moment is the mean distance gap that remains constant since the system is conservative. The covariance is nil, attesting that the invariant distribution of the process may have a product form.

In the car-following model given by generator (1) with $V(\Delta) = \Delta/T$, $r = \delta t/T$, $s = \delta t/T(1-\delta t/T)$ and

$$\lim_{t\to\infty} \mathbb{V}\Delta_i = \lim_{t\to\infty} \mathbb{E}\Delta_i^2 - (\mathbb{E}\Delta_i)^2 = D^2 \frac{\delta t}{T-\delta t} \qquad \forall i. \tag{3}$$

The variability of the vehicles distance gap tends towards 0 in the stationary state (i.e. the flow is homogeneous) when $\delta t \to 0$. This aspect is observed in the Newell car-following model with differential equation. It is well-known that the homogeneous state of this model is a stable equilibrium state when V is non-decreasing (see for instance [14]).

Invariant Distribution

The stationary measure of the process, denoted $\pi : E \mapsto [0,1]$, satisfies the invariant equation

$$\int_E \mathscr{L} f(\eta)\, \pi(\mathrm{d}\eta) = 0\,. \tag{4}$$

If one assumes that the stationary measure π admits a distribution ($\pi(\mathrm{d}\eta) = \pi(\eta)\prod_j \mathrm{d}\Delta_j$), the stationary equation is $\int_E \mathscr{L} f(\eta)\, \pi(\eta) \prod_j \mathrm{d}\Delta_j = 0$.

For the TARAP given by the generator (2), one obtains after substituting for all $i \in I_f$, Δ_{i-1} by $\Delta_{i-1} - (1-u)\Delta_i$ and Δ_i by Δ_i/u:

$$\int_E f(\eta) \prod_j \mathrm{d}\Delta_j \Bigg\{ \sum_{i\in I_f} \int_{\left(1+\frac{\Delta_{i-1}}{\Delta_i}\right)^{-1}}^1 p(u) \frac{\mathrm{d}u}{u} \times \\ \pi\left(\dots, \Delta_{i-2}, \Delta_{i-1} - \frac{1-u}{u}\Delta_i, \frac{\Delta_i}{u}, \Delta_{i+1}, \dots \right) \Bigg\} = \int_E f(\eta) \prod_j \mathrm{d}\Delta_j \Big\{ \sum_{i\in I_f} \pi(\eta) \Big\}.$$

This equality is satisfied for any function f depending on $I_f \subset \mathbb{Z}$ with $\mathrm{card}(I_f) < \infty$ if for all $\eta \in E$ the following equality using $x = \frac{1-u}{u}\Delta_i$ holds:

$$\sum_{i\in I_f} \int_0^{\Delta_{i-1}} p\left(\frac{\Delta_i}{x+\Delta_i}\right) \frac{\mathrm{d}x}{x+\Delta_i} \pi\left(\dots, \Delta_{i-2}, \Delta_{i-1} - x, \Delta_i + x, \Delta_{i+1}, \dots\right) = \sum_{i\in I_f} \pi(\eta).$$

Assuming that the invariant distribution of the process has product form of homogeneous in space marginal distribution $\tilde{\pi} : \mathbb{R}^+ \mapsto [0,1]$ such that $\int x\, \tilde{\pi}(x)\, \mathrm{d}x = D$, $\pi(\eta) = \prod_i \tilde{\pi}(\Delta_i)$, one obtains the equality (cf. [18] with p uniform):

$$\sum_{i\in I_f} \int_0^{\Delta_{i-1}} p\left(\frac{\Delta_i}{x+\Delta_i}\right) \frac{\mathrm{d}x}{x+\Delta_i} \frac{\tilde{\pi}(\Delta_{i-1} - x)\, \tilde{\pi}(\Delta_i + x)}{\tilde{\pi}(\Delta_{i-1})\tilde{\pi}(\Delta_i)} = \mathrm{card}(I_f).$$

This equality holds when one has for all $i \in I_f$:

$$\int_0^{\Delta_{i-1}} p\left(\frac{\Delta_i}{x+\Delta_i}\right) \frac{\mathrm{d}x}{x+\Delta_i} \frac{\tilde{\pi}\left(\Delta_{i-1}-x\right) \tilde{\pi}\left(\Delta_i+x\right)}{\tilde{\pi}(\Delta_{i-1})\tilde{\pi}(\Delta_i)} = 1. \tag{5}$$

In the case of the car-following model (1), this equation does not admit explicit solution for the marginal $\tilde{\pi}$. In the next paragraph we show that the process has an invariant product form distribution with gamma marginal solution for particular beta distributions of p.

Beta Distribution for *p*

We assume that p has a beta distribution on [0, 1] with parameters $m > 0$ and $n > 0$:

$$p(u) = \frac{1}{\beta(m,n)} u^{m-1}(1-u)^{n-1}\mathbb{1}_{[0,1]}(u)$$

with $\beta(m,n) = \int_0^1 u^{m-1}(1-u)^{n-1}\mathrm{d}u$.

To keep the meaning of the traffic model, one assumes that the expected value of p is equal to $1-\delta t/T$. One denotes $\sigma = \frac{1}{K}\frac{\delta t}{T}\left(1-\frac{\delta t}{T}\right)$ with $K > 1$ the variance of the p distribution. One obtains $m = (1-\delta t/T)(K-1)$ and $n = \delta t/T\,(K-1)$. The beta distribution tends towards a deterministic one concentrated on $1-\delta t/T$ when K tends towards infinity. It becomes bi-modal concentrated on $\{0, 1\}$ when K tends towards 1.

If we refer to formulas given in [7], the variability of the distance gap in stationary state for a beta distribution of p in stationary state is:

$$\lim_{t\to\infty} \mathbb{V}\Delta_i = D^2 \frac{T+\delta t(K-1)}{(T-\delta t)(K-1)} \qquad \forall i. \tag{6}$$

The variance tends towards infinity for $\delta t \to T$ or $K \to 1$. The distance gap variance tends towards $D^2/(K-1)$ when $\delta t \to 0$ and towards $D^2\delta t/(T-\delta t)$ when $K \to \infty$. Therefore, the variance tends towards 0 for $\delta t \to 0$ and $K \to \infty$ (that corresponds to the deterministic case of p described previously).

When we assume a beta distribution for p and a measure product form with marginal gamma distribution with parameter γ and θ:

$$\tilde{\pi}(x) = x^{\gamma-1}\frac{\exp(-x/\theta)}{\Gamma(\gamma)\theta^\gamma}\mathbb{1}_{[0,\infty)}(x)$$

with $\Gamma(\gamma) = \int_0^\infty u^{\gamma-1}e^{-u}\mathrm{d}u$, the condition of invariance (5) leads to

$$\int_0^{\Delta_{i-1}} \frac{1}{\beta(m,n)} \left(\frac{\Delta_i}{x+\Delta_i}\right)^{m-1} \left(\frac{x}{x+\Delta_i}\right)^{n-1} \frac{\mathrm{d}x}{x+\Delta_i}\left[(\Delta_{i-1}-x)(\Delta_i+x)\right]^{\gamma-1} = \left[\Delta_{i-1}\Delta_i\right]^{\gamma-1}$$

for all $i \in I_f$ By substituting x by $u \cdot \Delta_{i-1}$, one obtains after simplifications for all $i \in I_f$:

$$\frac{\Delta_i^{m-\gamma}\Delta_{i-1}^n}{\beta(m,n)} \int_0^1 (\Delta_i + \Delta_{i-1}u)^{\gamma-m-n}\, u^{n-1}(1-u)^{\gamma-1}\mathrm{d}u = 1. \tag{7}$$

If one considers the gamma distribution with $\gamma = m + n = K - 1$ (and $\theta = (K - 1)/D$ to be sure that the expected value is equal to D), the equation of invariance is:

$$\left(\frac{\Delta_{i-1}}{\Delta_i}\right)^n \frac{\beta(\gamma,n)}{\beta(m,n)} = \left(\frac{\Delta_{i-1}}{\Delta_i}\right)^n \frac{(\Gamma(\gamma))^2}{\Gamma(\gamma+n)\Gamma(\gamma-n)} = 1\,. \tag{8}$$

This equation is satisfied for any system state $\eta \in E$ in the limit case $n \to 0$. This limit is reached when $\delta t \to 0$ or if $K \to 1$. Yet, the invariant marginal gamma distribution depends on K but not on δt. For a given $K > 1$, the gamma distribution is the exact asymptotic invariant marginal distribution for $\delta t \to 0$. For given $\delta t > 0$ and $K > 1$, the product of gamma distributions is an approximation of the invariant measure of the system. The variance of the gamma distribution is equal to $D^2/(K-1)$ while the exact value given previously is $D^2\frac{T+\delta t(K-1)}{(T-\delta t)(K-1)}$ (cf. Eq. (6)). For any value of δt, the variability of the approximation is less that the exact one.

Some simulation experiments are undertaken to evaluate the precision of the gamma product approximation. 100 vehicles with a length of 5 m on a ring of length 2 km are considered. The targeted speed function is exclusively the pursuit one (i.e. $V_{\max} = +\infty$). The samples are obtained after a simulated time of 2 h which is sufficient to reach the stationary state. The initial configurations are uniform. The sample size for each value of the parameters is equal to $5 \cdot 10^5$ observations.

Figure 1 shows the distributions obtained. The lines represent the empirical distance gap distributions for various values of the parameter δt (varying from 1 to 0.05 s). The dotted lines are the asymptotic distributions $\Gamma(K-1, D/(K-1))$. Different values for K are compared (2, 10 and 50). The empirical distributions tend towards the asymptotic ones when $\delta t \to 0$ in all cases. For a given value of δt, the differences between the exact and the gamma distributions increase when K tends towards infinity.

The form of the distributions varies significantly in the limit case where δt tends towards 0 with the value of K. The form is exponential when K is less than 2 and monomial otherwise. The exponential case corresponds to the formation of platoon in stationary state in the system, while the flow is homogeneous when the distribution is monomial. On the Fig. 2, vehicles trajectories on a ring, from uniform initial conditions, are plotted for two values of the parameter K. The trajectories are perturbated and vehicles platooning occurs for K close to 1 (Fig. 2 at left). When K is high, p variability is low and the flow is homogeneous (Fig. 2 at right).

The formation of platoons observed when K tends towards 1 is due to the increase of noise in the system through the distribution p. For the deterministic

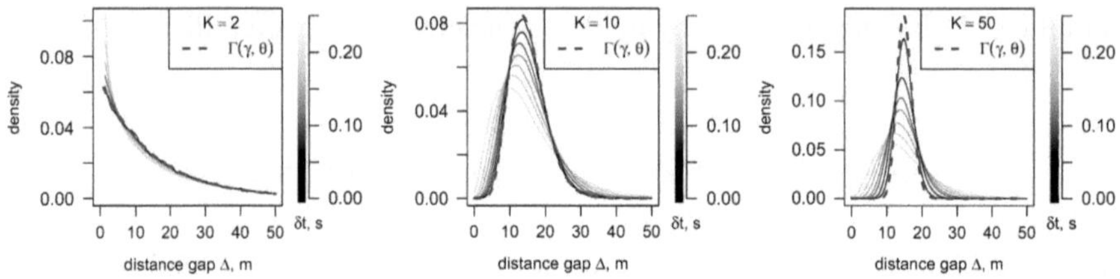

Fig. 1 Comparison of the asymptotic gamma marginal distribution with simulation data on a ring ($T = 1.2$ s, $V_{\max} = +\infty$ and $D = 15$ m)

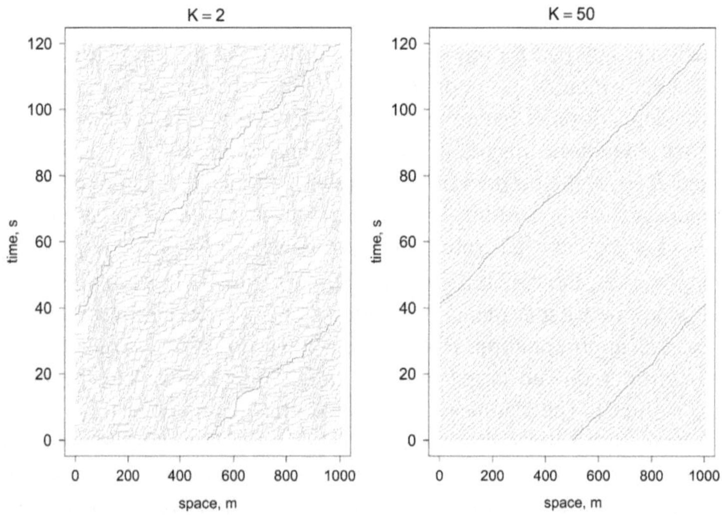

Fig. 2 Trajectories of 50 vehicles evolving on a ring ($\delta t = 0.01$ s, $T = 1.2$ s, $V_{\max} = +\infty$ and $D = 15$ m)

p distribution, the flow is homogeneous for δt small. In the next section, a model including a reaction time produces kinematic waves at the limit case when δt tends towards 0 and for the deterministic p distribution.

3 The Model Including a Reaction Time

The proposed model is a stochastic mapping of the delayed Newell car-following model. It is a stochastic mass transport process distinct from the TARAP.

3.1 Model Definition

As previously, the model is defined from an explicit eulerian discretisation of the differential equation with $T^r > 0$: $\frac{\mathrm{d}x_i}{\mathrm{d}t}(t) = V\big(\Delta_i(t - T^r)\big)$. If a linear development over the time is applied to the delayed distance gap to manipulate synchronous variable, the discretisation scheme for the motion of the vehicle i:

$$x_i(t + \delta t) = x_i(t) + \delta t \cdot V\big(\Delta_i(t) - T^r(v_{i+1}(t) - v_i(t))\big)$$

with the position x, the distance gap Δ to the predecessor and the speed v which is a new variable introduced in the system. Substituting the speed by the targeted speed function, a mass transport process corresponding to this discretisation can be characterised by the generator

$$\mathscr{L}f(\eta) = \sum_{i \in I_f} \frac{1}{\delta t}[f(\eta^i) - f(\eta)]\mathbb{1}_{\{\Delta_i > \delta t\, s_i\}} \tag{9}$$

with $s_i = V\big(\Delta_i - T^r(V(\Delta_{i+1}) - V(\Delta_i))\big)$ and $\eta^i = (\Delta^i_j)_j$ such that $\Delta^i_j = \Delta_j$ if $j \neq i$ and $j \neq i - 1$, $\Delta^i_i = \Delta_i - \delta t\, s_i$ and $\Delta^i_{i-1} = \Delta_{i-1} + \delta t\, s_i$.

With this form, the model is not a random average process even if V is linear since the jump size depends both on the distance gap and on the predecessor distance gap. If $V(\Delta) = \Delta/T$, one has the smallest sufficient condition on the time step parameter $\delta t \leq T/(1 + T^r/T)$.

3.2 Stationary State Description

The stationary state of the model including a reaction time is obtained by simulation. The simulation of this kind of stochastic process is easy and does not require to define a discretisation scheme. Each vehicle has an exponential clock giving the time of its next jump. The simulation is event driven by actualizing successively the vehicle with minimum jump time.

From the simulation results, two types of stationary states for the system on a ring are clearly identified for congested density levels, according to the value of the reaction time. the fist is a homogeneous stationary state with monomial and symmetric distributions of the vehicles distances gap. This state is observed when the reaction time is zero or not too large. The second is a heterogeneous stationary state with kinematic wave propagation and bi-modal distributions of vehicles distance gaps which occurs for sufficiently large reaction times.

In Fig. 3, vehicle trajectories are shown for a uniform initial configuration. The reaction time vanishes for the left system and the stationary state is homogeneous. For the right system the reaction time, equal to 1 s, is enough for convergence

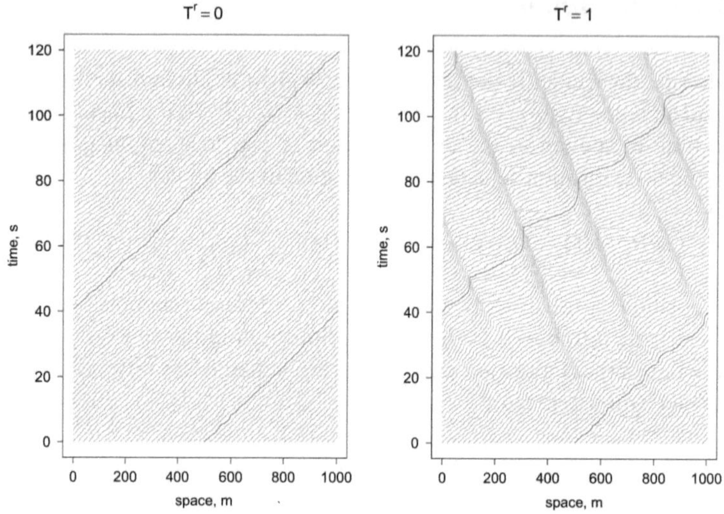

Fig. 3 Trajectories of 50 vehicles evolving on a ring ($\delta t = 0.01$ s, $T = 1.2$ s and $V_{\max} = +\infty$)

towards a heterogeneous stationary state. The formation of kinematic *stop-and-go* waves which propagate at constant speed can be observed. This phenomenon is well-known in real traffic. At a given density, the vehicles mean speeds are similar for the two stationary states although these states are different. The presence of kinematic waves is not observed in the model without reaction time, even if it produces locally vehicles platoons when the noise level is important.

Figure 4 presents the distribution of vehicle distance gaps and its mean and standard deviation for different values of the parameters. In the left part of the figure, distributions are obtained by simulations for different values of the time step parameter δt. One observes bi-modal distributions of the distance gap when δt is sufficiently close to 0. This reflects the presence of kinematic waves. At the center, the mean and standard deviation of the distributions are plotted for different values of the reaction time. A relation is observed, linking the reaction time and the targeted pursuit time with the emergence and propagation of waves in the stationary state. Various simulation experiments lead to the following condition for an homogeneous system in stationary state: $T > 2T^r$.

It is the same as the stability condition of the Newell delayed car-following model [14, 17]. It underlines a negative impact of the driver reaction time parameter on the homogeneity of the flow. To the right, one observes that the maximal speed parameter $V_{\max}$ induces a critical flow density threshold for kinematic waves to emerge. It refers to free and congested traffic states. For densities less that the critical density, vehicle speeds are close to the maximal value. Beyond the critical density threshold, vehicle speeds are regulated, with a mean less than the maximal value, and kinematic waves may emerge and propagate. Various simulation experiments show that the critical density is close to $1/(V_{\max}T + \ell)$ or equivalently for a given

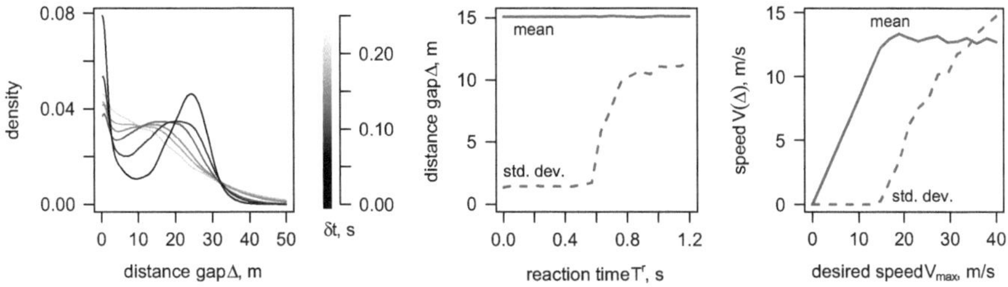

Fig. 4 Distributions in stationary state of the distance gap for different values of δt (*left*) and their mean and standard deviation for different values of, respectively, T^r and $V_{\max}$. ($\delta t = 0.01$ s, $T^r = 1$ s, $V_{\max} = 25$ m/s, $T = 1.2$ s and $D = 15$ m)

density with mean distance gap D, the critical maximal speed is $D/T = 12.5$ m/s in the figure.

4 Conclusion

We have presented results for a stochastic mass transport process calibrated by classical parameters of driver reaction time and targeted speed function, inspired by the deterministic car-following theory. The model, defined in continuous time, describes the evolution on a continuous space of the distances between vehicles without any collision. Its form allows analytical investigation in particular cases and simulations without requiring the use of a discretisation scheme.

When the reaction time vanishes, the introduction of randomness in the model induces the formation of vehicle platoons in the stationary state. Analytically, explicit approximations for the stationary distribution of the process are obtained. When the reaction time parameter is strictly positive, simulations exhibit a critical value beyond which kinematic stop-and-go waves propagate for congested densities in the stationary state. A relation linking the model parameters with the formation of waves is identified similar to the classical well-known stability conditions of car-following models.

The results underline the role of the two fundamental parameters of driver reaction time and targeted pursuit speed function. The reaction time, that is recognized and measured in real experiments, is identified as a negative factor on the homogeneity of the flow. Yet, it is known that the drivers are able to anticipate in order to deflect to the reaction time. They may increase the stability of the flow, but these aspects are not taken into account in the model.

Acknowledgements A. T. acknowledges support from VINCI/ParisTech Grant 8L1142 "Éco-Conception des ensembles bâtis et des infrastructures".

References

1. Schadschneider A, Chowdhury D, Nishinari K (2010) Stochastic Transport in Complex Systems. From Molecules to Vehicles. Elsevier, Amsterdam
2. Spitzer F (1970) Adv. Math. 5:246–290
3. Liggett T M (1985) Interacting Particle Systems. Springer
4. Andjel E (1982) Ann. Prob. 10(3):525–547
5. Evans M R, Hanney T (2005) J. Phys. A38:R195–R240
6. Pierson H M (1978) Adv. Appl. Prob. 10(3):613–632
7. Roussignol M (1980) Ann. Inst. Henri Poincaré B16(2):101–108
8. Cocozza-Thivent C (1985) Z. Wahr. Verw. Gebiete 70:509–523
9. Evans M (2007) in Traffic and Granular Flow '05. Schadschneider A, Pöschel T, Kühne R, Schreckenberg M, Wolf D E (eds). Springer. 447–459
10. Tordeux A, Roussignol M, Lassarre S. (2010) Traffic and Granular Flow'09 Proceedings. Springer.
11. Sopasakis A, Katsoulakis M (2006) SIAM 66(3):921–944
12. Kaupuzs J, Mahnke R, Harris R J (2005) Phys. Rev. E72(2):056125
13. Kanai M (2010) Phys. Rev. E82(6):066107
14. Tordeux A (2010) Ph.D. Thesis, Paris-Est University, Marne-la-Vallée, France
15. Zielen F, Schadschneider A (2003) J. Phys. A36(13):3709–3723
16. Newell G F (1961) Op. Res. 9(2):209–229
17. Bando M, Hasebe K (1995) Phys. Rev. E51(2):1035–1042
18. Krug J, Garcia J (2000) J. Stat. Phys. 99:31–55

Identifiability and Practical Relevance of Complex Car-Following Models

Gunnar Flötteröd, Peter Wagner, and Yun-Pang Flötteröd

Abstract This article looks at car-following models from a deliberately pragmatic perspective: What information about driver behavior can be extracted from a given data set without more or less speculative assumptions about underlying behavioral laws? The objective of this exercise is not to invalidate existing models but to obtain a better understanding of how much (complex) model structure can be revealed/validated from real data.

1 Introduction

The estimation of parameters of a microscopic traffic flow model appears at first glance to be a technically straightforward and well understood procedure. What is not that well understood is the question what is actually revealed by the calibration. Typically, the calibration exercise results in parameters that minimize some distance between model outputs and reality. Some of these parameters have immediate physical meanings: maximum speed, maximum acceleration, and the like. Other parameters are hard to interpret and hence are difficult to validate in hindsight, not even through simple plausibility checks. While sophisticated car-following model specifications abound, much of their added value lies in theoretically being able to explain certain (rare) phenomena, which, however, are of limited relevance if one is interested in estimating, say, a car-following model component for a complex network simulation from real data.

G. Flötteröd (✉)
KTH Royal Institute of Technology, Stockholm, Sweden
e-mail: gunnar.floetteroed@abe.kth.se

P. Wagner · Y.-P. Flötteröd
DLR German Aerospace Center, Cologne, Germany
e-mail: peter.wagner@dlr.de; yun-pang.floetteroed@dlr.de

V.V. Kozlov et al. (eds.), *Traffic and Granular Flow '11*,
DOI 10.1007/978-3-642-39669-4_5, © Springer-Verlag Berlin Heidelberg 2013

This article adopts an (almost naive) engineering perspective on the problem in that it starts the analysis by estimating a set of simple linear models from a given data set. If a linear model already explains the data well, there is little reason to complexify the model further. If, however, a linear model fails to explain certain aspects of the data, it still is possible to analyze the residuals in order to obtain data-driven hints of how to improve the model. Clearly, an argument against this approach is that a good model structure, based on physical and/or behavioral considerations, should also have a superior explanatory power. The counter-argument has in essence already been phrased: A simple model with only few, interpretable parameters may result in a slightly inferior fit, but its estimation may be more robust than for a complex model, and its simplicity and interpretability is likely to be a key feature in its practical application.

The remainder of this article investigates the above claims by constraining itself to utmost simple models and exclusively inferring model structure either from exact physical laws or from the data itself. Section 2 estimates and analyzes a set of linear car following models. Section 3 then discusses the implications of this article's findings, in particular with respect to more complex nonlinear model specifications.

2 Estimation and Analysis of Linear Models

The data used in this contribution has been recorded some years ago on a Japanese test track [2]. Nine drivers were following a lead driver, who implemented a certain speed protocol, changing between regimes of constant and varying speeds. Most acceleration values were moderate and in the range $[-3, 3]\,\mathrm{m/s^2}$. The drivers' kinematic states were tracked by means of a differential RTK-GPS system, which allows for centimeter accuracy in positions and centimeter-per-second precision in speeds.

Due to the real-time kinematics in the GPS recording, speeds are independently measured variables (essentially based on the Doppler shift in the signal frequency between vehicle and satellites) and are not derived via numerical differentiation from the positions. Altogether, data from eight different experiments of 25 min each is available, with different drivers and different vehicle sequences per experiment. Within this contribution, only the first experiment is analyzed. The full data set has been used in another study on car-following models and their calibration [1].

2.1 Considered Model Specifications

A number of increasingly complex linear models is calibrated from the data described above. The purpose of this is to understand how much information is contained in the data that can be captured by linear dynamics.

We start from the optimal velocity model (OVM) in discrete time t, which is written as

$$v(t) - v(t-h) = \alpha \cdot (v^*(t-h) - v(t-h)) + \varepsilon(t) \tag{1}$$

where v is the considered driver's speed, v^* is the instantaneous optimal velocity (yet to be defined), $\alpha \in (0, 1)$ defines the speed of the driver's adaptation towards v^*, h is the time step length, and ε is a temporally uncorrelated error term. (The last assumption is convenient in terms of model estimation but needs to be validated and possibly adjusted in future work [3].) Specifically, the following model instances are considered.

Speed synchronization model. This model assumes $v^*(t) = V(t)$, which means that the driver does nothing but follow the speed V of the leading car. The resulting linear model specification has only one independent parameter:

$$v(t) = a_1 v(t-h) + c_1 V(t-h) + \varepsilon(t) \tag{2}$$

where a_1 corresponds to $1 - \alpha$ and c_1 is constrained to be $1 - a_1$.

Newell-type model. This model assumes $v^*(t) = \frac{1}{\tau} g(t)$ with $\tau > 0$, which means that the desired speed is proportional to the distance g from the leading car. The resulting linear model specification has two parameters:

$$v(t) = a_1 v(t-h) + b_1 g(t-h) + \varepsilon(t) \tag{3}$$

where a_1 corresponds to $1 - \alpha$ and b_1 corresponds to $\alpha \frac{1}{\tau}$.

Combined OVM model. This model assumes $v^*(t) = \gamma \frac{1}{\tau} g(t) + (1-\gamma) V(t-h)$, $0 \leq \gamma \leq 1$, which defines the desired speed as a convex combination of lead car speed and distance. The resulting linear model specification has three parameters:

$$v(t) = a_1 v(t-h) + b_1 g(t-h) + c_1 V(t-h) + \varepsilon(t) \tag{4}$$

where a_1 corresponds to $1 - \alpha$, b_1 corresponds to $\alpha \frac{1}{\tau} \gamma$, and $c_1$corresponds to $\alpha(1 - \gamma)$.

Generalized OVM model. A natural extension of the combined OVM model is to allow for higher order dynamics in the speed adaptation process. This leads to the following linear model specification:

$$v(t) = \sum_{i=1}^{\dots} a_i v(t-ih) + \sum_{i=1}^{\dots} b_i g(t-ih) + \sum_{i=1}^{\dots} c_i V(t-ih) + \varepsilon(t). \tag{5}$$

Since this model allows for arbitrary linear combinations of past gaps and velocities of the modeled car and the lead car, it implicitly also captures speed differences as explanatory variables. Also, since $g(t) - g(t-h) = h \cdot (V(t-h) - v(t-h))$, a linear dependency between g, v, and V can be expected. Indeed, it turns out

Table 1 Estimated parameters for a single driver and the three simple linear models (2)–(4). The values in brackets indicate the statistical error in the parameter estimate, it relates to the last digit(s) of the parameter value, i.e. 0.9721 ± 0.0005 is written as 0.9721(5)

Model	$v(t-h)$	$g(t-h)$	$V(t-h)$	σ	LL	AIC
Speed -sync.	0.9721(5)	–	0.0279(6)	0.0565	23,761	−47,516
Newell	0.9968(1)	0.0021(1)	–	0.0595	22,917	−45,828
Combined	0.9719(6)	0.00127(9)	0.02620(6)	0.0561	23,865	−47,723

that the parameters $b_{i>1}$, $c_{i>1}$ cannot be identified. What worked, and therefore is exclusively considered in the following, is the simplification

$$v(t) = \sum_{i=1}^{\dots} a_i v(t-ih) + b_1 g(t-h) + c_1 V(t-h) + \varepsilon(t). \tag{6}$$

2.2 Description and Analysis of Estimation Results

A first set of model estimation results is shown in Table 1. All results are obtained with R's generalized linear model routine glm(). Also shown in this table are the following measures of fit:

- The average squared residual

$$\sigma^2 = \frac{1}{N}\sum_{i=1}^{N}\left(v_i^{\text{sim}} - v_i^{\text{data}}\right)^2 \tag{7}$$

  where N is the number of data points, v_i^{sim} is the ith simulated velocity, and v_i^{data} is its corresponding real data point;
- The log-likelihood

$$\text{LL} = \sum_{i=1}^{N} \ln\left(g\left(v_i^{\text{sim}} - v_i^{\text{data}}\right)\right) \tag{8}$$

  where in the given setting g is a standard normal distribution;
- The Akaike Information Criterion

$$\text{AIC} = 2k - 2\text{LL} \tag{9}$$

  where k is the number of parameters in the model. The AIC takes into account not only the likelihood but also the number of parameters needed to achieve

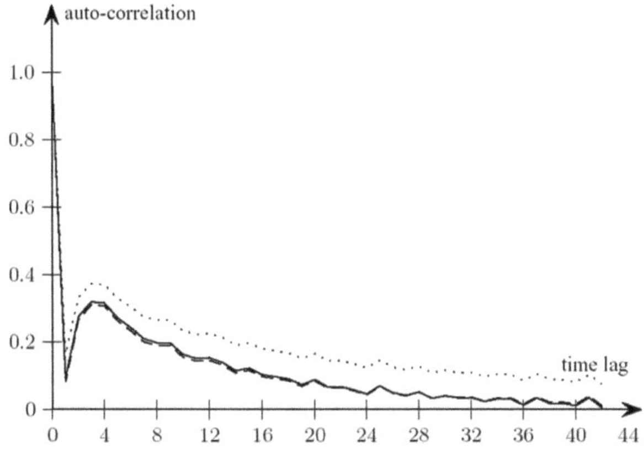

Fig. 1 Auto-correlation function of the residuals. *Solid*: Speed-synchronization, *dotted*: Newell, *dashed*: combined

the corresponding data fit: For two models with the same likelihood values, the model with the smaller AIC can be considered as "better".

Overall, the "combined" model yields both the best fit and the the smallest AIC, indicating its superiority over the two other models. All model parameters are within plausible ranges: the by far largest effect on the instantaneous velocity results from the previous velocity, which reflects the inertia of the driver/vehicle system. Small (but clearly significant) positive coefficients on g and V capture the distance-sensitivity of the driver as well as her adaptation to the preceding vehicle's speed.

Additional insights into the model dynamics can be obtained from Fig. 1, which shows the auto-correlation function of the residuals of the first three models. A first observation is that the Newell model performs substantially worse than the two other models, which is apparently due to the omission of the speed of the leading vehicle.

All models exhibit a relatively low autocorrelation at a lag of one, which then jumps to a larger value again and decays slowly. As a remedy to this "tail" in the auto-correlation, the "generalized" model (6) is implemented with more than one lag in the feedback of v on itself. A maximum lag of seven is chosen (after which the AIC becomes larger again). The resulting coefficients are shown in Table 2. One observes a clearly improved fit and AIC. Also, the obtained parameters are plausible again. The velocity v is fed back with a positive coefficient for small lags, representing inertia. For higher lags, the coefficients become negative, indicating overreaction. This could be (tentatively) interpreted as drivers being unable to react immediately but tending to overreact once their perception is updated.

The error autocorrelation function of the generalized linear model is flat, with only one peak at zero lag, indicating that all *linear* dynamics are captured well by the model.

Table 2 Parameter value(s) for the generalized linear model Eq. (6). All parameters but a_4 have significance levels well below 10^{-6}, the weakest ones being those for $g(t-h)$ and $v(t-5h)$

Model	$v(t-h)$	$g(t-h)$	$V(t-h)$	σ	LL	AIC
Generalized	0.813(8)	0.00037(7)	0.0148(6)			
Higher order	$a_2 = 0.26(1)$	$a_3 = 0.1621$	$a_4 = 0$ (not sign.)	0.0486	26,224	−52,429
Terms	$a_5 = -0.640$	$a_6 = -0.0964$	$a_7 = -0.0871$			

Table 3 Estimation results for the "combined" model, for all drivers

Driver	$v(t-h)$	$g(t-1)$	$V(t-h)$
1	0.9719	0.00127	0.02620
2	0.9748	0.00183	0.02230
3	0.9770	0.00254	0.01897
4	0.9740	0.00102	0.02375
5	0.9757	0.00307	0.01993
6	0.9822	0.00071	0.01648
7	0.9777	0.00159	0.01934
8	0.9679	0.00370	0.02586
9	0.9737	0.00081	0.02471

2.3 Investigation of Driver Heterogeneity and Time-Dependent Parameters

Table 3 shows the estimation results for all drivers individually, using the "combined" model (4). The is very little variation in the obtained parameter values between drivers: the results are extremely stable.

This remains the case when looking at the time-resolved parameter estimates for a single driver. For this, a window of a certain size w is moved along the data, and within each window the "combined" linear model is estimated. As the time window is moved along the data, a time-series of each of the parameters of the linear model is generated. For a window size of $w = 20$ s, the parameters of a single driver again turn out to be fairly stable, with the exception of the beginning of the time series, where the vehicles are standing still and not yet following each other; see Fig. 2 for an example.

2.4 Simulation Results

Now, the space-time plots of the real data are compared with those obtained by simulation of the linear models. To begin with, Fig. 3 shows the space-time plots of the original data for the leading and all nine following vehicles. This plot alone reveals little but the fact that the leading vehicle indeed varies its speed.

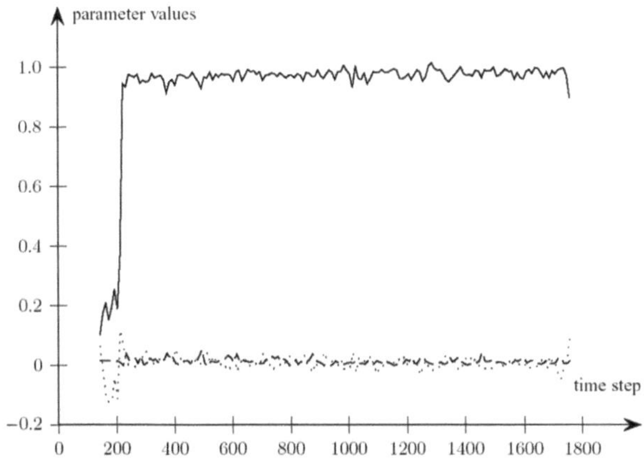

Fig. 2 Time dependence of the parameters of the generalized linear model for driver one. *Solid*: a_1, *dashed*: b_1, *dotted*: c_1

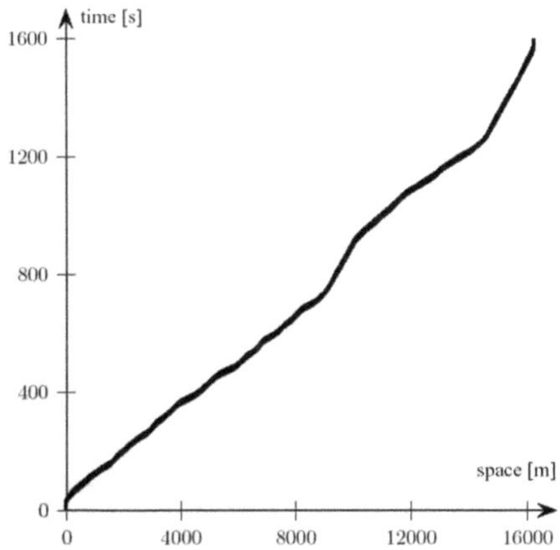

Fig. 3 Space-time plot of original data

Figure 4 (left) shows a "differential" perspective on the original data, where only the following vehicles are plotted relatively to the leading vehicle's coordinates. One clearly recognizes a widening and shortening of the platoon in reaction to the lead vehicle's driving pattern. Figure 4 (right) shows the corresponding trajectories as simulated by the "combined" model when following the (exogenously defined) lead vehicle. Although there are differences, the very simple linear model captures much of real data's dynamics quite well.

It turned out that a simulation of the higher-order model specified in Table 2 does not yield meaningful results: Without non-negativity constraints on the velocities and additional rules for collision avoidance, the trajectories do not make physical sense in that the space ordering of the vehicle is not maintained. An addition of

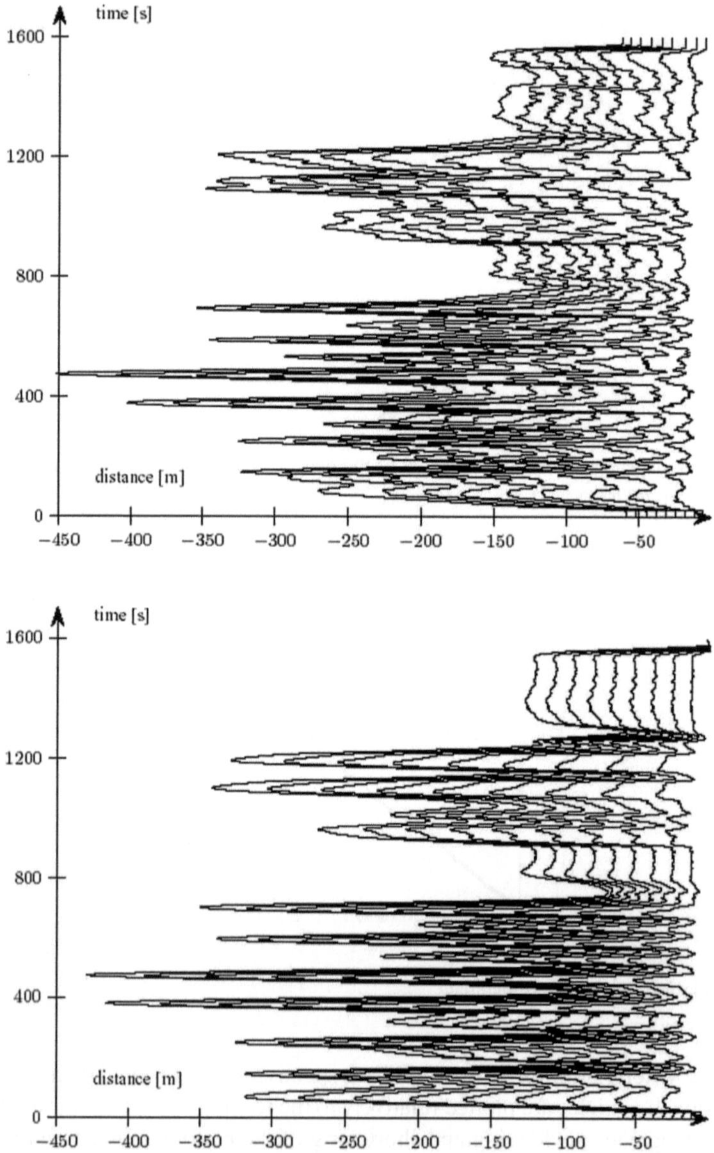

Fig. 4 Space-time plots of original data. X-coordinates are relative to the leading vehicle. *Top*: original data. *Bottom*: the "combined" model

these constraints leads to a configuration where all vehicles follow the lead vehicle with almost zero distance. Apparently, the increased number of parameters of this model enables a better fit, but it also reveals severe structural inconsistencies that most likely result from uncaptured nonlinearities.

3 Summary and Outlook

This study aims at an analysis of how much information can be obtained from real data about a car-following model's parameters. The results obtained from estimating a set of linear models from a rich data set lead to the following observations:

1. A linear model with only three parameters (the "combined" model) already captures much of the dynamics in the data set.
2. The linear model parameters can be estimated robustly and are stable (i) across the given driver population and (ii) across time.
3. Increasing the number of linear model parameters in order to capture higher-order dynamics (the "generalized" model) leads to some increase in fit. This, however, is counterbalanced by implausible simulation results, both without and with bounds on velocities and distances.

As a preliminary conclusion, it may be feasible to state that during most of the time, driver behavior is indeed linear. Further studies should therefore focus more selectively on those episodes in the data where nonlinearities actually take effect.

Acknowledgements We are grateful for the data-set, which has been made available to us by Prof. T Nakatsuji.

References

1. E. Brockfeld, R. D. Kühne, and P. Wagner. Calibration and validation of microscopic traffic flow models. *Transportation Research Records*, 1876:62–70, 2004.
2. G. S. Gurusinghe, T. Nakatsuji, Y. Azuta, P. Ranjitkar, and Y. Tanaboriboon. Multiple car following data using real time kinematic global positioning system. *Transportation Research Records*, 2003.
3. T. Söderström and P. Stoica. Instrumental variable methods for system identification. *Circuits, Systems, and Signal Processing*, 21(1):1–9, 2002.

5 Summary and Outlook

Latent Heat of a Traffic Model

Hans Weber and Reinhard Mahnke

Abstract We have studied the optimal velocity model (Bando et al., Jpn J Indust Appl Math 11:203, 1994; Phys Rev E 51:1035, 1995; J Phys I Fr 5:1389, 1995) for highway traffic. On a microscopic level, traffic flow is described by Bando's optimal velocity model in terms of accelerating and decelerating forces. We define an intrinsic energy of the model. We find a latent heat when the system undergoes a phase transition from single phase traffic (free flow) to a phase that contains two different, a dense and a dilute phase (congested or stop–and–go flow). Here we report on properties of the latent heat.

1 Introduction

We have studied a model for single lane highway traffic, the so called Bando Optimal Velocity Model (OVM) [1,2]. In the model a vehicle wants always to drive with its optimal velocity with respect to the distance to the vehicle ahead, the so called headway distance. In the model no overtaking is allowed. The model consists of a set of coupled differential equations one for each vehicle. We integrate out the equations of motion of the OVM by a Runge Kutta 4th order numerical method.

The phase diagram of the OVM consists of two phases. One is a high density phase where vehicles have a low velocity (congestion) and the other phase is a low density phase where vehicles run at nearly maximum velocity (free flow). A system can end up in one of these two phases in the entire system or it can end up in a mixed phase state. In the mixed phase state there will be two phase boundaries. The phase

H. Weber (✉)
Department of Physics, Luleå University of Technology, S–97187 Luleå, Sweden
e-mail: Hans.Weber@ltu.se

R. Mahnke
Institute of Physics, Rostock University, D–18051 Rostock, Germany
e-mail: reinhard.mahnke@uni-rostock.de

V.V. Kozlov et al. (eds.), *Traffic and Granular Flow '11*,
DOI 10.1007/978-3-642-39669-4_6, © Springer-Verlag Berlin Heidelberg 2013

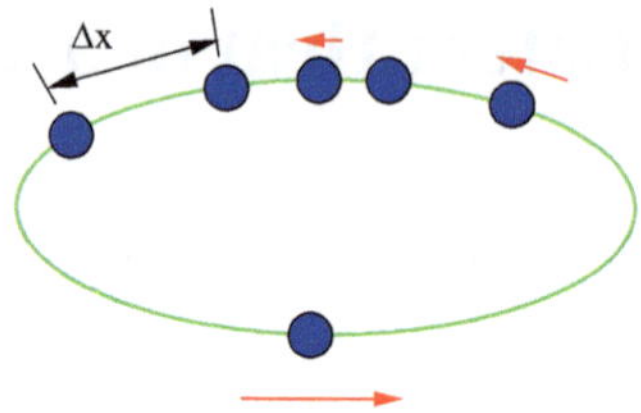

Fig. 1 Vehicles marked by *filled circles* driving on a closed loop. The *arrows* indicate the direction and magnitude of the velocity. Δx is the headway distance

boundaries are either at the head of the queue where vehicles leave the congested phase or at the tail where vehicles break and enter the queue (congested phase). If there are too few vehicles in the system a high density phase will not form and all vehicles will be in the low density phase. The OVM undergoes a dynamic phase transition.

Within the framework of the OVM an energy E of the model can be defined [3] and from the energy we can calculate a latent heat E_{gap} for the system of cars going from a phase of low density to one with a high density.

2 The Bando Optimal Velocity Model

The Bando OVM is a deterministic model for traffic flow. It consists of a one-dimensional circular road (single lane) with periodic boundary conditions, see Fig. 1.

The set of differential equations making up the Bando OVM dynamics are:

dimension | dimensionless

$$\frac{d}{dt}v_i = \frac{1}{\tau}\left(v_{opt}(\Delta x_i) - v_i\right) \qquad \frac{d}{dt}u_i = \left(u_{opt}(\Delta y_i) - y_i\right) \tag{1a}$$

$$\frac{d}{dt}x_i = v_i \qquad \frac{d}{dt}y_i = \frac{1}{b}u_i \tag{1b}$$

$$v_{opt}(\Delta x) = v_{max}\frac{(\Delta x)^2}{D^2 + (\Delta x)^2} \qquad u_{opt}(\Delta y) = \frac{(\Delta y)^2}{1 + (\Delta y)^2} \tag{1c}$$

$$b = \frac{D}{v_{max}\tau} \tag{1d}$$

The set of three equations to the right are the dimensionless version of the equations to the left. The velocity of the car i is v_i and its position is x_i. The optimal velocity function is $v_{opt}(\Delta x)$. The distance to the vehicle in front, the headway distance, is denoted by $\Delta x_i = x_{i+1} - x_i$ (bumper-to-bumper distance) and $c = N/L$ is the homogeneous density, where N is the number vehicles and L is the length of the road. Control parameters are maximal velocity of a vehicle v_{max}, an interaction distance D and a characteristic time τ. These three control parameters can be combined to a single parameter b in the dimensionless version.

Following [3] the acceleration of a vehicle cars can be split into two contributions

$$m\frac{d}{dt}v_i = F_{acc}(v_i) + F_{dec}(\Delta x_i)$$

where

$$F_{acc}(v_i) = \frac{m}{\tau}(v_{max} - v_i) \geq 0 \tag{2a}$$

$$F_{dec}(\Delta x_i) = \frac{m}{\tau}\left(v_{opt}(\Delta x_i) - v_{max}\right) \quad \in [-v_{max}m/\tau, 0] \leq 0 \tag{2b}$$

Adding together Eqs. (2a) and (2b) recovers Eq. (1a) from above

$$\frac{d}{dt}v_i = \frac{1}{\tau}\left(v_{opt}(\Delta x_i) - v_i\right) .$$

The decelerating force Eq. (2b) can be written as (using Eq. (1c))

$$F_{dec}(\Delta x_i) = v_{max}\frac{m}{\tau}\left(\frac{(\Delta x)^2}{D^2 + (\Delta x)^2} - 1\right) < 0 . \tag{3}$$

The decelerating force will always be less then zero but approach zero at infinite head away distance Δx and starting at $-v_{max}m/\tau$ at zero distance.

A potential energy V for the system of N vehicles can be defined as $V = \sum_{i=1}^{N}\phi(\Delta x_i)$ where $\phi(\Delta x_i)$ is the interaction potential of the i-th car with the car $(i+1)$ ahead, which is given by (Note $F_{dec}(\Delta x_i) \neq -\partial V/\partial x_i$ is a violation of Newton's 3rd law)

$$F_{dec}(\Delta x_i) = -\frac{\partial\phi(x_{i+1} - x_i)}{\partial x_i} = \frac{d\phi(\Delta x_i)}{d\Delta x_i}. \tag{4}$$

Integrating this equation gives

$$\phi(\Delta x) = v_{max}\frac{D\,m}{\tau}\left[\frac{\pi}{2} - \arctan\left(\frac{\Delta x}{D}\right)\right] \tag{5}$$

where the integrating constant is chosen such that $\phi(\infty) = 0$

The time derivative of the potential V becomes

$$\frac{dV}{dt} = \sum_{i=1}^{N}(v_{i+1} - v_i)F_{dec}(\Delta x_i) . \tag{6}$$

The time derivative for the kinetic energy $T = \sum_{i=1}^{N} \frac{mv_i^2}{2}$ is

$$\frac{d}{dt}\frac{mv_i^2}{2} = v_i m \frac{d}{dt} v_i = v_i \left(F_{acc}(v_i) + F_{dec}(\Delta x_i) \right) \, . \tag{7}$$

The time derivative of the total energy $E = T + V$ and the energy flux Φ obey the following balance equation

$$\frac{dE}{dt} + \Phi = 0 \tag{8}$$

where

$$\Phi = -\sum_{i=1}^{N} \left[v_i F_{acc}(v_i) + v_{i+1} F_{dec}(\Delta x_i) \right] \tag{9}$$

is the energy flux. It includes both input (from engine) and output (friction) of energy. Energy is not conserved but the driven system will reach a stationary state as time goes on.

3 Bando OVM Numerical Results

The system is started in a configuration close to the homogeneous state and as time goes on in the simulation the system ends up in one of two possible stationary states.

It can end up in a homogeneous flow with all vehicles traveling at the same headway distance. This solution is the fixed point $\Delta x_i = \Delta x_{hom}$, $v_i = v_{opt}(\Delta x_{hom})$ and all vehicles travel with the same velocity. This would give the dashed line in Fig. 2.

The other possibility is that the system reaches a limit cycle. In this solution there is one congested part and one free flow part in the system. Vehicles leave at a steady rate the head of the queue to enter the free flow regime and after a while they will reach the tail of the queue and enter the slowly moving regime. This would give the full line in Fig. 2. Note that there is only one queue in the stationary limit cycle. If there are more queues the system is still evolving and is not stationary. In the leftmost figure in Fig. 3 the reduction of the number of queues can clearly be seen as steps in the energy E as time increases.

Integrating out the equations (Bando OVM equation (1a–1d)) the energy of the system can be calculated. As the system evolves from its initial configuration the number of queues will be reduced till the system reaches the limit solution ($t \to \infty$). After that the energy E of the system will remain constant. In the leftmost Fig. 3 the evolution of the energy to a constant value can be seen clearly.

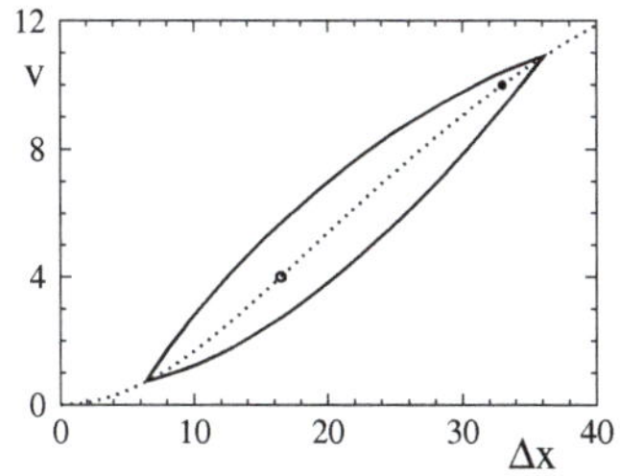

Fig. 2 Solution for $\rho = 0.0303\,\text{m}^{-1}$ *solid circle*. Solution for $\rho = 0.0606\,\text{m}^{-1}$ *open circle*, but this splits into a limit cycle

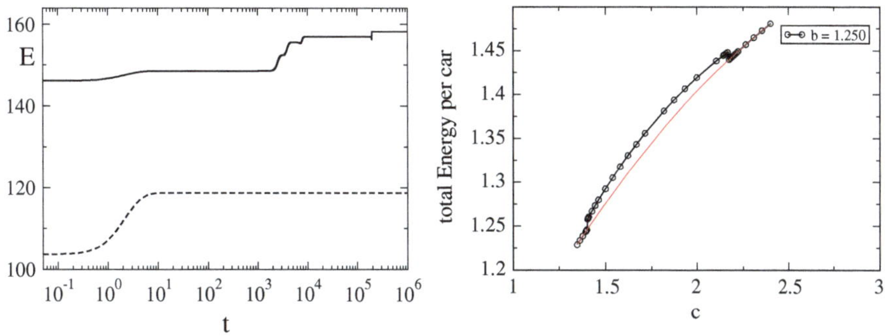

Fig. 3 In the *left figure* the *solid line* (larger density $\rho = 0.0606\,\text{m}^{-1}$) reaches the limit cycle over a series of smaller queues till finally only one queue is present. The *dashed line* small density $\rho = 0.0303\,\text{m}^{-1}$ reaches the fixed point a homogeneous system. In the *right figure* the energy as a function of the density of cars. The two joining points of the curves define the latent heat to go from the dense queued state to the free flow

For each particular combination of control parameters N, L and $b = \frac{D}{v_{max}\tau}$ a calculation is made. In the rightmost Fig. 3 one such result is shown combining several runs. In this figure there are two curves shown. Where the curves are on top of each other the system is in a homogeneous limit state. Where the two curves are separated the system is in the limit cycle. The difference in energy between the two joining positions of the curves is the energy as the system evolves from a dense homogeneous system to a dilute homogeneous system via a two phase regime. This energy difference represents a latent heat, here denoted by E_{gap}.

In the leftmost Fig. 4 the latent heat E_{gap} is presented as a function of the control parameter b in Eq. (1d). The shape of the curve suggests an analysis according to a simple power law:

$$E_{gap} = A(b_c - b)^{\alpha}. \tag{10}$$

where A is a constant and b_c is the value for b that gives a zero latent heat. In the rightmost Fig. 4 the result according to Eq. (10) is shown. As is apparent from

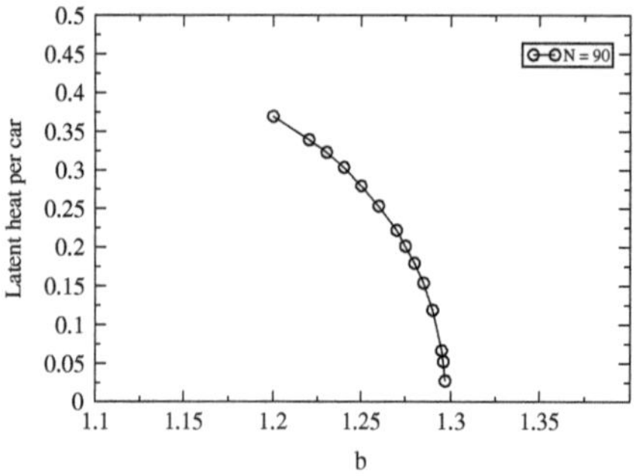

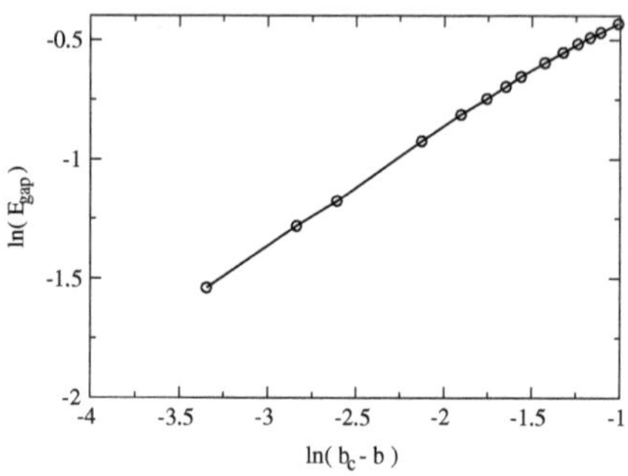

Fig. 4 In the *left figure* the latent energy is shown as a function of the control parameter b. In the *right figure* the same data as in the *left figure* is analyzed according to the scaling relation

the figure the power law seems to be fulfilled quit well. The value arrived for $\alpha = 0.4994$. The value used for $b_c = 1.29745$. The scaling is rather sensitive to small changes to b_c as the data will not join a straight line for small $b_c - b$.

4 Conclusions

We have shown how ideas from thermodynamics can be applied to such a many–particle system as traffic flow, based on a microscopic (car–following) description, in analogy to equilibrium physical systems like super saturated vapor forming liquid droplets.

We have calculated the latent heat of the model as the system changes from a low density phase to the a high density phase. We found the there is a scaling behavior in the latent heat for small $b_c - b$. Results found for the scaling exponent are $\alpha = 0.50$ and $b_c = 1.297$.

References

1. M. Bando, K. Hasebe, A. Nakayama, A. Shibata, Y. Sugiyama: Japan J. Indust. and Appl. Math. **11**, 203, 1994; Phys. Rev. E **51**, 1035, (1995)
2. M. Bando, K. Hasebe, K. Nakanishi, A. Nakayama, A. Shibata, Y. Sugiyama: J. Phys. I France **5**, 1389, (1995)
3. R. Mahnke, J. Kaupužs, J. Hinkel, H. Weber: Eur. Phys. J. B **57** 463–471, (2007)

Estimation of Transport Systems Capacity

Larisa Afanasyeva and Ekaterina Bulinskaya

Abstract A transport system capacity is introduced as maximal car flow density compatible with a desired quality of system performance. As an objective function one can choose mean car velocity or mean travel time dealing with highway capacity. Mean number of cars waiting before crossroads is useful to analyze the traffic lights capacity. Probability of a line of stationary or very slow moving traffic with length exceeding a given threshold can also serve for estimation of a transport system capacity. We consider three examples of transport systems (a highway without traffic lights with two car types, a single crossroads as well as hierarchical networks) and estimate their capacities.

1 Introduction

It is well known that in order to investigate a real-life process or system one has to construct an appropriate mathematical model. Interest in transport models is old enough, the first one was introduced by Greenshields, see [1], in 1935. Now it is impossible even to mention all the researchers who contributed to traffic modeling, see, e.g., [2–5] and references therein. Various methods such as cellular automata, statistical mechanics, mathematical physics or queueing theory are widely used.

One of the main problems is how to deal with traffic congestion, the condition on road networks that occurs as their use increases. It is characterized by slower speeds, longer trip times and increased vehicular queues, leading to loss of time, air pollution and many other bad consequences.

We define a transport system capacity as the maximal traffic intensity providing fulfillment of certain conditions concerning parameters characterizing its performance.

L. Afanasyeva (✉) · E. Bulinskaya
Lomonosov Moscow State University, Moscow, Russia
e-mail: afanas@mech.math.msu.su; ebulinsk@mech.math.msu.su

V.V. Kozlov et al. (eds.), *Traffic and Granular Flow '11*,
DOI 10.1007/978-3-642-39669-4_7, © Springer-Verlag Berlin Heidelberg 2013

The paper is organized as follows. In Sect. 2 we estimate the capacity of highway without traffic lights under assumption of two car types (quick and slow). The Sect. 3 deals with a single crossroads having traffic lights. A hierarchical transport network is treated in Sect. 4, whereas conclusions and further research directions are given in Sect. 5. Due to lack of space almost all proofs, as well as numerical results, are omitted.

2 Highway Without Traffic Lights (Two Types of Cars)

2.1 *Model Description*

Consider a one-lane highway without traffic lights. There are cars of two types (quick and slow) moving in the same direction. Let $\lambda_i\, dx\, dt$ be the probability that a car of type i appears in the interval $(x, x + dx)$ during the time interval $(t, t + dt)$, $i = 1, 2$. Velocity of a i-type car is V_i and $V_1 < V_2$. The distance covered by a car of type i on the highway is exponential random variable (r.v.) with parameter μ_i.

We make two assumptions:

- A car of type 2 (quick) on catching up with a car of type 1 (slow) begins to move with velocity V_1 until either of them leaves the road.
- The car size is not taken into account.

Let $X(t)$ be a stochastic process with values in the space of configurations $\mathcal{X} = \{(x_s, n_s, e_s)_{s=-\infty}^{s=+\infty}\}$, here x_s is the position of the sth group of cars ($x_s \in R^1$), n_s is the size of the sth group (cars number) and $e_s = 1$ if the group contains a slow car, $e_s = 2$ otherwise.

According to the model assumptions $X(t) = \{(x_s(t), n_s(t), e_s(t))_{s=-\infty}^{s=+\infty}\}$ is a homogenous Markov process with values in the space $\mathcal{X}$.

Theorem 1. *If* $\mu_i > 0$, $V_i > 0$, $i = 1, 2$, *then* $X(t)$ *is ergodic.*

A *proof* can be found in [2].

We suppose further on that $X(t)$ is stationary.

2.2 *Car Flow Densities*

Let I be a finite interval in R^1 and $|I|$ its length, whereas $\mathbb{I}(A)$ is indicator of the event A.

Definition 1. The density of cars moving with velocity V_i, $i = 1, 2$, at time t is given by

$$\delta_i = \lim_{|I|\to\infty} |I|^{-1} \sum_{s=-\infty}^{+\infty} n_s(t)\mathbb{I}(e_s(t) = i, x_s(t) \in I), \quad \text{a.s.},$$

the mean velocity on highway is

$$\tilde{V} = V_1 + (V_2 - V_1)\frac{\delta_2}{\delta_1 + \delta_2} \tag{1}$$

and the density of car flow is $\delta_0 = \delta_1 + \delta_2$.

The main result of this section is the following

Theorem 2. *The densities δ_i, $i = 1, 2$, have the form*

$$\delta_1 = \lambda_1\zeta_1 + \lambda_1\lambda_2 C(\mu_2 V_1(1 + \lambda_1 C))^{-1}, \quad \delta_2 = \lambda_2\zeta_2(1 + \lambda_1 C)^{-1} \tag{2}$$

where $\zeta_i = (V_i\mu_i)^{-1}$ and $C = \zeta_1\zeta_2(V_2 - V_1)\mu_2(\mu_1 + \mu_2)^{-1}$.

Proof. Let $p_j dx$ be the probability that interval $(x, x+dx)$ contains j cars of type 2 and one car of type 1, $j = 0, 1, 2, \ldots$, whereas $q_j dx$ the probability that interval $(x, x + dx)$ contains only j cars of type 2, $j = 1, 2, \ldots$. Put $\alpha = \sum_{j=0}^{\infty} p_j$ and $\beta = \sum_{j=1}^{\infty} q_j$.

Equations satisfied by p_j and q_j were established in [2]. They have the form

$$(\mu_1 V_1 + \beta(V_2 - V_1))p_0 = \mu_2 V_1 p_1 + \lambda_1, \tag{3}$$

$$(j\mu_2 V_1 + \mu_1 V_1 + \beta(V_2 - V_1))p_j = \mu_2 V_1(j + 1)p_{j+1} + (V_2 - V_1)\sum_{i=0}^{j-1} p_i q_{j-i}, \quad j > 0,$$

$$(\mu_2 V_2 + \alpha(V_2 - V_1))q_1 = 2\mu_2 V_2 q_2 + \mu_1 V_1 p_1 + \lambda_2, \tag{4}$$

$$(j\mu_2 V_2 + \alpha(V_2 - V_1))q_j = (j + 1)\mu_2 V_2 q_{j+1} + \mu_1 V_1 p_j, \quad j > 1.$$

These equations can be solved numerically, the algorithm was proposed in [2]. Since $\delta_1 = \sum_{j=0}^{\infty}(j + 1)p_j$ and $\delta_2 = \sum_{j=1}^{\infty} jq_j$, we introduce here probability generating functions $P(z) = \sum_{j=0}^{\infty} p_j z^j$ and $Q(z) = \sum_{j=1}^{\infty} q_j z^j$. Then we have

$$\delta_1 = P(1) + P'(1), \quad \text{whereas} \quad \delta_2 = Q'(1). \tag{5}$$

Moreover, in Eqs. (3) and (4) we have $\alpha = P(1)$ and $\beta = Q(1)$.

Multiplying the jth equation of (3) (and (4), resp.) by z^j, $j = 0, 1, \ldots$, ($j = 1, 2, \ldots$, resp.) and summing them, we get the following equations for generating functions

$$\mu_2 V_1(z - 1)P'(z) = P(z)[(V_2 - V_1)(Q(z) - Q(1)) - \mu_1 V_1] + \lambda_1, \tag{6}$$

$$\mu_2 V_2(z - 1)Q'(z) = \mu_1 V_1[P(z) - p_0] - P(1)(V_2 - V_1)Q(z) - \mu_2 V_2 q_1 + \lambda_2 z. \tag{7}$$

The left-hand sides of (6) and (7) are equal to zero for $z = 1$, the same is true of the right-hand sides. Thus, we get immediately $P(1) = \lambda_1 \zeta_1$. Furthermore, using the L'Hopital rule we obtain $P'(1)$ and $Q'(1)$ as follows

$$P'(1) = \lim_{z \to 1} \frac{P(z)[(V_2 - V_1)(Q(z) - Q(1)) - \mu_1 V_1] + \lambda_1}{\mu_2 V_1 (z-1)}$$
$$= (\mu_2 V_1)^{-1}[(V_2 - V_1)Q'(1)P(1) - \mu_1 V_1 P'(1)],$$
$$Q'(1) = \lim_{z \to 1} \frac{\mu_1 V_1 [P(z) - p_0] - P(1)(V_2 - V_1)Q(z) - \mu_2 V_2 q_1 + \lambda_2 z}{\mu_2 V_2 (z-1)}$$
$$= (\mu_2 V_2)^{-1}[\lambda_2 + \mu_1 V_1 P'(1) - \lambda_1 \zeta_1 (V_2 - V_1) Q'(1)].$$

Solving this system of equations for $P'(1)$ and $Q'(1)$ we get from (5) the desired expressions for densities δ_i, $i = 0, 1, 2$. □

2.3 *Highway Capacity*

Definition 2. The highway capacity is $\Lambda = \max\{\lambda_1 + \lambda_2 : \tilde{V} > V_1 + \Delta\}$, where $\Delta \in (0, V_2 - V_1)$ is specified.

Let the relationship between intensities of slow and quick cars be known, i.e. $\lambda_1 = a\lambda_2$. Then the densities δ_i are functions of λ_2, namely, $\delta_i(\lambda_2)$, $i = 1, 2$. Set $g(\lambda_2) = \delta_2(\lambda_2)[\delta_1(\lambda_2) + \delta_2(\lambda_2)]^{-1}$, hence, $\tilde{V}(\lambda_2) = V_1 + (V_2 - V_1)g(\lambda_2)$. Thus, according to Definition 2, the highway capacity is $\Lambda = (1+a)\widetilde{\lambda_2}$ with $\widetilde{\lambda_2} = \max\{\lambda_2 : \tilde{V}(\lambda_2) \geq V_1 + \Delta\}$, whence we get the following equation for getting $\widetilde{\lambda_2}$

$$(V_2 - V_1)g(\lambda_2) = \Delta. \tag{8}$$

From Theorem 2 one easily obtains

Corollary 1. *The function $g(\lambda_2)$ has the form*

$$g(\lambda_2) = \zeta_2[a\zeta_1 + \zeta_2 + \lambda_2 a C(a\zeta_1 + (\mu_2 V_1)^{-1})]^{-1}.$$

Since $g(\lambda_2)$ is continuous non-increasing, $g(\infty) = 0$ and $g(0) = \zeta_2(\zeta_2 + a\zeta_1)^{-1}$, Eq. (8) has a solution $\widetilde{\lambda_2}$ if $\Delta(V_2 - V_1)^{-1} < \zeta_2(\zeta_2 + a\zeta_1)^{-1}$. Therefore, for a fixed $\Delta \in (0, (V_2 - V_1))$, it follows that condition

$$a < \tilde{a} = \zeta_2(V_2 - V_1 - \Delta)(\zeta_1 \Delta)^{-1}$$

is necessary for existence of $\widetilde{\lambda_2}$. Using (2) it is easy to get the explicit form of solution

$$\widetilde{\lambda_2} = \frac{\zeta_2(V_2 - V_1)\Delta^{-1} - a\zeta_1 - \zeta_2}{a\zeta_1^2\zeta_2(V_2 - V_1)} \cdot \frac{\mu_1 + \mu_2}{\mu_1 + a\mu_2} \quad (9)$$

and highway capacity Λ.

It is clear that $\widetilde{\lambda_2}(a)$ given by (9) is a monotone decreasing function of a, it tends to infinity, as $a \to 0$, and equals zero for $a = \tilde{a}$. Moreover, this expression shows how to organize the traffic in order to increase the highway capacity. For example, building special exits for slow cars, we increase μ_1 thus decreasing ζ_1, and that results in Λ growth.

3 Crossroads

3.1 Model Description

Consider a crossroads with cars arriving in two perpendicular directions and introduce the following notation. Let $A_i(t)$ be a Poisson flow of cars arriving in the ith direction, λ_i intensity of the flow and $\{t_n^{(i)}\}_{n=1}^{\infty}$ a sequence of green light switching on times for the ith direction, $i = 1, 2$. Let $0 = t_1^{(1)} < t_1^{(2)} < \ldots$ then $\tau_n^{(1)} = t_n^{(2)} - t_n^{(1)}$, $\tau_n^{(2)} = t_{n+1}^{(1)} - t_n^{(2)}$, $n = 1, 2, \ldots$, are interswitching times. Their distribution functions are $G_i(x) = \mathsf{P}(\tau_n^{(i)} \le x)$ with means $\gamma_i^{-1} = \int_0^\infty x\,dG_i(x)$, $i = 1, 2$, whereas $\theta = \gamma_1^{-1} + \gamma_2^{-1}$ is the period length. Obviously, the intervals $\tau_n^{(i)}$ when the green light is switched on for the ith direction can be interpreted as a working state of the server, while the red interval can be considered as a repair state.

The time of passing crossroads by a car is usually supposed exponentially distributed, see, e.g., [6] or [7]. In practice the distribution of the crossing time for cars arriving to traffic lights during a green interval differs from that for cars already waiting before traffic lights. To take into account this difference we assume that a car arriving during a green interval passes the crossroads immediately if there are no waiting cars (the service time of such cars is zero). The other cars have exponential distribution with parameter ν. All the random variables involved in model description are mutually independent.

3.2 Ergodic Theorem

Let $X_i(t)$ be the car number in the ith direction before the traffic lights. Below we consider a stochastic process $X(t) = (X_1(t), X_2(t))$, that is, a continuous-time two-dimensional random walk. If this process is Markovian its ergodicity conditions are well known, see, e.g., [8]. However in our setting $X(t)$ is not a Markov process. Thus, we need to use another approach.

Theorem 3. *The limits*

$$p_j^{(i)} = \lim_{t\to\infty} \mathsf{P}(X_i(t) = j), \quad i = 1,2, \quad j = 0,1,2,\ldots, \tag{10}$$

with $p_j^{(i)} > 0$ *and* $\sum_{j=0}^{\infty} p_j^{(i)} = 1$ *exist iff*

$$\rho_i = \lambda_i \gamma_i \theta \nu^{-1} < 1. \tag{11}$$

The *proof* is based on the fact that in our case $X(t)$ is regenerative. Its regeneration points are the times $t_j^{(1)}$ with $X_i(t_j^{(1)} - 0) = 0$, $i = 1,2$, namely, there is no queue before the traffic lights in both directions and green light is switched on for the first direction. Existence of limits (10) follows immediately from the Smith theorem, see, e.g., [9]. More precisely, it is not difficult to establish that these limits form a probability distribution iff the process $X(t)$ is stochastically bounded. Conditions (11) are necessary and sufficient for stochastic boundedness. That can be proved by constructing two auxiliary processes, upper and lower bounds (in stochastic sense) of the process under consideration. The details are given in [5].

3.3 Estimation of Traffic Lights Capacity (Means as Criteria)

Definition 3. Traffic lights capacity in the ith direction $i = 1,2$, is the maximal intensity λ_i such that a chosen characteristic of performance quality does not exceed a certain threshold and the ergodicity condition (11) is fulfilled.

In particular, (11) provides the upper bound of capacity $\lambda_i < \nu(\gamma_i\theta)^{-1}$. It is possible to consider one of the following parameters as playing the main role in traffic lights performance (in a steady state):

1. Mean number $m_i(\lambda_i)$ of cars before the traffic lights in the ith direction,
2. Mean time $t_i(\lambda_i)$ of passing crossroads by a car.

Fixing the acceptable values of these parameters m^0 or t^0 the capacity can be obtained as solution of one of the equations

$$1.\ m_i(\lambda_i) = m^0 \quad \text{or} \quad 2.\ t_i(\lambda_i) = t^0. \tag{12}$$

Stationary Distribution Calculation

To implement the procedure of solving (12) we have to calculate the limits (10), since $m_i(\lambda_i) = \sum_{j=1}^{\infty} j p_j^{(i)}$.

Hence, begin by treating imbedded Markov chains $X_n^{(i)} = X_i(t_n^{(i)})$ where $t_n^{(i)}$ is the moment of the nth switching on of the green light in the ith direction, $n \geq 1$, $i = 1,2$. Condition (11) is supposed to be valid. Our aim is to propose an algorithm for calculating the stationary distribution of $\{X_n^{(i)}\}$, $i = 1, 2$, as well as steady-state expected numbers of waiting cars. It is possible to deal only with $\{X_n^{(1)}\}$ because the results for $\{X_n^{(2)}\}$ are obtained by putting everywhere λ_2 instead of λ_1, G_2 instead of G_1 and vice versa.

To obtain the transition probabilities for imbedded Markov chain we introduce an auxiliary birth-and-death process $Z(t)$ with absorbing state $\{0\}$, birth intensity λ_1 and death intensity ν. Put $\varphi_{kj}(t) = \mathsf{P}(Z(t) = j|Z(0) = k)$, $k, j > 0$. These functions satisfy the following system

$$\varphi_{k1}'(t) = -(\lambda_1 + \nu)\varphi_{k1}(t) + \nu\varphi_{k2}(t),$$
$$\varphi_{kj}'(t) = -(\lambda_1 + \nu)\varphi_{kj}(t) + \nu\varphi_{k,j+1}(t) + \lambda_1\varphi_{k,j-1}(t), \quad j > 1, \tag{13}$$

with initial conditions $\varphi_{kk}(0) = 1$ and $\varphi_{kj}(0) = 0$ for $j \neq k$. The solution of system (13) is obtained by applying the Laplace transform and its subsequent inversion. Thus we get the explicit form of $\varphi_{kj}(t)$ in terms of generalized Bessel functions of the first kind. For $j > 0$,

$$\varphi_{kj}(t) = e^{-(\lambda_1+\nu)t}(\lambda_1/\nu)^{(j-2)/2}(J_{|j-k|}(2\sqrt{\lambda_1\nu t}) - J_{j+k}(2\sqrt{\lambda_1\nu t}))$$

where

$$J_l(u) = \sum_{m=0}^{\infty} \frac{(u/2)^{l+2m}}{m!\Gamma(l+m+1)}.$$

Moreover, $\varphi_{k0}(t) = 1 - \sum_{j=1}^{\infty} \varphi_{kj}(t)$.

Next, denote by d_j the probability that j cars arrive during a red interval (in the first direction), therefore

$$d_j = \int_0^{\infty} e^{-\lambda_1 y}(\lambda_1 y)^j (j!)^{-1}\, dG_2(y), \quad j = 0, 1, \ldots.$$

Introduce also $b_{00} = 1$. Furthermore, let b_{kj} be the probability that at time of red light switching on there are j cars before traffic lights under condition that at time of the previous green light switching on there were k cars, namely, $b_{kj} = \mathsf{P}(X_1(t_n^{(2)}) = j|X_1(t_n^{(1)}) = k)$, $k, j \geq 0$. Then

$$b_{kj} = \int_0^{\infty} \varphi_{kj}(y)\, dG_1(y), \quad k, j > 0, \quad b_{k0} = 1 - \sum_{j=1}^{\infty} b_{kj}.$$

The transition probabilities for the imbedded Markov chain $\{X_n^{(1)}\}$ are given by

$$\mathsf{P}_{kj}^{(1)} = \mathsf{P}(X_{n+1}^{(1)} = j|X_n^{(1)} = k) = \sum_{m=0}^{j} b_{km}d_{j-m}, \quad k, j = 0, 1, \ldots.$$

Hence the stationary distribution $\{\pi_k\}$ of the imbedded Markov chain is obtained by solving the system

$$\pi_j = \sum_{k=0}^{\infty} \pi_k \mathsf{P}_{kj}^{(1)}, \quad \sum_{j=0}^{\infty} \pi_j = 1.$$

So it is easy to calculate the mean number of cars in a steady state (for imbedded Markov chain). Finally, the stationary distribution $\{p_j^{(1)}\}$ of the process $X_1(t)$ has the form

$$p_j^{(1)} = \theta^{-1} \sum_{k=0}^{\infty} \pi_k \int_0^{\infty} \varphi_{kj}(y)[1 - G_1(y)]\, dy$$

$$+\theta^{-1} \sum_{i=0}^{j} \sum_{k=0}^{\infty} \pi_k b_{ki} \int_0^{\infty} e^{-\lambda_1 y}(\lambda_1 y)^{j-i}[(j-i)!]^{-1}[1 - G_2(y)]\, dy. \tag{14}$$

Example 1. Assume interswitching intervals to be exponentially distributed. Consider the first direction. Let p_j (q_j, resp.) denote the probability of car queue before traffic lights in a steady state having length j and green (red, resp.) light being switched on for the first direction.

Generating functions for these probabilities are obtained by authors in [5]. The mean queue length $m_1(\lambda_1) = \sum_{j=1}^{\infty} j(p_j + q_j)$ has the form

$$m_1(\lambda_1) = \frac{\lambda_1\gamma_1(1 + (\nu - \lambda_1)(\gamma_1 + \gamma_2)^{-1})}{(\gamma_1 + \gamma_2)(\nu - \lambda_1)(\gamma_2(\gamma_1 + \gamma_2)^{-1} - \lambda_1\nu^{-1})}. \tag{15}$$

Function $m_1(\lambda_1)$, for $\lambda_1 \in (0, \nu(\theta\gamma_1)^{-1})$, is continuous and monotone increasing. Since $m_1(0) = 0$ and $\lim_{\lambda_1 \uparrow \nu(\theta\gamma_1)^{-1}} m_1(\lambda_1) = +\infty$, equation 1. in (12) has a unique solution $\widetilde{\lambda_1}$ which can be considered as traffic lights capacity for the first direction.

Without loss of generality put $\gamma_1 + \gamma_2 = 1$ and $x = \lambda_1\nu^{-1}$. Then it easily follows from (15) that $a_1x^2 - b_1x + m^0(1 - \gamma_1) = 0$, here $a_1 = m^0 + \gamma_1$, $b_1 = m^0(2 - \gamma_1) + \gamma_1(\nu + 1)$. Hence, $\widetilde{\lambda_1} = \nu(b_1 - \sqrt{b_1^2 - 4a_1m^0(1 - \gamma_1)})(2a_1)^{-1}$.

According to Little's formula, we calculate the mean passing time of crossroads as follows $t_1(\lambda_1) = \lambda_1^{-1}m_1(\lambda_1)$. Inserting this expression in equation 2. of (12) we establish that there exists a unique solution satisfying (11). It represents the traffic lights capacity according to this criterion.

3.4 Capacity According to Criterion of Level Crossing

Sometimes it is desirable that, for given $\varepsilon > 0$ and N_0, probability of waiting car queue length exceeding N_0 were less than ε.

Let $r_j^{(i)}$ be the probability that the number of cars waiting before traffic lights in the ith direction in a steady state is equal to j. The aim is to choose the maximal λ_i such that

$$\sum_{j=N_0+1}^{\infty} r_j^{(i)} < \varepsilon, \quad \lambda_i < \nu(\gamma_i\theta)^{-1}. \tag{16}$$

Using the results obtained in [2] it is possible to propose an algorithm for numerical estimation of traffic lights capacity in framework of this criterion.

For large N_0 one can also use the heavy-traffic asymptotics given by the following result proved in [5]. Put $\lambda_1^\delta = (1-\delta)\nu(\gamma_1\theta)^{-1}$ and let $X_1(\lambda_1^\delta)$ be the number of waiting cars in the first direction if the flow intensity is λ_1^δ.

Theorem 4. *If* $\mathsf{E}\tau_i^{2+\kappa} < \infty$, $i = 1, 2$, *for some* $\kappa > 0$ *then*

$$\mathsf{P}(\delta X_1(\lambda_1^\delta) > y) \to e^{-y/\sigma^2}, \quad \delta \to 0, \tag{17}$$

with $\sigma^2 = 1 + (\nu\gamma_1)(2\theta^2)^{-1}(\gamma_1^{-2}\mathsf{Var}\tau_1 + \gamma_2^{-2}\mathsf{Var}\tau_2)$, *here* τ_1 *and* τ_2 *are the green and red intervals respectively.*

Putting

$$\lambda_1 = (1 + N_0^{-1}\sigma^2 \ln\varepsilon)\nu(\gamma_1\theta)^{-1}, \tag{18}$$

one obtains from (17) that (16) is fulfilled, so for large N_0 the capacity is estimated by (18).

4 Transport Networks Analysis

4.1 Model Description

We study transport networks assuming that their nodes (vertices) are crossroads (with traffic lights) and arcs (edges) are the roads connecting the nodes. It is supposed that each node has two admissible motion directions. Moreover, the transport system under consideration is a superposition of two hierarchical networks S^+ and S^-. Hence, if S^+ contains a route from node A to node B then S^- has a route from B to A.

Time intervals when vehicular traffic in a given node is permitted in network S^+ (green light) correspond to the intervals when traffic is forbidden in S^- (red light) and vice versa. We establish conditions of stationary distributions existence for the processes describing the system performance under assumption that input flows are regenerative. This enables us to introduce the so-called traffic coefficients of nodes, to point out the most loaded ones and investigate the dependence of asymptotic behavior of car queues at crossroads on the traffic-lights performance. The estimates of system capacity are also provided.

We assume that the nodes of network S^+ (S^- resp.) consist of r (l resp.) levels (or classes). Thus, the cars arrive in the network through nodes of the first level, then they proceed through nodes of the second level etc. and leave the network through the nodes of the last level. Note that additional nodes with zero passing times at each level let consider the case with cars arriving (or leaving) the system on some intermediate level. The interswitching intervals for all the traffic lights are independent random variables with Erlang distribution. The crossroads passing times for the cars are exponentially distributed r.v.'s.

Consider S^+ and enumerate the nodes of each level in such a way that (i, j) means the jth node, $j = \overline{1, n_i}$, of the ith level, $i = \overline{1, r}$. The route is a sequence $I = (j_1, \ldots, j_r)$ where $j_i = \overline{1, n_i}$, $i = \overline{1, r}$. Let $d^{(i)}_{j_i, j_{i+1}}$ be the distance between the node j_i of level i and the node j_{i+1} of the level $i + 1$, $j_i = \overline{1, n_i}$, $j_{i+1} = \overline{1, n_{i+1}}$, and $v^{(i)}_{j_i, j_{i+1}}$ the admissible velocity for cars moving on this link (arc). Suppose a probability measure $\{\mathsf{P}(I), I \in J\}$ is defined on the set J of all the routes and an arriving car chooses a route I with probability $\mathsf{P}(I)$ independently of others.

4.2 Ergodicity Conditions

Thus, we consider an hierarchical transport system with input flow $A(t)$. It is a stochastic process $\{A(t), t \geq 0\}$, $A(0) = 0$, taking values $0, 1, 2, \ldots$ defined on a probability space $(\Omega, \mathcal{F}, \mathsf{P})$. The process has non-decreasing left-continuous trajectories with unit jumps.

Definition 4. The input flow $A(t)$ is regenerative if there exists an increasing sequence of r.v.'s $\{\theta_j, j \geq 0\}$, $\theta_0 = 0$, such that the sequence

$$\{\chi_j\}_{j=1}^{\infty} = \{\theta_j - \theta_{j-1}, A(\theta_{j-1} + t) - A(\theta_{j-1}), t \in [0, \theta_j - \theta_{j-1})\}_{j=1}^{\infty}$$

consists of i.i.d. random elements.

Note that for the most part flows used in queueing theory are regenerative. Doubly stochastic Poisson process, Markov modulated, semi-Markov flows and many others belong to this class, see, e.g., [3]. A very useful property of regenerative flows is their remaining regenerative (under some general assumptions) after passing a queueing system of any level of hierarchical networks. It plays important role in our reasoning.

We use the following notation. Let θ_i be the ith regeneration point, $\tau_i = \theta_i - \theta_{i-1}$ the ith regeneration interval, $\xi_i = A(\theta_i) - A(\theta_{i-1})$ the car number entering the network during this regeneration interval, $a = \mathsf{E}\xi_i$, $\mu = \mathsf{E}\tau_i$.

Intensity of input flow is given by $\lambda = \lim_{t \to \infty} A(t)/t = a/\mu$ a.s. The time of passing the traffic lights (i, k) has the mean ν_{ik}^{-1}, here i is the level of network and k is the number of the node on this level. We have assumed that this time has exponential distribution, however it is possible to consider an arbitrary distribution. The mean length of the green (red, resp.) interval for the traffic lights (i,k) is α_{ik}^{-1} (β_{ik}^{-1}, resp.), $k = \overline{1,n_i}$, $i = \overline{1,r}$.

Obviously, intensity of input in the node (i,k) is

$$\lambda_{ik} = \lambda\delta_{ik}, \quad k = \overline{1,n_i}, \ i = \overline{1,r}, \quad \text{with} \quad \delta_{ik} = \sum_{I:j_i=k} \mathsf{P}(I). \tag{19}$$

The traffic coefficient of the node (i,k) is defined as

$$\rho_{ik} = \lambda_{ik}(\alpha_{ik} + \beta_{ik})(\nu_{ik}\beta_{ik})^{-1}. \tag{20}$$

It is well known that Erlang distribution can be considered as a convolution of exponential distributions. Namely, if the interswitching times $\eta_{ik}^{(j)}$, $j = 1,2$, for the node (i,k) have Erlang distribution with parameters $(\gamma_{ik}^{(j)}, l_{ik}^{(j)})$, then $\eta_{ik}^{(j)} = \zeta_{ik}^{(j)}(1) + \ldots + \zeta_{ik}^{(j)}(l_{ik}^{(j)})$ where $\{\zeta_{ik}^{(j)}(m)\}_{m=1}^{l_{ik}^{(j)}}$ are i.i.d. r.v.'s having exponential distribution with parameter $\gamma_{ik}^{(j)}$. It follows immediately that

$$\alpha_{ik}^{-1} = \mathsf{E}\eta_{ik}^{(1)} = l_{ik}^{(1)}[\gamma_{ik}^{(1)}]^{-1}, \quad \beta_{ik}^{-1} = \mathsf{E}\eta_{ik}^{(2)} = l_{ik}^{(2)}[\gamma_{ik}^{(2)}]^{-1}.$$

We say that at time t the node (i,k) is in the mth phase, $m = \overline{1, l_{ik}^{(j)}}$, if $t \in (\zeta_{ik}^{(j)}(0) + \ldots + \zeta_{ik}^{(j)}(m-1), \zeta_{ik}^{(j)}(0) + \ldots + \zeta_{ik}^{(j)}(m))$ where $\zeta_{ik}^{(j)}(0) = 0$. Moreover, for $j = 1$ the green light is switched on, whereas for $j = 2$ the red light is switched on.

Introduce a stochastic process

$$X(t) = (q_{ik}(t), w_{ik}(t), e_{ik}(t), u_{ik}(t), i = \overline{1,r}, k = \overline{1,n_i}),$$

here $q_{ik}(t)$ is the number of cars waiting at the traffic lights (i,k), $w_{ik}(t)$ the number of cars on all the arcs going out of node (i,k), furthermore $e_{ik}(t) = 1$ (or 2) in case of green (red, resp.) light switched on and $u_{ik}(t)$ is the number of phase in the node (i,k) after the previous switching of lights.

Theorem 5. *The process $X(t)$ is ergodic iff*

$$\rho_{ik} < 1 \quad \textit{for all} \quad k = \overline{1,n_i}, \quad i = \overline{1,r}. \tag{21}$$

Sketch of proof. The process $X(t)$ is regenerative and its regeneration points are θ_j satisfying the following conditions $q_{ik}(\theta_j - 0) = 0$, $w_{ik}(\theta_j - 0) = 0$, $e_{ik}(\theta_j - 0) = 1$, $u_{ik}(\theta_j - 0) = 1$. We use the fact that input flow for each node is regenerative, a well-known Smith theorem and majorization procedure.

If for some node of network the ergodicity condition (21) is not fulfilled, that is, $\rho_{ik} \geq 1$, then $q_{ik}(t)$ will be stochastically unbounded, as $t \to \infty$.

4.3 Capacity of Transport Network

Conditions (21) enable us to estimate the maximal admissible intensity of car inflow in the network S^+ as follows

$$\lambda < \min_{k=\overline{1,n_i};\, i=\overline{1,r}} \nu_{ik}\beta_{ik}[(\alpha_{ik}+\beta_{ik})\delta_{ik}]^{-1}. \tag{22}$$

The same inequalities let us establish the network bottlenecks with ρ_{ik} defined by (20) very close to 1.

The objective function measuring the network performance we are going to consider is the mean time $T(\lambda)$ of passing the route by a car. Recall that λ is intensity of input flow to network.

Definition 5. The capacity of network is the maximal intensity of input flow λ satisfying the ergodicity condition (22) as well as inequality $T(\lambda) \le T_0$, where T_0 is the upper limit of desired network passing time by a car.

The question is how to determine T_0. At first we calculate the minimal mean time of passing the network by car. To this end consider a car arriving to empty network (that is, containing no other cars). Let $t(I)$ be the mean time of passing the route $I = (j_1, \dots, j_r)$, then

$$t(I) = \sum_{k=1}^{r} \frac{d^{(k)}_{j_k,j_{k+1}}}{v^{(k)}_{j_k,j_{k+1}}} + \sum_{k=1}^{r} \frac{\alpha_{kj_k}}{\alpha_{kj_k}+\beta_{kj_k}}(\beta^{-1}_{kj_k} + \nu^{-1}_{kj_k}). \tag{23}$$

The first sum in (23) is the mean time of passing of all the arcs connecting the nodes of the chosen route, whereas the second sum is the mean time of nodes (traffic lights) passing by a car.

The mean velocity of passing the route I is given by $v(I) = ||I||/t(I)$, here $||I|| = \sum_{k=1}^{r} d^{(k)}_{j_k j_{k+1}}$ is the length of the route.

Thus, the minimal mean time of route passing and the maximal mean velocity are

$$T_{min} = \sum_I t(I)\mathsf{P}(I), \quad v_{max} = \sum_I \mathsf{P}(I)||I||/t(I), \tag{24}$$

respectively. Obviously, the network capacity can be determined by the condition $v(\lambda) \le v_0$ for mean velocity instead of $T(\lambda)$. Here $v(\lambda)$ is the mean car velocity in case of input flow intensity λ and v_0 the admissible bound.

Expressions (24) let us establish the possible values of T_{min} and v_{max}. Clearly they depend only on network structure and its parameters and are independent of the inflow intensity. It is possible to take $T_0 = \vartheta T_{min}$ and $v_0 = v_{max}\vartheta^{-1}$ for some $\vartheta > 1$.

4.4 Approaches to Estimation of Network Capacity

For capacity estimation we have to calculate (or estimate) the mean time or mean velocity of network passing. Recall that intensity of car inflow to node (i,k) is given by (19). Suppose for simplicity that input to network is Poisson with parameter λ. Then the inflows to the nodes of the first level of network are also Poisson with parameters $\lambda\delta_{1j}$, $j = \overline{1,n_1}$.

The inflows to the nodes of the second level are not Poisson even for exponential traffic lights passing times. They are doubly stochastic Poisson processes. The previous level forms the random environment for these processes. However for large networks these inflows may be considered approximately Poisson as the sums of large number of regenerative flows, see, e.g., [10]. Hence, assume that for a chosen route $I = (j_1, \dots, j_r)$ we have Poisson inflows for all nodes with parameters $\lambda\delta_{1j_1}, \dots, \lambda\delta_{rj_r}$, respectively.

Consider some arc of the route, dropping all indices for brevity. Let d be the arc length and $\tilde{\lambda}$ the intensity of the Poisson input at the node beginning the arc. If $\tilde{\lambda}$ is small then a car does not disturb the others, that is, the mean car velocity does not depend on the intensity. Thus, to calculate the mean car velocity one can use the first term in (23). If $\tilde{\lambda}$ is large enough, a car on reaching a slow one either overtakes it or has to slow down. It is possible to use the model with two car types proposed in Sect. 1.

If the distance between the traffic lights (arc length) is not large we propose another model. Let $V_1 < V_2 < \dots < V_l$ be the set of possible velocities and p_j the probability that a car intends to move with velocity V_j, $j = \overline{1,l}$. The main assumption is that if after passing the traffic lights a car finds some other car on the arc it begins to move with the same velocity. Otherwise its velocity equals V_j with probability p_j.

To find the mean car velocity on the arc consider a queueing system $M|G|\infty$ with input intensity $\tilde{\lambda}$. Assume that the customer arriving to the empty system (that is, that beginning the busy period) has the service time d/V_j with probability p_j. All the customers arriving during this busy period have the same service time. We say that the busy period has type i if the service times are equal to d/V_i, $i = \overline{1,l}$. Denote by $T_b^{(i)}$ such a busy period. It is not difficult to establish that $\psi_i(s) = \mathsf{E}e^{-sT_b^{(i)}}$ has the form $(s+\tilde{\lambda})(se^{(s+\tilde{\lambda})d/V_i}+\tilde{\lambda})-1$, $Re\, s \geq 0$. It follows immediately that $\mathsf{E}T_b^{(i)} = \tilde{\lambda}^{-1}(e^{\tilde{\lambda}d/V_i} - 1)$. According to Little's formula, the mean number of customers served during this busy period is given by $\varkappa^{(i)} = e^{\tilde{\lambda}d/V_i} - 1$. Hence, the mean number of customers served during a busy period is $\varkappa = \sum_{i=1}^{l} p_i \varkappa^{(i)} = \sum_{i=1}^{l} p_i e^{\tilde{\lambda}d/V_i} - 1$. The probability of moving on arc with velocity V_i is equal to $p_i\varkappa^{(i)}/\varkappa$ and the mean velocity on the arc is

$$\bar{V} = \sum_{i=1}^{l} V_i (e^{\tilde{\lambda}d/V_i} - 1) p_i \left(\sum_{i=1}^{l} p_i e^{\tilde{\lambda}d/V_i} - 1 \right)^{-1}. \tag{25}$$

For the arc (j_k, j_{k+1}) of the chosen route denote by $p_i^{(k)}$ the probability of velocity V_i, $i = \overline{1,l}$, $k = \overline{1,r}$. Finally, the mean total time of passing the route $I = (j_1, \ldots, j_r)$ is given by the first sum in (23) with $\tilde{v}^{(k)}_{j_k j_{k+1}}(\lambda)$ instead of $v^{(k)}_{j_k j_{k+1}}$ where

$$\tilde{v}^{(k)}_{j_k j_{k+1}}(\lambda) = \frac{\sum_{i=1}^{l} V_i p_i^{(k)} (e^{\lambda \delta_{kj_k} d_{j_k j_{k+1}}/V_i} - 1)}{\sum_{i=1}^{l} p_i^{(k)} e^{\lambda \delta_{kj_k} d^{(k)}_{j_k j_{k+1}}/V_i} - 1}. \tag{26}$$

Now it is necessary to find the mean waiting time before the traffic lights. For the simple case of exponential interswitching times (with parameters $\gamma_{jk}^{(i)}$, $i = 1, 2$, for the node (j,k) with $i = 1$ corresponding to the green light and $i = 2$ for the red one) one can rewrite expression (15) for the mean waiting time before traffic lights in the form

$$m_{jk}(x_{jk}) = \frac{x_{jk}(1 - c_{jk})(1 + d_{jk}(1 - x_{jk})(1 - c_{jk}))}{(1 - x_{jk})(c_{jk} - x_{jk})} \tag{27}$$

with $x_{jk} = \lambda_{jk}/\nu$, $c_{jk} = [\gamma_{jk}^{(1)}\theta_{jk}]^{-1}$, $\theta_{jk} = [\gamma_{jk}^{(1)}]^{-1} + [\gamma_{jk}^{(2)}]^{-1}$, $d_{jk} = \nu/\gamma_{jk}^{(1)}$. The mean time of traffic lights (j,k) passing by a car is found by means of Little's formula, that is, $t_{(jk)} = \lambda_{jk}^{-1} m_{jk}(x_{jk})$.

Using (23) we get the mean time of route $I = (j_1, \ldots, j_r)$ as

$$t(I) = \sum_{k=1}^{r} \frac{d^{(k)}_{j_k j_{k+1}}}{\tilde{v}^{(k)}_{j_k j_{k+1}}(\lambda)} + \sum_{k=1}^{r} \frac{m_{jk}^{(k)}(x_{kj_k})}{\lambda \delta_{kj_k}}$$

with $\tilde{v}^{(k)}_{j_k j_{k+1}}(\lambda)$ given by (26) and $m_{jk}^{(k)}$ by (27).

Then the mean time of passing is

$$T(\lambda) = \sum_I t(I)\mathsf{P}(I).$$

Thus the problem is to find λ satisfying (22) and being the solution of equation

$$T(\lambda) = \vartheta T_{min} \quad \text{for a chosen} \quad \vartheta.$$

5 Conclusion and Further Research Directions

A transport system capacity was introduced as maximal car flow intensity compatible with a desired quality of system performance. As an objective function we have chosen

- The mean car velocity or mean travel time dealing with highway capacity,

- The mean number of cars waiting before crossroads useful to analyze the traffic lights capacity,
- Probability of a line of stationary or very slow-moving traffic (caused by traffic lights, road works, an accident or heavy congestion) with length exceeding a given threshold.

We have considered several examples of transport models, namely, a highway (without traffic lights) with two types of cars, a single crossroads with traffic lights, as well as hierarchical networks and estimated their capacities. For this purpose we investigated the limit behavior of such systems (proving ergodic theorems) in light- and heavy-traffic conditions. Markov, Poisson and doubly stochastic Poisson processes, renewal and regenerative ones are the main tools in our study.

There arise interesting problems for further investigation concerning the optimal choice of routes and optimization of traffic lights performance under general assumptions about the random variables involved in models description. The results will be published elsewhere.

Acknowledgements The research was partially supported by RFBR grant 13-01-00653.

References

1. Greenshields B D (1935) Proc Highway Res 14 : 448–477
2. Afanasyeva L G, Bulinskaya E V (2010) Proceedings of Moscow Institute of Physics and Technology, 2(4) : 6–21
3. Afanasyeva L, Bulinskaya E (2010) Automation and Remote Control, 71(7) : 1294–1307
4. Afanasyeva L, Bulinskaya E (2011) Communications in Statistics: Theory and Methods, 40(16) : 2830–2846
5. Afanasyeva L, Bulinskaya E (2012) Methodol Comput Appl Probab, DOI 10.1007/s11009-012-9291-x
6. Baykal-Gürsoy M, Xiao W, Ozbay K (2009) Eur J Oper Res, 195:127–138
7. Gaver D P Jr (1962) J Roy Stat Soc, B 24 : 73–90
8. Fayolle G, Malyshev VA, Menshikov MV (1995) Topics in the Constructive Theory of Countable Markov Chains, Cambridge University Press
9. Smith W L (1955) Proc Roy Soc A, 232 : 6–31
10. Grigelionis B (1963) Theory Probab and Appl. 8(2) : 177–182

Classification Approach to Traffic Flow Simulation

Yury V. Chekhovich and Nikita P. Ivkin

Abstract A new approach to a microscopic traffic flow simulation is presented. The approach is based on application of data mining techniques. Key moments of constructing process are considered and main advantages in comparison with classical models are provided.

1 Introduction

Simulating and predicting traffic macro parameters (such as velocities and flows) in heterogeneous traffic networks is an essential problem in transport management. In the majority of modern approaches to the given problem either heterogeneity of a simulated traffic flow isn't considered at all, or solution is offered just for several special cases. Taking into account heterogeneity means considering different types of agents in traffic flow (buses, trucks, private cars, etc.). But, the structure of heterogeneity is supposed to differ from one place to another and change in time. Thus, ready-made solutions give back bad quality of simulating. In this paper, we propose a new method of constructing microscopic models. This method allows taking into account different kinds of heterogeneities.

Y.V. Chekhovich (✉)
Computing Center of the Russian Academy of Science, Vavilova str. 40, Moscow, 119333, Russia
e-mail: chehovich@forecsys.ru

N.P. Ivkin
Moscow Institute of Physics and Technology (State University), Institutsky lane 9, Dolgoprudny, 141700, Russia
e-mail: ivkinnikita@gmail.com

V.V. Kozlov et al. (eds.), *Traffic and Granular Flow '11*,
DOI 10.1007/978-3-642-39669-4_8, © Springer-Verlag Berlin Heidelberg 2013

2 Key Ideas of Model Construction

2.1 Classification Approach

This paper is based on the classification approach of simulation modelling of the complex socially-technical systems offered in [1–3]. Obviously, the simulated object, namely a traffic flow, is a complex socially-technical system in the terms determined in mentioned articles: a traffic flow consists of a quantity of vehicles and drivers make decisions which completely define the movements of vehicles operated by them. Further we name pair <vehicle, driver> as a subject and, saying that the subject makes the decision, we mean that the decision is accepted by the driver.

The classification approach is based on three general assumptions:

1. Assumption on behaviour of system:
   - Each subject has the information about itself and about subjects from some "vicinity", that is "the situation description".
   - Analysing "the situation description" the subject makes the decision on the further behaviour.
   - Often the subject gets out the decision from finite, most likely "relatively small", set of decisions.
2. Assumption on behaviour of system:

   There is some measure of similarity (or even metrics) in the space of "descriptions of situations". It is supposed that the subject in situations of one class makes the same decisions and, thus, it allows classifying "situations".
3. Assumption on existence of subject types:

   Subjects which belong to one type in similar situations accept identical (or similar) decisions. This idea is effectively applicable only if there are many subjects and comparably less types.

2.2 Fitting Classification Approach to Traffic Micro-models

We propose to construct a model in terms of microscopic simulation. In a traffic flow "the situation description" for the simulated subject is the information about itself and about the nearest vehicles. This information can be presented in the form of vector of microscopic parameters. For example, for acceleration model [6] the vector of microscopic parameters consists of subject's speed, distance to a leader and leader's speed. Some other micro-parameters are presented in the Fig. 1: distance to a leader, distance to the neighbour car (L_{XA}), velocities of different neighbour cars and others.

The set of decisions is determined by possible changes in the movement mode undertaken by the subject. For acceleration model the multitude of decisions

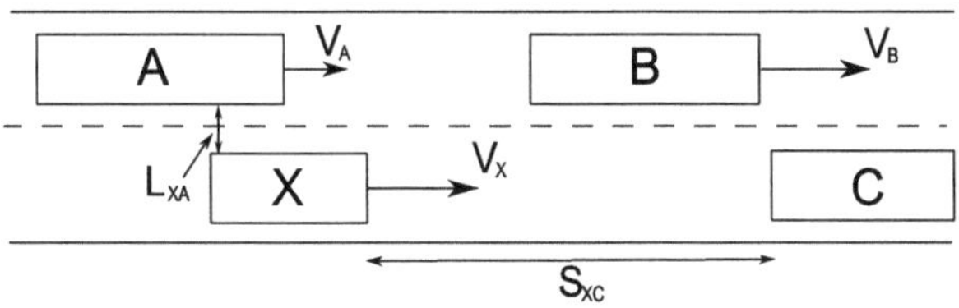

Fig. 1 Micro-parameters examples

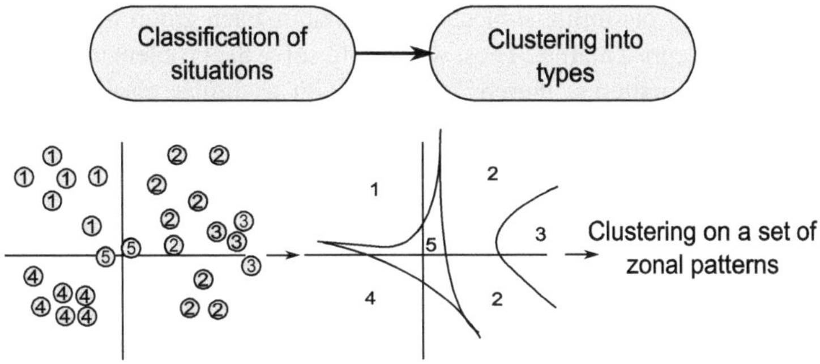

Fig. 2 Two stages of model constructing

consists of strong and weak accelerations, maintenance of speed, strong, weak and emergency braking. Such a poor model, apparently, fails any quality competition, but it is good for understanding key points.

When the formal description of situations space is received and the set of decisions is set, data for tuning our model should be presented in following view: <subject id, situation description—decision>. Further, the construction procedure of model divides into two stages (Fig. 2).

Stage One: Zonal Pattern of Behaviour for Each Subject

We propose to classify situations for each subject using provided sample of precedents. We need to introduce some measure of similarity in the space of situations descriptions. The most evident candidate is n-dimensional Euclidean metric or may be its weighted variant.

$$F(S_1, S_2) = \sqrt{\sum_{i=1}^{N} w_i (a_{1i} - a_{2i})^2}, \tag{1}$$

where S_1 and S_2—two situations, a_{ji}—micro-parameter number i that described situation S_j, w_i—weight of micro-parameter a_{ji} and N—number of given hboxmicro-parameters.

Then, we should use some algorithm of classification, for example, Support Vector Machine (SVM) [4]. The constructed classifier divides all space of situations into zones of various decisions (zonal pattern). Thus, decisions are evaluated for each vector of micro-parameters, that is for each situation.

Stage Two: Types Description

Dealing with a large sample of subjects described by zonal patterns, it's essential to search some groups of similar subjects. In simulation each group will be presented by some average representative. Thus, we need to solve the problem of clustering [4] on a set of zonal patterns, thereby subjects with a similar choice of decisions correspond to one group, that is to one type. To solve the problem of clustering we need to introduce measure of similarity on a set of zonal patterns. For example, we could divide a zonal pattern into finite number of small square parts and several infinite parts, then introduce the following function as measure of similarity between two zonal patterns, v and u:

$$\rho(v,u) = \sum_{i=1}^{n} w(v,u,i)\,[z(v,i) \neq z(u,i)]\,, \tag{2}$$

where $i = 1 \ldots n$—number of part (apparently it is supposed that all zonal patterns were divided identically), $z(v,i)$—function, that returns type of the zone, which occupy the biggest space in part number i of zonal pattern v, $w(v,u,i)$—weight function that depends on type of difference in part number i of zonal patterns v and u. We can use algorithm FOREL (FORmal ELement [4]) for clustering zonal patterns.

2.3 *Training Data Collecting*

To create a training sample mentioned above authors suggest to use the program stand simulating the movement of a car driven by an operator in a traffic flow. All situations descriptions and undertaken decisions are fixed by the program, and relying on the received sample of precedents the model is identified. More accurate tool of creation of precedent samples would be a real vehicle equipped with sensors which fix situations descriptions and undertaken decisions. The program stand allows to show general qualities of the offered model.

2.4 *Key Model's Features*

We propose to consider two classes of microscopic models: acceleration models and lane changing models. Several shortcomings of the classical approach, which are avoided in the offered model, are presented below:

- The majority of models are tuned on some "average" driver ("average" in terms of internal parameters) and often it isn't informative enough. The model constructed according to the classification approach assumes division of subjects into types, thus, the given lack isn't inherited. Even more, the number of types is evaluated from training sample, that give us an opportunity to take into account different types of heterogeneity.
- The majority of the program complexes based on microscopic modelling don't allow switching of models for simulating the same subject. Though in various situations this subject can behave differently. The given lack is also eliminated by construction.
- Historically, the majority of models were offered as complications and generalizations of already existent models. For example, here is a General Motors chain of models:

  1. $a_n(t) = \alpha \Delta V_n^{front}(t - \tau_n)$—Chandler et al. [5],
  2. $a_n(t) = \alpha \frac{(\Delta V_n^{front}(t-\tau_n))}{(\Delta X_n^{front}(t-\tau_n))}$—Gazis et al. [6],
  3. $a_n(t) = \alpha \frac{(V_n(t)^{\beta} \Delta V_n^{front}(t-\tau_n))}{(\Delta X_n^{front}(t-\tau_n)^{\gamma})}$—Gazis et al. [7],

  where a_n—the acceleration of a simulated car, ΔV_n^{front}—the relative speed of a leader, ΔX_n^{front}—the distance to a leader, V_n—the speed of a simulated car, τ_n—the time of driver's reaction, α, β, γ—internal parameters. Each subsequent model has new internal and external parameters. Therefore for each situation there is a question of choosing particular model. On the one hand, chosen model shouldn't be too simple, but on the other hand, too complex model involves an over-fitting problem. In the model constructed according to the classification approach, the problem of complexity turns into the problem of feature selection in classification. There already exist feature reduction methods for classification problems (features are the parameters of situations descriptions in our case) [4].

Thus, the offered approach allows eliminating the specified lacks. Also in the present paper it is proved that any reasonable classical model can be approximated by the model constructed according to the classification approach as precisely as needed.

3 Conclusion

We've proposed the application of the classification approach offered in [1, 2] to the traffic flow problem as an instrument of considering heterogeneity. We also have offered a program stand which can be used to demonstrate features of model identification.And finally, we've shown that the basic lacks of classical models are eliminated in the offered model. This branch of traffic modelling should be developed theoretically and, of course, should be compared experimentally with classical models.

References

1. Chekhovich Y.V.(2010) About identification of simulation models of difficult socially-technical systems by means of the aggregated data. Internationalization of processing of the information: 8th international conference. Republic Cyprus, Pathos, 17-24th oct, 2010: Collection of reports, MAX press. pp. 539–540.
2. Chekhovich Y.V.(2007) Application of algebraic approach to imitation modeling of complex socially-technical systems. Collection of reports of the third All-Russia scientifically-practical conference *Imitating modeling. Theory and practice*, St. Petersburg, Volume I, pp. 249–252.
3. Gasnikov A.V., Klenov S.L., Nurminsky E.A., Holodov Y. (2010) Introduction into mathematical modelling of transport streams. The manual: MIPT, Moscow
4. Hastie T., Tibshirani R., Friedman J. (2001) The Elements of Statistical Learning. Springer.
5. Chandler R., Herman R. and Montroll (1958) Traffic dynamics: Studies in car following. Operations Research 6, pp. 165–184.
6. Gazis D., Herman R. and Potts B.(1959) Car following theory of steady-state traffic flow. Operations Research 7, pp. 499–505.
7. Gazis D., Herman R. and Rothery R.(1961) Non-linear follow-the-leader models of traffic flow. Operations Research 9, pp. 545–567.

Multilane Vehicular Traffic Simulation Using the CA Theory

Ilya Furmanov, Natalia Churbanova, and Marina Trapeznikova

Abstract The paper deals with the development of microscopic mathematical models for the simulation of vehicular traffic flows taking into account real road geometry, i.e. the multilane structure, presence of crossroads, traffic lights etc. The proposed model is based on the cellular automata (CA) theory. Generalization of the Nagel-Schreckenberg model to the multilane case is presented. The model has been verified by the number of test problems. Practically important road situations have been predicted. Simulation of traffic on a crossroad demonstrates the influence of traffic lights regimes on the crossroad capacity. The road capacity depending on the entries/exits presence and on the distance between them is also investigated.

1 Model Description

The proposed microscopic model is based on the cellular automata (CA) theory. CA theory is an idealization of the physical system with discrete space and time, each of interacting units of this system has got a finite number of discrete states. For the description of vehicular traffic the CA concept is being developed since 1980s.

In the classical approach, a lane is represented by a one-dimensional lattice. Each cell of the lattice can be either empty or occupied by one particle, which represents a vehicle. Vehicles can skip from one cell to another (which must be empty) in one direction and cannot overtake one another. Since space and time are quantified, the speed and the acceleration take on only discrete values. In such models the particle movement is regulated by special laws of the cell state update incorporating stochastic observations. The update rules are identical for all cells and are applied to all cells in parallel. Therefore parallel computer codes can be developed for modeling to run on high-performance multiprocessors.

I. Furmanov · N. Churbanova (✉) · M. Trapeznikova
Keldysh Institute of Applied Mathematics RAS, 4 Miusskaya Square, Moscow, 125047, Russia
e-mail: nata@imamod.ru

V.V. Kozlov et al. (eds.), *Traffic and Granular Flow '11*,
DOI 10.1007/978-3-642-39669-4_9, © Springer-Verlag Berlin Heidelberg 2013

The CA rules feature the property of locality. In other words, to obtain the current state of the cell, it is necessary to know only the states of some of its neighbors called the cell vicinity. The cell length equals the length of the road interval occupied by a vehicle in the traffic jam that is the length of a vehicle and the distance between neighboring vehicles. Usually it is 7.5 m. The speed denotes how many cells the vehicle overpasses during a time step. The cell length, the maximal speed and the time step describe the model completely.

One of the well-known CA-based microscopic models is the Nagel—Schreckenberg model [1]. Its original variant is one-dimensional. In this model the speed v of each vehicle can take one of the integer values $v = 0, 1, \ldots, v_{max}$ (each vehicle can have its own v_{max}).

If x_n and v_n are the position and the speed of the current n-th vehicle, d_n is the distance between the current vehicle and the vehicle in front of it, then at each time step $t \rightarrow t + 1$ the algorithm of the vehicle arrangement update consists of the next stages:

1. Acceleration $v_n \rightarrow min(v_{n+1}, v_{max})$
2. Deceleration $v_n \rightarrow min(v_n, d_{n-1})$
3. Randomization $v_n \rightarrow max(v_{n-1}, 0)$ with some probability p
4. Vehicle movement $x_n \rightarrow x_n + v_n$

The first stage reflects the common tendency of all drivers to move as fast as possible, the second one guarantees avoiding collisions, the third one takes into account randomness in driver behavior, and the skip itself takes place on the fourth stage—each vehicle is moved forward according to its new velocity.

The Nagel-Schreckenberg model is a minimal model because it reproduces only primary features of real traffic flows. The present research takes this model as a basis to describe movements on relatively short road intervals with a high probability of traffic congestion. Aiming at simulation of multilane traffic, the authors have generalized the above model to the two-dimensional case. In this case the computational domain is the 2D lattice. The number of cells in the transverse direction corresponds to the number of lanes. Such a model allows vehicles to change lanes and to overtake one another. The algorithm of the cell state update is formed by two components:

1. Lane change (if it is necessary and possible);
2. Movement along the road by the rules of one-lane traffic (stages 1–4 above).

Change of lanes should happen during a time step. If there are more than two lanes in one direction, a conflict can occur when two vehicles from extreme lanes tend to the inner lane and try to occupy the same cell. The rule like the next one could help to resolve such a situation: vehicles change to the right only on even time steps and change to the left on odd steps.

In general rules and conditions for changing lanes are as follows:

1. Vehicle is located in the domain where lane change is allowed;

2. Lane change leads to increase of the speed (decrease of the density) or is necessary to reach the destination (to achieve the goal);
3. Target cell is empty;
4. Safety condition is satisfied—on the target lane the distance behind the vehicle is greater or equal to v_{max}, in front of the vehicle it is greater or equal to v_n, then the change takes place with some probability.

The algorithm proposed in the present work ensures the possibility to achieve the destination. For example, the side road exit or the appointed turn at traffic lights can be assumed as destinations in multilane traffic. In any case, starting from the certain time moment vehicles aim at the target lane and ignore the density and velocity values on it. However drivers cannot disregard the safety condition. If the destination is not far, vehicles change to the target lane at the first opportunity and do not quit it anymore. The situation is possible when a driver is not able to turn to the required lane up to the destination achievement. In such cases the vehicle has to stop near by the target lane and to wait for the opportunity of wedging itself in the lane. In doing so, it can disable forward movement of other vehicles on the current lane. General considerations help to evaluate the distance to the destination at which drivers start trying to change to the target lane. As a matter of fact, this distance depends on the flow density. In different problems it is 75–150 m. Thus the developed model keeps the destination parameter for each vehicle. Vehicle destinations are obligatory, they cannot be modified.

2 Test Predictions

In this work, results of test predictions are compared to results obtained on the same initial data with usage of macroscopic quasi-gas-dynamic (QGD) approach, described in [2]. Results obtained using CA approach are numerically close to those, obtained with the usage of the macroscopic model [3]. This fact leads us to conclusion that both models are consistent with the real road traffic movement.

2.1 Local Widening of the Road

The first test problem is vehicle movement on the road with local widening. The corresponding road configuration is shown in Fig. 1.

Figure 2 demonstrates the density field obtained with the use of the macroscopic model. The density for a given computational domain is shown. The density of vehicles falls at the wide part of the road but when the traffic flow shrinks back from three to two lanes, the speed falls significantly. Thus, the total time required to pass the given road interval grows as compared with the road without widening.

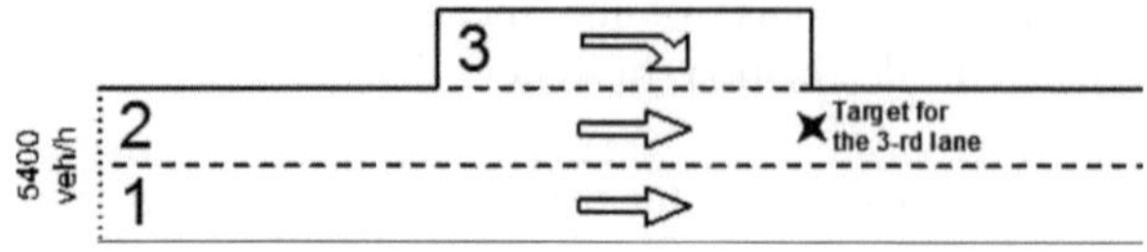

Fig. 1 Local widening: the problem statement

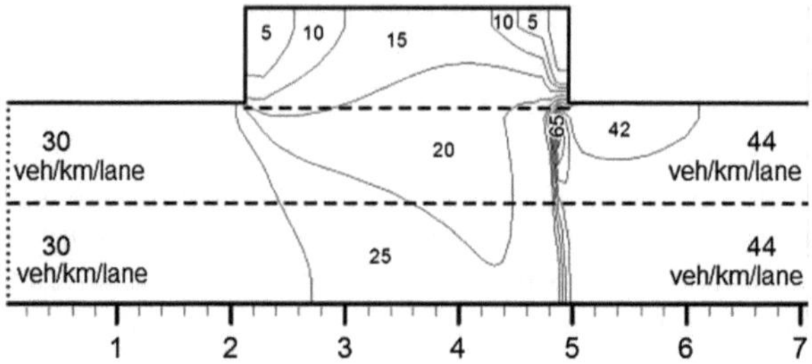

Fig. 2 Density field obtained via the macroscopic model

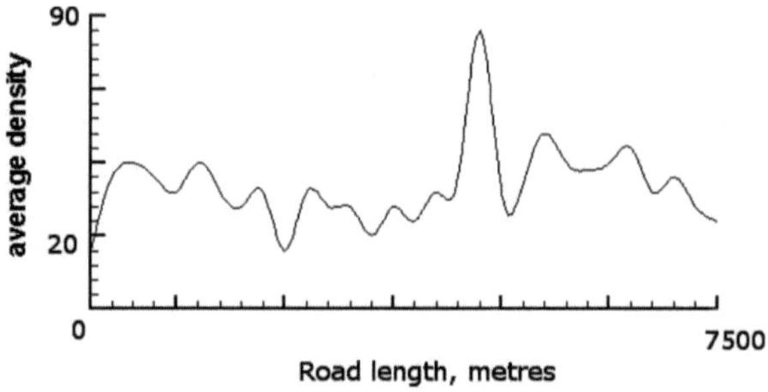

Fig. 3 Density field obtained via the microscopic model

The same test problem has been solved using the microscopic model. The average density is depicted in Fig. 3. One can observe the same tendency: the maximal density is located at the end of widening. Moreover, the density at the road exit exceeds the enter density.

2.2 *Crossroad Simulation*

The second problem to be considered is the simulation of traffic on a crossroad at different traffic lights regimes using the above microscopic model. The problem consists in obtaining the optimal traffic lights regime namely the signal durations to ensure the minimal time of stay on the crossroad for all traffic participants. Figure 4 illustrates the problem.

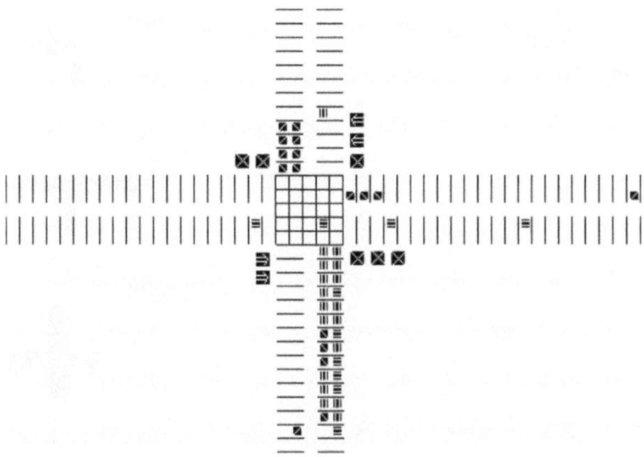

Fig. 4 Problem statement of traffic lights on the crossroad

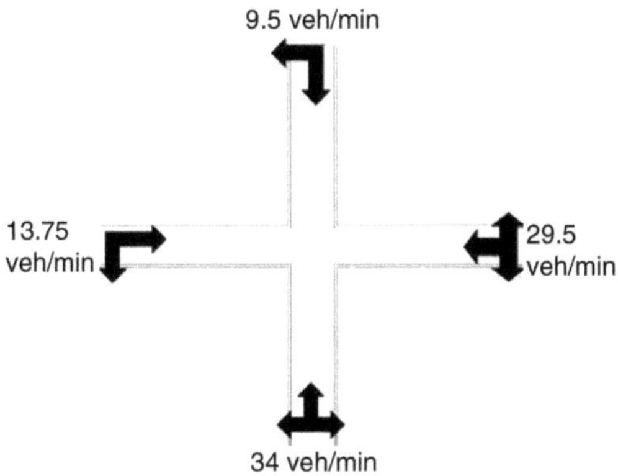

Fig. 5 Incoming flow rates and allowed directions on the crossroad

Lattice consists of four crossing roads each of them has four lanes: two in forward and two in reverse directions. The traffic flows are multicomponent, incoming flow rates are non-uniform. There are vehicles of four types according to its destination—to move forward or to turn to the right or to the left. Figure 5 shows incoming flow rates and allowed directions of movement on the crossroad.

Traffic lights has four operating modes which determine the order of traffic on the crossroad at the given time moment (see Fig. 6). Each mode has its own duration coinciding or not with other ones. All data correspond to some real crossroad.

Vehicles have more complicate behavior in the vicinity of the crossroad in comparison with straight intervals of the road. If the vehicle intends to turn it must decrease the speed till to allowed for turning. The vehicle must also take into account positions of vehicles in neighbor lanes and in other directions.

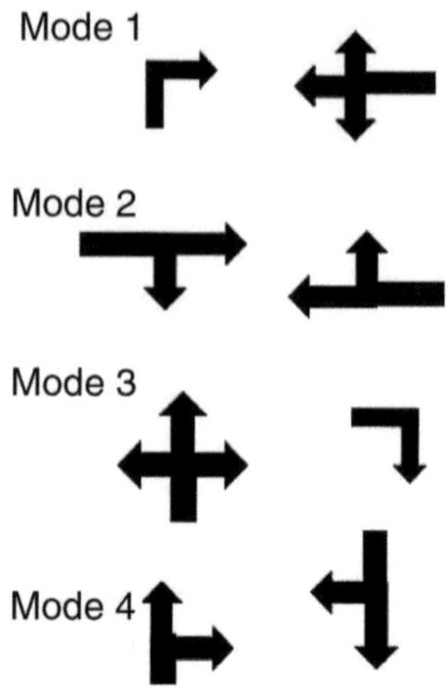

Fig. 6 Traffic lights operating modes allowing the given directions on the crossroad

Apart from steps, described in Sect. 1, following rules are also applied to vehicles in this simulation:

- Within 200 m before traffic lights the vehicle changes lane under its purpose according to the road laws.
- Additional speed decreasing takes place under the follows conditions:
  - If the vehicle is located near the turning point (at the turning point it stops);
  - If the traffic light is red;
  - If there is the collision threat on the crossroad.
- The vehicle turns if it is located in the turning point and has got the corresponding target.

Table 1 demonstrates results of predictions. Varying mode durations one can increase the crossroad capacity, i.e., the number of vehicles overpassing the crossroad for 10 min. Obtained results allow us to conclude the following. Equal mode durations do not provide the best capacity (see row 2 of Table 1). Mode 1 allows passing over the crossroad by vehicles with the maximal incoming flow rate. But excessive increase of the Mode 1 duration does not lead to the capacity increase (compare rows 2 and 4).

2.3 Comparative Analysis of Different Road Types

In the next set of problems, we perform a comparative analysis between ordinary roads without any entries/exits and roads of special configuration with multiple

Table 1 The traffic lights operating modes—simulation results

Mode 1 duration, s	Mode 2 duration, s	Mode 3 duration, s	Mode 4 duration, s	Capacity of the crossroad
90	90	45	45	490
60	60	60	60	545
90	60	90	60	545
45	45	60	60	575
45	45	90	90	575

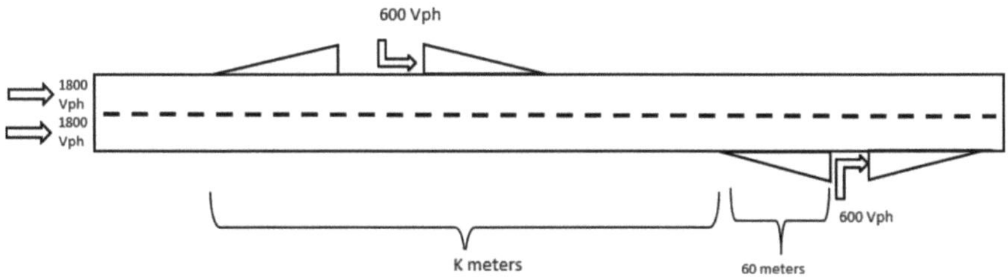

Fig. 7 Configuration of the road with multiple entries and exits

entries/exits. Reason for this test was to validate model and find optimal configurations of the road.

In the first problem of this set we compare the ordinary 3-lane road and the road with entries and exits placed at the both road sides at different distances (see Fig. 7). Both road types have the same length and input flows per a lane. All roads with multiple entries and exits have the same additional input flow per an entry and the same length of speed-up/slow-down lane. What we vary is the number of lanes in the road of special configuration (1, 2 or 3 lanes) and distance K between adjacent entries.

Test results show us that even the two-lane road of special configuration is almost twice as better as the ordinary road in terms of capacity (see Fig. 8).

We also found optimal distance between entries for the road of special configuration (Fig. 9). Note the fact that this distance is significantly different for one-lane roads and two- or three-lane roads. This is because in case of one lane additional entries, if situated too close to each other, add too much extra flux that cannot be digested by the system and results in traffic jams.

In the case of two- and three-lane roads the optimal distance between adjacent entries K is equal to 1.5 km vs. 3 km in the single lane case.

In the second problem related to roads of different configurations, we compare two proposed configurations of 75-km long road. The first road is the four-lane road with entries and exits located at every 3 km at the right side of the road. The second option is a two-lane road of special configuration, with exits and entries located at both sides of the road, 1.125 km away from each other, as on Fig. 7.

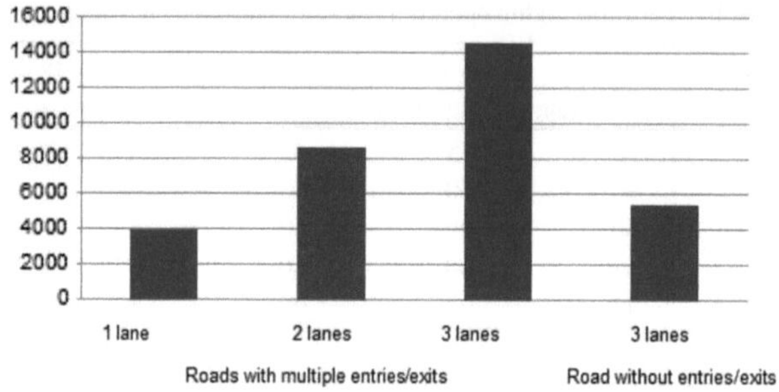

Fig. 8 Maximum capacity of the ordinary road vs. the roads with multiple entries and exits

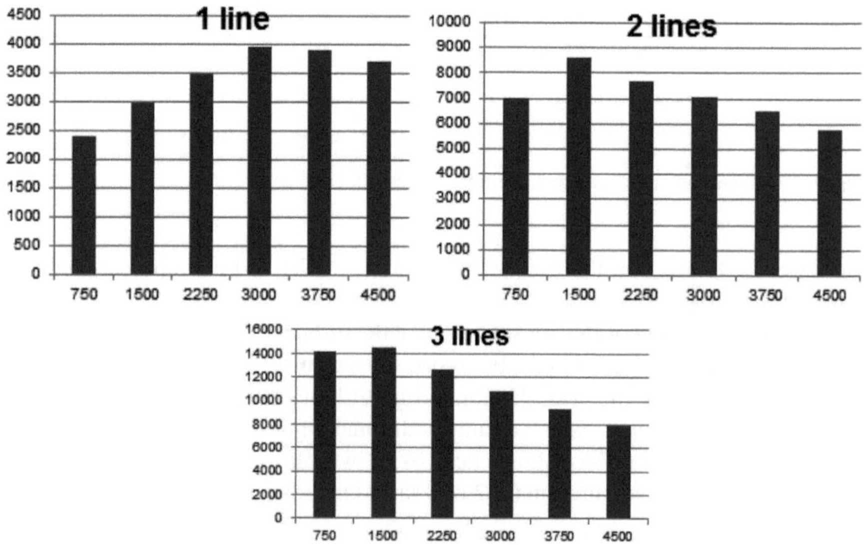

Fig. 9 Optimal distance between adjacent entries for 1-,2- and 3-lane roads

Table 2 Capacity (in vehicles per hour) of studied roads

Four-lane road, Ordinary vph	Three-lane road, Special vph
11,000	15,000

Test results show us that despite the road of special configuration has two times less lanes, overall capacity of it is higher, than one of the four-lane ordinary road (see Table 2).

3 Conclusions

- The model and algorithm developed can be implemented as a program package to be used in various engineering applications, including recommendations for the optimal motorway construction, road situation prediction and traffic flow control.
- High performance of modern supercomputers allows implementation of microscopic models with the detailed description of each vehicle behavior at large-scale computations. One of macroscopic models distinctive features—the possibility of large distances simulation—inheres now in microscopic models too.

References

1. Nagel K, Schreckenberg M (1992) A cellular automation model for freeway traffic. J. Phys. I France 2 (2221)
2. Sukhinova A, Trapeznikova M, Chetverushkin B and Churbanova N (2009) Two-dimensional macroscopic model of traffic flows. Mathematical Models and Computer Simulations 1(6):669–676
3. Chetverushkin B, Churbanova N, Furmanov I, Trapeznikova M (2010) 2D micro- and macroscopic models for simulation of heterogeneous traffic flows. In: CD-Rom Proc of ECCOMAS CFD 2010, Pereira J and Sequeira A (eds). IDMEC, Lisbon

Multi-agent Model of the Price Flow Dynamics

Vadim Malyshev, Anatoly Manita, and Andrei Zamyatin

Abstract On the real line initially there are infinite number of particles on the positive half-line., each having one of K negative velocities $v_1^{(+)}, \ldots, v_K^{(+)}$. Similarly, there are infinite number of antiparticles on the negative half-line, each having one of L positive velocities $v_1^{(-)}, \ldots, v_L^{(-)}$. Each particle moves with constant speed, initially prescribed to it. When particle and antiparticle collide, they both disappear. It is the only interaction in the system. We find explicitly the large time asymptotics of $\beta(t)$ – the coordinate of the last collision before t between particle and antiparticle.

1 Introduction

We consider one-dimensional dynamical model of the boundary between two phases (particles and antiparticles, bears and bulls) where the boundary moves due to reaction (annihilation, transaction) of pairs of particles of different phases.

Assume that at time $t = 0$ infinite number of $(+)$-particles and $(-)$-particles are situated correspondingly on R_+ and R_- and have one-point correlation functions

$$f_+(x, v) = \sum_{i=1}^{K} \rho_i^{(+)}(x)\delta(v - v_i^{(+)}), \qquad f_-(x, v) = \sum_{j=1}^{L} \rho_i^{(-)}(x)\delta(v - v_j^{(-)})$$

*Work of this author was supported by the Russian Foundation of Basic Research (grants 09-01-00761 and 11-01-90421)

V. Malyshev (✉) · A. Manita · A. Zamyatin
Faculty of Mechanics and Mathematics, Lomonosov Moscow State University, Leninskiye Gory, GSP-1, Moscow, 119991, Russia
e-mail: malyshev2@yahoo.com; manita@mech.math.msu.su; zamyatin@mech.math.msu.su

DOI 10.1007/978-3-642-39669-4_10, © Springer-Verlag Berlin Heidelberg 2013

Moreover for any i, j

$$v_i^{(+)} < 0, \qquad v_j^{(-)} > 0$$

that is two phases move towards each other. Particles of the same phase do not see each other and move freely with the velocities prescribed initially. The only interaction in the system is the following. When two particles of different phases find themselves at the same point they immediately disappear (annihilate). It follows that the phases stay separated, and one might call any point in-between them the phase boundary (for example it could be the point of the last collision). Thus the boundary trajectory $\beta(t)$ is a random piece-wise constant function of time.

One of the possible interpretations is the simplest model of one instrument (for example, a stock) market. Particle initially at $x(0) \in R_+$ is the seller who wants to sell his stock for the price $x(0)$, which is higher than the existing price $\beta(0)$. There are K groups of sellers characterized by their activity to move towards more realistic price. Similarly the (−)-particles are buyers who would like to buy a stock for the price lower than $\beta(t)$. When seller and buyer meet each other, the transaction occurs and both leave the market.

The main result of the paper is the explicit formula for the asymptotic velocity of the boundary as the function of $2(K + L)$ parameters – densities and initial velocities. It appears to be continuous but at some hypersurface some first derivatives in the parameters do not exist. This kind of phase transition has very clear interpretation: the particles with smaller activities (velocities) cease to participate in the boundary movement – they are always behind the boundary, that is do not influence the market price $\beta(t)$. In this paper we consider only the case of constant densities $\rho_i^{(+)}, \rho_i^{(-)}$, that is the period of very small volatility in the market. This simplification allows us to get explicit formulas. In [1] much simpler case $K = L = 1$ was considered, however with non-constant densities and random dynamics.

Other one-dimensional models (hardly related to ours) of the boundary movement see in [2, 3].

Main technical tool of the proof may seem surprising (and may be of its own interest) – we reduce this infinite particle problem to the study of a special random walk of one particle in the orthant R_+^N with $N = KL$. The asymptotic behavior of this random walk is studied using the correspondence between random walks in R_+^N and dynamical systems introduced in [4].

2 Model and the Main Result

2.1 Initial Conditions

At time $t = 0$ on the real axis there is a random configuration of particles, consisting of (+)-particles and (−)-particles. (+)-particles and (−)-particles differ also by the type: denote $I_+ = \{1, 2, \ldots, K\}$ the set of types of (+)-particles, and $I_- = \{1, 2, \ldots, L\}$ – the set of types of (−)-particles. Let

$$0 < x_{1,k} = x_{1,k}(0) < \ldots < x_{j,k} = x_{j,k}(0) < \ldots \tag{1}$$

be the initial configuration of particles of type $k \in I_+$, and

$$\ldots < y_{j,i} = y_{j,i}(0) < \ldots < y_{1,i} = y_{1,i}(0) < 0 \tag{2}$$

be the initial configuration of particles of type $i \in I_-$, where the second index is the type of the particle in the configuration. Thus all (+)-particles are situated on R_+ and all (−)-particles on R_-. Distances between neighbor particles of the same type are denoted by

$$\begin{aligned} x_{j,k} - x_{j-1,k} &= u_{j,k}^{(+)}, \quad k \in I_+, \quad j = 1, 2, \ldots \\ y_{j-1,i} - y_{j,i} &= u_{j,i}^{(-)}, \quad i \in I_-, \quad j = 1, 2, \ldots \end{aligned} \tag{3}$$

where we put $x_{0,k} = y_{0,i} = 0$. The random configurations corresponding to the particles of different types are assumed to be independent. The random distances between neighbor particles of the same type are also assumed to be independent, and moreover identically distributed, that is random variables $u_{j,i}^{(-)}, u_{j,k}^{(+)}$ are independent and their distribution depends only on the upper and second lower indices. Our technical assumption is that all these distributions are absolutely continuous and have finite means. Denote $\mu_i^{(-)} = Eu_{j,i}^{(-)}$, $\rho_i^{(-)} = \left(\mu_i^{(-)}\right)^{-1}, i \in I_-$, $\mu_k^{(+)} = Eu_{j,k}^{(+)}$, $\rho_k^{(+)} = \left(\mu_k^{(+)}\right)^{-1}, k \in I_+$.

2.2 *Dynamics*

We assume that all (+)-particles of the type $k \in I_+$ move in the left direction with the same constant speed $v_k^{(+)}$, where $v_1^{(+)} < v_2^{(+)} < \ldots < v_K^{(+)} < 0$. The (−)-particles of type $i \in I_-$ move in the right direction with the same constant speed $v_i^{(-)}$, where $v_1^{(-)} > v_2^{(-)} > \ldots > v_L^{(-)} > 0$. If at some time t a (+)-particle and a (−)-particle are at the same point (we call this a collision or annihilation event), then both disappear. Collisions between particles of different phases is the only interaction, otherwise they do not see each other. Thus, for example, at time t the j-th particle of type $k \in I_+$ could be at the point

$$x_{j,k}(t) = x_{j,k}(0) + v_k^{(+)} t$$

if it will not collide with some (−)-particle before time t. Absolute continuity of the distributions of random variables $u_{j,i}^{(-)}, u_{j,k}^{(+)}$ guaranties that the events when more than two particles collide, have zero probability.

We denote this infinite particle process $\mathbf{D}(t)$.

We define the boundary $\beta(t)$ between plus and minus phases to be the coordinate of the last collision which occured at some time $t' < t$. For $t = 0$ we put $\beta(0) = 0$. Thus the trajectories of the random process $\beta(t)$ are piecewise constant functions, we shall assume them continuous from the left.

2.3 Main Result

For any pair (J_-, J_+) of subsets , $J_- \subseteq I_-, J_+ \subseteq I_+$, define the numbers

$$V(J_-, J_+) = \frac{\sum_{i \in J_-} v_i^{(-)} \rho_i^{(-)} + \sum_{k \in J_+} v_k^{(+)} \rho_k^{(+)}}{\sum_{i \in J_-} \rho_i^{(-)} + \sum_{k \in J_+} \rho_k^{(+)}}, V = V(I_-, I_+) \qquad (4)$$

The following condition is assumed

$$\{V(J_-, J_+) \, : \, J_- \neq \varnothing, \, J_+ \neq \varnothing\} \cap \{v_1^{(-)}, \ldots, v_L^{(-)}, v_1^{(+)}, \ldots, v_K^{(+)}\} = \varnothing \, . \qquad (5)$$

If the limit $W = \lim_{t \to \infty} \frac{\beta(t)}{t}$ exists a.e., we call it the asymptotic speed of the boundary. Our main result is the explicit formula for W.

Theorem 1. *The asymptotic velocity of the boundary exists and is equal to*

$$W = V(\{1, \ldots, L_1\}, \{1, \ldots, K_1\})$$

where

$$L_1 = \max \left\{ l \in \{1, \ldots, L\} \, : \, v_l^{(-)} > V(\{1, \ldots, l\}, I_+) \right\}, \qquad (6)$$

$$K_1 = \max \left\{ k \in \{1, \ldots, K\} \, : \, v_k^{(+)} < V(I_-, \{1, \ldots, k\}) \right\}. \qquad (7)$$

Note that the definition of L_1 and K_1 is not ambiguous because $v_1^{(-)} > V(\{1\}, I_+)$ and $v_1^{(+)} < V(I_-, \{1\})$.

Now we will explain this result in more detail. As $v_K^{(+)} < 0 < v_L^{(-)}$, there can be three possible orderings of the numbers $v_L^{(-)}, v_K^{(+)}, V$:

1. $v_K^{(+)} < V < v_L^{(-)}$. In this case

$$K_1 = K, \quad L_1 = L, \quad W = V$$

2. If $v_K^{(+)} > V$ then $V < 0$ and $K_1 < K, \quad L_1 = L$. Moreover

$$W = V(\{1,\ldots,L\},\{1,\ldots,K_1\}) = \min_{k \in I_+} V(\{1,\ldots,L\},\{1,\ldots,k\}) < V < 0$$

3. If $v_L^{(-)} < V$ then $V > 0$ and $K_1 = K, \quad L_1 < L$. Moreover

$$W = V(\{1,\ldots,L_1\}, I_+) = \max_{l \in I_-} V(\{1,\ldots,l\}, I_+) > V > 0$$

2.4 *Another Scaling*

Normally the minimal difference between consecutive prices (a tick) is very small. Moreover one customer can have many units of the commodity. That is why it is natural to consider the scaled densities

$$\rho_j^{(+),\epsilon} = \epsilon^{-1} \rho_j^{(+)}, \quad \rho_j^{(-),\epsilon} = \epsilon^{-1} \rho_j^{(-)}$$

for some fixed constants $\rho_j^{(+)}, \rho_j^{(-)}$. Then the phase boundary trajectory $\beta^{(\epsilon)}(t)$ will depend on ϵ. The results will look even more natural. Namely, it follows from the main theorem, that for any $t > 0$ there exists the following limit in probability

$$\beta(t) = \lim_{\epsilon \to 0} \beta^{(\epsilon)}(t)$$

that is the limiting boundary trajectory.

2.5 *Example of Phase Transition*

The case $K = L = 1$, that is when the activities of (+)-particles are the same (and similarly for (−)-particles), is very simple. There is no phase transition in this case. The boundary velocity

$$W = \frac{v_1^{(+)} \rho_1^{(+)} + v_1^{(-)} \rho_1^{(-)}}{\rho_1^{(+)} + \rho_1^{(-)}} \tag{8}$$

depends analytically on the activities and densities. This is very easy to prove because the n-th collision time is given by the simple formula

$$t_n = \frac{x_n^{(+)}(0) - x_n^{(-)}(0)}{-v_1^{(+)} + v_1^{(-)}} \tag{9}$$

and n-th collision point is given by

$$x_n^{(+)}(0) + t_n v_1^{(+)} = x_n^{(-)}(0) + t_n v_1^{(-)}. \tag{10}$$

More complicated situation was considered in [1]. There the movement of $(+)$-particles has random jumps in both directions with constant drift $v_1^{(+)} \neq 0$ (and similarly for $(-)$-particles). In [1] *the order* of particles of the same type *can be changed* with time. There are no such simple formulas as (9) and (10) in this case. The result is however the same as in (8).

The phase transition appears already in case when $K = 2$, $L = 1$ and moreover the $(-)$-particles stand still, that is $v_1^{(-)} = 0$. Denote $\rho_1^{(-)} = \rho_0$, $v_i^{(+)} = v_i$, $\rho_i^{(+)} = \rho_i$, $i = 1, 2$. Consider the function

$$V_1(v_1, \rho_1) = \frac{\rho_1 v_1}{\rho_0 + \rho_1}.$$

It is the asymptotic speed of the boundary in the system where there is no $(+)$-particles of type 2 at all.

Then the asymptotic velocity is the function

$$W = V(v_1, v_2, \rho_1, \rho_2) = \frac{\rho_1 v_1 + \rho_2 v_2}{\rho_0 + \rho_1 + \rho_2}$$

if $v_2 < V_1$ and

$$W = V_1(v_1, \rho_1) = \frac{\rho_1 v_1}{\rho_0 + \rho_1}$$

if $v_2 > V_1$. We see that at the point $v_2 = V_1$ the function W is not differentiable in v_2.

3 Random Walk in R_+^N and Method of Proof

3.1 Associated Random Walk

One can consider the phase boundary as a special kind of server where the customers (particles) arrive in pairs and are immediately served. However the situation is more involved than in standard queuing theory, because the server moves, and correlation between its movement and arrivals is sufficiently complicated. That is why this analogy does not help much. However we describe the crucial correspondence between random walks in R_+^N and the infinite particle problem defined above, that allows to get the solution.

Denote $b_i^{(-)}(t)$ ($b_k^{(+)}(t)$) the coordinate of the extreme right (left), and still existing at time t, that is not annihilated at some time $t' < t$, $(-)$-particle of type $i \in I_-$ ($(+)$-particle of type $k \in I_+$). Define the distances $d_{i,k}(t) = b_k^{(+)}(t) - b_i^{(-)}(t) \geq 0$, $i \in I_-, k \in I_+$. The trajectories of the random processes $b_i^{(-)}(t)$, $b_k^{(+)}(t)$, $d_{i,k}(t)$ are assumed left continuous. Consider the random process $D(t) = (d_{i,k}(t), (i,k) \in I) \in R_+^N$, where $N = KL$.

Denote $\mathcal{D} \in R_+^N$ the state space of $D(t)$. Note that the distances $d_{i,k}(t)$, for any t, satisfy the following conservation laws

$$d_{i,k}(t) + d_{n,m}(t) = d_{i,m}(t) + d_{n,k}(t)$$

where $i \neq n$ and $k \neq m$. That is why the state space $\mathcal{D}$ can be given as the set of non-negative solutions of the system of $(L-1)(K-1)$ linear equations

$$d_{1,1} + d_{n,m} = d_{1,m} + d_{n,1}$$

where $n, m \neq 1$. It follows that the dimension of $\mathcal{D}$ equals $K + L - 1$. However it is convenient to speak about random walk in R_+^N, taking into account that only subset of dimension $K + L - 1$ is visited by the random walk.

Now we describe the trajectories $D(t)$ in more detail. The coordinates $d_{i,k}(t)$ decrease linearly with the speeds $v_i^{(-)} - v_k^{(+)}$ correspondingly until one of the coordinates $d_{i,k}(t)$ becomes zero. Let $d_{i,k}(t_0) = 0$ at some time t_0. This means that $(-)$-particle of type i collided with $(+)$-particle of type k. Let them have numbers j and l correspondingly. Then the components of $D(t)$ become:

$$d_{i,k}(t_0 + 0) = u_{j+1,i}^{(-)} + u_{l+1,k}^{(+)}$$

$$d_{i,m}(t_0 + 0) - d_{i,m}(t_0) = u_{j+1,i}^{(-)}, \quad m \neq k$$

$$d_{n,k}(t_0 + 0) - d_{n,k}(t_0) = u_{l+1,k}^{(+)}, \quad n \neq i$$

and other components will not change at all, that is do not have jumps.

Note that the increments of the coordinates $d_{n,m}(t_0+0) - d_{n,m}(t_0)$ at the jump time do not depend on the history of the process before time t_0, as the random variables. $u_{j,i}^{(-)}$ ($u_{j,k}^{(+)}$) are independent and equally distributed for fixed type. It follows that $D(t)$ is a Markov process. However, this continuous time Markov process has singular transition probabilities (due to partly deterministic movement). This fact however does not prevent us from using the techniques from [4] where random walks in Z_+^N were considered.

3.2 Ergodic Case

We call the process $D(t)$ ergodic, if there exists a neighborhood A of zero, such that the mean value $E\tau_x$ of the first hitting time τ_x of A from the point x is finite for any $x \in \mathcal{D}$. In the ergodic case the correspondence between boundary movement and random walks is completely described by the following theorem.

Theorem 2. *Two following two conditions are equivalent:*
(1) The process $D(t)$ is ergodic; *(2) $v_K^{(+)} < V < v_L^{(-)}$.*

All other cases of boundary movement correspond to non-ergodic random walks. Even more, we will see that in all other cases the process $D(t)$ is transient. Condition (5), which excludes the set of parameters of zero measure, excludes in fact null recurrent cases.

To understand the corresponding random walk dynamics introduce a new family of processes.

3.3 Faces

Let $\Lambda \subseteq I = I_- \times I_+$. The face of R_+^N associated with Λ is defined as

$$\mathcal{B}(\Lambda) = \{x \in R_+^N : x_{i,k} > 0, (i,k) \in \Lambda, x_{i,k} = 0, (i,k) \in \overline{\Lambda}\} \subseteq R_+^N \tag{11}$$

If $\Lambda = \emptyset$, then $\mathcal{B}(\Lambda) = \{0\}$. For shortness, instead of $\mathcal{B}(\Lambda)$ we will sometimes write Λ. However, one should note that the inclusion like $\Lambda \subset \Lambda_1$ is always understood for subsets of I, not for the faces themselves.

Define the following set of "appropriate" faces

$$\mathcal{G} = \left\{\Lambda : \overline{\Lambda} = J_- \times J_+, \ J_- \subseteq I_-, \ J_+ \subseteq I_+\right\}.$$

Lemma 1.

$$\mathcal{D} = \bigcup_{\Lambda_0 \in \mathcal{G}} (\mathcal{D} \cap \Lambda_0).$$

This lemma explains why in the study of the process $D(t)$ we can consider only "appropriate" faces.

3.4 Induced Process

One can define a family $\mathbf{D}(t; J_-, J_+)$ of infinite particle processes, where $J_- \subseteq I_-$, $J_+ \subseteq I_+$. The process $\mathbf{D}(t; J_-, J_+)$ is the process $\mathbf{D}(t)$ with $\rho_j^{(+)} = 0$,

$j \notin J_+$ and $\rho_j^{(-)} = 0, j \notin J_-$. All other parameters (that is the densities and velocities) are the same as for $\mathbf{D}(t)$. Note that these processes are in general defined on different probability spaces. Obviously $\mathbf{D}(t; I_-, I_+) = \mathbf{D}(t)$.

Similarly to $\mathbf{D}(t)$, the processes $\mathbf{D}(t; J_-, J_+)$ have associated random walks $D(t; J_-, J_+)$ in $R_+^{N_1}$ with $N_1 = |J_-||J_+|$. Usefulness of these processes is that they describe all possible types of asymptotic behavior of the main process $D(t)$.

Consider a face $\Lambda \in \mathcal{G}$, i.e., such face that its complement $\overline{\Lambda} = J_- \times J_+$ where $J_- \subseteq I_-$ and $J_+ \subseteq I_+$. The process $D_\Lambda(t) = D(t; J_-, J_+) = (d_{i,k}^\Lambda(t), (i,k) \in \overline{\Lambda})$ will be called an ***induced process***, associated with Λ. The coordinates $d_{i,k}^\Lambda(t)$ are defined in the same way as $d_{i,k}(t) = d_{i,k}^\Lambda(t)$, where $\overline{\Lambda} = \{\emptyset\}$. The state space of this process is $\mathcal{D}^{\overline{\Lambda}} = \mathcal{D}(R^{|\overline{\Lambda}|})$, where $|\overline{\Lambda}| = |J_-||J_+|$. Face Λ is called **ergodic** if the induced process $D_\Lambda(t)$ is ergodic.

3.5 Induced Vectors

Introduce the plane

$$\mathcal{R}(\Lambda) = \{x \in R^N : x_{i,k} = 0, (i,k) \in \overline{\Lambda}\} \subseteq R^N$$

Lemma 2. *Let Λ be ergodic with $\overline{\Lambda} = J_- \times J_+$, and $D_y(t)$ be the process $D(t)$ with the initial point $y \in \mathcal{B}(\Lambda)$. Then there exists vector $v^\Lambda \in \mathcal{R}(\Lambda)$ such that for any $y \in \mathcal{B}(\Lambda)$ $t \geq 0$, such that $y + v^\Lambda t \in \mathcal{B}(\Lambda)$, we have as $M \to \infty$*

$$\frac{D_{yM}(tM)}{M} \to y + v^\Lambda t$$

This vector v^Λ will be called the **induced vector** for the ergodic face Λ. We will see other properties of the induced vector below.

3.6 Non-ergodic Faces

Let Λ be the face which is not ergodic (non-ergodic face). Ergodic face Λ_1: $\Lambda_1 \supset \Lambda$ will be called outgoing for Λ, if $v_{i,k}^{\Lambda_1} > 0$ for $(i,k) \in \Lambda_1 \setminus \Lambda$. Let $\mathcal{E}(\Lambda)$ be the set of outgoing faces for the non-ergodic face Λ.

Lemma 3. *The set $\mathcal{E}(\Lambda)$ contains the minimal element Λ_1 in the sense that for any $\Lambda_2 \in \mathcal{E}(\Lambda)$ we have $\Lambda_2 \supseteq \Lambda_1$.*

3.7 Dynamical System

We define now the piece-wise constant **vector field** $v(x)$ in $\mathcal{D}$, consisting of induced vectors, as follows: $v(x) = v^{\Lambda}$ if x belongs to ergodic face Λ, and $v(x) = v^{\Lambda_1}$ if x belongs to non-ergodic face Λ, where Λ_1 is the minimal element of $\mathcal{E}(\Lambda)$. Let U^t be the **dynamical system** corresponding to this vector field.

It follows that the trajectories $\Gamma_x = \Gamma_x(t)$ of the dynamical system are piecewise linear. Moreover, if the trajectory hits a non-ergodic face, it leaves it immediately. It goes with constant speed along an ergodic face until it reaches its boundary.

We call the ergodic face $\Lambda = \mathcal{L}$ final, if either $\mathcal{L} = \emptyset$ or all coordinates of the induced vector $v^{\mathcal{L}}$ are positive. The central statement is that the dynamical system hits the final face, stays on it forever and goes along it to infinity, if $\mathcal{L} \neq \emptyset$.

The following theorem, together with Theorem 2, is parallel to Theorem 1. That is in all three cases of Theorems 1–3 describe the properties of the corresponding random walks in the orthant.

Theorem 3.

1. *If $D(t)$ is ergodic then the origin is the fixed point of the dynamical system U^t. Moreover, all trajectories of the dynamical system U^t hit* 0.
2. *Assume $v_K^{(+)} > V$. Then the process $D(t)$ is transient and there exists a unique ergodic final face $\mathcal{L}$, such that $v_{i,k}^{\mathcal{L}} > 0$ for $(i,k) \in \mathcal{L}$. This face is*

$$\mathcal{L}(L, K_1) = \{(i,k) : \ i = 1, \dots, L, \ k = K_1 + 1, \dots, K\}$$

   where K_1 is defined by (7). Moreover, all trajectories of the dynamical system U^t hit $\mathcal{L}(L, K_1)$ and stay there forever.
3. *Assume $v_L^{(-)} < V$. Then the process $D(t)$ is transient and there exists a unique ergodic final face $\mathcal{L}$, such that $v_{i,k}^{\mathcal{L}} > 0$ for $(i,k) \in \mathcal{L}$. This face is*

$$\mathcal{L}(L_1, K) = \{(i,k) : \ i = L_1 + 1, \dots, L, \ k = 1, \dots, K\}$$

   where L_1 is defined by (6). Moreover, all trajectories of the dynamical system U^t hit $\mathcal{L}(L_1, K)$ and stay there forever.
4. *For any initial point x the trajectory $\Gamma_x(t)$ has finite number of transitions from one face to another, until it reaches $\{0\}$ or one of the final faces.*

3.8 Simple Examples of Random Walks and Dynamical Systems

If $K = L = 1$ the process $D(t)$ is a random process on R_+. It is deterministic on $R_+ \setminus \{0\}$ – it moves with constant velocity $v^{(+)} - v^{(-)}$ towards the origin. When it reaches 0 at time t, it jumps backwards

$$D(t+0) = \eta$$

where η has the same distribution as $u_1^{(+)} + u_1^{(-)}$. The dynamical system coincides with $D(t)$ inside R_+, and has the origin as its fixed point.

If $L = 1, K = 2$ and moreover $v_1^{(-)} = 0$ then the state space of the process is $R_+^2 = \{(d_{11}, d_{12})\}$. Inside the quarter plane the process is deterministic and moves with velocity $(v_1^{(+)}, v_2^{(+)})$. From any point x of the boundary $d_{12} = 0$ it jumps to the random point $x + \eta_1$, and from any point of the boundary $d_{11} = 0$ it jumps to the point $x + \eta_2$, where η_1, η_2 have the same distributions as $(u_{j,1}^{(-)}, u_{j,1}^{(-)} + u_{j,2}^{(+)})$ and $(u_{j,1}^{(-)} + u_{j,1}^{(+)}, u_{j,1}^{(-)})$ correspondingly. The classification results for random walks in Z_+^2 can be easily transfered to this case; the dynamical system is deterministic and has negative components of the velocity inside R_+^2. When it hits one of the axes it moves along it. The velocity is always negative along the first axis, however along second axis it can be either negative or positive. This is the phase transition we described above. Correspondingly the origin is the fixed point in the first case, and has positive value of the vector field along the second axis, in the second case.

References

1. Malyshev V, Manita A (2009) Markov Processes Related Fields 15:575–584
2. Oshanin G, De Coninck J, Moreau M, Burlatsky S (1999) arXiv:cond-mat/9910243
3. Khorrami M, Aghamohammadi A (2003) Braz J Phys 33:421–430
4. Malyshev V (1993) Adv in Appl Probab 25:140–175

Part II
Pedestrian Traffic and Evacuation Dynamics

Calibration of Nomad Parameters Using Empirical Data

Winnie Daamen, Mario Campanella, and Serge P. Hoogendoorn

Abstract This paper describes the results of calibration of the Nomad pedestrian simulation model using empirical data from laboratory experiments. The results of the calibration show that complex walker models with large amounts of parameters can indeed be calibrated. The estimated average parameter values are shown and discussed, as well as their significance and correlation. Furthermore, pedestrian behaviors are investigated using the estimated parameters values in various conditions, among which flow configuration, pedestrian heterogeneity and traffic conditions.

1 Introduction

The microscopic pedestrian simulation model Nomad has (like other simulation models) a large amount of parameters. The calibration of such complex models is not a simple process given the reliability and correlation of parameters, as well as the problem of the information richness of the available data. Furthermore, pedestrian behaviours are complex and vary according to several factors such as walking area configurations, traffic conditions and pedestrian heterogeneity. Therefore, it is important to know to which situations a model can be applied for prediction. One way to estimate the general applicability of a walking model is to compare the parameter estimates when varying the different factors. Differences in the estimated parameter samples will reflect the inability of the model to correctly predict differences in the behaviours. If the parameter samples are significantly different then the model is not general enough and the samples reflect variations of pedestrian behaviours. As a positive consequence the estimated parameter samples

W. Daamen (✉) · M. Campanella · S.P. Hoogendoorn
Department of Transport & Planning, Delft University of Technology, Delft, The Netherlands
e-mail: w.daamen@tudelft.nl

V.V. Kozlov et al. (eds.), *Traffic and Granular Flow '11*,
DOI 10.1007/978-3-642-39669-4_11, © Springer-Verlag Berlin Heidelberg 2013

can therefore be used to investigate how pedestrians are behaving in the different situations.

This paper therefore presents the calibration results of Nomad using data from several controlled experiments. The parameters are simultaneously estimated for each individual pedestrian using trajectory data. The calibration results are then used to investigate the reliability, parameter correlation and general applicability of Nomad.

In the following, first a short description is given of the laboratory experiments of which the data are used. Then, the Nomad parameters that will be estimated are introduced, followed by the calibration results. Finally, we show the effect of the flow configuration, the population heterogeneity and the traffic conditions on the estimated parameters. We end with conclusions and recommendations for future research.

2 Laboratory Experiments

The data used for the model calibration come from the controlled experiments described in Daamen and Hoogendoorn [1]. Figure 1 shows the infrastructure layout and flow directions of the four configurations discussed in this paper: unidirectional flow, bidirectional flows, crossing flows and a narrow bottleneck. All experiments have been performed under normal walking conditions, while only in the narrow bottleneck experiment congestion occurred. Since the congestion occurred upstream of the corridor, the area of this experiment has been split into two: one (possibly) congested area upstream of the bottleneck and an area inside the bottleneck where capacity occurred. Trajectory data have been collected for 0.1 s and smoothed and interpolated to time-steps of 0.2 s to minimize numerical errors in the estimation process.

3 Microscopic Pedestrian Simulation Model Nomad

The microscopic pedestrian simulation model Nomad has been developed at the department Transport & Planning of the Delft University of Technology [2, 3]. Nomad is an agent-based model, covering the tactical and operational levels of human behaviour, including route choice behaviour, activity (area) choice behaviour, walking behaviour, waiting behaviour and behaviour in special infrastructure elements, such as revolving doors and turnstiles. The Nomad model is composed of linearly added components that correspond to specific pedestrian behaviours: path following, pedestrian and obstacle interaction. The parameters shown in Table 1 are included in the calibration process.

The parameters describing the lateral avoidance between pedestrians (A_l and R_l) are only estimated in the experiments in which these lateral movements

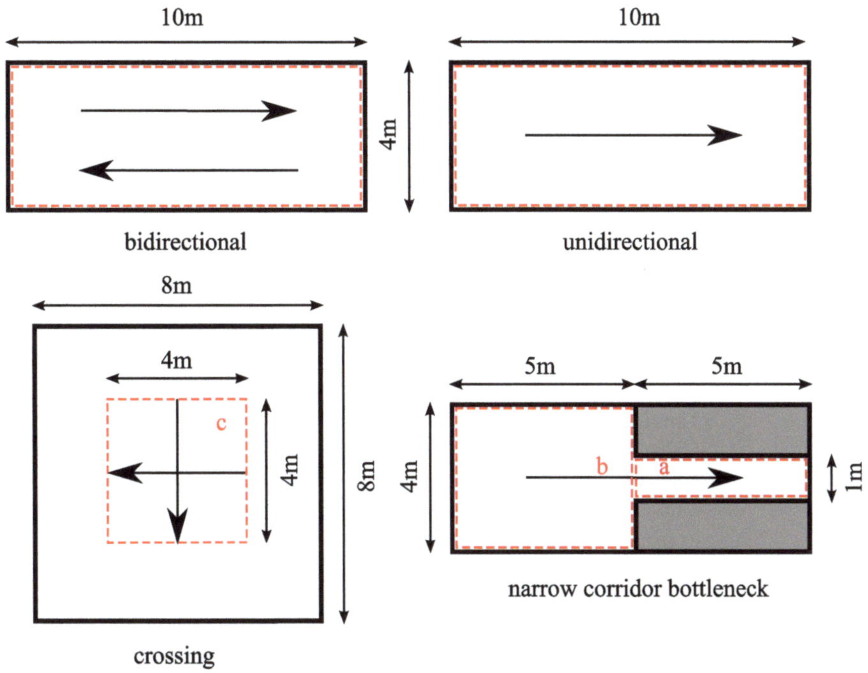

Fig. 1 Overview of laboratory experiment infrastructure configurations

occurred, that is, the bidirectional and the crossing flows. The parameter describing the interaction with obstacles is only estimated in the experiment with the narrow corridor, the only experiment where obstacles were present. All other parameters were estimated in all experiments.

Another important parameter in the Nomad model is the free speed V_0. Since this free speed is specific for each pedestrian, this speed is measured using the beginning of the trajectories (first 0.2 s), when the densities are low and it can be assumed that a pedestrian is walking in free flow conditions.

The parameters used to calculate the physical forces have been excluded from the calibration as previous tests have shown these parameters to be insensitive to calibration results. The parameters have been set to very large values to prevent compressions of pedestrians (K_0 = 1,000, K_1 = 1,000). The stochastic noise that accounts for modelling errors and behavioural variations was always set to zero, preventing it to influence the estimation. This way, the estimated parameters are optimal regardless of the size of the modelling errors and the behaviour variations can be measured by the differences of the parameters for different individuals.

Table 1 Overview of Nomad parameters estimated in this paper

Symbol	Explanation
T	Acceleration time (s), the time required to accelerate towards the free speed V_0 in the direction of the desired path. Small values of T will force pedestrians to walk very close to their desired path with their free speeds and any deviation from the path will generate large path following accelerations
A_0	Interaction strength (m/s^2), controls the intensity in which pedestrians are avoiding each other. Larger values of A_0 when other parameters are kept equal indicate an increase of importance of the avoidance accelerations due to other pedestrians
R_0	Interaction distance (m), controls how responsive the avoidance accelerations are to the distance between pedestrians. Small values of R_0 ($\sim$0.0 m) signify that only small distances between pedestrian cause avoidance accelerations
c_0^-	Transform the shape of the influence area behind pedestrians from circular (value = 1) to an ellipsoid. For values smaller than one the main axis of the ellipsoid is in the walking direction otherwise in the perpendicular direction
c_0^+	Transform the shape of the influence area in front of pedestrians from circular to an elongated ellipsoid similarly as c_0^-
ie_f	Influence area extension in the front (m), the largest distance at the front in which a pedestrian will provoke the avoiding behaviours
ie_b	Influence area extension in the back (m), the largest distance at the back in which a pedestrian will provoke the avoiding behaviours
A_1	Frontal interaction strength for pedestrians (m/s^2), controls the intensity of the extra lateral component of the avoidance accelerations when pedestrians are walking towards each other
R_1	Frontal interaction distance for pedestrians (m), controls how responsive the extra lateral avoidance accelerations are to the lateral distances of pedestrians walking towards each other
A_T	Anticipation time (s), the time in the future that pedestrians project the current locations of neighbouring pedestrians
A_W	Obstacle interaction strength (m/s^2), controls the extent in which pedestrians are avoiding obstacles. Larger values of A_W when other parameters are kept equal (*ceteris paribus*) indicate an increase of importance of the obstacle avoidance accelerations

4 Model Calibration Results

For the calibration results, samples of estimated parameters have been derived. A parameter set θ for an experiment consists of all the estimates of each parameter ($\theta_i = T, A_0, \ldots$). A single calibration thus results in an optimal parameter set θ^*, consisting of all samples $\theta_i^* \in \theta^*$. An estimated value is only part of the parameter set if it is significant, while values on the boundaries of the estimation interval are excluded from the set, since their optimality is not guaranteed. An overview of the

Table 2 Average values of all parameters for the five data sets

			Pedestrian interactions									
		Path follow	Ped avoid		Influence area				Lateral avoid		Antici-pation	Obstacle
	rad	T	A_0	R_0	c_0^-	c_0^+	ie_f	ie_b	A_I	R_I	A_T	A_W
bidir		0.32	2.02	1.24	0.94	1.01	1.30	1.98	7.08	0.69	0.53	
unidir		0.27	2.39	1.45	1.01	1.03	1.42	1.82			0.61	
cross		0.62	2.42	0.99	1.00	1.00	0.96	1.10	4.71	×	0.52	
bneckDown		0.63	4.51	1.28	1.07	1.06	1.04	1.51			0.56	9.43
bneckUp		0.57	2.84	1.06	0.99	0.96	1.36	1.14			0.60	9.02

estimated parameter values can be found in Table 2, where the × indicates that no significant estimations could be achieved.

Low values of the acceleration time T indicate a strong tendency of pedestrians to stay close to their desired paths. In general, the values of T are around values obtained in other calibrations [4, 5]. The unidirectional and the bidirectional flows resulted in significantly smaller values of T when compared with the other normal walking experiments, which is due to the lower densities in these experiments. In these relatively free situations, pedestrians have smaller probabilities to interact with other pedestrians and therefore only need small corrections to keep their desired path.

Larger values of the interaction strength A_0 alone do not indicate large avoidance accelerations because the final acceleration value depends on distances to other pedestrians and the interaction distance R_0. The relatively large values of A_0 and R_0 indicate that pedestrians in normal walking conditions are more reactive due to the presence of other pedestrians. This implies that when necessary pedestrians need to apply larger accelerations at larger distances to anticipate the longer distances between pedestrians (lower densities). The most reactive behaviours are encountered inside the narrow corridor (*bneckDown*), when they settle in a laterally displaced position that was identified by Hoogendoorn and Daamen [6] as the 'zipper-effect'. This arrangement puts pedestrians very close to each other and simultaneously close to the corridor walls, while the speeds inside the corridor are relatively high. With such high speeds, natural variations cause pedestrians to apply large accelerations to prevent collisions. The largest R_0 is estimated in the unidirectional flow (*unidir*) indicating that pedestrians are more sensitive to pedestrians further away than in the other experiments. This is largely due to the low densities in the experiment that make short distances very unlikely.

c_0^+ remained reasonably constant along the normal walking experiments with values around 1.0 indicating that pedestrians seem to be isotropic (scanning equally in all directions) contrary to previous findings [6]. However, further in this paper we show that for most walking experiments the slower the pedestrians walk the more they consider pedestrians walking near their walking paths (larger anisotropy). This is supported by the experiment *bneckUp* with the lowest average speeds presenting

Table 3 Percentages of significant parameters in each experiment (percentages below 5 % highlighted in bold)

		Pedestrian interactions									
Parameters	Path follow	Ped avoid		Influence area				Lateral avoid		Antici-pation	Obstacle
Experiments	T	A_0	R_0	c_0^-	c_0^+	ie_f	ie_b	A_1	R_1	A_T	A_W
bidir	49	58	10	66	62	62	68	7	**1**	57	
unidir	75	62	17	60	71	71	71			56	
cross	52	36	**3**	40	55	43	28	**2**	**0**	48	
bneckDown	72	91	32	72	81	72	79			61	28
bneckUp	66	75	21	55	66	58	53			53	13

the largest anisotropy in the frontal part of the influence area. The opposite happens in the back part of the influence area where pedestrians are clearly reacting to pedestrians immediately at their shoulders and lesser by those behind them. This is probably due to the large densities. As for c_0^+, the values of c_0^- are around 1 with the largest anisotropy for the bidirectional experiments (*bidir*) due to the self-organised presence of lanes that align the pedestrians and also makes them consider less the pedestrians in the nearby lanes.

In all but the last experiment pedestrians clearly consider more what happens in their back than in their front ($ie_b > ie_f$). This surprising result is counter-intuitive since pedestrians are certainly more attentive to what they see in front, especially in relatively uncomplicated traffic situations and low densities that usually do not require special attention to the surroundings. These results may reflect a deficiency of the model rather than pedestrian behaviours. Campanella et al. [4] already showed the necessity of the so-called 'push from behind' in the Nomad model to come up with good validation results. This effect is stronger in the experiments in which pedestrians are more aligned with each other (larger differences between ie_b and ie_f and larger values of ie_b).

The average value of A_T is around 0.55 s, whereas the lowest values are found for the bidirectional and crossing flows, indicating that the benefits of cooperation reduce the necessity of anticipation. However, these differences are too small to be conclusive.

4.1 Significance of Calibrated Parameter Values

Table 3 shows the percentages of significantly estimated parameters, which values were not on the limits of the interval used by the optimisation algorithm. Most parameters are seen to be significantly estimated for a large percentage of the trajectories: only 4 out of 46 significant parameters were constituted from less than 5 % of the total amount of available trajectories. This shows that the parameters are relevant to the model and that the calibration procedure is finding optimal values.

Table 4 Significant correlations between parameters (only correlations above an absolute value of 0.35 are shown)

Parameters		Pedestrian interactions									
		Ped avoid		Influence area				Lateral avoid		Antici-pation	Obstacle
Experiments		A_0	R_0	c_0^-	c_0^+	ie_f	ie_b	A_I	R_I	A_T	A_W
bidir	A_0							0.50			
	R_I			0.71							
unidir	R_0						0.41				
	c_0^+					0.42					
cross	A_0		−0.41					0.99			
	c_0^-						0.48				
	c_0^+					0.51					
bneckDown	A_0			0.39							0.84

The parameters describing lateral avoidance (A_I and R_I) show very low significance. To test the relevance of these parameters, a model with these parameters has been compared to a model without these parameters using the likelihood ratio test. These tests show a surprisingly high level of success, indicating that even though the parameter is not significant, any value different from zero (that is the equivalent of removing it) improves the simulated behaviour. The reason for the low percentage of significant parameters is the lack of a sufficient amount of moments when pedestrians are walking towards other pedestrians, which is the situation A_I and R_I can be estimated.

Correlation Between Calibrated Parameters

Table 4 shows that from a total of 190 possible pair combinations for all experiments, only 10 show significant correlations above an absolute value of 0.35. The low amounts of significant estimations for A_I, R_I and A_W did not allow for many valid pair comparisons between these and other parameters, thus only few correlations involve these parameters.

The only correlation between parameters from different components was between the obstacle component parameter A_W and the pedestrian avoidance parameter A_0 for the bottleneck experiment, suggesting that the three components have a large independence from each other. The bottleneck situation is in this case very specific, since pedestrians keep their relative distances to walls regardless of the local densities experienced by the pedestrians, thus pedestrian avoidance accelerations had to be large. Table 4 also shows that 50 % of the correlations included at least one parameter of the influence area, indicating that the variables shaping the influence area are not independent.

Table 5 Similarity statistics for the walking experiments

	bidir	*unidir*	*cross*	*bneckDown*	*bneckUp*
bidir	0.00	**1.19**	1.79	1.77	1.50
unidir		0.00	**1.84**	1.64	1.50
cross			0.00	1.47	1.40
bneckDown				0.00	1.50
bneckUp					0.0

5 Effect of Flow Configuration on Estimated Parameters

This section discusses the differences in pedestrian behaviour due to different flow configurations of the experiments. Table 5 shows the results of the similarity statistics *ST* that are the sum of the KS distances $D_{1,2}$ of every parameter sample $\theta^i \in \theta^*$ of a pair of experiments:

$$ST = \sum_{\theta^i \in \theta^*} D_{1,2}(\theta^i) \tag{1}$$

It is important to note that the KS statistics can be used regardless of if the KS test is passed. Once *ST* is measured for all pairs of experiments, the pairs with the smallest results can be identified and, based on the KS distances, it can be determined whether these experiments present similar pedestrian behaviours.

Table 5 shows that in general the experiments display a similar value of the similarity statistic. The unidirectional and bidirectional experiments show the smallest value of the statistic, indicating that these experiments are most similar, which can be explained by the occurrence of lane formation in the bidirectional flows, thus effectively turning these into several unidirectional flows. The largest value of the statistic occurs between the unidirectional flow and the crossing flows, which is also intuitive, but the difference is not significant.

6 Pedestrian Heterogeneity

In this section we investigate the heterogeneity of pedestrian behaviours by calculating the coefficient of variation *CV* for the parameter samples. Table 6 shows the coefficients for variation for all parameters in all experiments. For most of the samples the *CV* is large. Only the parameters c_0^- and c_0^+ presented relatively low averages of 0.23 along all experiments. The last column with the average *CV* per experiment was calculated using only the parameters that were common to all experiments. The overall average of *CV* is 0.71, in itself a large value that shows that there is a significant amount of heterogeneity in the population.

As the narrow corridor experiment has been split up into two separate areas, it is possible to see whether the behaviour of exactly the same pedestrians changed in

Table 6 Coefficient of variation for all estimated parameter samples

			Pedestrian interactions										
		Path follow	Ped avoid		Influence area				Lateral avoid		Antici-pation	Obstacle	
	rad	T	A_0	R_0	c_0^-	c_0^+	ie_f	ie_b	A_I	R_I	A_T	A_W	μ
bidir		0.92	1.54	1.49	0.22	0.24	0.68	0.58	0.95	1.21	0.51		**0.77**
unidir		0.75	1.20	1.34	0.23	0.24	0.51	0.50			0.48		**0.65**
cross		1.07	1.20	1.54	0.24	0.23	0.32	0.32	1.22	×	0.47		**0.67**
bneckDown		0.74	1.34	1.47	0.23	0.22	0.57	0.54			0.49	1.02	**0.70**
bneckUp		0.92	1.39	1.58	0.21	0.24	0.65	0.62			0.50	1.00	**0.76**

the upstream and the downstream part, the so-called intra-pedestrian heterogeneity. Table 6 shows the differences between the estimated parameters for each pedestrian i walking upstream of the narrow corridor θ_i^{up} and inside the narrow corridor θ_i^{in}. This difference *dif* is calculated as

$$dif = \frac{\left|\theta_i^{up} - \theta_i^{in}\right|}{\max\left(\theta_i^{up}, \theta_i^{in}\right)} \tag{2}$$

The results show that pedestrians show significantly different walking behaviour upstream of and inside the narrow bottleneck, which is already indicate by the relatively large similarity statistic (1.5) in Table 5. Especially the parameters relating to the influence area, and specifically the ones transforming its shape (c_0^- and c_0^+) show large differences, while the small *CV* values (see Table 6) indicate that the parameters are not much affected by external conditions. We can thus conclude that natural variations of pedestrian behaviour (intra-pedestrian heterogeneity) causes these differences.

The median values for the path following, pedestrian avoidance, anticipation time and obstacle avoidance parameters show median values well above 0.6. This confirms the hypothesis that the behaviours are significantly different and we can conclude that natural variations occur here as well (Fig. 2).

7 Influence of Traffic Conditions

In this section, we investigate how the parameters vary for traffic conditions. Here, we take the speed as the independent variable, since it is relatively easy to measure and it always reflects the local conditions. For each pedestrian, the average speed encountered during his entire trajectory is calculated and assigned to an interval of 0.2 m/s. The average values of the estimated parameters per interval are then compared to see whether pedestrian behaviours change with speed. Table 7 shows for which experiments which parameters are statistically different.

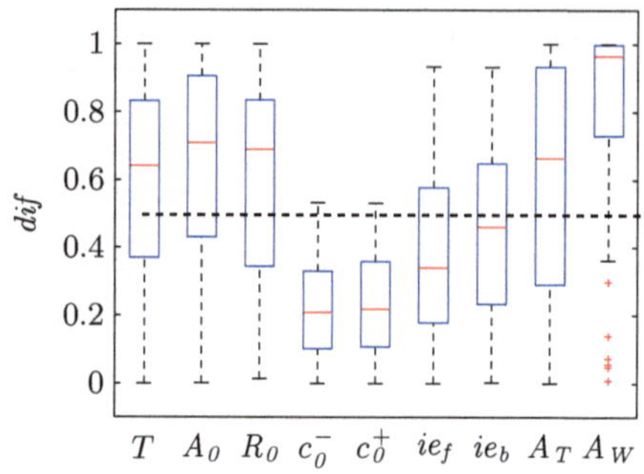

Fig. 2 Box-plot with normalised differences of parameters estimated for the same trajectories in the upstream and inside the narrow corridor

Table 7 Parameters for each experiment that have been shown a statistical difference for various speed intervals

	Path follow	Pedestrian avoidance		Influence area				Antici-pation	Obstacle
	T	A_0	R_0	c_0^-	c_0^+	ie_f	ie_b	A_T	A_W
bidir									
unidir		×				×			
cross	×	×							
bneckDown		×				×			
bneckUp	×	×			×		×	×	

Only the interaction strength appears to be statistically different for four out of five experiments, while the interaction distance, the shape transformation of the influence area behind pedestrians and the obstacle interaction strength does not seem to differ over speed. The samples of the frontal interaction strength and distance (A_1 and R_1) were too small to perform the test. Most significant difference were found for the upstream area of the narrow bottleneck experiment, which is also the experiment with congestion, and thus the largest speed range, whereas no significant differences were found for the bidirectional flow experiment, which might be caused by the self-organisation in the form of lane formation in which pedestrians simply follow other pedestrians. The relation between the significantly different parameters T, A_0 and are shown in more detail in Fig. 3, where the graphs on the left hand side show the experiments for which the parameter was significantly different, and in the graphs on the right hand side the parameter was not significantly different for the various speed intervals.

Figure 3a shows that the acceleration time follows a sigmoid curve for both the narrow bottleneck experiment (upstream part) and the crossing experiment, where the speed is constant for low and high speeds. For the other experiments the acceleration time is constant, and thus independent of the speeds. However, especially in the unidirectional and bidirectional experiments no high densities, and thus no low speeds, have been observed, so the sigmoid curve could still be possible

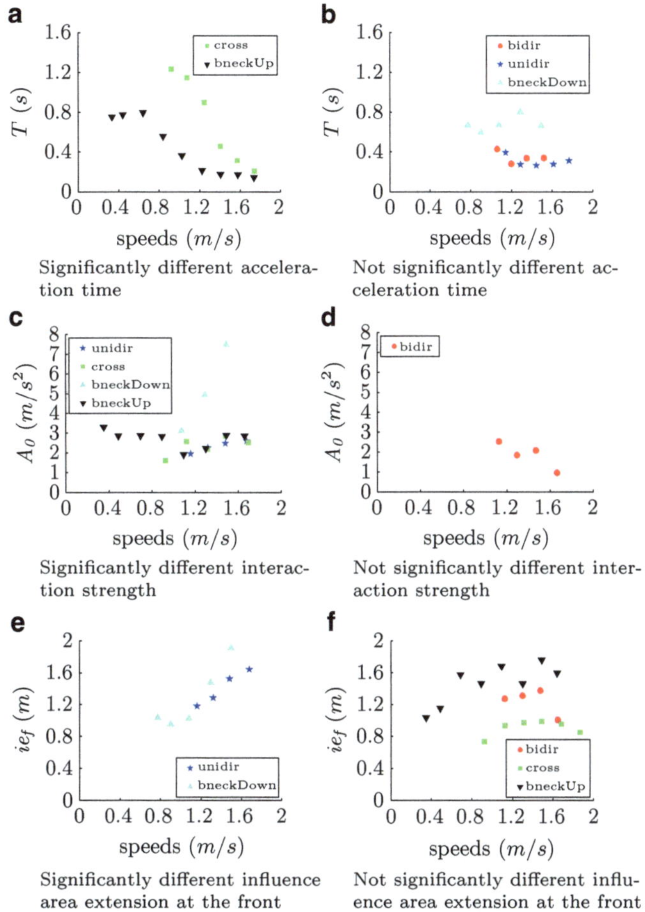

Fig. 3 Graphs with average parameter values of acceleration time (**a** and **b**), interaction strength (**c** and **d**) and influence area extension at the front (**e** and **f**) over speed

for these situations. The larger value of T is due to the more complex manoeuvres that are needed in congestion, especially when pedestrians also have to cross.

Figure 3c shows that A0 follows a U shaped curve, with a minimum value at speeds around 1.0 m/s for the narrow bottleneck experiment. For higher densities larger A_0 values are caused by the fact that pedestrians become more reactive and manoeuver more intensively, while for lower densities, pedestrians are compensating for the larger distances between pedestrians to avoid accelerations (more anticipation). Also Fig. 3e shows that pedestrians consider a longer horizon to perform their avoidance behaviours when walking faster: in more dense situations pedestrians are less likely to see pedestrians further away and even if they would they are more concerned with those nearby that are more likely to cause a conflict.

8 Conclusions and Recommendations

The main contribution of this paper is the extensive calibration of pedestrian behaviours using multiple sets of trajectories from several experiments, including different flow configurations, population characteristics and traffic conditions.

All parameters of the Nomad model could be calibrated, and are therefore necessary to be included in the model. Results of the calibrations from trajectories of different experiments show that flow configurations have a strong influence on pedestrian behaviours, and resulted in different parameters. Pedestrians in unidirectional flows behaved similar to pedestrians in bidirectional flows, showing that lane formation effectively separates the area in unidirectional regions. In congestion, pedestrian behaviour is quite different: in low densities pedestrians are more reactive due to the presence of other pedestrians, while in congestion pedestrians tend to follow other pedestrians and do not strain from their paths. The latter is also shown in the clear relation between some of the parameters and the (local) speed.

The clear differences between the parameters estimated for different situations show that a simulation model needs a specific parameter set to optimise prediction of pedestrian behaviour in a specific situation. However, simulation models are not applied for specific situations, but for a combination of situations, so future research aims at the assessment of the different parameter set for different situations.

References

1. W. Daamen, S. Hoogendoorn, Eur. Jrl. Transp. Infrastr. Res. **3**, 39 (2003)
2. S. Hoogendoorn, P. Bovy, Optimal Control Appl. Meth. **24**, 153 (2003)
3. S. Hoogendoorn, W. Daamen, M. Campanella, P. Bovy, in *6th Triennial Symposium on Transportation Analysis (TRISTAN), Thailand, 1–6* (2007)
4. M. Campanella, S. Hoogendoorn, W. Daamen, in *Proceedings of the 12th IFAC Symposium on Control in Transportation Systems (CTS09)* (2009)
5. S. Hoogendoorn, W. Daamen, in *Pedestrian Behaviour*, ed. by H. Timmermans (Emerald Group, 2010), p. 195
6. S. Hoogendoorn, W. Daamen, Transportation Science **39**, 147 (2005)

Estimating Model Parameters for Bottlenecks in Evacuation Conditions

Winnie Daamen and Serge P. Hoogendoorn

Abstract Emergency doors may be bottlenecks in the evacuation of a building. To assess and improve ways in which bottlenecks are used during evacuation conditions, knowledge regarding the microscopic and macroscopic phenomena at bottlenecks is required. Using data from laboratory experiments, parameters have been estimated using the microscopic simulation tool Nomad. The conclusion is that the pedestrian behavior shown upstream of bottlenecks in evacuation conditions is different from normal walking behavior. In the latter, especially in higher densities, pedestrians look for gaps which they can use to overtake other pedestrians when they have a higher free speed. In the experiments described here, pedestrians seem to have determined a path leading towards the door (the bottleneck), along which they only progress slowly, but where they stick to rather consistently. Overall, we could conclude that both types of behavior cannot be covered with a similar model type, but a multiregime model seems to be more appropriate, in which situations with low and high densities can be distinguished and dedicated models can be applied in the specific situations (regimes).

1 Introduction

Emergency doors may be bottlenecks in the evacuation of a building. In design guidelines requirements have been set to the door width, depending on the number of persons that rely on a specific door [1,2]. The Dutch guidelines are based on rather old experimental research with healthy students [3], while both the fire department and recent literature studies [4, 5] indicate that the capacities mentioned in the (Dutch) design guidelines are rather high. To see which capacities are realistic and to assess and improve ways in which bottlenecks are used during evacuation

W. Daamen (✉) · S.P. Hoogendoorn
Department of Transport & Planning, Delft University of Technology, Delft, The Netherlands
e-mail: w.daamen@tudelft.nl

V.V. Kozlov et al. (eds.), *Traffic and Granular Flow '11*,
DOI 10.1007/978-3-642-39669-4_12, © Springer-Verlag Berlin Heidelberg 2013

conditions, knowledge regarding the microscopic and macroscopic phenomena at bottlenecks is required. Therefore, experimental research has been performed to collect detailed data on pedestrian behaviour in and around doors during emergency conditions.

In this contribution we have estimated the parameters of the microscopic simulation tool Nomad to identify the ability of the tool to predict the microscopic pedestrian behaviour.

In the following, we first describe the set up and performance of the laboratory experiments. Then, a short description is given of the simulation tool Nomad, followed by the estimation results. The paper ends with conclusions and recommendations for further research.

2 Set Up and Performance of Laboratory Experiments

The evacuation experiments have been conducted on January 8th, 2009 in a large hallway. A wall was located perpendicular to the side wall of this hallway. A doorway opening was created in the wall, which was easy to adjust in width. An emergency exit sign was shown above the doorway to imitate the real situation, see left side of Fig. 1. A traffic signal was used which was turned to green when the participants could start walking, see left side of Fig. 1. Before every experiment, the group of participants lined up 5 m upstream of the wall behind a line that was put on the floor, see right side of Fig. 1. For an extensive description of the experimental set up, we refer to Daamen and Hoogendoorn [6].

The experimental variables consist of the composition of the population using the door, the door width and the conditions under which the door is used. We have used seven different population compositions, corresponding to characteristic buildings or situations, being school, station during peak hours, retirement home, work meeting, shopping centre, a disabled population and an average population. Due to ethical reasons, the disabled part of the population consisted of three blindfolded persons and three persons in wheelchairs. In total 75 children of 11 years old (blue caps), 90 adults (red caps) and 50 elderly persons (yellow caps) have participated in the experiments. This leads to populations of between 90 and 150 persons, which are large enough to cause congestion upstream of the door to observe capacities.

The conditions under which an emergency door is used may vary considerably. In the experiments, both the stress level of the participants and the sight are varied. Not much is known on how to introduce stress in an experiment. In the past two methods have been considered favourable: enforcing participants to hurry e.g. by rewarding participants according to their performance and exposing participants to noise. Here, we have chosen to use for the latter option by sounding the slow-whoop signal. In addition, the stress level of the participants is raised by a combination of the slow-whoop signal and stroboscope light. In total, participants have been exposed to three stress levels: none, a slow-whoop signal and a combination of a slow-whoop signal and stroboscope light. The sight is reduced by reducing illumination to a low level.

Fig. 1 Overview of the experiments

Two alternative light situations are considered: full lighting (200 lux) and dimmed (1 lux, corresponding to emergency lighting).

In the experiments, the opening width is varied between 50 cm (the minimal free passageway of an escape route in the Building decree for existing buildings) and 275 cm. In addition to an opening of 85 cm wide (minimal free passageway of an escape route in the Building decree for new estates) openings are a multiple of 55 cm. Furthermore, an opening of 100 cm is tested to see the correspondence with the normative capacity expressed as the number of persons passing an opening of 1 m wide in 1 min.

The final experimental variable relates to whether or not the outflow of pedestrians after passing the doorway is free. In reality, doors cannot always open 180°, but may be restricted. In the experiments we used a door being fixed at an opening of 90°. The total doorway width is not affected.

Ideally, all combinations of experimental variables should be investigated. Since this is not feasible due to time restrictions (the experiments should not last longer than a single day), for each experiment one variable is changed, while for the other variables the default value is maintained. By interpolation of the results of the various experiments, pronouncements can be made on the not performed experiments. The stress levels are varied for all experiments.

Each experiment has been performed multiple times to guarantee the reliability of the observations. To determine the number of repetitions, a total time of congestion of 3 min should be achieved. Since the time of congestion for wide doors is shorter than for narrow doors, more repetitions are performed for the wide doors. An overview of the experiments is shown in Table 1.

3 Microscopic Pedestrian Simulation Model Nomad

This section describes the microscopic pedestrian simulation model Nomad, developed at our department [7, 8]. To interpret the behaviour in the experiments, we estimate the individual pedestrian model parameters using the before mentioned trajectory data. We include inter-pedestrian differences expressed by the variability

Table 1 Overview of performed experiments

Experiment	Opening width [cm]	Population	Sight [lux]	Open door	Start time [hh:mm]
1	100	Average	200	No	9:58
2	220	Average	200	No	10:17
3	85	Retirement home	200	No	10:43
4	85	Average	200	No	10:58
5	165	Average	1	No	11:25
6	275	Average	200	No	11:52
7	85	Work meeting	200	No	12:49
8	85	Disabled	200	No	12:23
9	85	School	200	No	13:48
10	85	Average	1	No	14:08
11	50	Average	200	No	14:24
12	110	Average	200	No	14:39
13	85	Shopping centre	200	No	15:19
14	85	Average	200	Yes	15:40
15	165	Average	200	No	16:03
16	85	Station	200	No	16:24

in estimated model parameters. To do this, we use the maximum likelihood approach described in [9]. We estimate the anisotropic model, where pedestrians from the front have more impact on pedestrian behaviour than pedestrians from the back. The model predicts the two-dimensional acceleration vector $\mathbf{a_p(t)}$ as a function of the free velocity $\mathbf{v_p^0}$, the current speed $\mathbf{v_p(t)}$ and distance d_{pq} between pedestrians p and q as follows:

$$\mathbf{a_p(t)} = \mathbf{f_p}\left(\mathbf{v_p(t)}, \mathbf{r_p} - \mathbf{r_q}, \cdots\right) = \frac{\mathbf{v_p^0} - \mathbf{v_p(t)}}{T_p} - A_p \sum_q \in Q_p \mathbf{u_{pq}}(t) e^{-\frac{d_{pq}(t)}{R_p}} \tag{1}$$

where Qp denotes the set of pedestrians that influence pedestrian p, and where

$$d^*(t) = \frac{\mathbf{u}_{pq} \cdot \mathbf{v}_p(t)}{||\mathbf{v}_p(t)||} = \eta p \frac{\mathbf{w}_{pq} \cdot \mathbf{v}_p(t)}{||\mathbf{v}_p(t)||} \tag{2}$$

Here, ηp > 1 is a pedestrian specific factor that describes differences in pedestrian reaction to stimuli directly in front and stimuli from the sides of the pedestrians. In the estimations described in this section, ηp is chosen equal to 8. The desired walking direction $e_p^0 = \mathbf{v}_p^0 / \mathbf{V}_p^0$ V is derived from the data by taking a rough smoothing of the speeds.

The model has four pedestrian specific parameters, namely the free speed $\mathbf{V}_p^0 = ||\mathbf{v}_p^0||$, the acceleration time Tp, the interaction constant Ap and the interaction distance Rp that are to be estimated from the trajectory data.

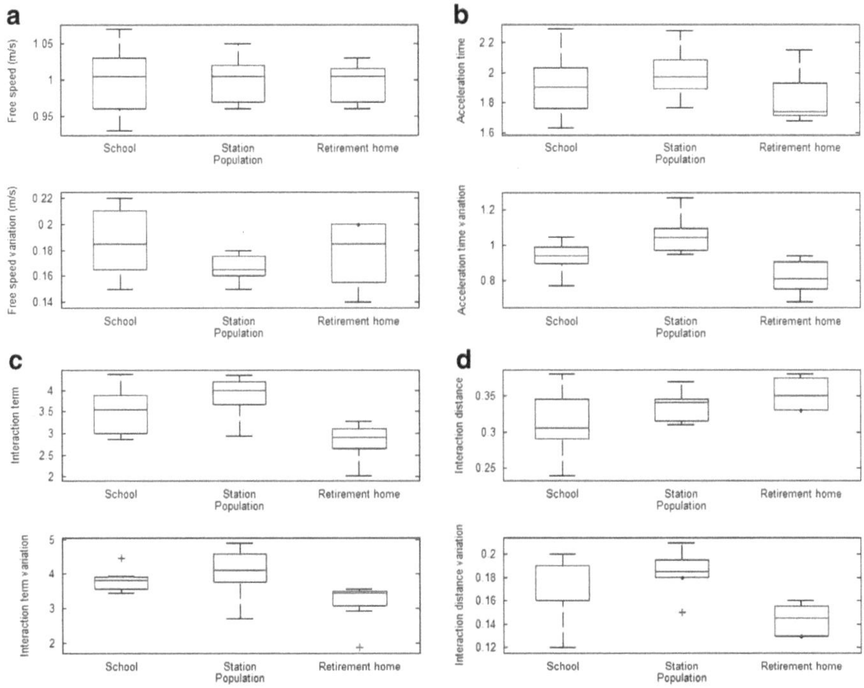

Fig. 2 Overview of the parameter estimation of three populations: school, work meeting and retirement home. (**a**) Free speed estimations. (**b**) Estimations for the acceleration time. (**c**) Estimations for the interaction term. (**d**) Estimations for the interaction distance

In previous analyses [6] we saw that both the school population and the population with the disabled pedestrians showed specific behaviour. In addition, we found out that heterogeneity might play an important role in pedestrian behaviour. We have chosen to estimate parameters for experiments with homogeneous yet very different populations, so the heterogeneity is between the populations and not within the population. For this, we have selected the school population (90 % children of 11/12 years old and 10 % adults), the station population (100 % adults) and the retirement home population (5 % children, 20 % adults and 75 % elderly of 65 years old or more). For the latter population, we have only estimated the trajectories for the elderly. For all populations, we have not estimated the trajectories of the first 25 pedestrians of each experiment, since these pedestrians did not have sufficient interactions and walked more in free flow conditions than in the congested conditions relevant for the experiments. The estimation results for the mentioned four parameters for these three populations are shown in Fig. 2.

Figure 2a shows that the free speed estimations are similar for children, adults and elderly. The estimates are lower than the free speeds usually observed in literature, which is due to the fact that pedestrians do not walk in free walking conditions, but in congested conditions. This might also be the reason that the estimated speeds are similar for the three different pedestrian types. The variation is higher for the school and retirement home population than for the station population.

The acceleration time indicates the time that pedestrians take to accelerate towards their free speed. The lower this time is, the more eager pedestrians are to walk with their free speed. This corresponds with behaviour where pedestrians are actively looking for gaps where they can overtake other pedestrians in order to pass the bottleneck as fast as possible. However, the values in Fig. 2b are very large, which means that pedestrians are not actively trying to walk with their free speed, but just continue with their current speed. It is remarkable that the acceleration time is lowest for the elderly pedestrians, whereas the macroscopic traffic characteristics (especially the speed variation) suggested that the children would have the highest preference to walk with their free speed.

The size of the interaction term (Fig. 2c) is rather hard to interpret in a behavioural way. The estimations also show that the values are not very reliable, since the standard deviation is almost as high as the average value of this parameter. This means that the strength of the interaction differs considerably between pedestrians, assuming a large heterogeneity between pedestrians. What can be seen is that the interaction between children and adults is stronger than the interaction between elderly, which is according expectations.

Finally, we look at the interaction distance (Fig. 2d), which indicates the distance that pedestrians prefer to keep to each other. This variable is directly related to density. Corresponding to other analyses, where we saw the highest densities for the children, children have the lowest interaction distance. However, the elderly have the highest interaction distance, whereas their density was higher than that for the adults. The findings for the macroscopic traffic characteristics do therefore not completely correspond to the findings on the microscopic level. If we compare the estimated parameters for these evacuation conditions near door openings with the parameters estimated in normal walking conditions described in [9], we see some remarkable differences. With a lower free speed and a twice as high acceleration time, the influence of the first term on the acceleration is much smaller than in normal walking conditions. This would imply that pedestrians are not eager to keep their free speed. The influence of other pedestrians on the acceleration (interaction) is also much lower, due to the lower value of the interaction term A, while the remaining part of the term remains similar since the interaction distance is comparable in the two situations.

4 Conclusions

These results as well as further observations during the estimation process bring us to the conclusion that the pedestrian behaviour shown in this experiment is different from normal walking behaviour. In the latter, especially in higher densities, pedestrians look for gaps which they can use to overtake other pedestrians when they have a higher free speed. This is also the way the Nomad model, and most other pedestrian simulation models, is set up. In the experiments described here, pedestrians seem to have determined a path leading towards the door (the bottleneck), along which they only progress slowly, but where they stick to rather consistently. This would require a completely different model formulation, in which only the size of the acceleration is determined, while its direction depends on the route choice. Overall, we could conclude that both types of behaviour cannot be covered with a similar model type, but a multiregime model seems to be more appropriate, in which situations with low and high densities can be distinguished and dedicated models can be applied in the specific situations (regimes).

References

1. *The Green Guide*, fourth edition edn. (HMSO London, 1997)
2. Ministry of Housing, Spatial Planning and the Environment. Bouwbesluit (in dutch), accessed on august 15, 2010 (2003). URL http://213.154.245.57/bouwbesluitonline/
3. I. Peschl, Bouw **26**, 62 (1971)
4. P. van Soomeren, H. Stienstra, J. Wever, G. Klunder, *Human behavior during evacuation from buildings* (2007)
5. M. Kobes, Ability to cope in fire: critical factors for safe escape from buildings. Tech. rep. (2008)
6. W. Daamen, S. Hoogendoorn, accepted for Fire Technology 2012 (2010)
7. S. Hoogendoorn, P. Bovy, Optimal Control Applications and Methods **24**, 153 (2003)
8. S. Hoogendoorn, W. Daamen, M. Campanella, P. Bovy, in *Preprint of the 6th Triennial Symposium on Transportation Analysis (TRISTAN), June, Thailand, 1–6* (2007)
9. S. Hoogendoorn, W. Daamen, R. Landman, *Microscopic calibration and validation of pedestrian models. Cross-comparison of models using experimental data* (Springer-Verlag, Berlin Heidelberg, 2005), pp. 253–266

Counterflow in Evacuations

Tobias Kretz

Abstract It is shown in this work that the average individual egress time and other performance indicators for egress of people from a building can be improved under certain circumstances if counterflow occurs. The circumstances include widely varying walking speeds and two differently far located exits with different capacity. The result is achieved both with a paper and pencil calculation as well as with a micro simulation of an example scenario. As the difficulty of exit signage with counterflow remains one cannot conclude from the result that an emergency evacuation procedure with counterflow would really be the better variant.

1 Introduction and Description

It is usually assumed that during evacuations counterflow [1–10] – or bi-directional flows – only occurs when occupants and rescue forces meet in opposite directions and that apart from that counterflows should not occur as they would inevitably imply that the time for evacuation is higher than necessary. This is surely true for a situation as shown in Fig. 1.

However, what if the exits have different capacities and are located at different walking distances from the two groups? What, furthermore, if the two groups have distinct walking speeds? Figure 2 shows such a situation.

There are two groups of occupants shown in blue and red each in their own room. Each group has exactly 100 members. The two groups differ as the blue agents have a desired walking speed of 2.5 ± 0.2 km/h (about 0.7 m/s) and the red agents 10.0 ± 1.0 km/h (about 2.8 m/s). Within the given boundaries the speeds are distributed equally among the population.

T. Kretz (✉)
PTV Planung Transport Verkehr AG, Stumpfstr. 1, D-76131 Karlsruhe, Germany
e-mail: Tobias.Kretz@ptv.de

V.V. Kozlov et al. (eds.), *Traffic and Granular Flow '11*,
DOI 10.1007/978-3-642-39669-4_13, © Springer-Verlag Berlin Heidelberg 2013

Fig. 1 Two groups of occupants (denoted by the *blue* and *red spheres*; walls are shown in *brown*) in a simple scenario with high symmetry: the exits (shown in *green* at the *right side*) are equally far away and have equal capacity. If the two groups have equal size in number and the occupants of both groups have comparable walking speeds it would not make sense to use the connecting corridor as it would imply a time delay

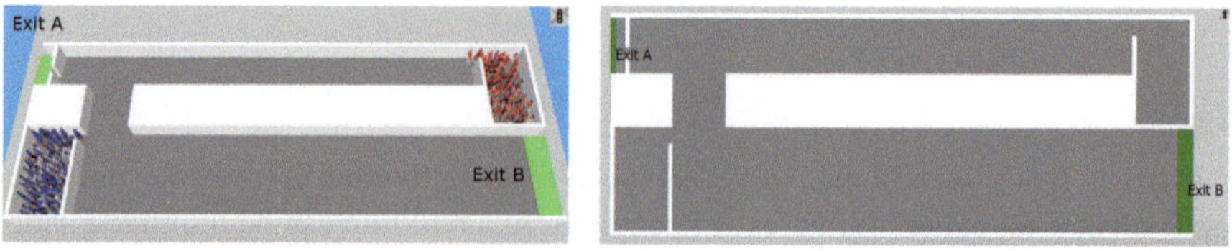

Fig. 2 3D and 2D view of the scenario: Two groups (*blue* and *red*) of occupants at the beginning of an evacuation process. The two available exits are marked in *light green* to the *upper left* and *lower right*. Note that here in front of exit A walls restrict the capacity

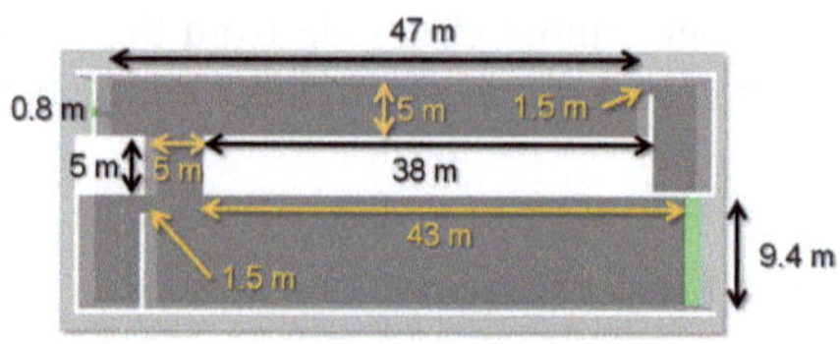

Fig. 3 Measures

The two exits differ in two important aspects: the one on the upper left (exit A) is the closer one for both groups of occupants, but it has a restricted capacity – the occupants need to pass a bottleneck before they reach the exit. The exit to the lower right (exit B) is more remote for both groups, but it has – considering the total number of occupants – nearly unlimited capacity. The geometric dimensions are shown in Fig. 3.

It is not difficult to see that the bottleneck in front of exit A will produce high total evacuation times as well as high average individual egress times if all occupants choose to use exit A. It therefore *might* make sense with regard to the user-equilibrium [11] as well as the system optimum that a number of occupants use exit B. However, as exit B is rather far away, the *costs* to use it in terms of travel time are lower for the faster red agents. From this, one can hypothesize that to optimize at least some of the egress performance measures it makes sense to accept counterflow in the small corridor which connects the two main corridors.

Table 1 Main property of the egress strategy as a consequence of exit choice on the group level

Red ⋱ Blue	Exit A	Exit B
Exit A	Shortest path	Separated
Exit B	Counterflow	Maximum capacity

Principally the distribution of pre-movement times as well as the individual choice of exits would produce a large number of variants. These have been reduced for this study to four by assuming a pre-movement time of zero and that all members of a group walk to the same exit.

This leaves four variants which are compared regarding the evacuation performance measure:

1. Shortest Path: blue group and red group both use exit A.
2. Maximum Capacity: blue group and red group both use exit B.
3. Separated: blue group uses exit B and red group uses exit A.
4. Counterflow: blue group uses exit A and red group uses exit B.

This is shown as a matrix in Table 1.

2 Paper and Pencil Calculation

To investigate this hypothesis we first do a simple calculation of the total evacuation times for both groups, i.e. the time when the last pedestrian of each group has left the building. For this the geometry of Fig. 3 is reduced to the network graph shown in Fig. 4. In addition to the variables and their values given there, what is needed for the following calculations are the walking speed of a member of the red group $v_r = 2.78\,\text{m/s}$, the walking speed of a member of the blue group $v_b = 0.69\,\text{m/s}$, the assumed specific flow $j = 1.31/\text{ms}$ (in accordance with [12]), the width of exit A $b_A = 0.8\,\text{m}$, the width of exit B: $b_B = 9.7\,\text{m}$, and the number $N = 100$ of the pedestrians in each of the groups. The widths of the doors of the rooms where the two groups start are equal: $b_r = b_b = 1.5\,\text{m}$.

With these definitions and values the egress times for the two groups can be calculated. We begin with the simplest case.

Case 1, Separated: In this case the bottleneck for the red group is exit A and the bottleneck for the blue group is the door of their starting room. Therefore the evacuation times result as

$$T_{red} = \frac{d_u + b_c + d_A}{v_r} + \frac{N}{jb_A} = 113.1\,\text{s} \tag{1}$$

$$T_{blue} = \frac{N}{jb_b} + \frac{b_c + d_l}{v_b} = 120.8\,\text{s} \tag{2}$$

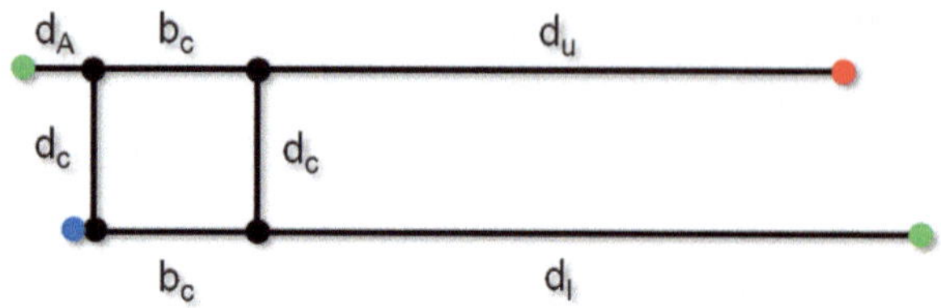

Fig. 4 Simplified network display for the paper and pencil calculation. The lengths of the links are $d_A = 4\,\text{m}$, $b_c = 5\,\text{m}$, $d_u = 38\,\text{m}$, $d_c = 5\,\text{m}$, $d_l = 43\,\text{m}$

Case 2, Maximum Capacity: We neglect that the members of the red group might slow down when walking around the corners and when they have to overtake the members of the blue group. For the blue group as well we neglect any negative impact from the fact that they now have to share the wide (lower) corridor with the red group. Therefore their evacuation time results identically as in Eq. (2).

$$T_{red} = \frac{N}{jb_r} + \frac{d_u + d_c + d_l}{v_r} = 82.2\,\text{s} \tag{3}$$

Case 3, Shortest Path: Here both groups queue before the narrow exit A. The evacuation times depend much on the assumption which group moves first through the bottleneck. If it was the red group, for the red group an evacuation time as in case 1 would result. If it was the blue group an evacuation time as in case 4 would follow. For the total evacuation time the worst case assumption is that both groups move alternating. This maximizes the time for both groups and gives – in this type of calculation – equal results.

$$T_{red} = T_{blue} = \text{Min}\left(\frac{d_u + b_c + d_A}{v_r}, \frac{d_c + d_A}{v_b}\right) + \frac{2N}{jb_A} = 205.3\,\text{s} \tag{4}$$

Case 4, Counterflow: The puzzling question in this case is with what factor one should take care for the fact of counterflow. For this simple calculation here we decide to do not at all so, i.e. assume that the counterflow is as efficient as if it was uni-directional. First the connecting corridor is relatively wide compared to the two doors of the starting rooms, second, from this simple calculation it does not follow how long the counterflow situation persists, and third with an assumed optimal flow from the result it can become apparent how large the loss in flow efficiency could be, before the conclusions of the calculation would have to be modified.

$$T_{red} = \frac{N}{jb_r} + \frac{d_u + d_c + d_l}{v_r} = 82.2\,\text{s} \tag{5}$$

$$T_{blue} = \frac{d_c + d_A}{v_b} + \frac{N}{jb_A} = 109.2\,\text{s} \tag{6}$$

Table 2 Summary of results of paper and pencil calculation (in seconds)

Red ⋱ Blue	Exit A	Exit B
Exit A	205.3 ⋱ 205.3	113.1 ⋱ 120.8
Exit B	82.2 ⋱ 109.2	82.2 ⋱ 120.8

Table 2 summarizes these results. It shows that the counterflow strategy performs best.

3 Simulation

Motivated by this result but still careful for having done many simplifications, we now simulate the scenario using VISWALK (the pedestrian module of VISSIM) [13–15]. The pedestrian dynamics model of VISWALK is the variant of the Social Force Model introduced as *elliptical variant II* in [16, 17] with minor modifications and extensions for specific situations.

At the beginning of the simulation the occupants are set into the simulation equally distributed over their room. Each of the four variants has been simulated 20 times with different random numbers for the stochastic term of the Social Force Model. Table 3 shows the average (over simulation runs) of averages of individual egress times. Figure 5 shows screen captures of the animated simulation runs. From the 20 simulation runs that were carried out those four were selected (respectively their random number generator seed value), which were closest to the average with respect to the average individual egress time for both the blue as well as the red agents.

These results show that for the blue as well as for the red group the case with counterflow leads to the smallest value of average individual egress times, with regard to this measure this therefore obviously is the system optimum. It can easily be seen that with Table 3 as cost matrix in an iterated game [18–21] the case with counterflow as well emerges as stable user equilibrium. With respect to the time when 95 % of the agents (i.e. 190 agents) have reached one of the two destination areas for the red group again the counterflow strategy would be clearly the best one, while within the variance of results for the blue group "counterflow", "separated", and "maximum capacity" are equally well suited, only "shortest path" is worse. If one neglects the statistical scatter of results and only considers the mean value then the "Maximum Capacity" strategy would be the user equilibrium (Table 4).

For total evacuation times (Table 5) – only considering mean values and neglecting the standard deviations – the counterflow strategy remains the best one for the blue group, the sum of times is best for the maximum capacity strategy which is also the user equilibrium strategy. For the blue group the separated strategy scores best. It is not surprising that with increasing number of pedestrians within the same geometry maximum capacity at some time becomes the best strategy.

Table 3 Average of averages of individual egress times in seconds

Red ⋱ Blue	Exit A	Exit B
Exit A	104.9 ± 3.8 ⋱ 124.5 ± 3.1	62.7 ± 2.2 ⋱ 112.6 ± 1.1
Exit B	47.9 ± 0.5 ⋱ 87.1 ± 1.8	49.8 ± 0.7 ⋱ 114.7 ± 1.1

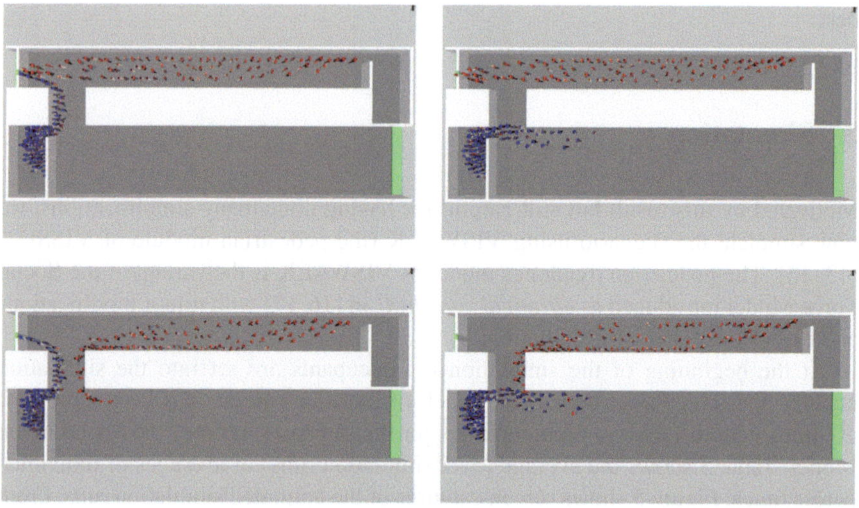

Fig. 5 Situation after 20 s for all four variants (the four figures are placed according to the matrix of Table 1). The corresponding animations are available online at http://youtu.be/-cXH6ExUTg4

Table 4 Average of simulation times when 95 % of all agents had reached one of the two destination areas

Red ⋱ Blue	Exit A	Exit B
Exit A	221.2 ± 4.9 ⋱ 189.3 ± 4.6	107.5 ± 4.2 ⋱ 143.5 ± 1.6
Exit B	58.1 ± 0.6 ⋱ 146.5 ± 2.8	61.3 ± 1.2 ⋱ 145.4 ± 1.6

Table 5 Average of total evacuation times (all agents have left the simulation)

Red ⋱ Blue	Exit A	Exit B
Exit A	229.5 ± 4.9 ⋱ 206.1 ± 6.3	116.1 ± 4.6 ⋱ 150.1 ± 2.1
Exit B	59.8 ± 0.6 ⋱ 158.0 ± 3.4	63.5 ± 1.3 ⋱ 151.9 ± 1.9

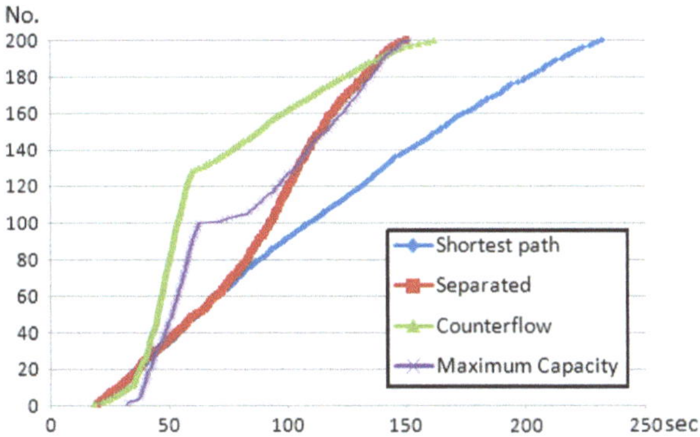

Fig. 6 Number of agents that have arrived at their destination in dependence of the simulation time (evacuation graph). Shown are not averages, but the time dependence of those simulation runs from which the Fig. 5 were taken and which with their average individual egress time are closest to the average of all simulation runs. This plot shows most clearly that for a wide range the counterflow strategy is the best one

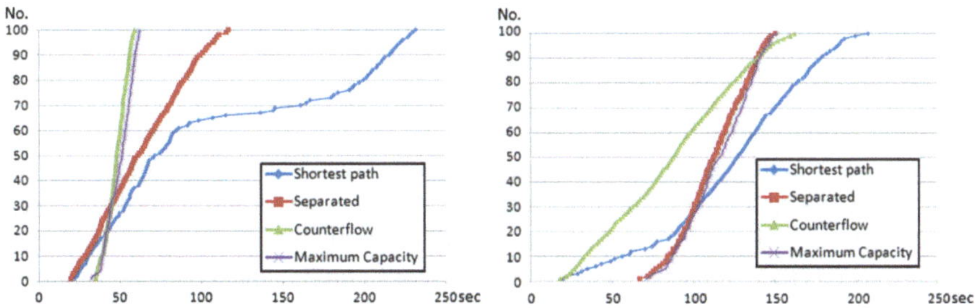

Fig. 7 *Left:* Number of *red agents* that have arrived at their destination in dependence of the simulation time (evacuation graph). This plot shows that in the beginning it is faster for members of the *red group* to use exit A (as they are the first to arrive there), however if a *red agent* is not among the very first of his group then exit B is the better choice. The delay in the "shortest path" strategy curve exists as in this run many blue agents passed exit A before the *red agents* joined. *Right:* Number of *blue agents* that have arrived at their destination in dependence of the simulation time (evacuation graph). From this plot it becomes clear that the counterflow strategy benefits the *blue agents* most

Finally, Figs. 6 and 7 show the number of agents that have arrived at their destination area in dependence of the simulation time.

4 Discussion and Summary

It has been shown that situations exist where the quickest evacuation can be achieved when the egress procedure includes a counterflow. Both, calculation as well as simulation, were carried out with simplifications. If route choice were not bound to whole groups, but could be carried out and optimized individually clearly results would change. However, probably still *parts* of the blue group would find themselves in a situation of counterflow with *parts* of the red group, which would imply that the main claim of the contribution would still be valid. Route choice based on travel time has recently been in the focus of a number of studies [22–32]. Such methods could be used to compute route choice individually.

For this study the corridor width was chosen wide enough such that the friction from counterflow between the two groups is small. However, the results as shown in Table 3 are clear enough to allow for increased counterflow friction. As different models of pedestrian dynamics yield strongly different friction for identical inflow and corridor width, the case with small friction is the one where models agree most.

The impact of this study for application is currently very small. Present day emergency egress signage is static and group independent. To implement the solution as proposed, an emergency egress signage would have to tell "if you are a fast walker walk this way, else the other way", which to date is too complicated. An application could be possible when individual routing information could be given via mobile devicesq to each occupant individually in a comprehensible way [33].

What can nevertheless be learned from this result is to mistrust simple truths. There could be other situations, where it is not so unrealistic to place an adequate emergency egress signage. Second, the fundamental condition for the results as presented is that the two groups have different walking speeds. Therefore this is another example that the phenomena of pedestrian dynamics depend much on the width of the distribution of walking speeds, as it is the case for the stability of lanes in counterflows. Most obviously the example is a hint that with largely scattered walking speeds there are new phenomena to be considered in the search for quickest travel time solutions.

References

1. P.M. Simon and H.A. Gutowitz. A cellular automaton model for bi-directional traffic. *Phys. Rev. E*, 57:2441, 1998.
2. M. Muramatsu, T. Irie, and T. Nagatani. Jamming transition in pedestrian counter flow. *Physica A*, 267:487–498, 1999.
3. V.J. Blue and J.L. Adler. Cellular Automata Microsimulation of Bi-Directional Pedestrian Flows. *Transp. Res. Rec.*, 1678:135–141, 2000.
4. S.A.H. AlGadhi, H.S. Mahmassani, and R. Herman. A Speed-Concentration Relation for Bi-Directional Crowd Movements with Strong Interaction. In Schreckenberg and Sharma [34], pages 3–20.
5. Y. Tajima, K. Takimoto, and T. Nagatani. Pattern formation and jamming transition in pedestrian counter flow. *Physica A*, 313:709–723, 2002.
6. M. Isobe, T. Adachi, and T. Nagatani. Experiment and simulation of pedestrian counter flow. *Physica A*, 336:638–650, 2004.
7. T. Kretz, A. Grünebohm, M. Kaufman, F. Mazur, and M. Schreckenberg. Experimental study of pedestrian counterflow in a corridor. *J. Stat. Mech.*, P10001, 2006.

8. T. Kretz, M. Kaufman, and M. Schreckenberg. Counterflow Extension for the F.A.S.T.-Model. *Lect. Notes Comp. Sc.*, 5191:555–558, 2008.
9. A. Schadschneider, W. Klingsch, H. Klüpfel, T. Kretz, C. Rogsch, and A. Seyfried. Evacuation Dynamics: Empirical Results, Modeling and Applications. In R.A. Meyers, editor, *Encyclopedia of Complexity and System Science*, page 3142. Springer, 2009.
10. A. Schadschneider, H. Klüpfel, T. Kretz, C. Rogsch, and A. Seyfried. Fundamentals of Pedestrian and Evacuation Dynamics. In A. Bazzan and F. Klügl, editors, *Multi-Agent Systems for Traffic and Transportation Engineering*, pages 124–154. Information Science Reference, USA, 2009.
11. M.G.H. Bell and Y. Iida. *Transportation network analysis*. Wiley, 1997.
12. IMO. Interim Guidelines for Evacuation Analyses for New and Existing Passenger Ships. Technical Report MSC/Circ. 1238, IMO, 2007.
13. T. Kretz, S. Hengst, and P. Vortisch. Pedestrian Flow at Bottlenecks – Validation and Calibration of VISSIM's Social Force Model of Pedestrian Traffic and its Empirical Foundations. In M. Sarvi, editor, *ISTS08*, 2008.
14. M. Fellendorf and P. Vortisch. Microscopic Traffc Flow Simulator VISSIM. *Fundamentals of Traffic Simulation*, pages 63–94, 2010.
15. PTV. *VISSIM 5.30 User Manual*. Karlsruhe, DE, 2010.
16. A. Johansson, D. Helbing, and P.K. Shukla. Specification of the Social Force Pedestrian Model by Evolutionary Adjustment to Video Tracking Data. *Adv. Compl. Sys.*, 10(4):271–288, 2007.
17. D. Helbing and A. Johansson. Pedestrian, Crowd and Evacuation Dynamics. In R.A. Meyers, editor, *Encyclopedia of Complexity and System Science*, volume 16, page 6476. Springer, 2009.
18. D.M. Kreps. Game theory and economic modelling. *OUP Catalogue*, 1992.
19. C. Camerer and Russell Sage Foundation. *Behavioral game theory: Experiments in strategic interaction*, Vol. 9. Princeton Univ. Press, 2003.
20. M.J. Osborne. *An introduction to game theory*, 2004.
21. T. Kretz. A Round-Robin Tournament of the Iterated Prisoner's Dilemma with Complete Memory-Size-Three Strategies. *Complex Systems*, 19(4):363–389, 2011.
22. T. Kretz. Pedestrian Traffic: on the Quickest Path. *J. Stat. Mech.*, P03012, 2009.
23. E. Kirik, T. Yurgel'yan, and D. Krouglov. The shortest time and/or the shortest path strategies in a ca ff pedestrian dynamics model. *Journal of Siberian Federal University. Mathematics & Physics*, 2(3):271–278, 2009.
24. T. Kretz. The use of dynamic distance potential fields for pedestrian flow around corners. In *ICEM 2009*. TU Delft, 2009.
25. T. Kretz. Applications of the Dynamic Distance Potential Field Method. In Dai, S. et al., editor, *TGF '09*, 2009. submitted.
26. T. Kretz. The Dynamic Distance Potential Field in a Situation with Asymmetric Bottleneck Capacities. *Lect. Notes Comp. Sc.*, 6350:480–488, 2010.
27. J. Ondřej, J. Pettré, A.H. Olivier, and S. Donikian. A synthetic-vision based steering approach for crowd simulation. *ACM TOG*, 29(4):1–9, 2010.
28. D. Dressler, M. Groß, J.P. Kappmeier, T. Kelter, J. Kulbatzki, D. Plümpe, G. Schlechter, M. Schmidt, M. Skutella, and S. Temme. On the use of network flow techniques for assigning evacuees to exits. *Procedia Engineering*, 3:205–215, 2010.
29. M. Höcker, V. Berkhahn, A. Kneidl, A. Borrmann, and W. Klein. Graph-based approaches for simulating pedestrian dynamics in building models. In *ECPPM 2010*, 2010.
30. M. Moussaïd, D. Helbing, and G. Theraulaz. How simple rules determine pedestrian behavior and crowd disasters. *PNAS*, 108(17):6884, 2011.
31. A.U. Kemloh Wagoum, A. Seyfried, and S. Holl. Modelling the dynamic route choice of pedestrians to assess the criticality of building evacuation. *Adv. Compl. Sys.*, 15: 1250029, 2012.
32. T. Kretz, A. Große, S. Hengst, L. Kautzsch, A. Pohlmann, P. Vortisch. Quickest Paths in Simulations of Pedestrians. *Adv. Compl. Sys.*, 14:733, 2011.

33. H. Klüpfel, A. Seyfried, S. Holl, M. Boltes, M. Chraibi, U. Kemloh, A. Portz, J. Liddle, T. Rupprecht, A. Winkens, W. Klingsch, C. Eilhardt, S. Nowak, A. Schadschneider, T. Kretz, and M. Krabbe. HERMES – Evacuation Assistant for Arenas. In *Future Security*, 2010. eprint.
34. M. Schreckenberg and S.D. Sharma, editors. *Pedestrian and Evacuation Dynamics*. Springer, 2002.
35. R.D. Peacock, E.D. Kuligowski, and J.D. Averill, editors. *Pedestrian and Evacuation Dynamics*. Springer, 2011.

Using a Telepresence System to Investigate Route Choice Behavior

Tobias Kretz, Stefan Hengst, Antonia Pérez Arias, Simon Friedberger, and Uwe D. Hanebeck

Abstract A combination of a telepresence system and a microscopic traffic simulator is introduced. It is evaluated using a hotel evacuation scenario. Four different kinds of supporting information are compared: standard exit signs, floor plans with indicated exit routes, guiding lines on the floor and simulated agents leading the way. The results indicate that guiding lines are the most efficient way to support an evacuation but the natural behavior of following others comes very close. On another level the results are consistent with previously performed real and virtual experiments and validate the use of a telepresence system in evacuation studies. It is shown that using a microscopic traffic simulator extends the possibilities for evaluation, e.g. by adding simulated humans to the environment.

1 Introduction

To accurately plan and evaluate safety measures in buildings and transportation systems, knowledge of the behavior of pedestrians in emergency situations is required. There are two major ways to acquire this information, conducting experiments or performing computer simulations. The problem with experiments is the immense effort that is necessary to recreate an appropriate situation. This is even more difficult in the planning phase when a prototype would have to be constructed. In addition many situations cannot be evaluated experimentally because the safety of the test subjects cannot be guaranteed. Simulations on the other hand are usually

T. Kretz (✉) · S. Hengst
PTV Planung Transport Verkehr AG, Stumpfstr. 1, D-76131 Karlsruhe, Germany
e-mail: tobias.kretz@ptv.de; stefan.hengst@ptv.de

A.P. Arias · S. Friedberger · U.D. Hanebeck
Intelligent Sensor-Actuator-Systems Laboratory (ISAS), Karlsruhe Institute of Technology (KIT), Karlsruhe, Germany
e-mail: Antonia.Perez@kit.edu; simon.friedberger@kit.edu; Uwe.Hanebeck@kit.edu

V.V. Kozlov et al. (eds.), *Traffic and Granular Flow '11*,
DOI 10.1007/978-3-642-39669-4_14, © Springer-Verlag Berlin Heidelberg 2013

based on relatively simple models of pedestrian behavior and cannot capture the complexities of human thought processes and decision making. A common solution is to conduct simple experiments and simulations and attempt to derive generally applicable rules from them [1]. The generality makes these rules necessarily crude; an example would be the required exit width depending on the number of people in a building. Kobes et al. [2] propose the usage of so called serious games to combine the benefits of simulations and experiments. Their system allows a test subject to try to escape from a simulated emergency situation using common video game techniques. The problem with this approach is that the user doesn't actually move even though proprioception (i.e. self-perception of motion) has been shown to improve navigational abilities. We therefore introduce a combination of an extended range telepresence system and a pedestrian simulation. The effectiveness of the system is shown by comparing different utilities for finding fire exits. Such different utilities, like escape exit signs or guiding lines, can be seen and followed with varying simplicity. As for example guiding lines are visually very intrusive and may be considered unaesthetic a quantitative evaluation of the usefulness of different signage is desirable.

2 Combined System

This section presents the components of our experimental setup; the extended range telepresence system and the pedestrian simulation software.

2.1 Extended Range Telepresence

The extended range telepresence system allows a user to feel present in a remote or virtual environment (called the target environment) by locally reproducing perception for the user and remotely reproducing actions by the user. To achieve this, the user is wearing a head-mounted display (HMD) capable of displaying the target environment in 3D and playing back sound (cf. Fig. 1). The head-mounted display is fitted with additional sensors that allow its position and orientation in the room to be tracked. When the user takes a step forward, this movement will be registered and transferred to the target environment and what the user sees will change accordingly. This process is further complicated by an algorithm called Motion Compression [3–5], which allows the target environment to be much larger than the user environment. The path of the user is curved to require less space while keeping length and turning angles of the paths in both environments identical. It has been shown that users do not notice slight changes in curvature [6]. The image the user sees is then rotated slightly to steer him on the calculated user path.

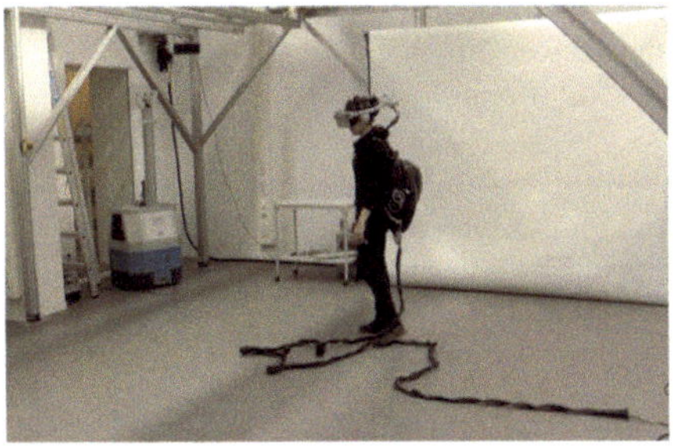

Fig. 1 User in the telepresence system, wearing HMD and backpack computer for processing

Fig. 2 Screenshot from the VISSIM microscopic traffic simulator

2.2 *VISSIM*

We have connected the pedestrian and vehicle simulation software VISSIM [7, 8] to the extended range telepresence allowing us to simulate environments that include virtual agents (cf. Fig. 2). These virtual (simulated) agents react to the telepresent user as if he was a simulated agent allowing him to become part of the simulation and interact with it.

2.3 *Benefits*

This setup has several benefits apart from proprioception. Three dimensional vision with the head mounted display allows to naturally judge distances and increases realism. The user position is tracked, which allows the simple creation of detailed records of his movement. The orientation of the head-mounted display is tracked as well which can be used to extract coarse focus of attention information, e.g., whether a fire exit is visible. Using a pedestrian simulation enables us to populate the simulated world with other humans without requiring further test subjects. This allows conducting experiments concerning the behavior of individuals in large crowds quickly and cheaply.

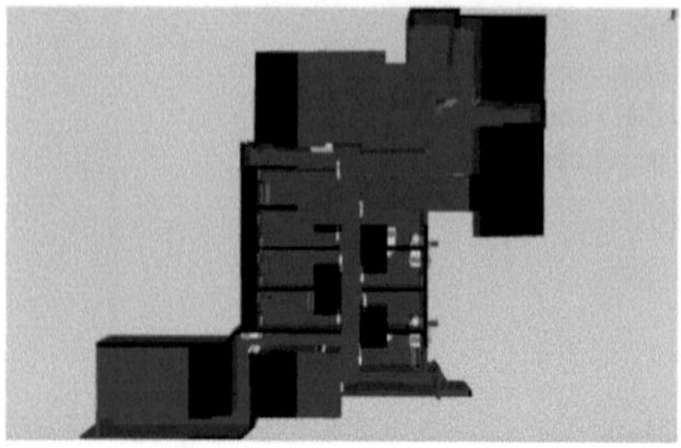

Fig. 3 Orthogonal view of the 3D scenario

3 Route Choice Behavior in a Hotel Evacuation

3.1 Scenario

The combined VISSIM-Telepresence system is used to study the influence of different signage for finding fire exits in a virtual hotel scenario (Fig. 3), it has the same layout as the scenario used by Kobes et al. in [2, 9–11]. The layout was enhanced with textures and furniture in order to realistically reproduce a typical hotel scenario.

Using this scenario has two advantages: First, the layout of the chosen hotel is classified as complex [10] and the choice of the nearest exit is not trivial. Second, this scenario reproduces a real hotel and data of a real case study performed in the hotel are available in [10], so that we can compare our results with the real data.

3.2 Experiment Description

Preliminary experiments were conducted to check the user behavior concerning the virtual walls. In principle, the user could consciously decide to move through the virtual walls. Different to an experiment designed to be done in front of a screen and controlled by keys, mouse or joystick, the telepresence system has no mechanism to prevent this. The preliminary experiments showed that the visual information that the user receives via the head-mounted display is sufficient to navigate in the virtual environment without colliding with the virtual walls (Fig. 4 shows exemplary trajectories of these experiments). However, without supporting information (cf. Fig. 5) most test subjects were unable to find the nearest exit in these experiments.

A case study was designed to investigate the influence of the signage in finding the nearest exit in case of a hotel evacuation. As part of this study, we compared exit choice, travel times, walking distances, and walking speeds towards exits under following conditions:

There is a guiding line on the floor (Fig. 6a). There are other persons (simulated agents) walking to the exit (Fig. 6b). There is standard escape exit signage above

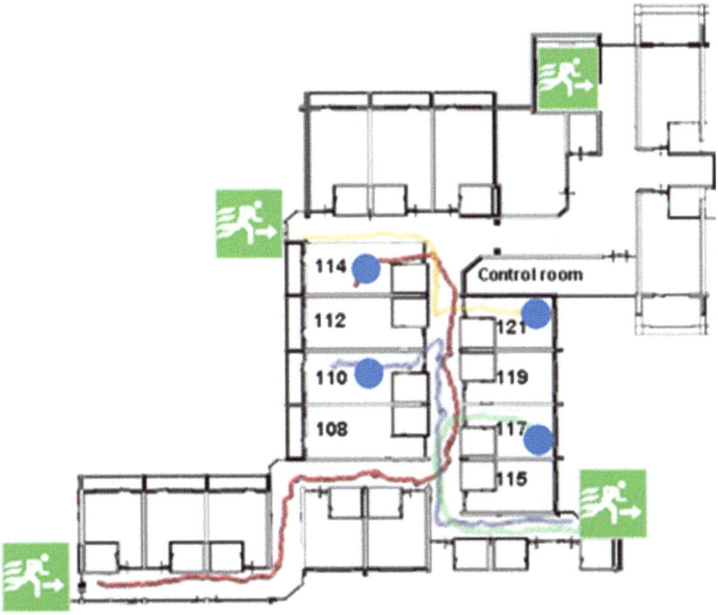

Fig. 4 Example trajectories (Base picture adapted from [9])

Fig. 5 Screenshot of evacuation scenario without supporting information

head (Fig. 6c). There is an evacuation floor plan in the room where the evacuation starts (Fig. 6d).

3.3 Experiment Participants

We introduced 20 participants, all male, between 21 and 32 years old to the scenario. The participants had the opportunity to familiarize themselves with the

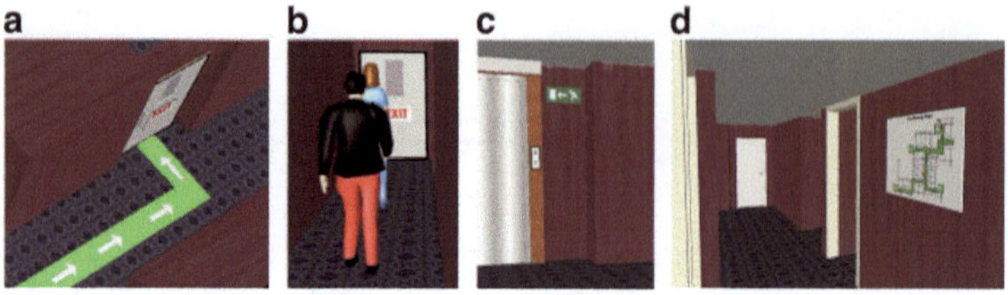

Fig. 6 Screenshots of scenarios with supporting information: (**a**) guiding lines, (**b**) simulated agents, (**c**) escape exit signs, and (**d**) floor plan

telepresence system in a simpler scenario. All participants tested the four conditions. The participants started each test run at different positions to avoid learning effects. Moreover, 5 participants started with condition 1, 5 with condition 2, etc., in a way that we could analyze the performance under different signage conditions with and without learning effects separately. The participants were instructed to leave the building as fast as possible as they would be a real evacuation.

3.4 Performance Measures

In order to evaluate the efficiency of the supporting information, exit choice, travel times, walking distances, and walking speeds towards exits were recorded. In order to quantify the subjective preference of the participants for each type of signal, a questionnaire was used.

4 Results

The participants were not able to find the nearest exit without supporting information. Therefore, the next results only report the performance measures for the scenarios with supporting information.

4.1 Objective Measures

4.1.1 Nearest Exit Choice

The evaluation of the exit choice in Fig. 7 shows that the guiding lines are the most efficient signage to find the nearest exit. By using other signage the nearest exit is only chosen in about 50 % of the test runs.

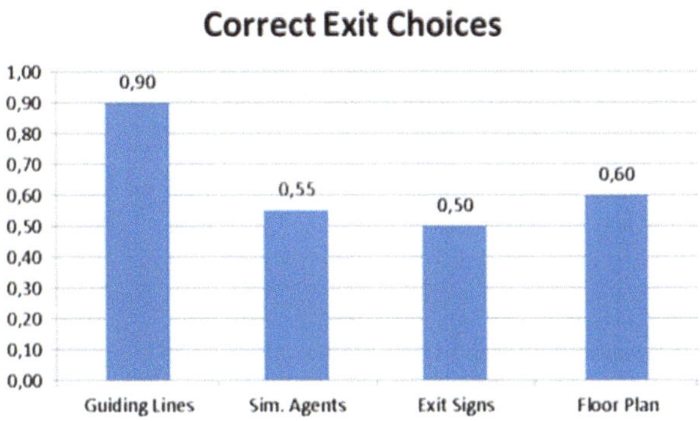

Fig. 7 Percentage of correct exit choice for each signage condition

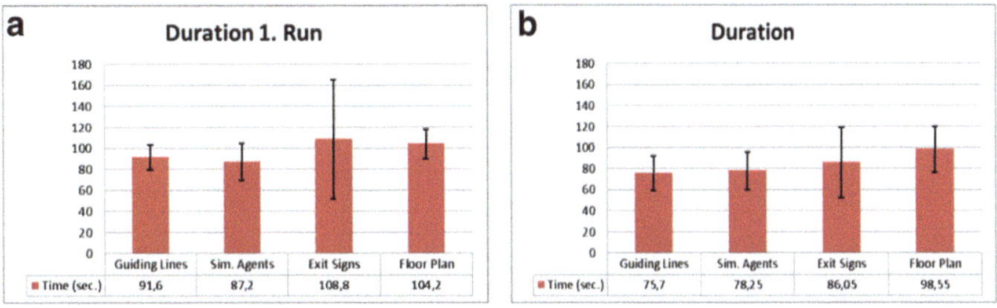

Fig. 8 (**a**) Average duration for each signage condition using only the first runs of the participants. (**b**) Average duration using all runs of the participants

4.1.2 Travel Time

The time needed to find the exit in the evacuation scenario is shown in Fig. 8a. When only the first test runs are evaluated (i.e., the participants do not know the hotel scenario in advance), the guiding lines and the presence of other pedestrians lead to the shortest travel times with average values of 91.6 s and 87.2 s, respectively. Escape exit signs and the escape floor plan have the longest travel times. Note that there is a higher standard deviation across the participants when using the exit signs, whereas the floor plan leads to longer times for most participants.

The guiding lines and the presence of other pedestrians also lead to faster evacuations when regarding the average time duration for all test runs Fig. 8b, although the duration of the evacuation using the exit signs and the floor plan is as expected shorter when the user knows the building in advance.

4.1.3 Walking Distance

Figure 9a shows the average of the covered distances to the exits using only the first runs of the participants. The guiding lines and the presence of other pedestrians

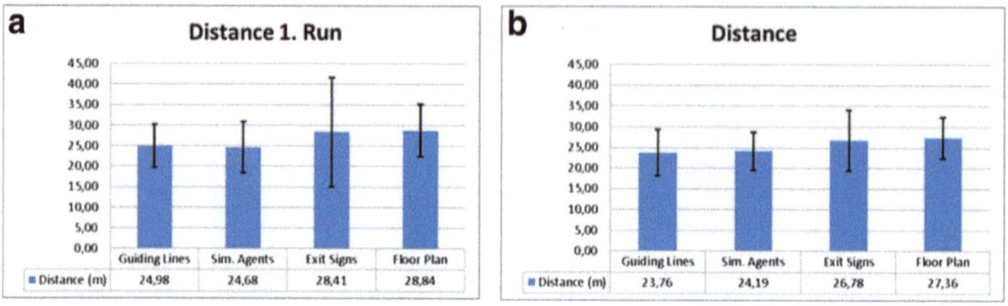

Fig. 9 (**a**) Average walking distance using only the first runs of the participants. (**b**) Average walking distance using all runs of the participants

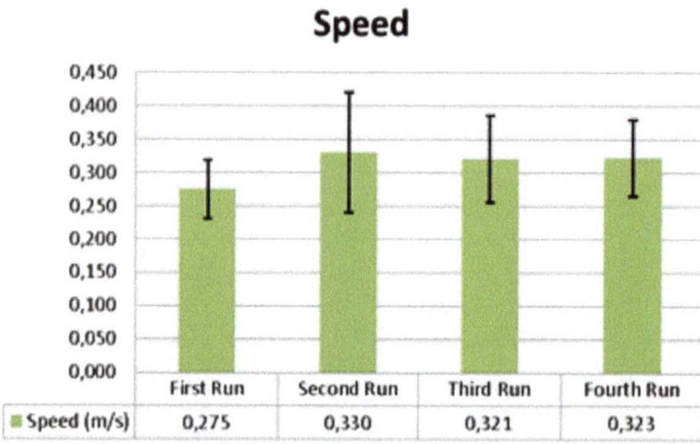

Fig. 10 Average velocity of participants at each run

again lead to shorter walking distances than the exit signs and the floor plan. The same trend is observed when considering all test runs of the participants (Fig. 9b).

4.1.4 Walking Speed

The learning effect is clearly observed by regarding the average velocity of the participants at each test run in Fig. 10. The average velocity in the first test run is significantly lower than in the other runs. However, no significant difference is observed between the second run and the next runs.

4.2 Subjective Measures

The evacuation scenario was found to be modeled realistically by almost all (19) participants. Most participants found the guiding lines to be the most efficient signal.

Table 1 Questionnaire used to evaluate the preference of the participants

	Guiding lines	Exit signs	Sim. agents	Floor plan
Which aid seemed to provide the fastest way out?	9	7	2	2
Which aid seemed the most useful?	17	2	1	0
Which aid would you prefer in case of fire?	19	5	2	0
(Multiple answers possible)				

Moreover, in case of fire, 19 participants would prefer the guiding lines and 7 participants would prefer a combination of guiding lines and exit signs (Table 1).

4.3 Discussion

In our route choice behavior study, the guiding lines turned out to be the most efficient signage method in order to find the nearest exit. This signage condition achieved shorter times and walking distances than exit signs or a floor plan hanging on the wall. The evaluation of the questionnaires also showed that participants have a clear preference for this signage condition, especially in the case of a fire evacuation. These results are in agreement with studies reported in [9, 12, 13] that indicate that photoluminescent low-level exit path markings are likely to be more effective compared to conventional escape route signs. The presence of other pedestrians turned out to be very beneficial for the evacuation in our study leading to exit times close to those of guiding lines. However, following other agents did not always guide to the nearest exit for the user who may have started in a different room. In order to validate our extended range telepresence system as an adequate tool to perform such a route choice study, a comparison of our results with real data is necessary. For this purpose, we use the results from the case study in the hotel scenario presented in [10]. In our experiments not all participants chose the nearest exit, neither did they in the real experiments. The mean value of the covered distance to the chosen exit is 48.8 m in the real experiment, with a minimum value of 13.5 m and a maximum value of 83.2 m [10]. The mean value of the covered distance in our experiments (considering only the first run) is 26.7 m, with minimum and maximum values of 13.5 m (following simulated pedestrians) and 54.5 m (following the exit signs), respectively. The mean and the maximum covered distance in extended range telepresence are rather lower than in the real experiments. However, all the values are within the range of distances achieved in the real experiments. The mean value of the walking velocity in the real experiments [10] is 1.03 m/s, which is higher than the mean velocity in our experiments. This difference may be due to the user being afraid of running outside the borders of the user environment or damaging the carried equipment. This difference is irrelevant for our evaluation of route choice as there are no differences in speed limiting factors along the different exit paths.

5 Summary

A combination of a telepresence system and a microscopic traffic simulator has been introduced and its efficacy for evaluating evacuation scenarios has been shown. As a first test scenario, the evacuation of a hotel using different kinds of signage has been evaluated. The results indicate that low-level exit path markings are the most efficient way of guiding people to an emergency exit but also that following others is efficient as well. These results are consistent with previously performed real and virtual experiments, which validates the use of our telepresence system in evacuation studies, and also shows the extended possibilities of using a pedestrian simulation software to add virtual agents.

Acknowledgements This work was supported by the research project "The Pedestrian Simulation VISSIM within a Telepresence System" within the Central Innovation Programme for Small and Medium-sized Enterprises (ZIM) of the German Federal Ministry of Economics and Technology (BMWi).

References

1. U. Chattaraj, A. Seyfried, and P. Chakroborty. Comparison of pedestrian fundamental diagram across cultures. *Adv. Compl. Sys.*, 12(3):393–405, 2009.
2. M. Kobes, N. Oberijé, and K.G.T. Morsche. Serious gaming for behavioral assessment and research in case of emergency. an evaluation of experiments in virtual reality. In *Proceedings of Conference: Simulation - Concepts, Capability and Technology (SimTecT 2009)*, 2009.
3. N. Nitzsche, U.D. Hanebeck, and G. Schmidt. Extending telepresent walking by motion compression. In *1. SFB-Aussprachetag Human Centered Robotic Systems (HCRS 2002)*, pages 83–90, 2002.
4. N. Nitzsche, U.D. Hanebeck, and G. Schmidt. Motion compression for telepresent walking in large target environments. *Presence: Teleoperators & Virtual Environments*, 13(1):44–60, 2004.
5. P. Rößler, U.D. Hanebeck, and N. Nitzsche. Feedback controlled motion compression for extended range telepresence. In *Proceedings of IEEE Mechatronics & Robotics (MechRob 2004), Special Session on Telepresence and Teleaction*, pages 1447–1452, 2004.
6. N. Nitzsche, U.D. Hanebeck, and G. Schmidt. Motion compression for telepresent walking in large-scale remote environments. In *Proceedings of SPIE, AeroSense Symposium*, 2003.
7. M. Fellendorf and P. Vortisch. Microscopic Traffc Flow Simulator VISSIM. *Fundamentals of Traffic Simulation*, pages 63–94, 2010.
8. PTV AG. *VISSIM 5.30 User Manual*. Karlsruhe, 2010.
9. M. Kobes, I. Helsloot, B. de Vries, N. Oberijé, and N. Rosmuller. Fire response performance in a hotel. behavioural research. In *Interflam 2007*, p. 1429, 2007.
10. M. Kobes, N. Oberijé, and M. Duyvis. Case studies on evacuation behaviour in a hotel building in bart and in real life. In Klingsch et al. [14], pages 183–201.
11. M. Kobes, I. Helsloot, B. de Vries, J.G. Post, N. Oberijé, and K. Groenewegen. Way finding during fire evacuation; an analysis of unannounced fire drills in a hotel at night. *Building and Environment*, 45(3):537–548, 2010.
12. G. Proulx, B. Kyle, and J. Creak. Effectiveness of a photoluminescent wayguidance system. *Fire Technology*, 36:236–248, 2000.
13. M.J. Quellette. Visibility of exit signs. *Progressive Architecture*, 36:39–42, 1993.
14. W.W.F Klingsch, C. Rogsch, A. Schadschneider, and M. Schreckenberg, editors. *Pedestrian and Evacuation Dynamics 2008*. Springer Berlin Heidelberg, 2010.

A Cellular Automaton Approach for Lane Formation in Pedestrian Counterflow

Stefan Nowak and Andreas Schadschneider

Abstract The formation of lanes is a well known emergent behavior in pedestrian counterflow as well as in some other physical systems. Nevertheless, not much is known quantitatively which is related to the fact that the empirical situation is quite poor. Here we analyze lane formation in a two-dimensional cellular automaton for pedestrian dynamics. To quantify the lane structure, we make use of a laning order parameter which has been used previously to detect lanes in colloidal systems. The main purpose of our work is to determine a phase diagram in dependence on the coupling constants and the particle density. Furthermore, we compare the results of our simulation with experimental data.

1 Introduction

In recent years, pedestrian dynamics has become more and more attractive for scientists from different fields. This is not very surprising because there are many applications which require a profound knowledge of crowd dynamics. It can help to make forecasts of evacuation scenarios as well as to plan the layout of buildings and venues. Furthermore, pedestrian dynamics is very interesting from a physical point of view. Human crowds are many-body systems which show many interesting collective effects like jamming at bottlenecks, density waves, flow oscillations and lane formation [1]. The latter is also known from other driven systems [2], e.g., from complex plasmas [3], molecular ions [4] and colloidal suspensions [5–7].

There are many models which are able to reproduce the collective phenomena mentioned above [8, 9], but quantitative studies are rare. Especially, lane formation in counterflow is often used as a validation to check whether a model is realistic. However, there are very few studies that go beyond the statement that lanes

S. Nowak (✉) · A. Schadschneider
Institut für Theoretische Physik, Universität zu Köln, 50937 Köln, Germany
e-mail: sn@thp.uni-koeln.de; as@thp.uni-koeln.de

V.V. Kozlov et al. (eds.), *Traffic and Granular Flow '11*,
DOI 10.1007/978-3-642-39669-4_15, © Springer-Verlag Berlin Heidelberg 2013

are present in the system, even though simulations of bi-directional pedestrian movement are made very often [9–24].

Comparisons with empirical data are even made less frequently. This is related to the fact that the empirical situation is not very satisfying. Even for quite simple scenarios, like the fundamental diagram in a corridor, there is no consensus [1,25]. But the situation has been improved recently within the Hermes project [26, 27] where several large scale experiments were performed [28–31] with high accuracy [32] and automatized evaluation [33].

The goal of this article is to make quantitative statements about lane formation with help of an order parameter adopted from the analysis of colloidal suspensions [7]. The results of the simulations will also be compared with experimental data.

2 Definitions

2.1 Floor Field Cellular Automaton

The Floor Field Cellular Automaton (FFCA) [14,34] is defined on a square lattice of size $L \times W$. Each cell corresponds to an area of $0.4 \times 0.4\,\text{m}^2$ and can either be empty or contain a type-A particle, a type-B particle or a wall. The two different types of particles represent pedestrians with opposite walking direction: Type-A particles go to the right, type-B particles to the left. Their movement is defined by update rules that transfer the system from one timestep to the next. One timestep corresponds to about 0.25 s. The update rules are defined by the probability

$$p_{ij} = \frac{1}{Z} \exp\left(k_\text{S} S_{ij} + k_\text{D} D_{ij}\right) \xi_{ij} \tag{1}$$

that a certain particle goes to the cell (i, j). The factor $\xi_{ij} \in \{0, 1\}$ ensures that movement only takes place at allowed cells, i.e., $\xi_{ij} = 0$ for cells which are occupied by other particles or walls and cells which are not nearest neighbors of the origin-cell. Z is a normalization constant which ensures that $\sum_{ij} p_{ij} = 1$. The actual direction of motion is determined by the so called floor fields S_{ij} and D_{ij}. Movement is preferred into the direction of larger fields. The coupling constants k_S and k_D control the strength of the fields.

The *static floor field* (SFF) S_{ij} contains the information about the desired direction of movement, i.e., $S_{ij}^\text{A} = 1$ for cells to the right of a type-A particle, $S_{ij}^\text{A} = -1$ for cells to the left and $S_{ij}^\text{A} = 0$ otherwise. The corresponding field of a type-B particle is given by $S_{ij}^\text{B} = -S_{ij}^\text{A}$.

The *dynamic floor field* (DFF) D_{ij} is used to model longer ranged interactions. Each type-X particle that moves from a cell (i, j) to another cell increases D_{ij}^X by 1. It can be interpreted as a virtual trace that acts attracting to other particles. The DFF has also its own dynamics, namely diffusion and decay. At the end of each timestep it is updated according to

$$D_{ij}(t+1) = (1-\delta)\Big[D_{ij}(t) + \frac{\alpha}{4}\Delta D_{ij}(t)\Big] \tag{2}$$

where

$$\Delta D_{ij}(t) = D_{i,j+1}(t) + D_{i,j-1}(t) + D_{i+1,j}(t) + D_{i-1,j}(t) - 4D_{i,j}(t)$$

is a discretization of the Laplace operator.

Since all particles choose a destination cell synchronously, a *conflict* occurs when $n > 1$ particles try to move to the same cell. In this case, with probability $(1-\mu)$, the conflict is solved, i.e., one of the particles is chosen at random to move. For simplicity we assume that each particle is chosen with equal probability $1/n$. With probability μ the conflict is unsolved, i.e., all particles stay where they are. μ can be interpreted as a *friction* parameter [35].

In summary, the update rules are given as follows:

1. All particles chose a destination cell according to Eq. (1)
2. Conflicts are solved
3. All particles move to their destination cell. The dynamic floor field of the origin cells is increased by one
4. Diffusion and decay of the dynamic floor field according to Eq. (2)

2.2 Order Parameter for Lane Formation

In order to quantify the lane structure in the system, we use an order parameter which is based on the order parameter that was introduced in Ref. [7] to detect lanes in a colloidal suspension. Other attempts for lane detection are given in Refs. [14, 36, 37].

Considering the discreteness of space in our system, the order parameter can be written as

$$\Phi = \frac{1}{N}\sum_{n=1}^{N}\left(\frac{N_{i_n}^A - N_{i_n}^B}{N_{i_n}^A + N_{i_n}^B}\right)^2 \tag{3}$$

where i_n is the vertical position of the n-th particle, N_i^A (N_i^B) denotes the number of particles of type A (B) at row i and N is the total number of particles. This value is close to 0, if there is a homogenous mixture of type-A and type-B particles, and equal to 1, if each row contains only one type of particle. Note that the value of Φ is in general larger than zero, even if all particles are distributed at random. The mean value of Φ in that case is denoted by Φ_0 and can be large although there is no lane structure in the system, especially for small densities. We take this into account by defining a reduced order parameter

$$\tilde{\Phi} = \frac{\Phi - \Phi_0}{1 - \Phi_0}. \tag{4}$$

3 Results

3.1 Determination of Phases

In this section the number of particles of type A and B is equal, i.e., $\rho_A = \rho_B = \rho/2$. The parameters are given by

$$k_S = 2.5, \quad \alpha = 0.3, \quad \delta = 0.1, \quad W = 10, \quad L = 100 \tag{5}$$

if not mentioned otherwise. The boundary conditions are periodic in walking direction and open perpendicular to it, i.e., there is a virtual wall at rows 0 and $W + 1$. We use the von Neumann neighborhood, i.e., each cell has four nearest neighbors. As a first result one observes that in general the probability P_{jam} that the system forms a gridlock increases both with an increasing density and a decreasing coupling constant k_D (Fig. 1a). This is not very surprising since an increasing density means a larger number of type-A and type-B particles which can block each other. Additionally, there is less space to avoid such collisions. An increasing k_D means that it becomes more likely for a particle to follow another particle of the same kind and thus collisions of type-A and type-B particles are avoided. Note that the jamming transition to a gridlock state was examined frequently [15–19], but there is no empirical evidence that such a transition occurs in reality [31]. Therefore, the occurrence of gridlocks at too low densities should be seen as an indication for unrealistic behavior of the model.

For the analysis of the order parameter $\tilde{\Phi}$ and in the following only those simulations are considered that did not evolve into a gridlock. One can see in Fig. 1b that a certain minimal density ρ and coupling constant k_D is necessary to form lanes. If one of these parameters is reduced, the interaction between particles becomes too weak. For large densities ($\rho > 0.225$) and small k_D no data are available because the system always evolves in a gridlock. If k_D is increased at those large densities one observes that no configuration with a medium value of $\tilde{\Phi}$ is stable, i.e., the system is either in a gridlock state or in a state with almost perfect lanes.

3.2 Properties of Lanes

Figure 2a shows the distribution of densities ρ_i in the different rows $i = 1, \ldots, W$. If the particles were equally distributed over the system (disordered configuration),

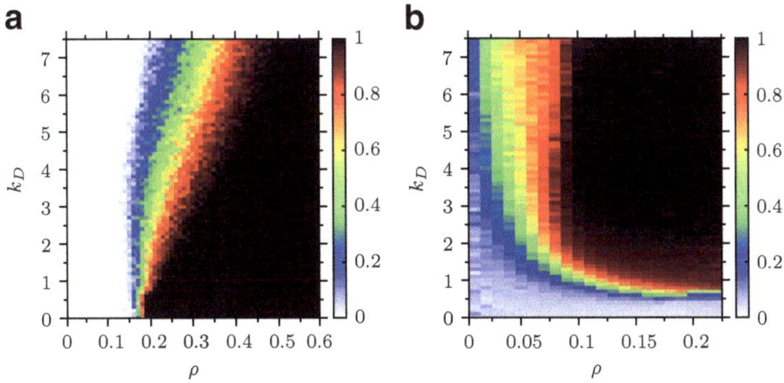

Fig. 1 Jam probability and order parameter in dependence of the density ρ and the coupling constant k_{D}. (**a**) Jam probability P_{Jam}. (**b**) Reduced order parameter $\tilde{\Phi}$

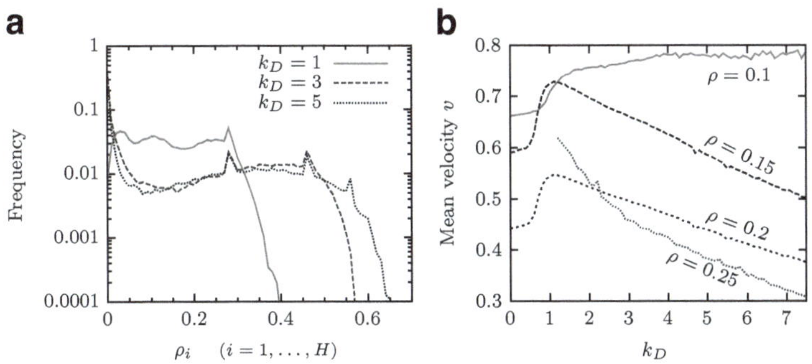

Fig. 2 (**a**) Distribution of densities in the different rows. The global density is $\rho = 0.15$. (**b**) Mean velocity in dependence on the coupling constant k_D. For both figures only those simulations were taken into account which did not evolve in a gridlock

one would expect a sharp peak in the distribution at the position of the global density ρ. But since the DFF leads to an attractive interaction between particles, the number of rows which are not occupied as well as the number of rows with larger density grows. This can be observed in the distribution as a sharp peak at $\rho_i = 0$ and a larger support for increasing coupling constant k_{D}. The latter means that the mobility of particles and thus the average velocity is reduced.

Another feature of the distribution is the existence of a characteristic lane density which does neither depend on the coupling constant k_{D} nor on the global density ρ. This becomes visible as peaks in the distribution. The positions of the peaks are at $\rho_1^* = 0.28$, $\rho_2^* = 0.46$ and $\rho_3^* = 0.56$. Note that not all peaks have to be present at a given set of parameters. It turns out that, if there is a peak located at ρ_i^* for some $i \in \{1, 2, 3\}$, there are also peaks at ρ_j^* for all $j < i$. This was tested for different coupling constants $k_{\mathrm{D}} \in [1, 7]$ and for different (global) densities $\rho \in [0.1, 0.3]$. For instance, at $\rho = 0.15$ there is only one peak at $k_{\mathrm{D}} = 1$, two peaks at $k_{\mathrm{D}} = 3$ and three peaks at $k_{\mathrm{D}} = 5$.

Since lane formation is a way of separating different types of particles which leads to less opposing traffic one can expect that lanes increase the mean velocity v in the system. But on the contrary, lane formation requires a sufficiently large value of the coupling constant k_{D} which can lead to large local densities in a row. As mentioned above, this decreases the velocity. Therefore, there is an optimal value of k_D and one has a maximum in $v(k_{\mathrm{D}})$ (Fig. 2b). But note that not for all densities such a maximum is present. For instance, at $\rho = 0.1$ the global densities does not suffice to create a local density which is larger than 0.5. Hence, each particle has on average one empty cell in front and the velocity does not decrease. If the density is too large one has also no maximum because there are no data available for small k_D due to a large jam probability.

4 Additional Mechanisms for Lane Formation

The basic model as described above shows interesting effects, but it is not sufficient to reproduce experimental data. The main reason is that the tendency of the system to form a gridlock is too large. Therefore, we introduce a few additional mechanisms which are able to prevent jams.

4.1 *Anticipation*

When pedestrians are in a counterflow they usually try to avoid collisions by estimating the prospective route of pedestrians with opposite walking direction. This behavior can be imitated by introducing an additional "Anticipation Floor Field" [38]. For particles of type A it is defined as

$$A_{ij}^{A} = \sum_{j'} \lambda^{d_A(j,j')} \tag{6}$$

where the sum is over all type-A particle positions in the i-th row. $\lambda \in (0, 1)$ is a parameter which controls the range of anticipation. $d_A(j, j')$ is the minimal number of cells which have to be passed if a particle of type A goes from a cell (i, j') to the cell (i, j), only taking steps towards its desired walking direction. The field A_{ij}^B for particles of type B is defined analogously. This definition of A_{ij} causes a large value of A_{ij}^X ($X \in \{A, B\}$) if there are many particles of type X which are going to tread the cell (i, j) in the nearby future. Particles of the other type Y should avoid this cell. This can be achieved by reducing the transition probability p_{ij} by an additional factor $\exp(-k_{\mathrm{A}} A_{ij}^X)$ in Eq. (1), where k_{A} is the coupling constant to the anticipation floor field A_{ij}^X.

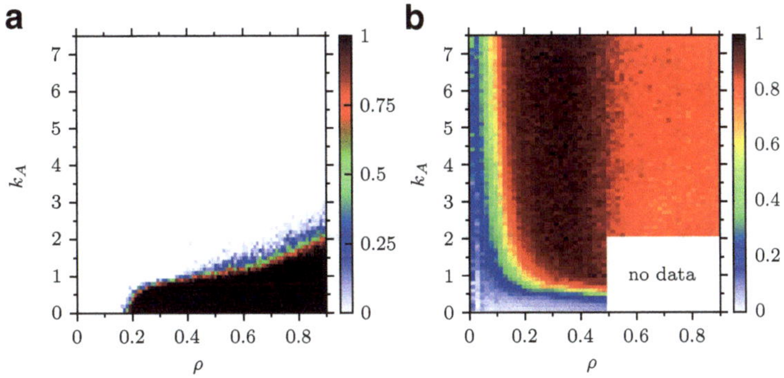

Fig. 3 Order parameter and jam probability in dependence of the density ρ and the coupling constant k_A. The dynamic floor field is turned off ($k_D = 0$). (**a**) Jam probability P_{Jam}. (**b**) Reduced order parameter $\tilde{\Phi}$

As Fig. 3b shows, the anticipation mechanism is able to create lanes, too. But the more important fact is that the jam probability is strongly reduced compared to the system without anticipation (cf. Fig. 3a).

4.2 Swapping

In reality pedestrians are not incompressible squares like in the model but more or less compressible bodies. They are also able to turn sideways such that the contact area becomes smaller, i.e., they are able to walk past each other even if the available space is smaller than two cells (= 0.8 m). We take this into account by providing a certain probability π that two opposing particles can swap their position. Swapping is applied in each timestep after the "normal" movement of particles, but only to those pairs of particles that did not move previously in that timestep and if their combined neighborhood contains at least one empty cell.

4.3 Politeness Factor

The "politeness factor" [39] takes into account that pedestrian usually try to avoid to get too close to other pedestrians, i.e., it is a kind of repulsive force. It is not necessary for lane formation or the prevention of gridlocks, but it is important for quantitative predictions of the fundamental diagram [40]. It can be easily implemented by adding an additional politeness factor $\exp(-k_P \tilde{N}_{ij})$ in the transition probabilities (1). Here we define $\tilde{N}_{ij} = \max(0, N_{ij} - N_{kl})$ for a particle on the cell (k,l), N_{ij} is the number of nearest neighbors of the cell (i,j) and k_P controls the strength of politeness.

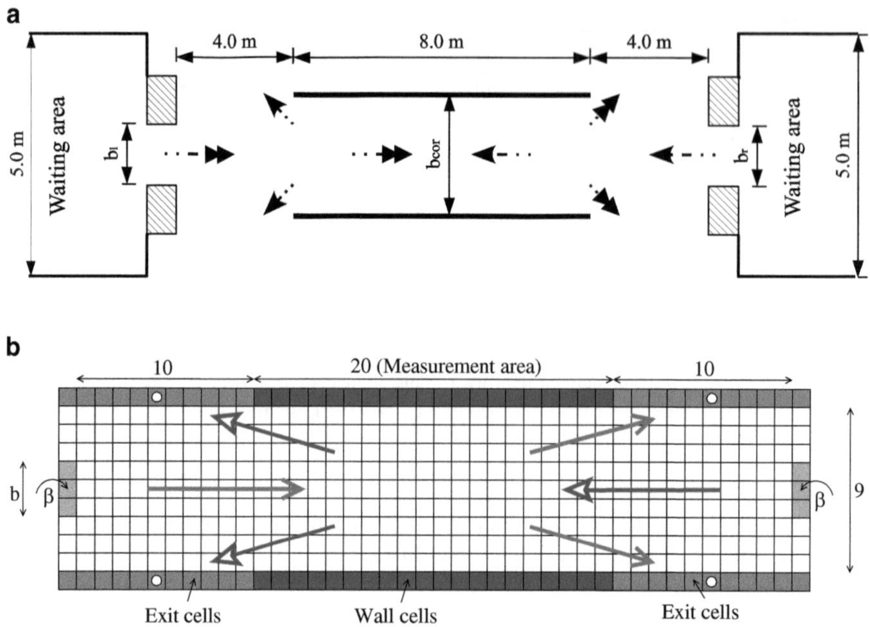

Fig. 4 Pictures of the experimental setup (**a**) in reality (Taken from [31]) and (**b**) in the simulation

5 Comparison with Experimental Data

5.1 Experiment

Experiments were performed within the Hermes project [26, 27] in the fairground Düsseldorf [31]. A schematic picture of the experimental setup can be found in Fig. 4a. There were several experiments with varying width of the corridor, but here we will only consider the experiments with $b_{\text{cor}} = 3.6$ m. In order to vary the shape of the lanes in the counterflow, there are two versions of the experiment. In the first version, the participants get no instruction about which of two possible exits (left or right) they have to choose. One observes a complete segregation of the two opposing streams. The lanes are formed immediately after the beginning of the experiment and stay stable the whole time. This regime is called "Stable Separated Lanes" (SSL). In a second version, they were told to choose an exit according to a number they got before the experiment: Odd numbers exit to the left, even numbers to the right. One still observers lane formation in this version, but the lanes are unstable and their number increases. This regime is called "Dynamical Multi-Lanes" (DML). A surprising result is that in both cases the fundamental diagram (Fig. 5) looks very similar, at least up to a density of 2 Persons/m^2. Unfortunately, there have been no experiments for larger densities in the SSL regime. From the fundamental diagram

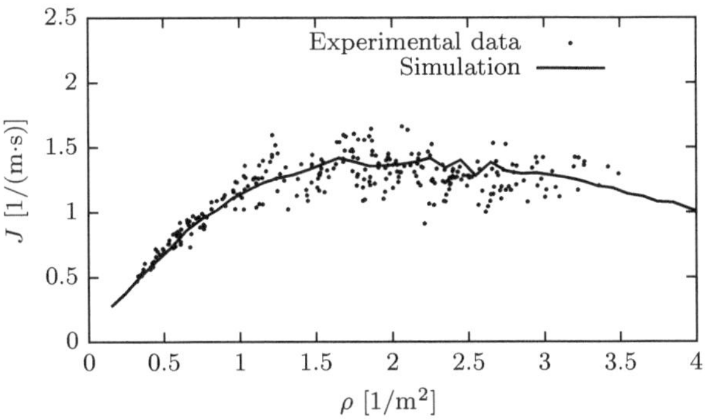

Fig. 5 Empirical fundamental diagram vs. simulations in the DML regime

of the DML regime one can see that densities up to 3.5 Persons/m^2 are possible without a noticeable decrease of the flow. This corresponds to a density of $\rho = 0.56$ in the model, i.e., no gridlocks should occur for smaller densities.

5.2 Simulation

The "experimental" setup in the simulation can be found in Fig. 4b. Particles are inserted into the system at every timestep on each of the b entrance cells with probability β. If they reach one of the exit cells, they are removed from the system. We assign the static floor field according to the negative Euclidean distance to the four cells which are indicated with white circles in Fig. 4b. In the DML regime there are two different SFFs S_{ij}^{L} and S_{ij}^{R} corresponding to the left and right exit. Particles which enter the system choose one of them at random. In the SSL regime there is only one SFF $S_{ij} = \max\{S_{ij}^{\mathrm{L}}, S_{ij}^{\mathrm{R}}\}$.

First of all, we try to adjust the parameters of the model such that the experimental fundamental diagram is reproduced. The best agreement was found with the following parameters: $k_{\mathrm{S}} = 2.5$, $k_{\mathrm{D}} = 2$, $k_{\mathrm{A}} = 2$, $k_{\mathrm{P}} = 3.2$, $\mu = 0.5$, $\pi = 0.6$, $\alpha = 0.25$, $\delta = 0.15$, $\lambda = 0.8$. The resulting plot is shown in Fig. 5.

From visualisations of the simulations one can also clearly see the difference in the dynamics of the SSL and the DML regime. Like in the experiment, in the simulated SSL regime the particles form a configuration of two separated lanes which is stable for a long time. The only difference is that it takes some longer time to form the lanes while in the experiment lanes are formed immediately.

In the DML regime one can observe the same instability of lanes as in the experiment. It seems that the dynamics in the inside of the corridor is at least similar to the behavior in the experiment. The dynamics in the entrance/exit area still needs some improvement. In the simulation particles go often to the center and wait for a gap in the opposing stream, which they cannot cross easily due to anticipation

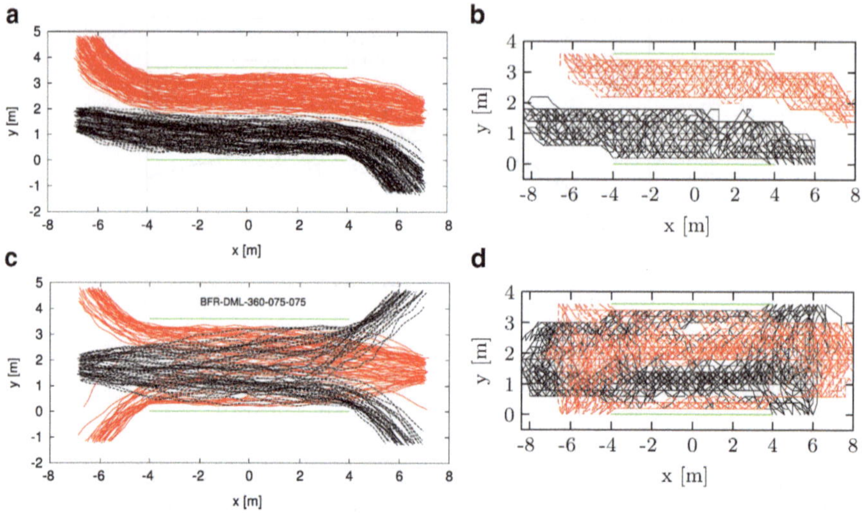

Fig. 6 Trajectories in the experiment and the simulation. (**a**) SSL regime, experiment. (**b**) SSL regime, simulation. (**c**) DML regime, experiment. (**d**) DML regime, simulation

(cf. Fig. 6d). In the experiment participants usually take a more direct way out of the corridor (cf. Fig. 6c). The discrepancy is related to the fact that anticipation and swapping only works well when the walking direction is parallel to the corridor.

6 Conclusion and Outlook

We have discussed the phenomenon of lane formation in a cellular automaton model. With help of an order parameter one can easily estimate the range of model parameters where lane formation can be observed in the system. The basic floor field model can reproduce lane formation, but it turns out that the tendency to form a gridlock is too large to be realistic. For this reason additional mechanisms have to be introduced. If they are included the model is able to reproduce the experimental data quite well.

In future work one can try to improve the model further, especially to deal with more complex scenarios. It would also be nice if there were quantitative empirical data on the lane structure, for instance in terms of the order parameter. This could help to make the model calibration even more sophisticated.

Acknowledgements This work was supported by the project Hermes funded by the Federal Ministry of Education and Research (BMBF) Program on “Research for Civil Security – Protecting and Saving Human Life” and the Bonn-Cologne Graduate School of Physics and Astronomy. We thank Jun Zhang and Armin Seyfried for providing the empirical data.

References

1. A. Schadschneider, W. Klingsch, H. Klüpfel, T. Kretz, C. Rogsch, and A. Seyfried. In R. A. Meyers, editor, Encyclopedia of Complexity and Systems Science, pages 3142–3176. Springer, 2009.
2. B. Schmittmann and R. K. P. Zia. Phys. Rep., 301:45–64, 1998.
3. K. R. Sütterlin, A. Wysocki, A. V. Ivlev, C. Räth, H. M. Thomas, M. Rubin-Zuzic, W. J. Goedheer, V. E. Fortov, A. M. Lipaev, V. I. Molotkov, O. F. Petrov, G. E. Morfill, and H. Löwen. Phys. Rev. Lett., 102:085003, 2009.
4. R. R. Netz. Europhys. Lett., 63:616, 2003.
5. J. Dzubiella, G. P. Hoffmann, and H. Löwen. Phys. Rev. E, 65:021402, 2002.
6. M. E. Leunissen, C. G. Christova, A.-P. Hynninen, P. Royall, A. I. Campbell, A. Imhof, M. Dijkstra, R. van Roij, and A. van Blaaderen. Nature, 437:235–240, 2005.
7. M. Rex and H. Löwen. Phys. Rev. E, 75(051402), 2007.
8. A. Schadschneider, A. Kirchner, and K. Nishinari. Lecture Notes in Computer Science, 2493:239–248. Springer, 2002.
9. D. Helbing, P. Molnar, I. J. Farkas, and K. Bolay. Environment and Planning B, 28:361–383, 2001.
10. V. J. Blue and J. L. Adler. Transportation Research Record, 1678:135–141, 1999.
11. F. Weifeng, Y. Lizhong, and F. Weicheng. Physica A, 321:633–640, 2003.
12. W. G. Weng, T. Chen, H. Y. Yuan, and W. C. Fan. Phys. Rev. E, 74:036102, 2006.
13. Y. F. Yu and W. G. Song. Phys. Rev. E, 75:046112, 2007.
14. C. Burstedde, K. Klauck, A. Schadschneider, and J. Zittartz. Physica A, 295:507–525, 2001.
15. M. Muramatsu, T. Irie, and T. Nagatani. Physica A, 267:487–498, 1999.
16. Y. Tajima, K. Takimoto, and T. Nagatani. Physica A, 313:709–723, 2002.
17. K. Takimoto, Y. Tajima, and T. Nagatani. Physica A, 308:460–470, 2002.
18. H. Kuang, T. Song, X.-L. Li, and S.-Q. Dai. Chin. Phys. Lett., 25:1498, 2008.
19. T. Nagatani. Physics Letters A, 373:2917–2921, 2009.
20. T. Kretz, M. Kaufman, and M. Schreckenberg. Lecture Notes in Computer Science, 5191:555–558. Springer, 2008.
21. W.G. Weng, S.F. Shen, H.Y. Yuan, and W.C. Fan. Physica A, 375:668–678, 2007.
22. M. Isobe, T. Adachi, and T. Nagatani. Physica A, 336:638–650, 2004.
23. W. J. Yu, R. Chen, L. Y. Dong, and S. Q. Dai. Phys. Rev. E, 72:026112, 2005.
24. Y.Q. Jiang, T. Xiong, S.C. Wong, C.W. Shu, M. Zhang, P. Zhang, and W.H.K. Lam. Acta Mathematica Scientia, 29:1541–1555, 2009.
25. A. Schadschneider and A. Seyfried. In H. J. P. Timmermans, (eds.), Pedestrian Behavior: Models, Data Collection and Applications, pages 27–44. Emerald Group Publishing, 2009.
26. Hermes project. http://www.fz-juelich.de/jsc/hermes.
27. A. Schadschneider, C. Eilhardt, S. Nowak, A. Seyfried. This proceedings.
28. A. Seyfried, B. Steffen, W. Klingsch, and M. Boltes. J. Stat. Mech., 2005:P10002, 2005.
29. A. Seyfried, M. Boltes, J. Kähler, W. Klingsch, A. Portz, T. Rupprecht, A. Schadschneider, B. Steffen, and A. Winkens. Pedestrian and Evacuation Dynamics 2008, pages 145–156. Springer, 2010.
30. J. Zhang, W. Klingsch, A. Schadschneider, and A. Seyfried. J. Stat. Mech., 2011:P06004, 2011.
31. J. Zhang, W. Klingsch, A. Schadschneider, and A. Seyfried. Preprint. J. Stat. Mech., 2012:P02002, 2012.
32. B. Steffen and A. Seyfried. Physica A, 389:1902–1910, 2010.
33. M. Boltes, A. Seyfried, B. Steffen, and A. Schadschneider. In W. F. Klingsch, C. Rogsch, A. Schadschneider, and M. Schreckenberg, (eds.), Pedestrian and Evacuation Dynamics 2008, pages 43–54. Springer Berlin Heidelberg, 2010.
34. A. Kirchner and A. Schadschneider. Physica A, 312:260–276, 2002.
35. A. Kirchner, K. Nishinari, and A. Schadschneider. Phys. Rev. E, 67:056122, 2003.
36. K. Yamori. Psychological Review, 105(3):530–557, 1998.

37. S. P. Hoogendoorn and W. Daamen. In Traffic and Granular Flow 2003, pages 373–382. Springer, 2005.
38. Y. Suma, D. Yanagisawa and K. Nishinari. Physica A, 391:248–263, 2012.
39. T. Kretz, and M. Schreckenberg. Lecture Notes in Computer Science, 4173:712–715. Springer, 2006.
40. A. Schadschneider, C. Eilhardt, S. Nowak, and R. Will. In R. D. Peacock et al. (eds), Pedestrian and Evacuation Dynamics, pages 557–566. Springer, 2011.

A Methodological Approach to Adjustment of Pedestrian Simulations to Live Scenarios: Example of a German Railway Station

Maria Davidich and Gerta Köster

Abstract Pedestrian stream simulations serve to predict the flow of a crowd. Applications range from planning safer buildings, performing risk analysis for public events to evaluating the clever placement of advertisement. The usability of a simulator depends on how well it reproduces real behavior. Unfortunately very little data from live scenarios has been available so far to compare simulations to. Calibration attempts have relied on literature values or, at best, laboratory measurements. This paper is based on live video observations at a major German railway station. We present a methodological approach to extract key data from the videos so that calibration of the simulation tool against live video observations becomes possible. The success of the approach is demonstrated by reproducing the real scenario in a benchmark simulator and comparing the simulation with the live video observations.

1 Introduction

Real time simulations of pedestrian behavior collect live data and calibrate the simulator against the actual scenario. After that predictions on the evolution of the scenario are made. For a useful prediction it is essential that the simulator is capable of reproducing real situations. An accurate reproduction of known data, without being an exact proof, indicates that the simulator may be trusted. This is best tested beforehand by comparing simulation data to data that was extracted from live observations and stored.

M. Davidich (✉)
Siemens AG, Otto-Hahn-Ring 6, 81739 München, Germany
e-mail: maria.davidich@siemens.com

G. Köster
University of Applied Sciences, Lothstr. 64, 80335 München, Germany
e-mail: gerta.koester@hm.edu

V.V. Kozlov et al. (eds.), *Traffic and Granular Flow '11*,
DOI 10.1007/978-3-642-39669-4_16, © Springer-Verlag Berlin Heidelberg 2013

Unfortunately such live data is not readily available and most calibration attempts are still based on literature data or laboratory experiments. They focus on some specific phenomenon, such as the reproduction of the classic fundamental diagram by Weidmann [1] and Schadschneider et al. [2]. This is very helpful and fully justified when one seeks to better understand an isolated phenomenon. However, one cannot expect to reproduce a complete scenario with all the aspects that define this scenario. For this a more holistic attempt is necessary. The authors are not aware of any publications on comprehensive calibration.

In this paper we use data gathered through video cameras installed at a major German railway station to demonstrate step by step, how data collection, data analysis and calibration of simulation against data from live scenario are combined to ensure high quality predictions of pedestrian flows.

The results also shed some more light on to which extent the use of the classic literature values on density-flow relationships compiled in [1] is adequate to calibrate the scenario we investigate. In particular, we look at the free-flow velocities of pedestrians and the fundamental diagram. We finally demonstrate the success of the proposed methodology: We apply it to a one of our sample videos. Then we feed the results into a benchmark pedestrian stream simulator and compare the evolution of the density of the simulation experiment to the measurements in several observation areas.

2 Gathering Data for a Scenario: A German Railway Station

Several cameras filmed the crowd flow in a part of a German railway station from the bird's view. The trajectories of individual pedestrians in time and space must be extracted manually from the videos using a tool that allows to "click" positions on the video. At this point we have finished analyzing two of the videos, each of which has a duration of at least 1.5 min. The number of pedestrians walking on each video is about 400 persons. The area covered on each video includes several platforms and a part of the station's main hall.

All trajectories within the complete area of video observation were extracted and analyzed. Some trajectories are partially obscured from the camera view or the view is distorted by distance. Obscured or distorted trajectories are only used to gather source-target statistics. The detailed analysis of velocities and flows is conducted exclusively on the visible and undistorted parts of the trajectories to keep measurement errors small. Figure 1 shows a schematic picture of an area of observation of a measurement experiment, highlighting the fact that one has to deal with hidden areas.

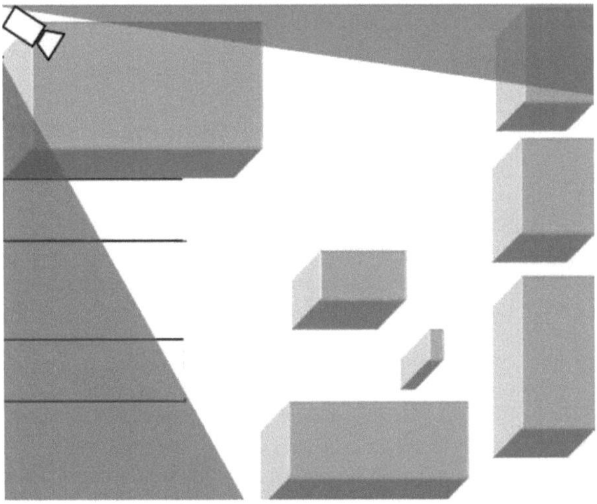

Fig. 1 Schematic representation of the measurement experiment: The *white area* corresponds to the field of vision of a camera mounted at a location in the *upper left* corner of the picture. There are platforms on the *left*, obstacles in the *middle* and a wall on the *right side*

3 The Benchmark Simulation Tool

There are many different models for pedestrian movements, all of them having their own merits. For surveys we refer to [3–7] with descriptions of a large number of approaches for modeling pedestrian movements.

Our own simulation tool is a cellular automaton. The simulator has been described in earlier publications by the authors and by their colleagues, namely in [8]. Details on calibration according to fundamenatal diagrams are given in [9] and the path finding method based on geodesics is described in [10, 11]. We will therefore restrict the description to the minimum necessary to understand the paper.

In a cellular automaton the area of observation is divided in a lattice of cells. Each cell at each time step has a status: either empty or occupied by either a person, or an obstacle, or a source or a target. Virtual persons enter and leave the scenario through sources and targets. The cells are updated by rules which together form the automaton. The cell diameter is usually set to 53 cm to accommodate an average sized European male (Fig. 2).

The core of the model is contained in the automaton, that is, the set of rules according to which the cell states are updated when the simulation steps forward in time. In many aspects, the model is similar to other cellular automaton models based on potentials [2, 12–15].

We imagine that attractive forces act between targets and pedestrians, whereas obstacles and other pedestrians repulse pedestrians. These forces between pedestrians, targets and obstacles are expressed through suitable scalar functions: the

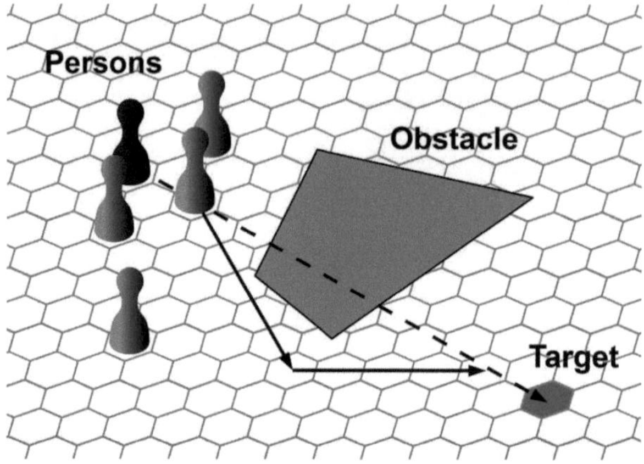

Fig. 2 Pedestrians move on a grid with hexagonal cells towards a target. Persons, targets and obstacles occupy cells. Positions are updated sequentially in each simulation step, so that collision is impossible

potentials. Virtual persons seek to minimize the overall potential when stepping ahead. Obstacles are successfully skirted because the target potential is given through the arrival time of a traveling wave front [10]. A sequential update scheme makes collisions impossible. Furthermore, each person has an individual speed that the person tries to achieve – and indeed does achieve when the path is free: the free-flow velocity. Our model enriches the basic ideas by a number of sub-models to compensate the most relevant shortcomings [8].

4 Methodology: Realistic Calibration According to Video Data

The next step is to adjust the simulations to video data. In order to achieve this, we need to define sets of data and parameters that capture the most important phenomena. Some data, such as the location and form of obstacles, is extracted from the videos and then directly fed into the simulation. We also extract parameter estimates by statistically analysing the data. For some parameters, we may also need to find statistical distributions. This is the case for the free-flow velocity, where literature suggests a normal distribution [1]. In some cases, we may have to measure a dependency instead of a simple parameter, such as the relationship between density and velocity or equivalently density and flow in a crowd.

We propose to extract the following information, which we consider necessary for the reconstruction of live scenarios, from the video footage and, if available, floor plans. We would like to stress that in some scenarios additional input, like the average size of pedestrians, may be necessary or at least beneficial.

- **The topology of the area of interest** in two dimensions. This is direct input data.
- **The positions of sources and targets within a scenario**, that is, the locations where people come from and to where they go to. This is again direct input data.
- **Statistical information on the distribution of trajectories between sources and targets.** This data is necessary to direct virtual pedestrians from sources to targets in a way that fits the scenario.
- **A schedule of pedestrian appearances and disappearances.** How many persons per second appear or disappear at each location? This data is necessary to feed virtual pedestrians into the simulation so that it fits the scenario and to adequately remove them from the simulation.
- **The scenario specific distributions of free-flow velocities**.
  It is very important to use the measured free-flow velocity distribution that is correct for a particular scenario. Only then our simulation will produce consistent and quantitatively correct data: e.g. during the rush hour we observe mostly fit walkers with a clear purpose in mind. They tend to walk fast towards their goals. Only minutes later, when the last commuter train has arrived and its passengers have left the station, we may observe a more relaxed crowd of casually strolling tourists with no particular goal or time schedule in mind.
  At 17:26 p.m. we measured a mean free-flow velocity of 1.04 m/s with a standard deviation of 0.51 m/s at the railway station. This differs significantly from the benchmark data for a Gaussian distribution about a mean of 1.34 m/s with a standard deviation 0.26 m/s in [1]. The histogram and quantile plot of the measured distributiona are depicted in Fig. 3 and in Fig. 4. In the benchmark simulations presented in Sect. 5 we decided to use a normal distribution with the mean and standard deviation from the data including some very slow and very fast pedestrians.
- **Measured data from which the density-flow relationship (fundamental diagram) valid for the current scenario can be derived.** The density-flow relationship observed in the video also deviates from the fundamental diagram provided by Weidmann (see Fig. 5). The method to calibrate the pedestrian simulation against a given density-flow relationship that we use in the benchmark simulator is described in detail in [9].

5 Results

As soon as the steps from Sect. 4 are completed and the adjustments against measured data have been performed, a comparative simulation can be started. We compare simulation results and video measurements for data that was recorded during the rush hour in the afternoon. In the benchmark scenarios a train arrives on a platform (on the left side of Fig. 1), passengers exit the train and move to different destinations, such as the entrances to the subway, food stalls, elevators and other platforms. Accordingly, the density is low at first. Then the bulk of the passengers appear on the video and a higher pedestrian density is measured.

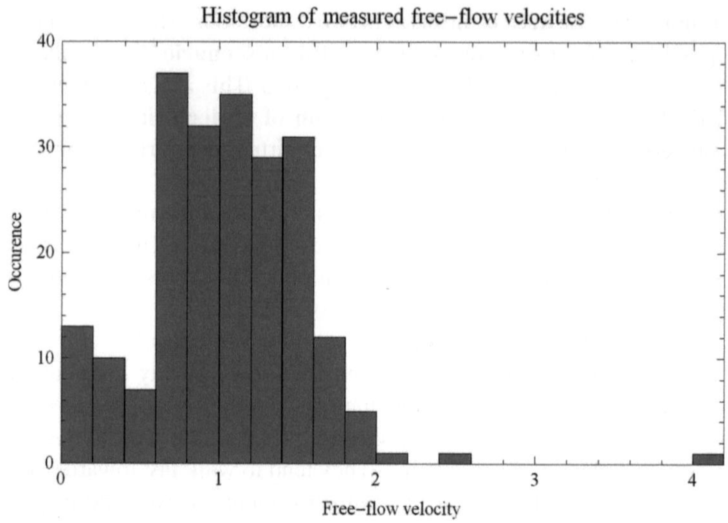

Fig. 3 Histogram of free-flow velocities at 17:26 p.m. at a German railway station. Only the 214 trajectories with a free path in the direction of movement are considered. The mean free-flow velocity is 1.04 m/s and the standard deviation is 0.51 m/s

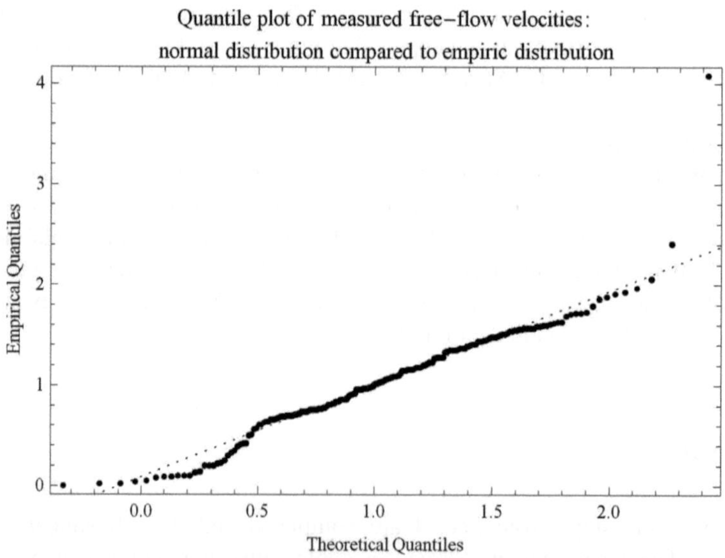

Fig. 4 Quantile plot of free-flow velocities comparing the distribution of the free-flow velocities measured at 17:26 p.m. at a German railway station to a normal distribution. Only the 214 trajectories with a free path in the direction of movement were considered

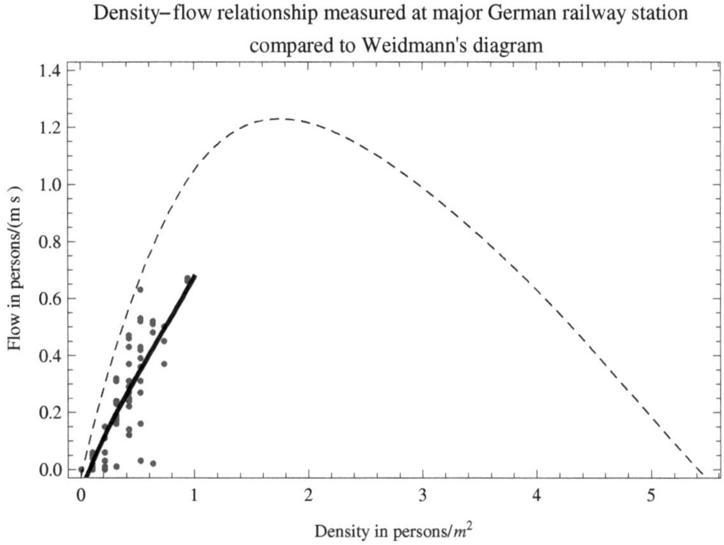

Fig. 5 Measured density-flow relationship at a German railway station at 17:26 p.m. on a workday compared to Weidmann's diagram (*dashed line*). Densities above 1 person/m^2 did not occur. The *solid line* is a smooth approximation of the measured data

Finally, the pedestrian density decreases slowly. After 3 min no more passengers come from the platform. Only pedestrians that come from other trains or entrances are seen in the main hall.

When comparing measurements to simulation results one must find suitable quantities that allow such a comparison. The sources, destinations and speeds of the virtual pedestrians in our simulations match measurements statistically, that is, the distributions coincide. Individually no such match can be expected and, as a consequence, individual trajectories cannot be compared. We need an aggregated quantity instead. We pick the density of the crowd as it evolves with time in an area of observation. The density cannot only be measured quite easily in both cases, but is also of immediate interest, because densities above a certain threshold would be an indicator for impending danger.

Figure 6 shows a comparison of simulated and measured densities in a time span of 3 min. Solid lines correspond to the video footage, dashed lines to the simulation. The prediction from the simulator qualitatively reproduces the scenario quite well, that is, the peak density occurs in the area of observation, at the correct time, for the correct duration and in the correct order of magnitude.

However, the result somewhat overestimates the densities. Part of the differences can be explained by the influence of chance. The simulation is subject to random input as far as the velocities of the virtual pedestrians and their chosen trajectories are concerned. Only the distributions of the input and measurement parameters coincide. Therefore, for each new seed and for each new simulation the results differ and one cannot ever expect a complete match with the measurements.

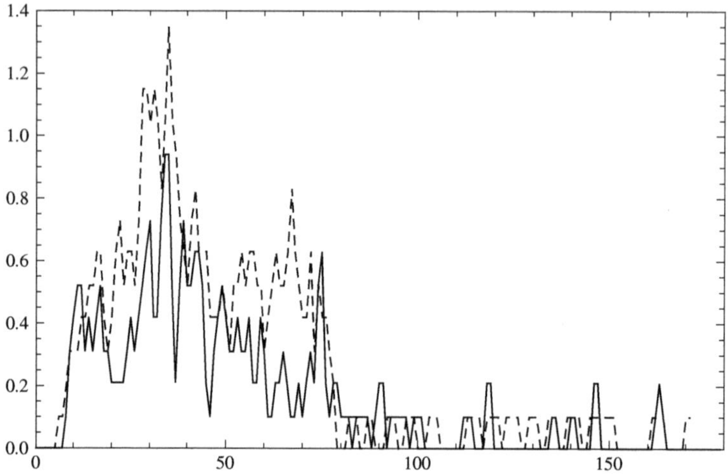

Fig. 6 Comparison of measured densities from video footage (*solid line*) and simulated densities (*dashed lines*) when pedestrians are fed into the scenario with the source target distribution, the free-flow velocities and the density-flow relationship measured at the start of the scenario. One hundred eighty seconds (3 min) are simulated

We also suspect that the real pedestrians coordinate their movements better than the virtual pedestrians. The virtual pedestrians are quite 'short sighted' and take steps to avoid collision only when the actually 'feel' the potential of the other pedestrians. Real pedestrians are more likely to plan ahead. This is a typical disadvantage of high speed pedestrian stream simulators that need to restrict influences to a near field, so-called greedy algorithms, to keep computation times low.

The important question is whether this systematic overestimation is acceptable. In our case, we are interested in a warning system for potentially dangerous densities. Therefore we believe that slight overestimates can be tolerated, whereas underestimations would be inacceptable.

6 Discussion and Next Steps

In this paper, we proposed a methodological approach to adjust pedestrian simulations to live scenarios. We applied the methodology to a complex live scenario at a major German railway station. The most important aspects and characteristics that are taken into account and should be adjusted are: the topology of a scenario, the positions of sources and targets and the statistical distribution of trajectories between sources and targets, the current schedule of pedestrian appearances and disappearances, the current distribution of free-flow velocities and the density-flow relationship. Some of the parameters are direct input parameters, others, like the density-flow relationship, must be used as target functions for parameter adjustments.

The analysis of the data from our live scenario revealed significant differences to known literature data [1], which underlines the importance of scenario specific measurements as input data and calibration to measured relationships, especially if predictive simulations are attempted. The success of the proposed approach has been tested by comparing the density evolution of the simulated data to the measured data in an area of observation. The simulation predicts the density evolution rather well at least qualitatively and also, to some extent, quantitiatively.

References

1. U. Weidmann. *Transporttechnik der Fussgänger*, volume 90. Schriftenreihe des IVT, Zürich, March 1993.
2. A. Schadschneider, W. Klingsch, H. Klüpfel, T. Kretz, C. Rogsch, and A. Seyfried. Evacuation dynamics: Empirical results, modeling and applications. *arXiv:0802.1620*, February 2008. Encyclopedia of Complexity and Systems Science (Editor-in-Chief: R.A. Meyers), pages 3142–3176 (Springer 2009).
3. S. Gwynne, E.R. Galea, M. Owen, P. J. Lawrence, and L.A. Filippidis. A review of the methodologies used in the computer simulation of evacuation from the built environment. *Building and Environment*, 34:842–855, 1999.
4. K. Teknomo. Application of microscopic pedestrian simulation model. *Transportation Research F*, 9:15–27, 2006.
5. Xiaoping Zheng, Tingkuan Zhong, and Mengting Liu. Modeling crowd evacuation of a building based on seven methodological approaches. *Building and Environment*, 44(3):437–445, 2009.
6. Kazuhiro Yamamoto, Satoshi Kokubo, and Katsuhiro Nishinari. Simulation for pedestrian dynamics by real-coded cellular automata (RCA). *Physica A: Statistical Mechanics and its Applications*, 379(2):654–660, June 2007.
7. Miho Asano, Takamasa Iryo, and Masao Kuwahara. Microscopic pedestrian simulation model combined with a tactical model for route choice behaviour. *Transportation Research Part C: Emerging Technologies*, 18(6):842–855, December 2010.
8. G. Köster, D. Hartmann, and W. Klein. Microscopic pedestrian simulations: From passenger exchange times to regional evacuation. In *Operations Research Proceedings 2010*, Munich, Germany, 2010.
9. M. Davidich and G. Köster. Towards automatic and robust adjustment of human behavioral parameters in a pedestrian stream model to measured data. In R. D. Peacock, E. D. Kuligowski, and J. D. Averill, editors, *Pedestrian and Evacuation Dynamics*, Gaithersburg, MD, USA, 2010.
10. D. Hartmann. Adaptive pedestrian dynamics based on geodesics. *New Journal of Physics*, 12(4):043032, April 2010.
11. G. Köster, M. Seitz, F. Treml, D. Hartmann, and W. Klein. On modelling the influence of group formations in a crowd. *accepted: Contemporary Social Sciences, Special Issue: Crowd in the 21st Century*, 2012.
12. H. L. Klüpfel. *A Cellular Automaton Model for Crowd Movement and Egress Simulation*. PhD thesis, 2003.
13. C. Burstedde, K. Klauck, A. Schadschneider, and J. Zittartz. Simulation of pedestrian dynamics using a two-dimensional cellular automaton. *Physica A*, 295:507–525, 2001.

14. H. W. Hamacher and S. A. Tjandra. Mathematical modelling of evacuation problems: A state of art. Technical Report 24, Fraunhofer-Institut für Techno- und Wirtschaftsmathematik ITWM, Kaiserslautern, 2001.
15. C. Kinkeldey and M. Rose. Fußgängersimulation auf der basis rechteckiger zellularer Automaten. In K Kaapke and A. Wulf, editors, *Forum Bauinformatik 2003: Junge Wissenschaftler forschen*. Aachen, 2003.

Optimal Density in a Queue with Starting-Wave

Akiyasu Tomoeda, Daichi Yanagisawa, Takashi Imamura, and Katsuhiro Nishinari

Abstract The propagation speed of people's reaction in a relaxation process of a queue, so-called *starting-wave*, has an essential role for pedestrians and vehicles to achieve smooth movement. For example, a queue of vehicles with appropriate headway (density) alleviates the traffic jams, since the delay of reaction to start is minimized. In the previous study (Tomoeda et al., Fifth international conference on pedestrian and evacuation dynamics. Springer), it was found that the fundamental relation between the propagation speed of starting-wave and density is well approximated by the power law function. We have revealed the existence of optimal density, where the travel time of last pedestrian in a queue with the starting-wave to pass the head position of the initial queue is minimized. This optimal density inevitably plays a significant role to achieve smooth movement of crowds.

A. Tomoeda (✉)
Meiji Institute for Advanced Study of Mathematical Sciences, Meiji University, Nakano-ku, Tokyo, Japan

CREST, Japan Science and Technology Agency, Nakano-ku, Tokyo, Japan
e-mail: atom@isc.meiji.ac.jp

D. Yanagisawa
College of Science, Ibaraki University, Bunkyo, Mito, Ibaraki, Japan

T. Imamura
Research Center for Advanced Science and Technology, The University of Tokyo, Komaba, Meguro-ku, Tokyo 153-8904, Japan

K. Nishinari
Research Center for Advanced Science and Technology, The University of Tokyo, 4-6-1, Komaba, Meguro-ku, Tokyo, 153-8904, Japan

V.V. Kozlov et al. (eds.), *Traffic and Granular Flow '11*,
DOI 10.1007/978-3-642-39669-4_17, © Springer-Verlag Berlin Heidelberg 2013

1 Introduction

Various kinds of self-driven many-particles (SDP) systems, such as pedestrian dynamics, vehicular traffic and traffic phenomena in biology have attracted a great deal of attention in a wide range of fields during the last few decades [2–4]. Most of these complex systems are interesting not only from the point of view of natural sciences for fundamental understanding of how nature works but also from the points of view of applied sciences and engineering for the potential practical use of the results of the investigations. Especially, the interdisciplinary investigations for the dynamics of jamming phenomena in SDP systems, so-called *Jamology*, have been progressed by developing sophisticated mathematical models considered as a system of interacting particles driven far-from equilibrium [5–11]. These contributions to analyze the mechanism of jamming formation tell us that one of the most important factor to cause the jamming phenomena is a *sensitivity*, which indicates the time delay of reaction of particles to the stimulus. As an example, if the reaction of drivers are extremely sensitive, they can avoid the traffic jam by adjusting their behavior immediately to their front car's movement. The reaction time of pedestrians is similarly important toward smooth movement of crowd. Moreover, we would like to point out that the wave of successive reaction in a queue, so-called *starting-wave*, plays a significant role for the waiting time in a queuing system of pedestrians and vehicles, since quick-start in walking accomplishes the more smooth movement of crowds and jams. In this paper we have investigated the optimal density to minimize the travel time of last pedestrians in a queue to pass the head position of the initial queue by taking into account the propagation speed of pedestrians' reaction, which is characterized by the power law [1].

This paper is organized as follows. In the next section, let us introduce the fundamental relation between propagation speed of starting-wave and density. In Sect. 3, the existence of optimal density which minimizes the travel time in a line and the effect of the partition number of groups are shown by analytical calculation. Moreover, we show the existence of optimal density by actual experiments of pedestrians. Finally, Sect. 4 is devoted to the concluding discussions.

2 Propagation Speed of Starting-Wave

In this section, let us introduce the fundamental relation between propagation speed of starting-wave and density obtained from both numerical simulations and experimental measurements.

Our mathematical model is built on the stochastic cellular automaton, which recently prevails to modeling the complex systems [12]. Let us imagine that the passage is partitioned into $L \in \mathbb{N}$ identical cells that each cell can accommodate at most one particle (pedestrian) at a time. Note that, in the following, we refer to "particle" as a representation of a pedestrian in a model and "pedestrian" as a person itself. The length of each cell corresponds to 0.5 m by considering the

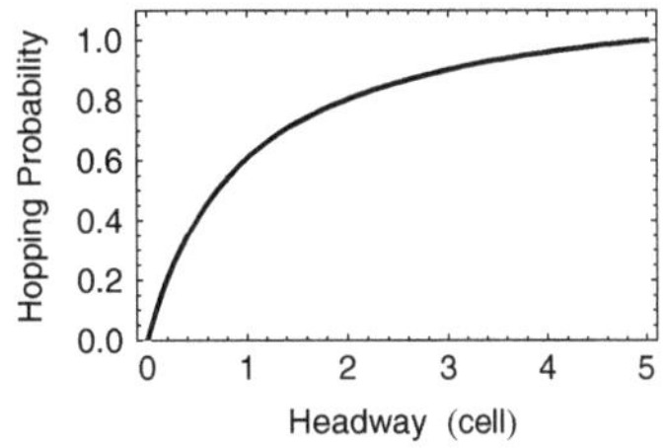

Fig. 1 Optimal Velocity function for walking model

reasonable volume exclusion effect of pedestrians. Moreover, a total number of N $(2 < N \leq L)$ indicates the particles which are placed at equal distance $\bar{h}$ cell. In our model, the parameters $(N, L, \bar{h})$ satisfies the equally-spaced condition as follows:

$$(N-1)(\bar{h}+1)+1 = L. \quad (1)$$

The update rules of our cellular automaton model are as follows: first of all, only the particle at the right boundary (head of a queue) moves forward. Then the following particle (only second particle in the queue) can also move forward. After the second particle moves forward, the next particle can also start to move in sequence. These rules of pedestrians' walking are applied in *parallel* to all particles. Note that, in this study, in order to investigate the propagation speed of successive reactions, all of the following particles cannot move forward before the starting-wave reaches to themselves. Therefore, unlike the usual stochastic cellular automaton model such as ASEP [13–15], ZRP [16–18], in our model, only if the next cell is empty and predecessor had already moved, following particles can move forward with probability $p(h)$ which depends on their headway distance h. This hopping probability of particles $p(h)$, which indicates the velocity of particles, is given in analogy with the idea of Optimal Velocity (OV) function, which is often introduced into the mathematical model for vehicular traffic as a desired velocity of drivers depending on headway distance (or density) [5, 19]. This function is motivated by the common expectation that drivers have their desired velocity and adjust to their suitable driving behavior. That is, the velocity must be reduced and become small enough to prevent crashing into the preceding particle when the headway becomes short. Whereas, when the headway becomes longer, particles can move with their maximum velocity, which corresponds to the legal velocity in vehicular traffic.

We approximate the experimental data of pedestrians' walking on a circular passage way [20] as a linear type function for the simplicity, which is widely used in the traffic flow modeling built on the fluid dynamics [19, 21]. The constant value $V_{\max}$ and $\rho_{\max}$ indicate the maximum velocity of particles and the density at complete stand-still, respectively. This linear relation gives us the OV function by translating the headway distance into the reciprocal of the density as $h \sim 1/\rho$. In order to apply the OV function obtained from the experimental data to the stochastic cellular automaton model, the OV function is scaled so that the function from the origin and the free hopping probability satisfies $p = 1$, if $h = 5$, which is based on the maximum headway (5 cells) in our experiment as shown in Fig. 1.

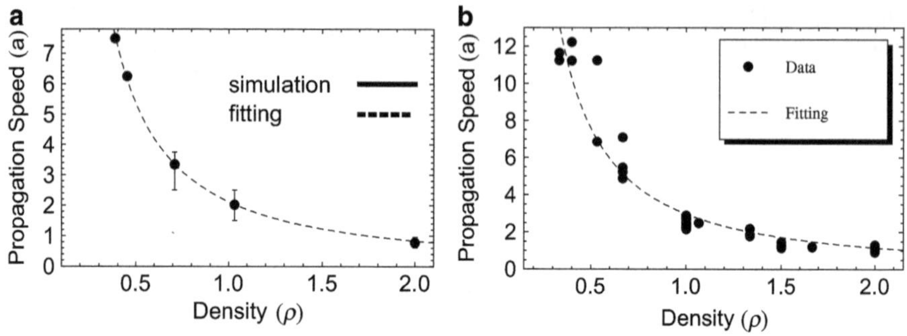

Fig. 2 Simulation results (**a**) and experimental measurements (**b**) between propagation speed of starting-wave and density. The *dashed curve* shows the fitting function based on the form (2)

On the other hand, in our experiments, we have measured the propagation speed of starting-wave under several densities which are decided by the initial number of pedestrians along a line. At first, we make a long straight passage ($L \in \{5, 10, 15, 20\}$ m) and put marks with a distance of 0.5 m between them. As an initial condition of pedestrians, all pedestrians stand in line with same headway distance. After that, the leader of queue starts to walk by the cue and then we measure the elapsed time until the last pedestrian starts to walk. Thus, we have obtained the propagation speed of successive reaction which is derived from the length of initial queue divided by the elapsed time under each given density.

As a significant result of both numerical simulations and experimental measurements, we have found the power law in the relation between propagation speed of starting-wave and the initial density of pedestrians as shown in Fig. 2.Taking into account this power law, we have assumed the simple relation between propagation speed and density of pedestrians in analogy with the sonic speed of aerodynamics as the form

$$a(\rho) = \alpha \rho^{-\beta} \tag{2}$$

where ρ and a are the density and the propagation speed of starting-wave, relatively. α and β indicate positive parameters. By fitting this simulation result, we have obtained the parameter values $\beta \sim 4/3$, which shows the good agreement with the experimental measurements.

3 Optimization of Initial Distribution

Now, let us apply this fundamental relation (2) to the optimization of initial distribution for a long queue, for example, teemingly number of the athletes in marathon. This subject is quite important for organizers to achieve the smooth movement, since the athletes located rearward position have some unavoidable delay to pass the head of the initial queue. Moreover, if there is an optimal density to minimize the delay time, the control of density reduces the waste of waiting time in a queue. If athletes stand in line with large headway (low density), the starting-wave propagates fast, but the queue becomes long. Whereas, if they stand in line with small headway (high density), the starting-wave propagates slowly, but the queue becomes short. Which situation decreases the delay to pass the start line? This problem which includes the effect of starting-wave has not been investigated. Thus, the optimal initial distribution for a queue is investigated here.

We set the problem as follows:

Which density minimizes the required time for the last to pass the head position of the initial queue?

We generally set N as the total number of pedestrians. T_n and L indicate the travel time T of n-th pedestrian and the length of initial queue, respectively. The initial equally-spaced density ρ_0 is calculated by $\rho_0 = N/L$.

3.1 Existence of Minimum T_n

The travel time T_n for given initial density ρ_0 is calculated as

$$T_n = \frac{L}{a(\rho_0)} + \frac{L}{v(\rho)}. \tag{3}$$

First term and second term of right-hand-side in (3) indicate the required time to start walking and the one to pass the head position of the initial queue after start walking, respectively. Substituting the relation $L = N/\rho_0$, (3) is translated into

$$T_n = N\Big(\frac{1}{\rho_0 a(\rho_0)} + \frac{1}{\rho_0 v(\rho)}\Big). \tag{4}$$

If the velocity $v(\rho)$ is constant v, (4) reduces to

$$T_n = N\Big(\frac{1}{\rho_0 \alpha \rho_0^{-\beta}} + \frac{1}{\rho_0 v}\Big), \tag{5}$$

by substituting the relation (2). This relation (5) is shown in Fig. 3.From this figure, we have found the existence of optimal density which minimizes the travel time in the case of constant walking velocity. Moreover, the density which satisfies the extreme value $dT_n/d\rho_0 = 0$ is calculated by

$$\rho_0^{\beta} = \frac{\alpha}{v(\beta - 1)}. \tag{6}$$

In order to exist the value of ρ, β satisfies $\beta > 1$, since the parameter α, β, and v are all positive value. As noted before, the value of $\beta \sim 4/3$ is larger than 1.

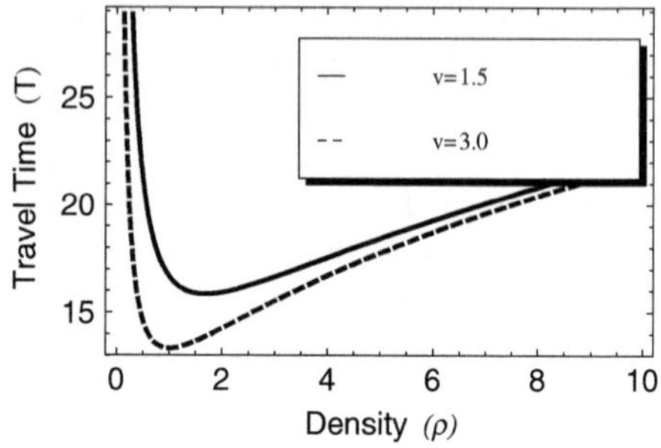

Fig. 3 Numerical plots of (4) against the constant velocity $v = 1.5, 3.0$

Therefore, it is found that an optimal density does exist for given parameter α, β and v. Note that, as shown in (6), the optimal density becomes lower as the velocity becomes faster.

3.2 *Effect of Divided Groups*

We consider the several divided groups which are often introduced to avoid the large crowd for their safety. For example, *Wave Starts* by several groups is prevailed in New York City Marathon to alleviate congestion at the start.

As an common expectation, the best movement for these divided groups is to connect smoothly with each other, that is, the leader of following group catches up with the last of former group without loss of the velocity. Under this expectation, therefore, one satisfies the condition

$$\frac{\frac{L-(m-1)d}{m}}{a(\rho_0)} = \frac{d}{V(\rho)}, \tag{7}$$

where L, m and d are corresponds to the total length, the number of partitions for groups and the distance between successive groups, respectively. ρ_0 corresponds to the initial density in a group and is same for all groups. Once the density is given, the propagation speed of starting-wave is constant value $a(\rho_0) = const.$

Here, we assume that the leader of the group walks with constant velocity $\bar{V}$. The required time T for the last of the following group to pass the head of queue is given by

$$T = \frac{\frac{L-(m-1)d}{m}}{a(\rho_0)} + \frac{L}{\bar{V}} = \frac{d+L}{\bar{V}}. \tag{8}$$

The message we receive from this equation is the travel time T does not depend on the number of partitions m under above assumptions. Thus, it is enough to consider the optimization in only one group to minimize the travel time.

Table 1 The travel time data for each density from experiments. The density is measured in pedestrians per a meter

Density ρ	1.0	2.0	3.0	4.0
Travel time T	33.6	26.2	27.8	27.3

3.3 Experimental Results

Here, we measure a required time after the leader starts to walk by the cue and before the last pedestrian passes the head position of the initial queue for four situations; $\rho = 1.0, 2.0, 3.0$, and 4.0.

The experimental results of the travel time T are shown in Table 1. We have found that the optimal initial density, which minimizes the travel time, does exist at density $\rho = 2.0$. If the density is small, the pedestrian moves longer distance, since the distance to the head position of queue becomes large. Whereas, if the distance to the head position of queue is small, it is difficult for the pedestrian to move without perspectives, since the density becomes large. In other words, longer queue or higher density causes the lost time for smooth start. It is common expectation that there is an optimal density to minimize the travel time, since both opposite extreme situations make travel time longer. We have verified this expectation by both analytical calculations built on mathematical modeling and substantiative experiments.

4 Conclusion

In this contribution, we have investigated the propagation speed of pedestrians' reaction in relaxation process of a queue which is called starting-wave. The faster the starting-wave propagates, the more smooth the crowd moves.

We have revealed the existence of optimal density, where the travel time of last pedestrians to pass the start line for the initial queue is minimized by both analytical calculations and experiments. If the walking velocity is very slow, the value of optimal density is detached from reality. However, the optimal density is suitable to apply to the real situation, for example, a queue of athletes starts to move in marathon, since the optimal density becomes lower as the walking velocity increases. This optimal density inevitably plays a significant role to design not only the initial queue of pedestrians but also the traffic intersections and signals.

Acknowledgements We thank Kozo Keikaku Engineering Inc. in Japan for the assistance of the experiments. The author (AT) is supported by the Meiji University Global COE Program "Formation and Development of Mathematical Sciences Based on Modeling and Analysis". We acknowledge the support of Japan Society for the Promotion of Science and Japan Science and Technology Agency.

References

1. A. Tomoeda, D. Yanagisawa and K. Nishinari, Fifth International Conference on Pedestrian and Evacuation Dynamics, Springer.
2. D. Chowdhury, L. Santen and A. Schadschneider (2000), Phys. Rep. **329**, 199.
3. D. Helbing (2001), Rev. Mod. Phys. **73**, 1067.
4. A. Schadschneider, D. Chowdhury and K. Nishinari (2010), *StochasticC Transport in Complex Systems: From Molecules to Vehicles*, Elsevier.
5. M. Bando, et al. (1995), Phys. Rev. E, **51**, 1035.
6. B. S. Kerner and P. Konhauser (1993), Phys. Rev. E, **48**, R2335.
7. M. Kanai, et al. (2005), Phys. Rev. E **72**, 035102.
8. D. Yanagisawa, et al. (2009), Phys. Rev. E, **80**, 036110.
9. A. John, et al. (2009), Phys. Rev. Lett., **102**, 108001.
10. K. Nishinari, et al. (2005), Phys. Rev. Lett. **95**, 118101.
11. A. Tomoeda, et al. (2007), Physica A, **384**, 600.
12. S. Wolfram (1984), Nature **311**, 419.
13. B. Derrida (1998), Phys. Rep. **301**, 65.
14. N. Rajewsky, L. Santen, A. Schadschneider and M. Schreckenberg (1998), J. Stat. Phys. **92**, 151.
15. G. M. Schütz, in *Phase Transitions and Critical Phenomena*, vol. 19 (Acad. Press, 2001), G. M. Schütz (2003), J. Phys. A: Math. Gen. **36** R339.
16. F. Spitzer (1970), Adv. Math. **5**, 246.
17. M. R. Evans and T. Hanney (2005), J. Phys. A: Math. Gen. **38**, R195.
18. M. Kanai (2007), J. Phys. A **40**, 7127.
19. B. D. Greenshields (1935), in *Proceedings of the Highway Research Board*, Washington, D. C., **14**, 448.
20. A. Seyfried, et al. (2005), J. Stat. Mech., 10002.
21. H. J. Payne (1971), in *Mathematical Models of Public Systems*, edited by G. A. Bekey, 51.

Arch-Shaped Equilibrium Solutions in Social Force Model

Tsubasa Masui, Akiyasu Tomoeda, Mayuko Iwamoto, and Daishin Ueyama

Abstract In the present paper, we investigate arch-shaped equilibrium solutions in the social force model proposed by Helbing and Molnar (Phys Rev E 51:4282, 1995) and Helbing et al. (Nature 407:487, 2000). The social force model is a system of ordinary differential equations, which describe the motion of the pedestrians under a panic situation. In the simulation of the social force model, we observe an intermittent appearance of arch-shaped structures (i.e. the "Blocking clusters" Parisi and Dorso, Physica A 354:606, 2005; Physica A 385:343, 2007; Frank and Dorso, Physica A 390:2135, 2011) around an exit which block up the flow of pedestrians. To understand such a dynamic behavior, we study arch-shaped equilibrium solutions around an exit under the simplest configuration.

T. Masui (✉)
Graduate School of Science and Technology, Meiji University, 1-1-1 Higashi Mita, Tama-ku, Kawasaki, Kanagawa, 214-8571, Japan
e-mail: ce06209@gmail.com

A. Tomoeda
Meiji Institute for Advanced Study of Mathematical Sciences, Meiji University, Nakano-ku, Tokyo, Japan

CREST, Japan Science and Technology Agency, Nakano-ku, Tokyo, Japan

M. Iwamoto
Graduate School of Advanced Mathematical Sciences, Meiji University, Nakano-ku, Tokyo, Japan

D. Ueyama
Meiji Institute for Advanced Study of Mathematical Sciences, Meiji University, Nakano-ku, Tokyo, Japan

Graduate School of Advanced Mathematical Sciences, Meiji University, Nakano-ku, Tokyo, Japan

V.V. Kozlov et al. (eds.), *Traffic and Granular Flow '11*,
DOI 10.1007/978-3-642-39669-4_18, © Springer-Verlag Berlin Heidelberg 2013

1 Introduction

Investigations of pedestrian dynamics have been attractive not only for engineers but also for scientists over recent decades, since it shows the large variety of interesting collective effects and self-organization phenomena that can be observed [6–8]. Especially, understanding of the characteristics of pedestrian behaviors is important and helpful to avoid the jamming state of panicking pedestrians in emergency escape situations [9–12]. This jamming phenomenon occurs due to an arching effect when many pedestrians rush through an exit to leave a room at the same time and often it brings on tragic disasters [13].

Sometimes, the kinds of crowd stampede are triggered in life-threatening situations. The evacuation dynamics from a room is one of the important subjects for our safety in the design of buildings. It is well known that there are blocking effects around an exit under a panic situation. Several models can reproduce blocking effects of pedestrians around an exit. The correlation of the "Blocking clusters" due to the blocking effect and the "Faster is slower effect", which states that the faster the pedestrians wish to move the slower they can escape from the room under a panic situation [3–5]. Here, in order to achieve a smooth flow around an exit by resolving the blocking effect, the "Blocking clusters" is investigated from a perspective of the equilibrium solutions in the social force model.

2 The Equilibrium Arch-Shaped Solutions

Let us consider the social force model proposed by Helbing and Molnar [1] and Helbing et al. [2]. The model equations are the following,

$$m_i \frac{d\mathbf{v}_i(t)}{dt} = m_i \frac{v_i^0 \mathbf{e}_i(t) - \mathbf{v}_i(t)}{\tau_i} + \sum_{j(\neq i)} \mathbf{f}_{ij} + \sum_{w} \mathbf{f}_{iw}, \tag{1}$$

where m_i is the mass of i-th particle,[1] $\mathbf{v}_i$ and $v_i^0\mathbf{e}_i$ are the current and the desired velocities, respectively. $\mathbf{e}_i$ is the unit vector pointing to the desired target (particles inside the room have their targets located at the closest position over the line of the exit) and τ_i is a constant related to the relaxation time of the particle to achieve v_0. In the model equation, $\sum_{j(\neq i)} \mathbf{f}_{ij}$ and $\sum_w \mathbf{f}_{iw}$ are respectively the sum of the forces from the other particles and the walls, which come from the psychological force, the physical repulsion force and the sliding friction force. The functions to describe these forces have been developed in [1, 2]. In the simulation of (1), we observe arch-shaped structures around an exit intermittently as in Fig. 1. These arch-shaped

[1] "Particle" is used to represent "pedestrian" in the model.

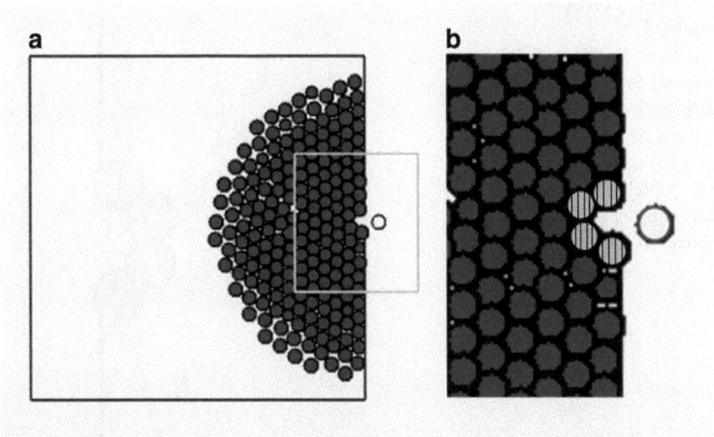

Fig. 1 (**a**) An arch-shaped structure in the simulation of the social force model [1, 2]. In the case of high desired velocity, such arch-shaped structures appear intermittently and cause the "Faster is slower effects" [3–5]. (**b**) Magnification of (**a**). Four particles (*striped circle*) around an exit form an arch-shaped structure. The *white circle* indicates the particle who has just escaped from the room

structures, so-called the "Blocking clusters", are made in a self-organized manner and they block up the flow of the particles for a moment [3–5].

Such a blocking effect is obviously undesirable for an escape situation, therefore, we focus on the arch-shaped structures around an exit in this contribution. From the mathematical viewpoint, the arch-shaped structure can be assumed as an equilibrium solution of the system. This equilibrium solution of the system satisfies the following equation,

$$m_i \frac{v_i^0 \mathbf{e}_i}{\tau_i} + \sum_{j(\neq i)} \mathbf{f}_{ij} + \sum_{w} \mathbf{f}_{iw} = 0. \tag{2}$$

For the equilibrium situation, the sliding friction can be negligible. However, it is still hard to obtain the equilibrium solution of the system due to its high degree of freedom, i.e. 200 particles in the simulation of Fig. 1. In the first step of our investigations, we develop the simplest configuration for the arch-shaped solution; say the system made from a few particles. The particles are split up into two groups, i.e. the particles in the arch-shaped structure (Arch) and the outside of the arch-shaped structure (Out). In Fig. 1b, (Arch) particles correspond to the striped circles and (Out) particles correspond to the gray filled circles. As we have mentioned earlier, let us assume this situation as the steady state here, since all particles nearly remain still due to the arch-shaped structures. Moreover, we have found that the symmetric arch consists of four particles which appears frequently in our simulations. Then, we consider the symmetric arch for simplicity. The external forces from (Out) particles acting on the arch are approximated by introducing a

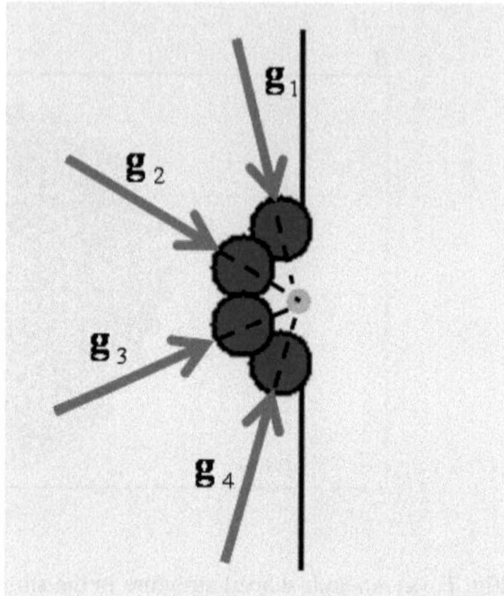

Fig. 2 Schematic description of the simplified configuration (3). In this figure, there are four particles that make an arch-shaped structure. Each particle in the arch-shaped structure gets the force $\mathbf{g}_i$, which represents the force from other particles in the original configuration

function $\mathbf{g}_i = g(n)\mathbf{e}_i$, which is a sum of the forces from (Out) particles to the i-th particle of (Arch) and n is a number of (Out) particles.

Thus, the equation for i-th particle in the arch-shaped structure (Arch) is

$$m\frac{v^0\mathbf{e}_i}{\tau} + \sum_{j(\neq i)} \mathbf{f}_{ij} + \sum_{w} \mathbf{f}_{iw} + \mathbf{g}_i = 0. \tag{3}$$

Here, we assume all particles have same size (0.3 m in radius), weight (80 kg), relaxation time(0.5 s) and the desired velocity (4.9 m/s). As in Fig. 2, we assume the force from (Out) particles direct to the center of the exit. The function $g(n)$ is obtained by the numerical simulation of full system (1) for a given number of the particles where four particles around the exit are pined arch-shaped form. It is fitted by square root function to the numerical data for a present study as follows,

$$g(n) = 800\left(\frac{\sqrt{0.18n}}{0.6} - 1\right), \tag{4}$$

where n is a number of the particles. As shown in Fig. 3, (4) shows a good agreement with the numerical results, which is estimated by the numerical simulations based on the full system (1).

Further decreasing the degree of freedom, we assume symmetry of the motion of particles to the x-axis. Then, we employ Mathematica's *NSolve* function to obtain arch-shaped equilibrium solutions numerically.

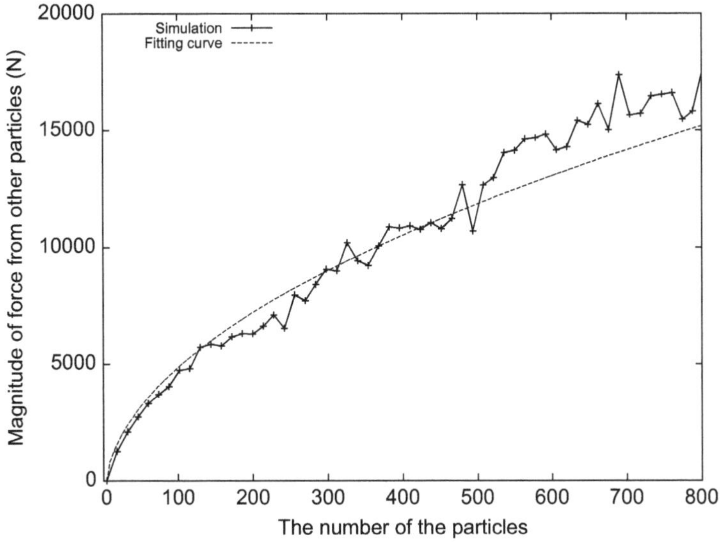

Fig. 3 Magnitude of the force from the other particles to arch-shaped particles. *Horizontal axis* denotes the number of the particles and the *vertical axis* describes the magnitude of the force. A *dashed curve* is square root function fit to numerical data

3 Results

The examples of the obtained arch-shaped equilibrium solutions are shown in Fig. 4. Here we take a number of the particles in 200. Then $|\mathbf{g}_i| = g(200)$ is used in (3). Fig. 4a represents one of the arch-shaped equilibrium solutions of (3) for 1.2 m exit size. Fig. 4b is one of the solutions for 1.8 m exit size. It is noted that the shape of the arch-shaped structures depends on the parameters. The numerical results show that those arch-shaped equilibrium solutions of (3) are not stable.

4 Summary

In this paper, we obtained the equilibrium arch-shaped solution of the social force model under a very simplified configuration. However, there are ongoing topics on the relationship between the arch-shaped equilibrium solutions of (3) and the one in (2). We know that there is a difference between them, especially the assumption for $\mathbf{g}_i$ is not appropriate, say $\mathbf{g}_i$ is not directed to the center of the exit in a general configuration.

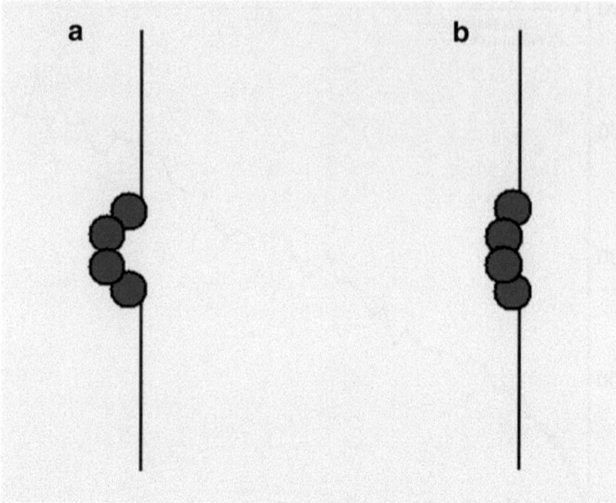

Fig. 4 The arch-shaped equilibrium solutions for 200 particles for different width of the exit. (**a**) 1.2 m (**b**) 1.8 m

Acknowledgements We are grateful to the Meiji University Global COE Program "Formation and Development of Mathematical Sciences Based on Modeling and Analysis" for the support. The author (AT) would like to thank Japan Science and Technology Agency for the support.

References

1. D. Helbing, P. Molnar, Social force model for pedestrian dynamics, Phys. Rev. E, 51 (1995) 4282.
2. D. Helbing, I. Farkas, T. Vicsek, Simulating Dynamical Features of Escape Panic, Nature 407 (2000) 487.
3. D. R. Parisi, C. O. Dorso, Microscopic dynamics of pedestrian evacuation, Physica A, 354 (2005) 606.
4. D. R. Parisi, C. O. Dorso, Morphological and dynamical aspects of the room evacuation process, Physica A, 385 (2007) 343.
5. G. A. Frank, C. O. Dorso, Room evacuation in the presence of an obstacle, Physica A, 390 (2011) 2135.
6. D. Helbing, Traffic and related self-driven many-particle systems, Rev. Mod. Phys., 73 (2001) 1067.
7. M. Schreckenberg, S. D. Sharma, Pedestrian and Evacuation Dynamics, Springer, Berlin (2002).
8. A. Schadschneider, D. Chowdhury, K. Nishinari, Stochastic Transport in Complex Systems: From Molecules to Vehicles, Elsevier, (2010).
9. D. Yanagisawa, K. Nishinari, Mean-field theory for pedestrian outflow through an exit, Phys. Rev. E, 76 (2007) 061117.
10. D. Yanagisawa, A. Kimura, A. Tomoeda, R. Nishi, Y. Suma, K. Ohtsuka, K. Nishinari, Introduction of frictional and turning function for pedestrian outflow with an obstacle, Phys. Rev. E, 80 (2009) 036110.

11. T. Kretz, A. Grunebohm, M. Schreckenberg, Experimental study of pedestrian flow through a bottleneck, J. Stat. Mech. (2006) P10014.
12. A. Seyfried, O. Passon, B. Steffen, M. Boltes, T. Rupprecht, W. Klingsch, New insights into pedestrian flow through bottlenecks, Transp. Science, 43 (2009) 395.
13. R. A. Smith J. E. Dickie, Engineering for Crowd Safety, Elsevier, Amsterdam (1993).

Effect of Rhythm on Pedestrian Flow

Daichi Yanagisawa, Akiyasu Tomoeda, and Katsuhiro Nishinari

Abstract We have calculated a fundamental diagram of pedestrians by dividing the velocity term into two parts, length of stride and pace of walking (number of steps per unit time). In spite of the simplicity of the calculation, our fundamental diagram reproduces that of traffic and pedestrian dynamics models in special cases. Theoretical analysis on pace indicates that rhythm which is *slower* than normal walking pace in free-flow situation increases flow if the fundamental diagram of flow is convex downward in high-density regime. In order to verify this result, we have performed the experiment by real pedestrians and observed improvement of pedestrian flow in congested situation by slow rhythm. Since slow rhythm achieves large pedestrian flow without dangerous haste, it may be a safety solution to attain smooth movement of pedestrians in congested situation.

D. Yanagisawa (✉)
College of Science, Ibaraki University, Bunkyo, Mito, Ibaraki, Japan
e-mail: daichi@mx.ibaraki.ac.jp; http://yana.sci.ibaraki.ac.jp/

A. Tomoeda
Meiji Institute for Advanced Study of Mathematical Sciences, Meiji University, Nakano-ku, Tokyo, Japan

CREST, Japan Science and Technology Agency, Nakano-ku, Tokyo, Japan
e-mail: atom@isc.meiji.ac.jp; http://dow.mydns.jp/

K. Nishinari
Research Center for Advanced Science and Technology, The University of Tokyo, 4-6-1, Komaba, Meguro-ku, Tokyo, 153-8904, Japan
e-mail: tknishi@mail.ecc.u-tokyo.ac.jp; http://park.itc.u-tokyo.ac.jp/tknishi/

V.V. Kozlov et al. (eds.), *Traffic and Granular Flow '11*,
DOI 10.1007/978-3-642-39669-4_19, © Springer-Verlag Berlin Heidelberg 2013

1 Introduction

Pedestrian dynamics has been studied vigorously over last decades [1, 2], and many collective phenomena typified by arching and lane formation are analyzed by macroscopic models as well as the microscopic models such as the social force model [3], the lattice gas model [4], and the floor field model [5]. Fundamental diagram, which depicts the relation between density and flow (or velocity) is important to know the characteristics of pedestrian flow, so that it is also investigated actively by both theoretical [6, 7] and experimental [8, 9] approaches. Experimental and observational data have been accumulated through the coordinated efforts of the researchers at the ped-net.org homepage [10]. Due to the variety of the experimental and observational conditions, the fundamental diagrams also vary; however, there is a common feature accepted widely to the researchers. As density increases velocity decreases, whereas flow increases until it reaches maximum and then decreases.

Important goals of pedestrian-dynamics research along with elucidation of collective phenomena are development of solutions to ease congestion and contribution to the safety [11]. As the fundamental diagrams indicate, congestion occurs at high-density regime, and pedestrians are exposed to stressful situation, i.e., difficulty to move in the desired direction and conflicts with others. Therefore, we will develop a method to increase pedestrian flow in the congested situation in this paper.

We obtain the fundamental diagram by considering the effect of length of stride and pace of walking (number of steps per unit time) instead of microscopic interaction between pedestrians. It reproduces the fundamental diagrams of basic models for traffic and pedestrian dynamics, and indicates that when pedestrians walk in a constant pace irrespective to the density, the flow increases in the high-density regime even if the pace is *slower* than that in the normal-walking condition. We have performed the real experiment and verified that the pedestrian flow increases when their walking pace is controlled by slow rhythm.

This paper is organized as follows. In the next section, we simply calculate the fundamental diagram and discuss the relation between the existing models. The effect of pace on the pedestrian flow is also studied. Increase of pedestrian flow is theoretically predicted in Sect. 3 and it is verified by real experiments in Sect. 4. The final section is devoted to summary and proposition of future works.

2 Simple Theoretical Fundamental Diagram

2.1 Introduction of Stride Function

Let us start calculating the fundamental diagram theoretically without considering complex microscopic interaction. We consider one-dimensional periodic circuit, whose length is L, as in Fig. 1. The width of the circuit is as wide as that of one pedestrian, so that pedestrians walk in a line without overtaking. Most fundamental

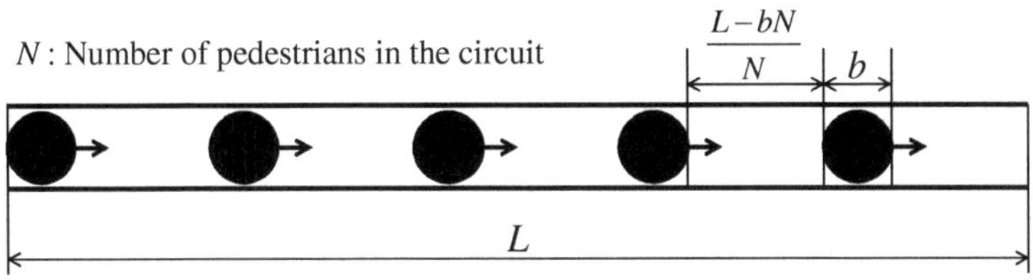

Fig. 1 Schematic view of the situation considered in our calculation. N pedestrians, which are depicted as *black* particles, walk in the circuit. The length of each pedestrian and the circuit are described as b and L. The boundary is periodic, so that pedestrians move to the leftest end after they pass the rightest end. Due to the assumption of the homogeneous distribution, the headway, i.e., the distance between pedestrians, are simply calculated as $(L - bN)/N$

diagrams are depicted by quasi-one-dimensional flow, where overtaking is acceptable, and the variety of them is partially attributed to overtaking. Thus, this condition contributes to reduce variety and focus on simple situation. The similar pedestrian flow is studied in Ref. [8, 12]. Furthermore, we assume that N homogeneous pedestrians, whose length are b, distribute *homogeneously* in the circuit without considering complex microscopic interaction as shown in Fig. 1. This simplification deprives the model of interesting and important phenomena such as instability of flow and the propagation of stop waves; however, it still preserves the characteristics of previous successful models and let us obtains the method for increasing flow.

Due to the homogeneous distribution of pedestrians, the density and headway are calculated as follows:

$$\rho = N/L, \tag{1}$$

$$h = (L - bN)/N, \tag{2}$$

thus, the headway h is described as the function of ρ as $h(\rho) = (1 - b\rho)/\rho$.

Since we have simplified the distribution of pedestrians, we consider the velocity V of individual pedestrian in detail by dividing it to two parts as

$$V(h) = S(h)P(h), \tag{3}$$

where S (stride function) and P (pace function) denote length of a stride and pace of walking (number of steps per unit time), respectively.

The explicit formulation of the stride function is intuitively determined as follows. It is plausible to assume that there is maximum length of stride for pedestrians, which is given as s, in the low-density regime. When the density becomes large, pedestrians are no more able to walk with their maximum stride. Since we assume homogeneous distribution, the pedestrians can maximally proceed the headway h in one step; however the effect of personal space, which is usually larger than the exact pedestrians length, prevents the pedestrians from contact with their predecessors. This phenomenon is introduced by the parameter $k \in (0, 1]$, and the stride function is described as

$$S(h) = \begin{cases} kh & (0 \leq h \leq h_c), \\ s & (h_c < h < \infty), \end{cases} \tag{4}$$

where h_c is calculated from the equation $kh_c = s$, and obtained as $h_c = s/k$ as well as $\rho_c = k/(kb + s)$.

Because of the homogeneity of pedestrians and their distribution, the flow Q is computed by using hydrodynamic equation and individual velocity V as

$$Q(\rho) = \rho V(h) = \rho S(h)P(h) = \begin{cases} s\rho P(h) & (0 \leq \rho \leq \rho_c), \\ k(1 - b\rho)P(h) & (\rho_c < \rho \leq \rho_j), \end{cases} \tag{5}$$

where $\rho_j = 1/b$ is the maximum density. Headway $h = 0$ at $\rho = \rho_j$.

2.2 Correspondence with Traffic and Pedestrian Dynamics Models

Before the detailed analysis on the pace function, we would like to consider the correspondence between our model and the previous sophisticated models for traffic and pedestrian dynamics in the case $P(h) = 1 =$ const. to investigate the characteristics of our model.

Firstly, Eq. (5) represents the flow of the totally asymmetric simple exclusion process (TASEP) [13, 14] with deterministic movement, in other words the elementary cellular automata with the rule 184 [15], which is one of the fundamental model for traffic, pedestrian, and molecular motor dynamics, in the case $b = s = k = 1$ as shown in Table 1. The TASEP with stochastic movement is also applied to many models for self-driven particles; however, the equation $b = s$ indicates that the maximum stride and the length of pedestrians are same, which is not very realistic. Therefore, our model obtains the detailed effect of stride in return for the loss of stochastic dynamics.

Secondly, the zero-range process (ZRP) [16, 17] with piecewise-linear hopping-probability function $u(h)$ is also reproduced when the reciprocal of the critical headway h_c^Z, where the hopping probability reached 1 in the $u(h)$, corresponds to the parameter k as in Table 1. Moreover, Eq. (5) represents the fundamental diagram of the Nagel-Schreckenberg model (NaSch) [18] without random-braking rule if we substitute the parameter v_m^N, which is the number of maximum proceeding cells in one time step, to the maximum stride s. These are successful cellular-automata models for highway traffic, and many extended models are developed. Velocity of vehicles in these models changes according to their headway, i.e, the number of empty cells in front of them, and this characteristic enables us to obtain realistic

Table 1 Parameters which reproduce the expression of flow of the previous models in the special cases. The parameter h_c^Z is the headway where the hopping probability reached 1 in the piecewise-linear hopping-probability function $u(h)$ in the ZRP, v_m^N denote the maximum velocity, i.e., the number of the maximum proceeding cells in one time step in the NaSch, and v_m^O represents the maximum velocity in the OVM

Model	b	s	k
ASEP	1	1	1
ZRP	1	1	$1/h_c^Z$
NaSch	1	v_m^N	1
OVM	b	v_m^O	1

fundamental diagrams. The correspondence between our model and the above two models implies that the effect of important relation between headway and velocity does not harmed by the homogeneous assumption.

Finally, we would like to confirm that our model is a reduction of the optimal velocity model (OVM) [19], which is a successful microscopic continuous model for highway traffic. When the parameter of driver's sensitivity tends to infinity, and a piecewise-linear optimal velocity function with the size of vehicles b and the maximum velocity v_m^O is introduced, the OVM shows homogeneous distribution of the vehicles similar to our model, thus, our assumption of the homogeneity corresponds to that of agile pedestrians. Since inertia of pedestrians is much smaller than vehicles, the homogeneous assumption does not greatly deviate from the real pedestrian dynamics. From the discussion above, we have verified that our model is one of the minimum models, which can be reduced from many previous one-dimensional models. Pedestrians are agile enough in our model, and only the change of stride against headway is considered in detail.

2.3 *Influence of Pace Function on Pedestrian Flow*

In this subsection, we focus on the pace function. If the density is low and a pedestrian does not interact with each other, it is feasible to assume that pedestrians walk with constant pace. However, contrary to the stride function case, it is difficult to obtain the explicit formulation of the pace function with some intuitive assumptions in the high-density regime, thus, we consider simple linear function and investigate how the change of pace affect on the flow. Mathematical formulation of the pace function is given as follows:

$$P(h) = \begin{cases} p - a(h_c - h) & (0 \le h \le h_c), \\ p & (h_c < h < \infty), \end{cases} \tag{6}$$

where p is a pace in the free-flow situation and a represents the influence of the density on pace. If $a > 0$ ($a < 0$) the pace of pedestrians decrease (increases) when the density increases, and if $a = 0$ pedestrians keep walking with the constant pace regardless of the density. Note that $a \leq p/h_c = kp/s$ since $P(h) \geq 0$. Substituting Eq. (6) for $P(h)$ in Eq. (5), the expression of the flow is described as

$$Q(\rho) = \begin{cases} sp\rho & (0 \leq \rho \leq \rho_c), \\ k[p - a(h_c - h(\rho))](1 - b\rho) & (\rho_c < \rho \leq \rho_j), \end{cases} \tag{7}$$

and, the maximum of the flow is calculated as

$$Q_{\max} = \begin{cases} sp\rho_c & \text{at } \rho = \rho_c & (a \leq a_c), \\ kp_j - 2kab\left(1 - \sqrt{1 - \frac{p_j}{ab}}\right) & \text{at } \rho = \left[b\sqrt{1 - \frac{p_0}{ab}}\right]^{-1} & (a > a_c), \end{cases} \tag{8}$$

where

$$\begin{cases} p_j = P(\rho_j) = p - ah_c, \\ a_c = -\dfrac{bp}{h_c(b + h_c)}. \end{cases} \tag{9}$$

Since second derivative of the flow are computed as

$$\frac{d^2Q(\rho)}{d\rho^2} = \frac{1}{\rho^3}\frac{d^2V(h)}{dh^2} = \begin{cases} 0 & (0 < \rho < \rho_c), \\ 2ka/\rho^3 & (\rho_c < \rho < \rho_j), \end{cases} \tag{10}$$

the convexity in the region $\rho > \rho_c$ becomes

$$\begin{cases} \text{convex downward} & (a > 0), \\ \text{convex upward} & (a < 0). \end{cases} \tag{11}$$

Therefore, the convexity of the flow changes at $a = 0$. Plots of the flow for various a are depicted in Fig. 2a. We can confirm that the parameter a, i.e., the pace function, affects on the density which achieves the maximum flow and the convexity as we have analyzed above.

3 Increase of Pedestrian Flow by Slow Rhythm

If pedestrians walk with rhythm, in other words if we can control the walking pace by rhythm using a device such as a metronome, the rhythm exactly corresponds to the pace, i.e., the parameter p. Therefore, fast and slow rhythm increase and decrease the flow, respectively, as we can verify from Eq. (7). In this case, the

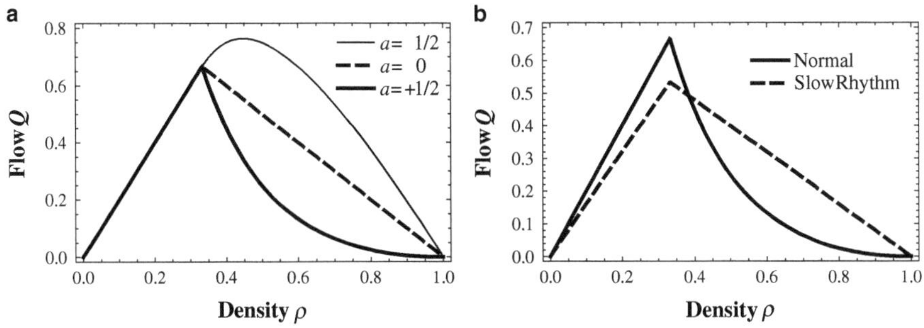

Fig. 2 (**a**) Theoretical fundamental diagrams for various a in the case $b = 1$, $s = 2$, and $p = 1$. We see that they are convex downward (upward) when $a > 0$ ($a < 0$). (**b**) Theoretical fundamental diagrams of the normal walking ($p = p^N = 1, a = 0.5$) and the slow-rhythmic walking ($p = p^R = 0.8, a = 0$) in the case $b = 1$ and $s = 2$. The crossing is observed at $\rho_s = 5/13$, and the rhythmic walking, whose flow is smaller than that of the normal walking in the low-density regime due to the slow pace, i.e., $p_R < p_N$, achieve larger flow in the high-density regime

parameter $a = 0$ because pedestrians walk with the constant rhythm irrespective to the density.

Suppose that pedestrians in the normal situation walk with the pace $p = p_N$, and it decreases in the high-density regime, i.e., $a > 0$. Then the surprising figure as shown in Fig. 2b is depicted. Due to the convexities of the fundamental diagrams in the normal and rhythmic walking cases, i.e., convex downward and linear, the rhythm which is *slower* than the normal walking pace, increases the flow in the high-density regime. The crossing is achieved at

$$\rho_s = \frac{\rho_c}{1 - \rho_c(p^N - p^R)/a}, \tag{12}$$

where $\rho_c < \rho_s < \rho_j$, if the condition $p_j^N < p^R < p^N$ is satisfied.

This phenomenon may give a solution to ease congestion in the real word. The significant advantage is that any excessive haste does not necessary at all. The flow increases by just keeping the slow walking pace. Thus, pedestrians do not consume extra energy or conflict with others by moving aggressively in pedestrian jam.

4 Experiment

4.1 Experimental Condition

We have performed an experiment with real pedestrians to verify the theoretical result obtained in the previous section. Schematic view of the circuit used in our experiment is depicted in Fig. 3a. Inner and outer radius are $r_i = 1.8$ and $r_o = 2.3$ [m], respectively, and the participants of the experiment, who are male university

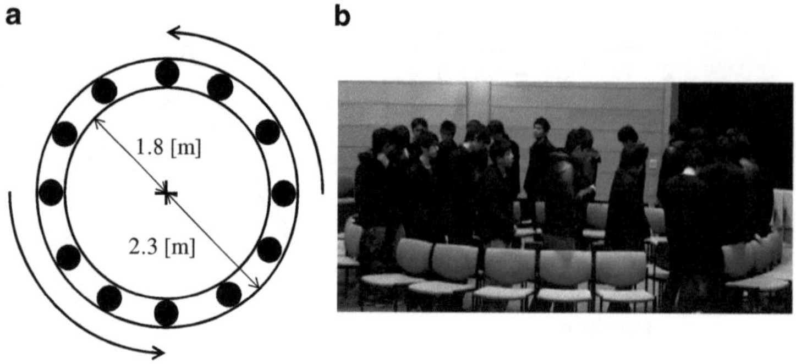

Fig. 3 (**a**) Schematic view of the circuit used in the experiment. The *black* particles represent the participants of our experiment. (**b**) Snap shot of the experiment (Normal walking, $N = 24$ ($\rho = 1.86$ [persons/m]))

students between 18 and 39 years old, walk the circuit in the counter-clockwise direction. Snap shot of the experiment is shown in Fig. 3b.

We have executed 11 kinds of density conditions. The number of the participants in the circuit in each condition is $N = \{1, 3, 6, 9, \cdots, 30\}$. The condition $N = 1$ and $N = 3$ are tried three times, and the other conditions are tried once. Each trial is more than 1 min. The density is calculated as

$$\rho = \frac{N}{L} = \frac{N}{\pi\,(r_i + r_o)}. \tag{13}$$

Two kinds of walking were performed in the experiment. In the first case, we did not give any specific instructions to the participants, so that they walked normally in the circuit. In the latter case, the participants were instructed to walk with the sound from the electric metronome, whose rhythm is 70 beat per minutes (BPM). Note that we did not uniform which foot to move first.

4.2 Experimental Result

Figure 4 shows the fundamental diagram obtained from our experiment. Firstly, we see that the flow is larger in the normal case than the rhythmic case at the leftest plot, i.e., $N = 1$. Hence, the pace 70 [BPM] is slower than the normal-walking pace of the participants. Secondly, we observe that the flow in the normal case is convex downward in the high-density regime as we have assumed in the theoretical analysis. Thus, the walking pace decreases by the influence of the predecessors. Thirdly, the flow decreases linearly in the rhythmic case, which follows our intuition. Finally, since the theoretical assumptions of the convexity are satisfied in the experimental

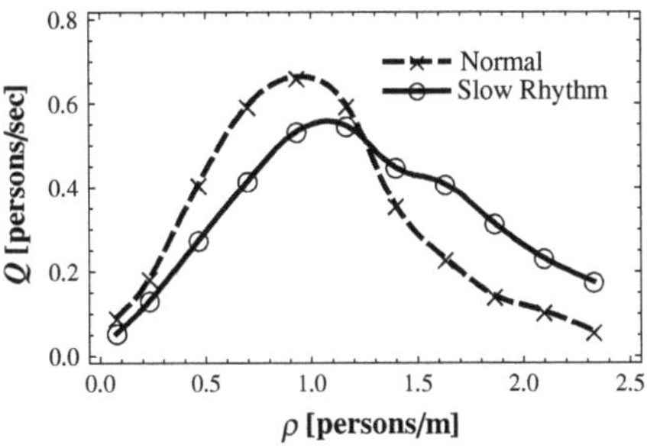

Fig. 4 Fundamental diagram of the normal walking and the rhythmic walking (70 [BPM]). Data plots represented by the *cross* and *circle* markers are interpolated by second-order *curves*. The values at $\rho = 0.07$ and 0.23 [persons/m], i.e., $N = 1$ and 3, are average of the three trials. We clearly observe the crossing and the rhythmic walking achieve larger flow in the high-density regime

flows, the crossing appears after the both flows achieve maximum, and the flow of the rhythmic case exceeds that of the normal case. Therefore, we have succeeded to verify that slow rhythm improves the pedestrian flow.

5 Conclusion

In this paper, we have calculated the fundamental diagram of pedestrians by dividing the velocity term into the stride and pace parts. It reproduces the fundamental diagram of the basic models, which are applied for traffic and pedestrian dynamics. Analyzing the effect of pace, we have surprisingly discovered that the rhythm which is slower than the normal pace in the free-flow situation increases the flow in the congested situation. This phenomenon is also verified by our real experiment and expected to be applied to ease congestion and increase flow in the high-density regime.

One of the important future works related to this study is theoretical analysis on the effect of rhythm in higher-dimensional case, i.e, quasi-one and two dimensional cases. In the quasi-one-dimensional case, where the width of the circuit is larger than one pedestrian, fast pedestrians are able to pass slow ones. When we consider bi-directional flow and two dimensional cases, we need to consider conflicts between pedestrians. Therefore, investigation of the effect of rhythm may not be as simple as one-dimensional case in this paper; however, success of the study broadens the range of application of the slow-rhythmic flow.

In our experiment, we have instructed the participants to walk with the rhythm from the electronic metronome; however, it is not certain that whether pedestrians in the real world walk with the rhythm as in the experiment. Therefore, research on the relation between the improvement of flow and the ratio of pedestrians walking with the rhythm is another important future work to apply the rhythmic-walking method to the real world. When pedestrians' movement is synchronized to the rhythm without explicit instruction, a smooth-flow society will be achieved by our slow-rhythmic method.

Acknowledgements We would like to appreciate the staffs of our experiment described in Sect. 4 for helping us. This work is financially supported by the Japan Society for the Promotion of Science and the Japan Science and Technology Agency.

References

1. Dirk Helbing. Traffic and related self-driven many-particle systems. *Rev. Mod. Phys.*, 73(4):1067–1141, 2001.
2. Andreas Schadschneider, Debashish Chowdhury, and Katsuhiro Nishinari. *Stochastic Transport in Complex Systems*. Elsevier, 2010.
3. Dirk Helbing and Péter Molnár. Social force model for pedestrian dynamics. *Phys. Rev. E*, 51(5):4282–4286, 1995.
4. Masakuni Muramatsu, Tunemasa Irie, and Takashi Nagatani. Jamming transition in pedestrian counter flow. *Physica A*, 267:487–498, 1999.
5. C. Burstedde, K. Klauck, A. Schadschneider, and J. Zittartz. Simulation of pedestrian dynamics using a two-dimensional cellular automaton. *Physica A*, 295:507–525, 2001.
6. Daniel R. Parisi, Marcelo Gilman, and Herman Moldovan. A modification of the social force model can reproduce experimental data of pedestrian flows in normal conditions. *Physica A*, 388:3600–3608, 2009.
7. Mohcine Chraibi, Armin Seyfried, and Andreas Schadschneider. Generalized centrifugal-force model for pedestrian dynamics. *Phys. Rev. E*, 82(4):046111, 2010.
8. Armin Seyfried, Bernhard Steffen, Wolfram Klingsch, and Maik Boltes. The fundamental diagram of pedestrian movement revisited. *J. Stat. Mech.*, 2005:P10002, 2005.
9. J Zhang, W Klingsch, A Schadschneider, and A Seyfried. Transitions in pedestrian fundamental diagrams of straight corridors and t-junctions. *J. Stat. Mech.*, 2011:P06004, 2011.
10. Online-Database at ped-net.org homepage (http://www.pednet.org/) and the references there in, 2008.
11. Daichi Yanagisawa, Ayako Kimura, Akiyasu Tomoeda, Ryosuke Nishi, Yushi Suma, Kazumichi Ohtsuka, and Katsuhiro Nishinari. Introduction of frictional and turning function for pedestrian outflow with an obstacle. *Phys. Rev. E*, 80(3):036110, 2009.
12. Daniel Jezbera, David Kordek, Jan Kříž, Petr Šeba, and Petr Šroll. Walkers on the circle. *J. Stat. Mech.*, 2010:L01001, 2010.
13. Frank Spitzer. Interaction of markov processes. *Adv. Math.*, 5(2):246–290, 1970.
14. B. Derrida. An exactly soluble non-equilibrium system: The asymmetric simple exclusion process. *Physics Reports*, 301(1–3):65–83, 1998.
15. Stephen Wolfram. *Cellular Automata and Complexity: Collected Papers*. 1994.
16. M. R. Evans. Phase transitions in one-dimensional nonequilibrium systems. *Braz.J.Phys.*, 30(1):42–57, 2000.
17. Masahiro Kanai. Exact solution of the zero-range process: fundamental diagram of the corresponding exclusion process. *J. Phys. A: Math. Theor.*, 40:7127–7138, 2007.
18. Kai Nagel and Michael Schreckenberg. A cellular automaton model for freeway traffic. *J. Phys. I France*, 2:2221–2229, 1992.
19. M. Bando, K. Hasebe, A. Nakayama, A. Shibata, and Y. Sugiyama. Dynamical model of traffic congestion and numerical simulation. *Phys. Rev. E*, 51(2):1035–1042, 1995.

Social Field Model to Simulate Bidirectional Pedestrian Flow Using Cellular Automata

Jorge D. González, M. Luisa Sandoval, and Joaquín Delgado

Abstract A collective phenomenon appearing in the simulation of bidirectional pedestrian flow in corridors is dynamic multi-lane (DML) flow. We present a cellular automata model that reproduces this behavior. We propose to incorporate a social distance emulating a territorial effect through a social field, similar to the dynamic floor field of Burstedde et al. (Physica A 295:507–525, 2001). This model also considers a vision field allowing a pedestrian to collect information from cells in front of him/her and to get the weighted social parameter as well; the importance of this parameter in the formation of dynamic lanes is that it helps a pedestrian to choose the lane that contains the highest concentration of persons walking in the same direction. We present numerical simulations in corridors with bidirectional flow and the fundamental diagram for unidirectional flow.

1 Introduction

In recent decades, the study of the dynamics of pedestrians has witnessed a great development, mainly through cellular automata (CA) modeling due to its efficiency in describing model complex systems.

Several models have been developed to simulate pedestrian traffic based on cellular automata [1–9]. In order to model the change of walking lane, these models have used different types of probabilities or parameters: either fixed in advance or dynamical. The latter are calculated according to the conditions that exist around the pedestrians. An important factor to determine the dynamic parameters is to define properly the shape of the area that influences the behavior of pedestrians.

J.D. González · M.L. Sandoval (✉) · J. Delgado
Departamento de Matemáticas, Universidad Autónoma
Metropolitana Iztapalapa, A. P. 55–534, México, C.P. 09340, D.F., México
e-mail: jd_nova@hotmail.com; mlss@xanum.uam.mx; jdf@xanum.uam.mx

V.V. Kozlov et al. (eds.), *Traffic and Granular Flow '11*,
DOI 10.1007/978-3-642-39669-4_20, © Springer-Verlag Berlin Heidelberg 2013

Few cellular automata models take into account social distance between pedestrians [1, 10]. We introduce the idea of social distance through a social field, according to the dynamic floor field of Burstedde et al. [1], which resembles the territorial effect [11].

Our goal is to combine a dynamic parameter with the overlapping of different fields of influence in order to obtain numerical simulations that generate dynamic lanes and to avoid body contacts. As will be shown, with the addition of the social field, we were able to reproduce Weidmann's fundamental diagram [12].

This paper presents the modeling of bidirectional pedestrian flow in a corridor, where in order to change the walking lane, we define a vision field for each pedestrian and a weighted social parameter. In Sect. 2 the proposed model is described and in Sects. 3 and 4 the numerical simulations and conclusions are presented, respectively.

2 Description of the Model

Our microscopic model based on CA is intended to simulate bidirectional pedestrian traffic along a corridor. The CA is defined on a rectangular grid of size $I_1 \times I_2$ contained in $\mathbb{R}^2$. Each cell is a square of $40 \times 40\,\text{cm}^2$. Pedestrian enter and exit through (lateral) left and right boundaries of the corridor. Impenetrable boundary conditions are imposed at the top and bottom borders of the corridor, and two types of boundary conditions on the lateral boundaries are considered: open and periodic.

There are three basic elements that a microscopic pedestrian model must take into account: to mitigate conflicts, forward and lateral motions (as in change of lane) [3]. For this reason a pedestrian is only allowed to move as shown in Fig. 1.

A collective phenomenon, the dynamic multi-lane (DML) flow is formed by groups of pedestrian moving along directional lanes. Pedestrians generate a lane by avoiding others coming from the opposite direction and by following the one just ahead of him/her and going in the same direction [3]. A factor enabling self-organization of collective phenomena in pedestrian behavior is the social distance or social field [1, 10, 11].

Formation of dynamical lanes can be understood as follows: a typical pedestrian chooses to minimize its energy by walking along currents of people in its own direction. Thus when he/she walks along, he/she has to worry about keeping its own distance from the pedestrian in front of him/her. Without forming lanes, pedestrians will confront each other frequently and would have to deviate from its own route avoiding physical contact, thus expending more energy.

2.1 Social Field

The territorial effect arises changes in pedestrian behavior, which are guided by social fields. Helbing and Molnár [11] takes this idea to model territorial behavior

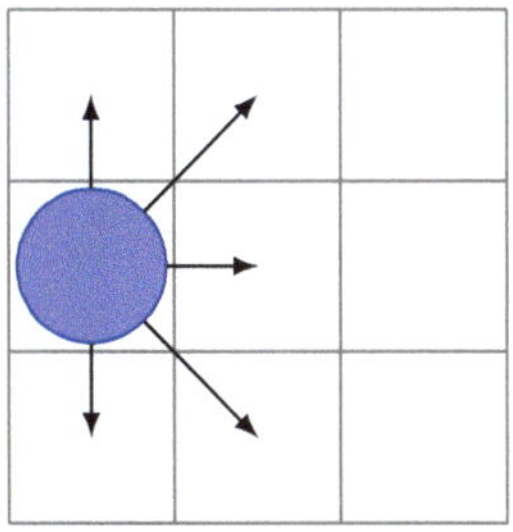

Fig. 1 Movements allowed in the model

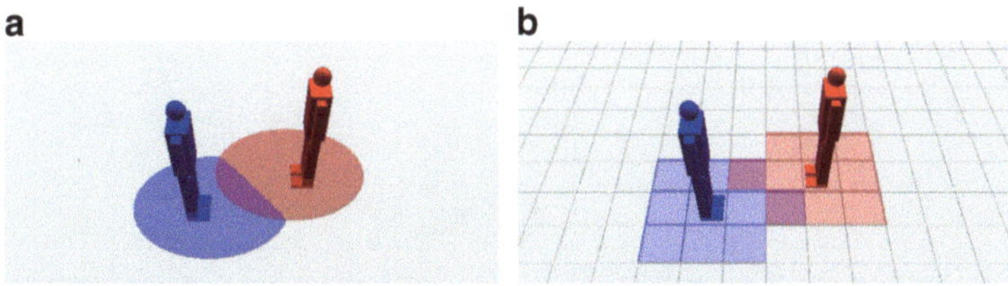

Fig. 2 Social fields: (**a**) continuous and (**b**) discrete

by means of repulsion forces, in such a way that persons tend to move away from others or objects to avoid collisions or hurting. Social field resembles an individual space where each person feels comfortable. Thus the social field acts in a similar manner as the repulsive force of the social force model.

Whenever a person invades the social space of another, the last tries to move in a direction that diminishes the discomfort, acting as a repulsive force among persons. In this way the social distance is kept among individuals. Social field also augments the presence of the individual in such a way that he/she is perceived among others. Figure 2 shows two kinds of social fields: continuous and discrete. The shape of a discrete field is similar to the continuous field due to the fact that it is elliptical. A discrete value of the social field is assigned to each cell. In the case that the fields intersect, the values at the intersected cells are summed up as shown in Fig. 3, so the intensity or presence of the individual is increased.

2.2 *Vision Field*

The majority of pedestrians look at all what happens in front of them, as they walk by. In this way they can avoid collisions with either objects or another persons. The vision field reflects the attention that a pedestrian pays to situations that happens in front of him/her, giving them more attention than those which happen besides or back of him/her. On the other hand, the vision scope of pedestrians becomes important from certain distance, where he/she can change direction to avoid collision with obstacles. These characteristics permit the pedestrian to decide, to continue walking straight ahead or deflect.

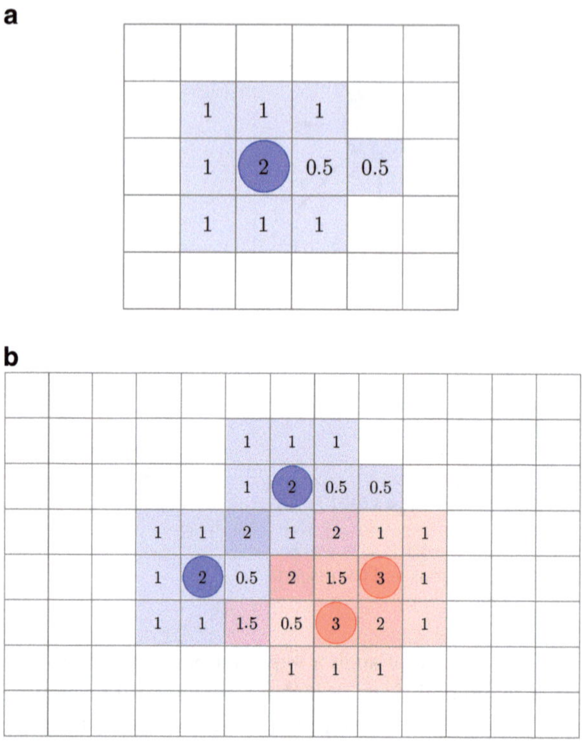

Fig. 3 (**a**) Social field configuration and (**b**) intersections of social fields

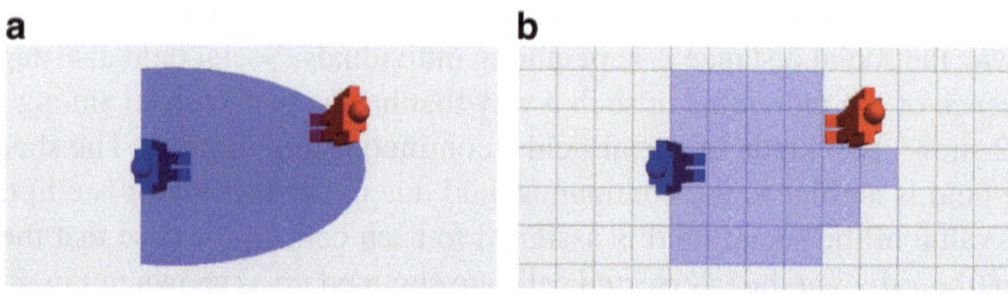

Fig. 4 Vision fields: (**a**) continuous and (**b**) discrete

Figure 4 shows two types of vision fields: continuous and discrete. The vision scope of a pedestrian will be equal to its individual velocity, thus he/she will have a vision of the route along time. Figure 5 shows different scopes of vision field.

2.3 *Weighted Social Parameter*

Hao Yue et al. [4] introduce dynamical parameters in order to control pedestrian motion. Similarly we introduce the so called weighted social parameter. The idea of

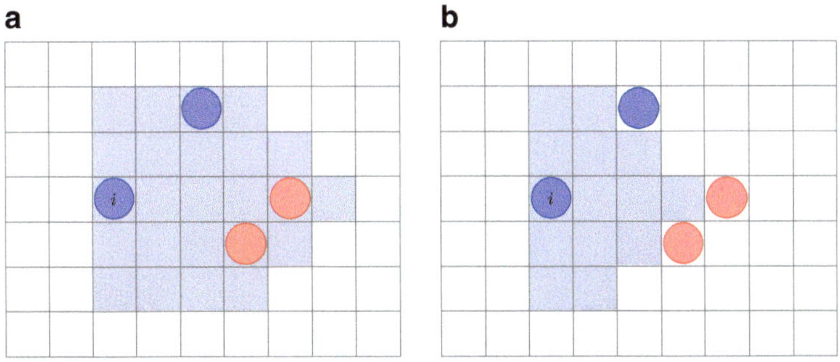

Fig. 5 Different scopes of vision field: (**a**) $v^i = 5$ and (**b**) $v^i = 3$

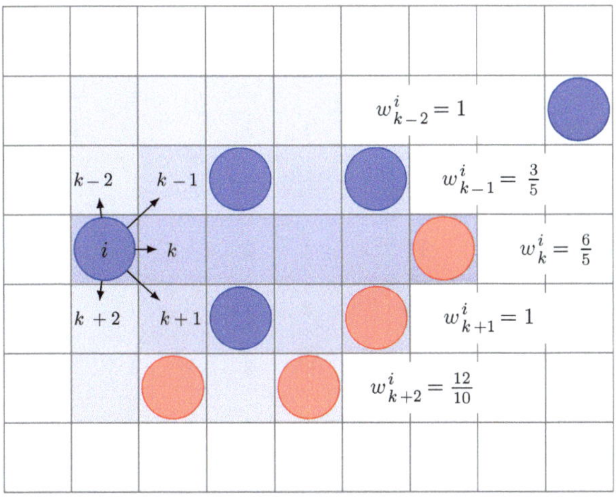

Fig. 6 Weights per lane for each walking lane

this parameter is to help a pedestrian decide on the lane with the highest pedestrian concentration walking in the same direction. In this way formation of dynamical lanes is more likely. Let us introduce two intermediary parameters: the weight and the social value per lane.

The weight per lane for lane i is defined by

$$w_k^i = 1 + \sum_{j=1}^{np_k} \frac{d_j}{l_k \times v^i} \tag{1}$$

where np_k is the number of persons in lane k, v^i is the number of cells within the vision scope of a pedestrian, $l = 1$ for walking lanes $k-1$, k and $k+1$, and $l = 2$ for walking lanes $k-2$ and $k+2$; $d_j = +1$ if pedestrian j walks in the opposite direction as pedestrian i and $d_j = -1$ for those walking in the same direction. The composition of walking lanes is shown in Fig. 6; cells of the same color belong to the same walking lane.

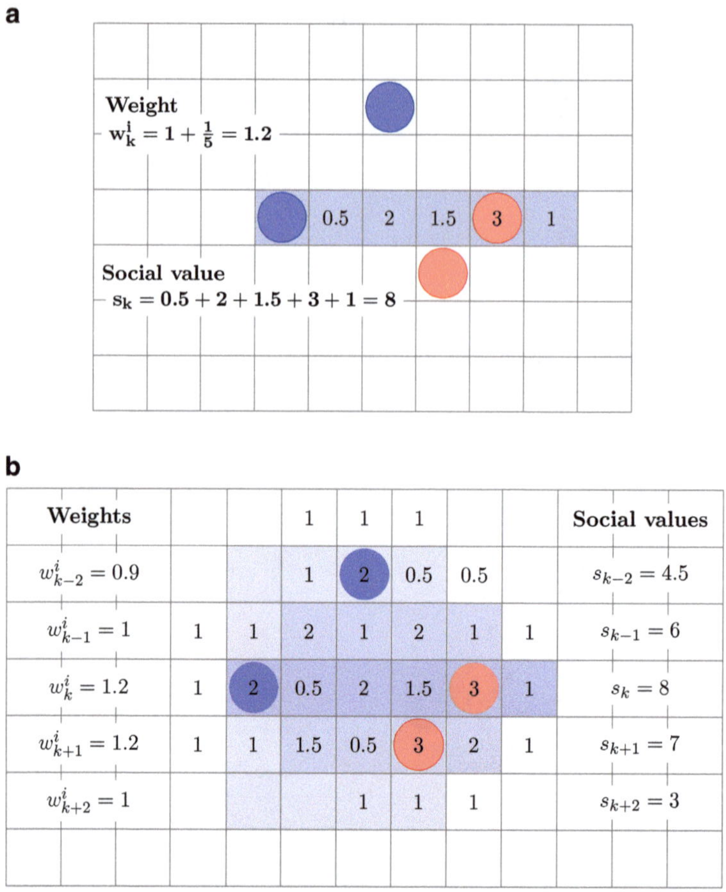

Fig. 7 Computation of weights and social values per lane for: (**a**) the lane k and (**b**) the five lanes

The social value of lane k is obtained by summing the social field of each cell that form the walking lane, see Fig. 7 for an example. The weighted social parameter for each lane is computed by multiplying the weight by the social value per lane. That is, $vp_k = w_k^i s_k$.

2.4 No Crossing Paths

The restriction of "no crossing paths" applies to models that use maximum speeds greater than one per cell in each time-step: The pedestrian moves ahead as long as there is no other pedestrian blocking his/her path, see Fig. 8. We have incorporated this restriction since it was shown to work well [13].

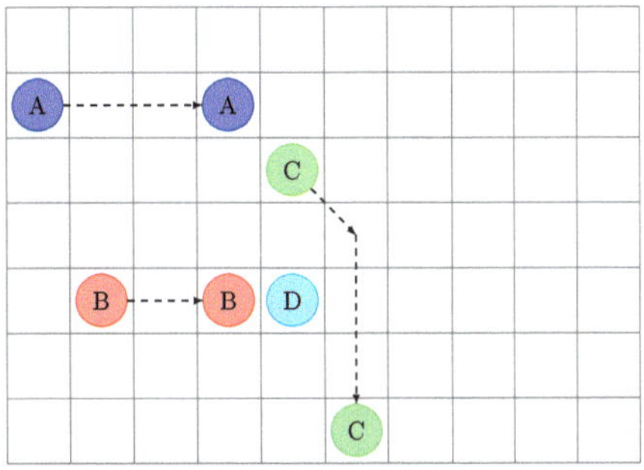

Fig. 8 Restriction of no crossing paths

2.5 Pedestrian Motion

Pedestrian motion is achieved in two stages: first, the whole path of each pedestrian is computed, then each pedestrian position is updated. Each time-step is sectioned in time sub-steps. Pedestrian motion applies to each time sub-step until completing the whole time-step or the individual velocity is attained. The updating of pedestrian positions is made in parallel, that is, in each time step a pedestrian moves to the last cell of the computed path.

Following is the sequence of pedestrian motion:

I. Computation of pedestrian path

1. Compute the vision scope of the pedestrian.
2. Compute vp_k the weighted social parameter of walking lane k.
3. If the weighted social parameter is less than 1, then go to step 6. Otherwise go to the step 4.
4. Compute the weighted social parameter for adjacent walking lanes $k-1$, $k+1$, $k+2$ and $k-2$, as shown in Fig. 7.
5. Compare and choose the free cell with the least weighted social parameter. That is, $z = \{z \in [k, k-1, k+1, k+2, k-2] \mid vp_z = \min[vp_k, vp_{k-1}, vp_{k+1}, vp_{k+2}, vp_{k-2}], Cell(z) = 0\}$.
6. Move and update pedestrian social field, go to step 7.
7. In case there is no free cell, pedestrian stays in the same cell.
8. End.

3 Numerical Simulations

We consider two types of boundary conditions in a corridor of size $10 \times 50\,\text{m}^2$ for bidirectional pedestrian flow: (a) open and (b) periodic. For both boundary conditions, the common parameters are: time step $\Delta t = 2$ s which is divided in five subintervals of equal length. Each pedestrian is assigned a velocity taken from a normal distribution with mean 1.34 m/s and variance 0.26 m/s. In the following

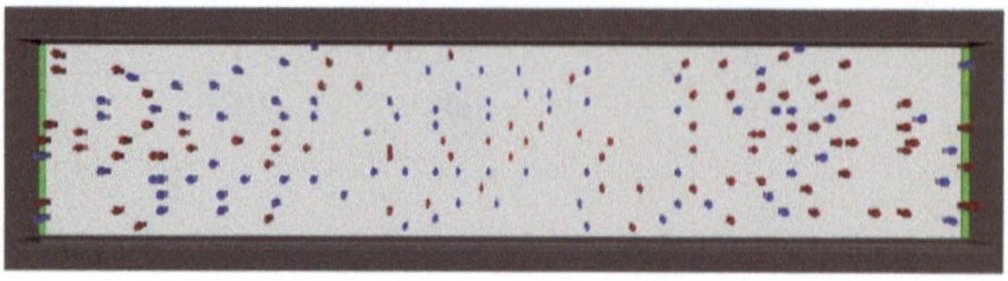

Fig. 9 Numerical simulation with open boundary conditions

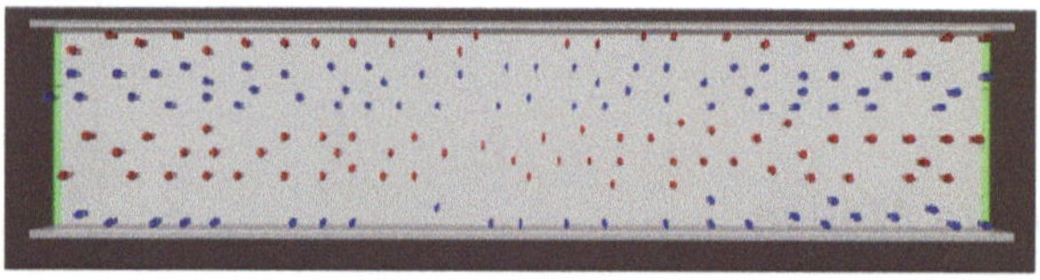

Fig. 10 Numerical simulation with periodic boundary conditions, time: 1,197 s and 160 pedestrians

figures, pedestrians in blue move from left to right, those in red move in the opposite direction.

For open boundary conditions, 0–5 pedestrians are introduced at the left and right boundaries every two time-steps. Figure 9 shows the numerical simulation in this case for 167 pedestrians. Dynamical multi-lane flow is clearly apparent in our simulation.

For periodic boundary conditions, Fig. 10, observed that most of the simulations show a self-organization by formation of four flow currents, two of them larger than the others.

3.1 Fundamental Diagram

We compute a fundamental diagram of velocity vs. density from our CA model. Data were generated as follows: A corridor of $3.6 \times 40\,\mathrm{m}^2$ size was simulated. Each simulation was run for 500 s. After a transient period of 60 s when the flow has been stabilized, data (mean velocity and density) were computed in a sub-region of the corridor of size $3.6 \times 5.6\,\mathrm{m}^2$.

Figure 11 shows: (a) Weidmann's empirical curve (continuous), (b) computed data with social field model (red dots) and (c) computed data with modified social force model (blue asterisks) [14]. Remarkably, social field model reproduces Weidmann's curve even for high densities (greater as $4.8\,\mathrm{p/m}^2$) at a low computational cost.

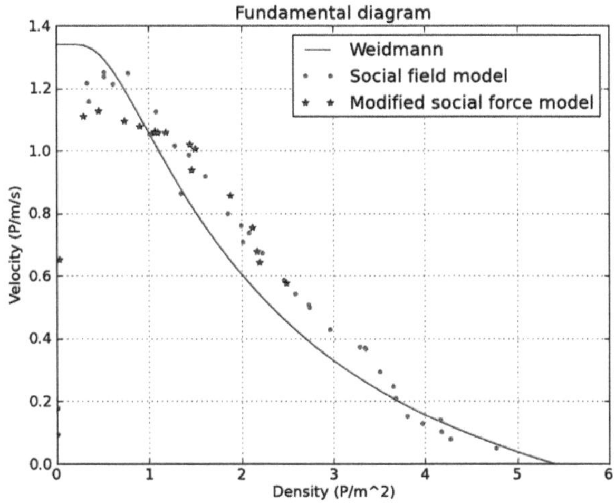

Fig. 11 Fundamental diagram

4 Conclusions

We have proposed a CA model that incorporates three main features: a social field, a vision field and a weighted social parameter. Our results reproduce dynamical multi-lanes as observed empirical data published elsewhere.

Our social field model fits quite well to Weidmann's empirical curve up to high densities such as 4.8 p/m^2.

The advantages of a CA modeling are its simplicity and its computational speed which are very convenient for real time simulation.

Acknowledgements The authors were supported by the FOMIX grant 120375, "Desarrollo de modelos matemáticos para mejorar la operación de la Red del STC".

References

1. Burstedde C., Klauck K., Schadschneider A., and Zittartz J., Simulation of pedestrian dynamics using a two-dimensional cellular automaton. *Physica A*, 295:507–525, 2001.
2. Gipps P.G. and Marksjö B., A micro-simulation model for pedestrian flows. *Mathematics and computers in simulation*, 27:95–105, 1985.
3. Blue V.J. and Adler J.L., Cellular automata microsimulation for modeling bi-directional pedestrian walkways. *Transportation Research Part B*, 35(3):293–312, 2001.
4. Hao Yue, Hongzhi Guan, Juan Zhang, and Chunfu Shao, Study on bi-direction pedestrian flow using cellular automata simulation. *Physica A*, 389:527–539, 2010.
5. Varas A., Cornejo M.D., Mainemer D., Toledo B., Rogan J., Muñoz V., and Valdivia J.A., Cellular automaton model for evacuation process with obstacles. *Physica A*, 382:631–642, 2007.
6. Fang Weifeng, Yang Lizhong, and Fan Weicheng, Simulation of bi-direction pedestrian movement using a cellular automata model. *Physica A*, 321:633–640, 2003.

7. Li Jian, Yang Lizhong, and Zhao Daoliang, Simulation of bi-direction pedestrian movement in corridor. *Physica A*, 354:619–628, 2005.
8. Weng W.G., Chen T., Yuan, and Fan W.C., Cellular automaton simulation of pedestrian counter flow with different walk velocities. *Phys. Rev. E*, 74:036102, 2006.
9. Narimatsu K., Shiraishi T., and Morishita S., Acquisiting of local neighbour rules in the simulation of pedestrian flow by cellular automata, *Lecture Notes in Computer Science*, 3305:211–219, 2004.
10. Was J., Gudowski B., and Matuszyk P.J., Social distances model of pedestrian dynamics, *Lecture Notes in Computer Science*, 4173:492–501, 2006.
11. Helbing D. and Molnár P., Social force model for pedestrian dynamics. *Phys. Rev. E*, 51:4282–4286, 1995.
12. Weidmann U., *Transporttechnik der Fußgänger*. Schriftenreihe des IVT 90, Zürich, Switzerland, 1992.
13. Klüpfel H., *A cellular automaton model for crowd movement and egress simulation*. PhD thesis, University of Duisburg-Essen, Germany, 2003.
14. Parisi D.R., Gilman M., and Moldovan H., A modification of the social force model can repreoduce experimental data of pedestrian flows in normal conditions. *Physica A*, 388:3600–3608, 2009.

Some Empirical Studies on Evacuation from a Hall

Ujjal Chattaraj, Partha Chakroborty, and Armin Seyfried

Abstract Studies on egress time and spatio–temporal progression of pedestrians inside a hall are important for design of exits of halls. In this study experiments on evacuation from a hall are conducted to understand the impact of exits and the geometry of the flow space on pedestrian flow. The width of the door openings as well as number, shape, size and positioning of obstacles are varied to change the nature of the goals and the geometry of the flow space. Results from this study explain how evacuation time from an enclosed space varies with number of persons inside the flow space and nature of exits present in the flow space as well as geometry of the space. Results also show how pedestrians distribute themselves inside the flow space while evacuating due to the above mentioned variations in the flow space. These results may help in designing enclosed space geometry and exits.

1 Introduction

Any space provided for human circulation (for example, airport terminals, sidewalks, shopping malls, fair grounds, etc.,) involves movement of pedestrians. Efficient design of facilities catering to pedestrian movement can be achieved

U. Chattaraj (✉)
Department of Civil Engineering, National Institute of Technology Rourkela, Rourkela 769 008, India
e-mail: chattaraju@nitrkl.ac.in

P. Chakroborty
Department of Civil Engineering, Indian Institute of Technology Kanpur, Kanpur 208 016, India
e-mail: partha@iitk.ac.in

A. Seyfried
Jülich Supercomputing Centre, Forschungszentrum Jülich GmbH, 52425 Jülich, Germany

Computer Simulation for Fire Safety and Pedestrian Traffic, Bergische Universität Wuppertal, Pauluskirchstraße 11, 42285 Wuppertal, Germany
e-mail: a.seyfried@fz-juelich.de; seyfried@uni-wuppertal.de

V.V. Kozlov et al. (eds.), *Traffic and Granular Flow '11*,
DOI 10.1007/978-3-642-39669-4_21, © Springer-Verlag Berlin Heidelberg 2013

only if one understands pedestrian flow. Empirical studies available in literature on pedestrian flow can be broadly classified as (i) studies on speed, density and their interrelationship and (ii) studies on the various different phenomena that can be observed in pedestrian dynamics. In this section, first, studies related to speed, density and their interrelationship are discussed. Later studies related to different pedestrian dynamics phenomena are presented.

Over the years various studies on speed–density (or flow–density or speed–flow) relationships (also known as fundamental relationship) of pedestrian streams have been reported [1–8]. The results however varied substantially primarily due to differences in the ways the data was collected and represented. Seyfried et al. [7] have tried to develop an experiment scenario which tries to capture only the impact of density on speed. Others have studied such relations in the specific case of experiment/observation without nullifying the impacts of influencing factors like entrance and exit condition of the corridor, width of the corridor, overtaking, pedestrians moving side by side, etc. [1–6, 8]. Morrall et al. [9] and Chattaraj et al. [10] have studied whether cultural differences impact the fundamental diagram.

Some studies have concentrated on determining capacity and specifying guidelines for Level of Service (LOS). Notable among these are the work by Polus et al. [11], Hoogendoorn and Daamen [12] and Seyfried et al. [13]. Polus et al. [11] have tried to determine LOS definitions in terms of nature of flow (free flow, unstable flow, dense flow and jammed flow) for uniform width sidewalks. Hoogendoorn and Daamen [12] have studied capacity of bottlenecks. Seyfried et al.'s [13] study relates exit widths to immediate upstream capacities.

There are yet other studies which relate to speed of pedestrians only. For example, Henderson and Lyons [14] observed male and female pedestrians in the same homogeneous mix follow different speed distributions. A similar, but more restrictive, remark was also made by Polus et al. [11], who observed that speeds of male pedestrians are far greater than female pedestrians. Young [15] has also done some speed studies on pedestrians in airport terminals.

Many studies relate to empirical observations on interesting phenomena that occur in pedestrian flow. Many authors, for example, have studied the spontaneous formation of lane-like structures in primarily bi–directional flow. Isobe et al. [16] have observed pattern formation and jamming transition (occurrence of jam when the density exceeds certain threshold value) in pedestrian counter flow. Kretz et al. [17] have plotted frequency distribution of number of lanes formed for bi–directional pedestrian flow. Hoogendoorn and Daamen [18] have studied lane formation and cluster formation for bi–directional pedestrian flow. Zhang et al. [19] have studied ordering in bidirectional pedestrian streams and its influence on the fundamental diagram. They found that the maximum of the specific flow in bidirectional streams is significantly lower than that in unidirectional streams. In another study Hoogendoorn and Daamen [12] have observed zipper effect (staggered positioning of pedestrians when the width of the corridor is in excess than that required for single file movement but not sufficient for two pedestrians moving side by side) at bottlenecks. Oscillations at bi–directional bottlenecks (alternate passing of pedestrians from one direction blocking pedestrians from the opposite direction) with emphasis on alternate passing time and frequency distribution of time headway was studied by Helbing et al. [20]. Helbing et al. [8] have observed upstream moving (back–propagating) stop and go shock waves forming in pedestrian streams.

It is evident from the literature reviewed that till now reasonable number of experimental studies have been done on speed, density and their interrelationship. But, till now there is no experimental study available in literature to understand how pedestrian flow parameters (speed and density) change spatially and temporally, especially, in response to geometric and other factors of the flow space. These phenomena motivated the present study. It can be thought that pedestrian movement is broadly classified as either movement inside enclosed spaces with few entry–exit points or movement through corridors. In this paper results from experiments on evacuation from a hall are presented. Results on lateral and longitudinal variations of density and speed along a corridor having geometric variations along it are presented in [21].

This category of experiments is on evacuation from a hall with two doors. The width of the door openings as well as shape, size and positioning of obstacles are varied to change the nature of the goals and the geometry of the flow space. As stated earlier, these experiments are conducted to understand the impact of goals and the geometry of the flow space on pedestrian flow. Various experiments designed to investigate the progression of pedestrians while evacuating from a hall have been conducted as a part of this study [22].

2 Experimental Set–Up and Procedure

These experiments on evacuation are conducted in an indoor hall with two doors and paved floor. The lighting in the hall was very good during the experiments. Figure 1 shows a schematic of the experimental set–up. The size of the goals and obstacles can be varied. Different combinations of the goal sizes and obstacles lead to different experiments. In all the experiments the pedestrians are initially in a waiting zone as shown in Fig. 1. At their initial position they keep 0.4 m gap from the person immediately ahead and immediately behind. They also keep 0.4 m gap from the persons on either sides. This 0.4 m is approximate body size of a human being. After the instruction to start is given the pedestrians move through the flow space at their comfortable speed and go out of the flow space through the goal (door) of their choice. Each experiment is conducted three times. Data is collected by video recording by overhead camera. In order to determine the position of pedestrians on the flow space a grid (0.4×0.4 m) is constructed using thin but highly visible wires at a height of 1.65 m (this is approximately equal to the average height of Indian people [23]) from the ground. Once the grid is constructed the fixed cameras record this grid. Once the grid is recorded neither the camera position nor the camera angle is altered till all the experiments are over. The grid is removed before the experiments begin. The video recordings of the experiments are projected on a 53 in. television for extraction of data. Before beginning to extract data from

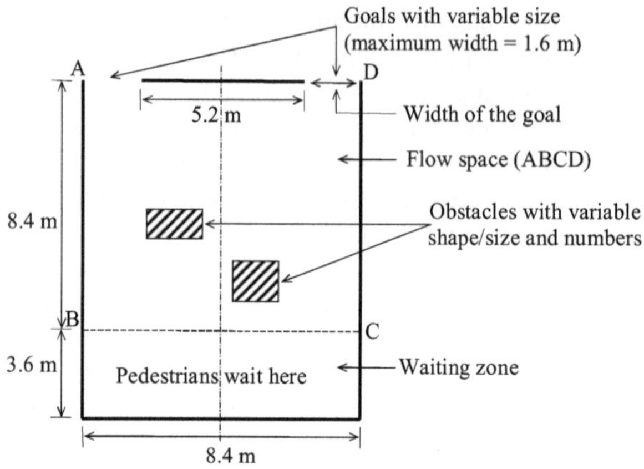

Fig. 1 Schematic of the flow space for evacuation from a hall

the recording, the grid recorded by the camera is painstakingly recreated by using removable marker to mark each line of the grid on the television screen. These lines on the screen constitute a virtual grid on which the motion of pedestrians recorded on tapes are played back. For every experiment, at every instant of time 't' the cells (i, j) that are occupied are noted. The set of experiments conducted here are as follows:

(i) Both doors are fully open; no obstacle present in the flow space
(ii) Left door is fully open and right door is half open; no obstacle present in the flow space
(iii) Both doors are half open; no obstacle present in the flow space
(iv) Left door is fully open and right door is closed; no obstacle present in the flow space
(v) Left door is half open and right door is closed; no obstacle present in the flow space
(vi) Both doors are fully open; a rectangular obstacle in the form of a barrier as shown in Fig. 2a is in the flow space
(vii) Both doors are fully open; an obstacle near the initial position of the pedestrians as shown in Fig. 2b is placed in the flow space
(viii) Both doors are fully open; a rectangular obstacle near the left door as shown in Fig. 2c is placed in the flow space.

In all the three cases where obstacles are present, height of the obstacle is kept low so that the doors are visible. Further, each experiment is done with two levels of initial density; in one set of experiments 14 people are used and in the other 35 people are used. The subjects are all male in the age group of 20–30. They are not familiar with one another.

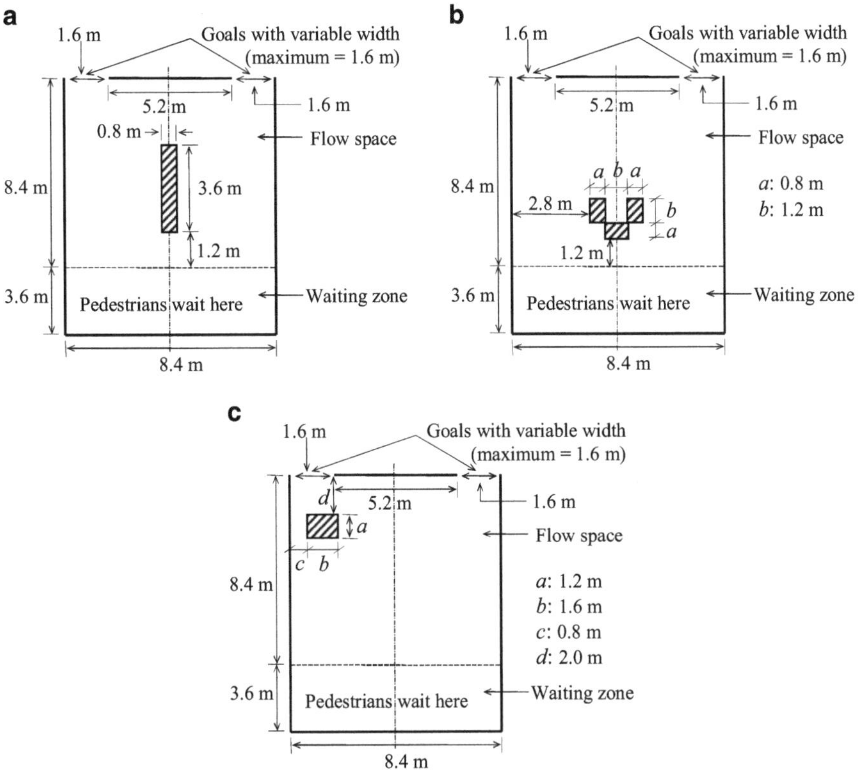

Fig. 2 Sketch of flow space for evacuation when both the doors are fully open and (**a**) a rectangular obstacle in the form of a barrier (**b**) an obstacle near the initial position of the pedestrians and (**c**) a rectangular obstacle near the *left door*

2.1 The Observed and Computed Parameters

For easy understanding of the data, the whole flow space is divided into zones. Data on the indicator variable $O^t_{i,j}$ (where, (i, j) indicates the cell co–ordinate and t the time instance), is collected from the video recording. In order to extract meaningful information the data on $O^t_{i,j}$ is aggregated. To represent the data, two parameters, zonal density and relative zonal density are defined with regard to time and their temporal profile is also studied to understand the progression of pedestrians. These parameters are defined as follows:
Zonal density of zone z at time t:

$$ZD^t_z = \sum_{(i,j)\in z} O^t_{i,j} \tag{1}$$

and Relative zonal density of zone z at time t:

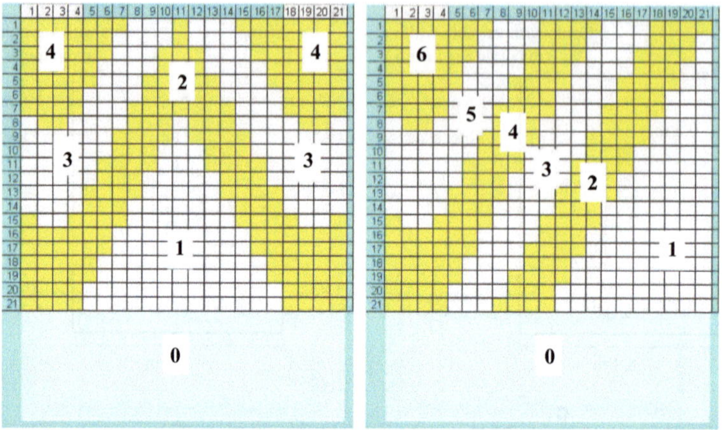

Fig. 3 Zoning pattern of the flow space when (**a**) both the doors are open (full or half) (**b**) *left door* is open (full or half); *right door* is closed

$$RZD_z^t = \frac{ZD_z^t}{N_0} \tag{2}$$

where, N_0 is the total number of pedestrians participating in the experiment.

Zones used here are shown in Fig. 3a, b. Zone 0 is outside the flow space and represents the waiting area for pedestrians. These definitions for zones are motivated by the positions of the goals and the hypothesis on how these goals might affect pedestrian motion. While defining these zones, cells which are assumed to get similarly impacted by the goals are grouped together. Number of zones and their positions are primarily user–defined and therefore qualitatively fixed. For example, Zone 4 in Fig. 3a (or Zone 6 in Fig. 3b) represent cells which have large attraction for pedestrians. Similarly, Zone 3 in Fig. 3a (or Zone 5 in Fig. 3b) represent cells which are reasonably attractive for pedestrians and so on.

3 Results and Discussions

Figure 4 shows the temporal profiles of RZD_z^t for the case when both the doors are fully open and there is no obstacle in the flow space for different number of persons. There are similar other figures for all the rest seven cases (which are not presented here due to space restriction). In the figures all the zones from 0 to 4 (or 0 to 6) are the same as shown in Fig. 3a, b; whereas, Zone E indicates the zone where pedestrians go after evacuation through the door(s), i.e., the space outside the flow space. From all the figures two points can be readily observed:

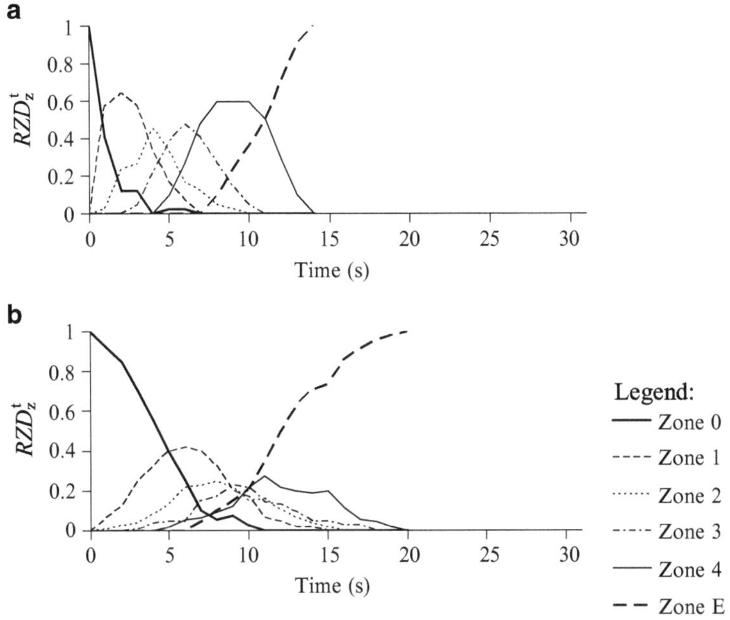

Fig. 4 Temporal profile of RZD_z^t when both the doors are fully open and there is no obstacle in the flow space for number of persons (**a**) 14 (**b**) 35

(i) As the initial number of pedestrians increase the total evacuation time increases because of the fact that at higher density pedestrians' movements get obstructed frequently, and
(ii) The RZD_z^t are sharper with higher peaks when number of pedestrians is lower. This indicates that at higher densities there is more uniformity in the way pedestrians distribute themselves over the flow space.

The results are further used to study the impact of width of opening and the impact of choice of goals (number of doors) on the pedestrian flow. The results are also used to study the impact of obstacle positioning, size and shape on pedestrian flow.

3.1 Study on Impact of Width of Opening

In this study five different experiments with different door openings are conducted. If one adds the width of the doors that are open then these five cases collapse into four; for the cases when both the doors are fully open, the total door opening is 3.2 m, for the cases when left door is fully open and right door is closed, the total door opening is 2.4 m, for the cases when (a) both the doors are half open, and (b) left door is fully open and right door is closed the total door opening is 1.6 m and for

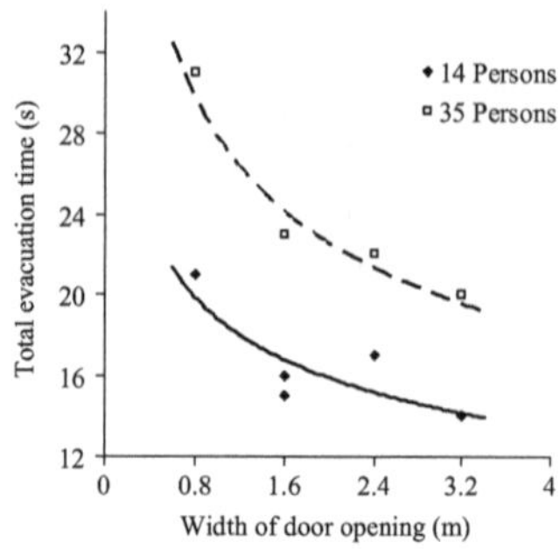

Fig. 5 Total evacuation time versus width of door opening for 14 and 35 persons

Table 1 Total evacuation time (in second) for different placement, size and shape of obstacles

	Number of persons	
Case	14	35
No obstacle	14	20
Barrier obstacle	15	22
Composite obstacle	14	20
Obstacle near left door	14	19

the cases when left door is half open and right door is closed the total door opening is 0.8 m. Figure 5 shows a plot of the total evacuation time (time required by all the pedestrians in the flow space to evacuate the flow space) versus width of door opening for experiments with 14 and 35 persons. From the figure it can be seen that:

(i) The total evacuation time versus width of door opening plot for 35 persons is above that for 14 persons,
(ii) Total evacuation time rises sharply when the width of door opening reduces beyond a certain value (from the figure the value seems to be around 1.6 m),
(iii) As expected the rise is sharper when more number of persons participating in the experiment,
(iv) The width of 1.6 m is obtained in two ways; in one case only one door is fully open and in the other case both doors are half open; the results show there is not much change in the total evacuation time thereby showing that the humans choose openings judiciously.

3.2 *Study on Impact of Obstacle Positioning, Size and Shape*

When the three different sets of obstacle positioning, size and shape are compared with the no obstacle case, it is seen that the total evacuation time of different cases differ randomly and minimally from one another. This suggests that for the densities studied here pedestrians adjust their speeds such that the obstacle does not impact the time they take to evacuate the flow space.

Table 1 shows total evacuation time for different placement, size and shape of obstacles.

4 Conclusions

Experiments on pedestrian motion in a closed space with varying number and width of exit locations and different obstacle positioning, size and shape yield among other things information on:

(i) How pedestrians choose a goal and how this affects their movement,
(ii) The impact of obstacles on pedestrian movements, and
(iii) How exit geometry (width) affects pedestrian evacuation process.

It needs to be mentioned that these types of experiments help to understand pedestrian motion inside enclosed spaces, which is essential for their design.

References

1. Hankin BD, Wright RA (1958) Passenger Flow in Subways. Operational Research Quarterly, Vol 9, No 2, pp 81–88
2. Oeding D (1963) Verkehrsbelastung und Dimensionierung von Gehwegen und anderen Anlagen des Fußgängerverkehrs. Forschungsbericht 22, Technische Hochschule Braunschweig
3. Older SJ (1968) Movement of Pedestrians on Footways in Shopping Streets. Traffic Engineering and Control, Vol 10, pp 160–163
4. Navin FPD, Wheeler RJ (1969) Pedestrian Flow Characteristics. Traffic Engineering, Vol 39, pp 30–36
5. Mori M, Tsukaguchi H (1987) A New Method for Evaluation of Level of Service in Pedestrian Facilities. Transp. Res. A 21, pp 223–234
6. Weidmann U (1993) Transporttechnik der Fußgänger. Schriftenreihe des IVT, No 90, ETH Zürich
7. Seyfried A, Steffen B, Klingsch W, Boltes M (2005) The Fundamental Diagram of Pedestrian Movement Revisited. J. Stat. Mech., P10002
8. Helbing D, Johansson A, Al-Abideen HZ (2007) Dynamics of Crowd Disasters: An Empirical Study. Phys. Rev. E 75, pp 046109
9. Morrall JF, Ratnayake LL, Seneviratne PN (1991) Comparison of CBD Pedestrian Characteristics in Canada and Sri Lanka. Transportation Research Record 1294, pp 57–61
10. Chattaraj U, Seyfried A, Chakroborty P (2009) Comparison of Pedestrian Fundamental Diagram Across Cultures. Advances in Complex Systems, Vol 12, No 3, pp 393–405
11. Polus A, Joseph JL, Ushpiz A (1983) Pedestrian Flow and Level of Service. Journal of Transportation Engineering, ASCE, 109, pp 46–56
12. Hoogendoorn SP, Daamen W (2005) Pedestrian Behavior at Bottlenecks. Transportation Science, Vol 39, No 2, pp 147–159
13. Seyfried A, Passon O, Steffen B, Boltes M, Rupprecht T, Klingsch W (2009) New Insights into Pedestrian Flow Through Bottlenecks. Transportation Science, Vol 43, pp 395–406
14. Henderson LF, Lyons DJ (1972) Sexual Differences in Human Crowd Motion. Nature, 240, p. 353
15. Young SB (1999) Evaluation of Pedestrian Walking Speeds in Airport Terminals. Transportation Research Record 1674, pp 20–26
16. Isobe M, Adachi T, Nagatani T (2004) Experiment and Simulation of Pedestrian Counter Flow. Physica A, Vol 336, pp 638–650
17. Kretz T, Gruenebohm A, Kaufman M, Mazur F, Schreckenberg M (2006) Experimental Study of Pedestrian Counterflow in a Corridor. J. Stat. Mech., P10014

18. Hoogendoorn SP, Daamen W (2004) Self-organization in Pedestrian Flow. Traffic and Granular Flow '03, pp. 373–382
19. Zhang J, Klingsch W, Schadschneider A, Seyfried A (2012) Ordering in Bidirectional Pedestrian Flows and its Influence on the Fundamental Diagram. J. Stat. Mech., P02002
20. Helbing D, Buzna L, Johansson A, Werner T (2005) Self Organized Pedestrian Crowd Dynamics: Experiments, Simulations and Design Solutions. Transp. Sc., 39, pp 1–24
21. Chattaraj U, Chakroborty P, Seyfried A (2010) Empirical Studies on Pedestrian Motion Through Corridors of Different Geometries. Proc. 89th Annual Meeting Transp. Res. Board, Washington, D.C., USA
22. Chattaraj U (2011) Understanding Pedestrian Motion: Experiments and Modelling. PhD Thesis, I.I.T. Kanpur, Kanpur, India
23. Brennan L, Mcdonald J, Shlomowitz R (1995) The Variation in Indian Height. Man in India, Vol 75, No 4, pp 327–337

Understanding and Simulating Large Crowds

S.M.V. Gwynne and A.A. Siddiqui

Abstract Simulation tools are often used to establish pedestrian and evacuee performance. The accuracy and reliability of such tools are dependent upon their ability to qualitatively and quantitatively capture the outcome of this performance. This paper investigates the relationship between the representation of low-level agent actions and the generation of reliable emergent, high-level conditions that can then be used to better understand the conditions that may develop in large crowds and mitigate against them. Data has been collected concerning the movement of pilgrims during the Hajj. This paper presents a simple framework for categorizing these real-world observations and then translating them into the simulated environment – extracting key information from the data collected to configure the simulation tool as required. Several scenarios are simulated using the buildingEXODUS model to test the impact of representing these observations to a greater or lesser degree. This enables the importance of low-level behaviours upon emergent conditions to be investigated, even when simulating large numbers of pilgrims attending the Hajj; i.e. in large crowds. The relationship between low-level agent actions and the high-level emergent conditions is investigated using analytical and simulation tools. This paper should help future researchers (1) identify and extract key factors from crowd data and then (2) appropriately configure simulation tools to represent agent behaviour and the subsequent emergent conditions produced (e.g. congestion, flow patterns, etc.).

S.M.V. Gwynne (✉)
Hughes Associates, UK

Fire Safety Engineering Group, University of Greenwich, Greenwich, UK
e-mail: sgwynne@haifire.com

A.A. Siddiqui
Fire Safety Engineering Group, University of Greenwich, Greenwich, UK
e-mail: A.A.Siddiqui@gre.ac.uk

V.V. Kozlov et al. (eds.), *Traffic and Granular Flow '11*,
DOI 10.1007/978-3-642-39669-4_22, © Springer-Verlag Berlin Heidelberg 2013

1 Introduction

The Hajj represents one of the largest annual human gatherings. Pilgrims congregate from around the world to perform religious rites within a fixed area and a specified period of time. This gathering is made more complex by the orchestrated movement of this population between locations and the performance of rituals at pre-determined times. As such, it provides a unique opportunity to (predictably) observe movement under a wide range of conditions: from low-density unmotivated movement, to high-density, driven movement.

The Hajj has been subject to a number of recent serious incidents; for instance, between 1994 and 2006 there were five major incidents leading to 1,053 fatalities and 1,295 injuries [1, 2]. These were produced through a combination of the population size (several million people) and population densities produced (significantly beyond $4\,\mathrm{p/m^2}$), the physical and environmental constraints, and the procedures employed. This suggests the need for close scrutiny of the event, irrespective of the general value of the data extracted. As such, the event demands close examination both to determine (1) ways in which pilgrim safety can be protected, and (2) lessons that can be learned and applied more generally.

A number of studies of the Hajj movement have been conducted [2], primarily characterising the flow-density relationships generated. In addition, there have been several attempts at simulating the pedestrian conditions present during the Hajj primarily to investigate mitigation opportunities [2–13]. However, little, if any, analysis has been performed relating the behaviours of the individual hajji (those individuals performing the Hajj) to the high-level conditions produced, the ability to simulate these behaviours within computational tools, and the improved fidelity in the overall results produced due to the representation of these behaviours. Recent work (examining other incidents) has been conducted investigating the relationship between low-level behaviours and high-level conditions [14] albeit in a simplistic manner.

Simulation tools are often used to establish pedestrian performance [15]. The accuracy and reliability of such tools are dependent on their ability to qualitatively and quantitatively capture the outcome of pedestrian performance; i.e. whether agents perform the expected acts and take the expected amount of time to complete them. In the past, it was often assumed that as crowds got larger so it became more reasonable to reduce the detail in which pedestrian behaviour was represented. This assumption was based as much on necessity as on theoretical understanding; i.e. given the computational expense of simulating large crowds. However, there was certainly a widely held opinion that low-level agent behaviours became less important, as the crowd became larger: that the omission of low-level behaviours had a negligible impact on the high-level outcomes produced. This confused the impact of an individual performing an action in isolation with the cumulative impact of many individuals performing an action (or set of actions). In addition, it is often assumed that it is possible to understand individual actions by isolating them from the surrounding procedural, physical and organizational influences – the external context of an event. By doing so, the individual response is divorced from the original context in which it occurred, and conclusions drawn from the results produced may be inappropriately associated primarily with the individual response rather than the procedural/organizational influences underlying them [16–22]. This may lead to misunderstanding and scape-goating.

This paper is an attempt to demonstrate the importance of understanding low-level pedestrian behaviours, even in large crowds: establishing the conditions under which they occurred, the factors that influenced them and the conditions that were produced by them. These behaviours can be represented in the assessment of pedestrian/evacuee performance. For this to happen, a sufficiently detailed and comprehensive data-set is required, locating coincident factors and responses, allowing a contextual understanding of events (at the individual, procedural and emergent levels), and enabling these events to be represented within simulation tools to a higher degree of fidelity, if possible. To do this, an analytical framework has been developed to categorize the observations made, and to relate these categories (e.g. relating low-level pedestrian actions observed and the constraining factors present at the time these actions were performed) with the high-level conditions that emerged. Observations have been made by the second author during his own pilgrimage. These observations enabled the actions of hajjis to be recorded, along with the surrounding influential factors and the resultant high-level conditions that emerged. These observations were made across the full pilgrimage and were categorized accordingly. A smaller event within the Hajj was examined in greater detail and is used as a case study during this paper. This event (the Sa'ee) was selected as numerous observations were made and also given the self-contained nature of the event itself. A summary of the observations and conditions recorded during this event are presented in this paper. This analytical framework is demonstrated by focusing on the Sa'ee observations, and then applying a simulation tool to explore the relationship between individual actions and emergent conditions. A similar approach has been adopted elsewhere to investigate evacuee behaviour during fires incidents [23, 24].

Three sets of simulations have been conducted using the buildingEXODUS model to examine the impact of representing hajji performance at different levels of detail: (a) hajji capabilities based on the model's default travel speeds, (b) observed hajji travel speeds, and (c) observed hajji travel speeds and the inclusion of several observed behaviours. These were conducted assuming a population of 15,000 hajjis. The intent of these sets of simulations was to determine the ability of the model to represent the observed conditions in a credible manner and enable a comparison between the high-level conditions produced and the conditions observed in the actual event. This should allow several questions to be addressed:

1. Is the simulation model employed able to represent the low-level behaviours observed?
2. Do the low-level behaviours represented have a comparable impact on the high-level conditions to those observed in the actual event?

3. Are the sets of simulated emergent conditions produced sufficiently different from each other to demonstrate the sensitivity of the model to the underlying assumptions made?
4. Is it important for the accuracy of the emergent conditions to represent low-level pedestrian behaviours, even in large-scale crowd movement?

2 Hajj Overview

The Hajj involves the annual pilgrimage of Muslims to Makkah, Saudi Arabia. During this physically demanding event, pilgrims perform a series of rituals in and around Makkah. The exact manner and order in which these actions are performed is influenced by the branch of Islam to which the individual belongs, religious rites, operational needs and the geographical and environmental conditions present. The nature of Hajj therefore means that pedestrians have different objectives and expectations even though they are using the same routes within the same time period. This variation complicates the management of the pilgrim movement, produces unbalanced loading of locations and routes, and increases the likelihood that pedestrian incidents occur. Given the scale of the event, the requirements of those involved, and the conditions produced, the Hajj presents an opportunity for safety scientists to (1) make observations regarding a range of pedestrian conditions produced, and (2) from these observations, improve our understanding and subsequently improve the conditions for the *hajjis*.

This cultural and geographical diversity has enormous implications for the management of the pilgrims and their subsequent behaviour. In 2010 (1431 in Hijiri), 2.8 million pilgrims were recorded as having attended the Hajj: 1 million were Saudis, while 1.8 million were non-Saudis [1–14]. This represents a 40 % increase on the overall figure recorded in 2004, posing significant difficulties in planning pilgrim movement and also in the provision of services. The majority of those performing the Hajj come from overseas and may well not be familiar with the spatial layout of the area or with the services available. In addition, a large number of volunteers, security personnel and those on business also arrive into the area during the Hajj period. Catering for people of different races, demographics, language/education skills, and cultural practices is a challenge for the organisers (i.e. the Ministry of Hajj of Saudi Arabia), especially given that the organisers are attempting a delicate balance: (a) ensuring the security and heritage of the site; (b) the safety of those involved; and critically (c) the (religious) experience of the pilgrims. The relative importance of these three components is not necessarily the same as comparable non-religious events; however, the need to balance competing priorities is typical of the management of large numbers of people.

The procedural requirements associated with religious rites place low-level obligations upon the individual hajji; i.e. religious rites have individual implications. Although the visiting hajji will be familiar with the high-level religious rites required of them (from their religious observances), many will be unfamiliar with the conditions that the mass performance of these rites produce and the impositions that are subsequently placed upon them.

Although the Hajj is unique, it is formed from a collection of activities and conditions that (when reduced to their constituent components) are comparable to other pedestrian events – it is only the combination of the activities and scale of the event that is unique. This enables the observations produced here to be of general value, and also enables the existing theories and tools available to aid in the mitigation of conditions at the Hajj. The data collected and the observations made can be used in a number of ways: (1) to inform the development of behavioural theories; (2) to inform the implementation of these theories within existing egress models; (3) to validate/verify the quantitative performance of these models; (4) to configure these models for use in examining large crowds; (5) to validate/verify the qualitative performance of these models. This paper primarily focuses on the last two applications.

3 General Approach

Observations were made during the Hajj. An analytical framework was developed to enable an understanding of the underlying (low-level) factors present and the (high-level) conditions that emerged because of them. This framework enabled the observations to be clearly categorized and, where possible, relationships derived. The relationship between these factors and subsequent conditions were then explored using the simulation tool.

3.1 Observation

A large number of observations were made during the 2008 Hajj and then catalogued; e.g. associated with particular locations, times and events. This included over 250 catalogued still photographs, several hours of video footage (collected directly by the second author or collated from other resources), and hundreds of manual observations. Each piece of material was interrogated to extract as many qualitative observations as possible. This helped build a picture of the relationship between the key factors at specific times and locations.

These observations focused on qualitative aspects of the pilgrimage; i.e. whether actions/events took place, the manner in which they took place, the hypothesized impact of these actions/events. Pictorial evidence, video evidence and manual observations were collected in support of this data collection effort. A limited number of quantitative measurements were also made, primarily relating to the population densities present, the travel speeds achieved and the completion times for specific events. These are supported by observations from other sources [1–14], and were found to be broadly representative. Real-world observations were made across the Hajj phases at two distinct levels: the low-level actions of the hajji and high-level emergent conditions that were the result of the combined hajji activities, environmental conditions and structural constraints. In addition, information was gathered on the procedural influences present; i.e. the set of religious, safety/security/operational, and cultural objectives that influenced the hajji actions. This allowed the authors to speculate on the objectives and motives underlying hajji actions; however, it should be noted that the inferred objectives are only based on known external factors – no survey was conducted. The information gathered also allowed the authors to hypothesise on the causal relationship between the combined hajji actions and the subsequent production of emergent conditions.

3.2 Analytic Framework

An analytical framework has been developed to help categorize the data collected, to enable relationships between low-level and high-level observations to be made, and for real-world observations and simulated results to be compared. This comparison enables us (1) to establish whether the simulation of low-level actions actually improves the accuracy and sophistication of simulated high-level conditions, and (2) to suggest which of the real low-level factors had the biggest impact on the high-level conditions produced and when they might have occurred. It has both forensic and testing applications. This approach was developed by first deriving the key factor groupings that might influence performance, and the level at which they might operate. The potential responses were then identified along with the constraints that influence response selection and inhibit performance. These factors, responses and constraints were derived from examining Hajj-specific data, but also from looking at broader sources in emergency and non-emergency literature [25–40]. The framework developed is presented in Fig. 1.

The framework to the observations made was applied in several stages. Stage [1] of the framework (shown in Fig. 1), was first applied to the full set of Hajj data. The observations made of the Hajj movement were grouped into the following categories:

1. Low-Level Actions [LLA] – actions that could be attributed to individual hajji (primarily from direct observation)
2. Procedural Factors [PrF] – factors that are not tied to an individual, but instead existed prior to the event and which had an impact on the motivations and objectives of the pilgrim population; e.g. religious rites, safety/operational procedures, services, etc. (combination of direct observation and secondary material)
3. Physical Constraints [PhC] – the physical and environmental constraints that influenced pedestrian movement; e.g. nature of the terrain, configuration/dimensions of the space, temperatures, etc. (combination of direct observation and secondary material)

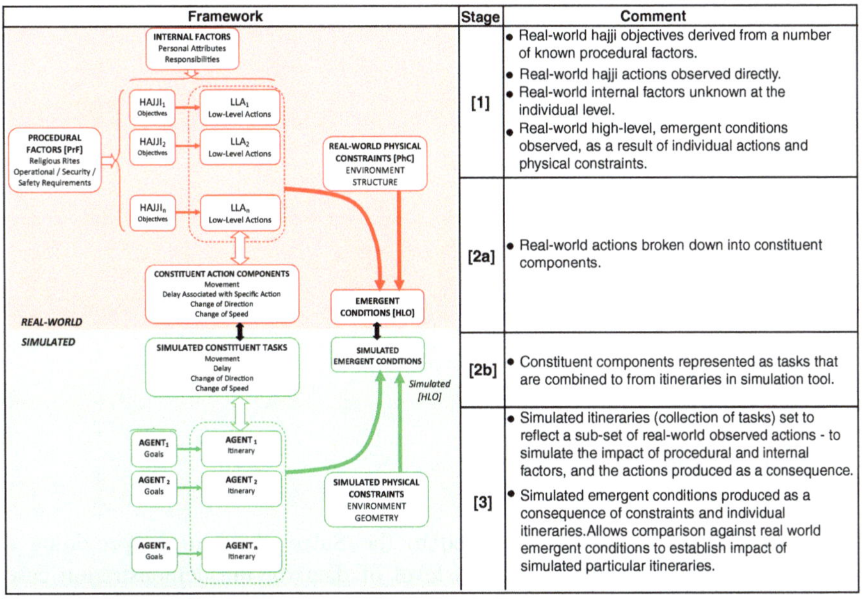

Fig. 1 Analytical framework

4. High-Level Outcomes [HLO] – emergent conditions produced through the combination of procedural factors, physical constraints and low-level actions (primarily from direct observation)

This categorisation enabled sufficiently complex, but manageable, sub-sets to be identified. A sub-set of the overall Hajj data-set was selected (see next section) for the application of Stages [2,3] of the analytical framework; i.e. decomposition of actions and representation within a simulation tool. This data was categorized in the manner discussed previously (i.e. a dictionary of key terms produced in Stage [1]). Each of the observed low-level actions was then decomposed into its basic constituent components in Stage [2a]: movement, delay in movement associated with specific actions, change of direction, and change of speed. All of the actions represented in the dictionary could be crudely represented within a simulation model through the recombination of these constituent components. This was particularly useful in translating the real world actions into simulated tasks selected to be represented within the model. It was also useful in gaining a simple overview of the core components of each action. These components were then translated into tasks within a simulation tool (in Stage [2b]). These were then grouped to form representative scenarios (in Stage [3]; e.g. geometrical, environmental conditions), allowing the functionality of the tool to be established and the relationships between the factors represented tested.

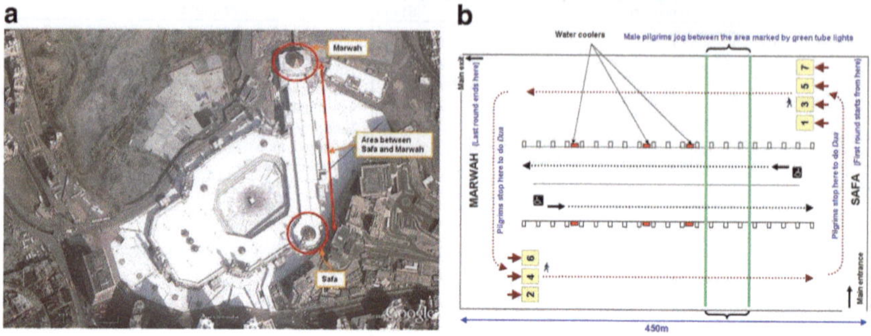

Fig. 2 (**a**) Location of the Sa'ee; (**b**) Sa'ee procedure

3.3 *Case Study: Sa'ee*

The analytical framework was applied to the Sa'ee observations providing a manageable (but sufficiently complex) level of data for the demonstration case presented here. The Sa'ee movement was selected as

1. A large number of observations were made leading to a sizeable data-set;
2. The Sa'ee event is relatively self-contained allowing the analysis to focus on specific events while managing the impact of external events;
3. The conditions present, the underlying procedural/behavioural factors and the outcomes produced were sufficiently complex to be of interest.

Sa'ee Description and Observations

Direct observations of the Sa'ee were made in two forms: (1) the observer detached from the Sa'ee, and (2) the observer embedded in the flow movement during the performance of the Sa'ee. These two perspectives allowed numerical observations to be collected at the individual (speed) and population (flow) level, and allowed descriptive observations to be made (experienced directly and observed). A large selection of still photographs, video footage and manual observations were collected. This was supported by existing observations made by other researchers, secondary material or material collected describing the religious requirements of the hajji during the Sa'ee, and the operational procedures in place [1–14]. This amounted to approx. 300 separate sources from which numerical and descriptive observations could be made.

During the Sa'ee, hajji have to make three circuits (seven 'rounds') between Safa and Marwah (the remains of two small hills now located inside the Grand Mosque in Makkah, Fig. 2a) in a manner indicated in the religious stories associated with this rite (Fig. 2b). The pilgrims start Sa'ee from Safa and while facing in the direction of the Ka'ba, make Dua (i.e. to supplicate themselves, delaying their movement). The hajji then head towards the Marwah while reading/reciting Duas (Fig. 2). Occasionally, this is performed in groups, with a leader reciting text while others in the group repeat it. Male pilgrims are required to increase their speed in the area marked by green lights, but walk at the normal pace elsewhere. Female pilgrims generally walk at a normal pace throughout. When the pilgrims reach Marwah, they again stop and make Dua, as they did at Safa. This is considered one round, as is the return movement from Marwah to the Safa. A total of seven rounds of movement are required to perform the Sa'ee, with the last round ending at Marwah.

The Sa'ee takes place over a path 450 m in length (between Safa and Marwah) and 40 m in width (in some places the presence of obstacles reduce the available width to approximately 24 m). The distance travelled during the seven rounds amounts to approximately 3 km [1, 41]. The two points and the path between them are inside the Grand Mosque (see Fig. 2a). Since 2009, the capacity of the Sa'ee has increased from an expected 44,000 to 118,000 pilgrims per hour, given the expansion of the area involved (from 24,400 to 87,000 m^2) and the number of floors involved (currently four covered floors and an additional uncovered floor) [1]. The temperature inside the structure is approximately 25 °C, which is significantly cooler than the external environment. This is aided by the use of ceiling and wall fans. The terrain is typically flat with firm footing (generally marble flooring), except for a slight incline at each end (at the approach to the Marwah and Safa) and some standing water around the water dispensers.

Anyone who is unable to walk or is impaired can use a wheelchair. Those (males) in wheelchairs are still required to accelerate in the designated area between the green lights and to stop at Safa and Marwah so that they can make Dua as is expected of the rest of the hajji population. Dedicated lanes are provided for wheelchair users located in the central area (marked in Fig. 2b). These dedicated lanes are provided on three of the five levels. These pilgrims are usually assisted by someone in pushing the wheelchairs. However, occasionally wheelchair users are also seen moving around the general area. Water coolers are provided at designated locations between Safa and Marwah for pilgrims: typically at six locations along the inner boundary of the route. These locations are usually monitored by dedicated maintenance staff whose responsibility it is to ensure that spilled water is removed to avoid people from slipping and falling.

As with the Hajj process in general, each of the observations made was closely scrutinized to identify elements that might influence performance. Examples are shown in Fig. 3. At this stage they were in the form of raw observations, associated with each other only in that they were contemporaneous; i.e. taken from the same source. These direct observations formed the basis for the detailed analysis to be performed. The observations were categorized according to the groupings suggested in the analytical framework, informing the identification of low level actions, physical constraints, procedural factors and high-level outcomes, complementing the information derived from secondary sources (derived from documentary and internet resources [1–14]).

Fig. 3 Sa'ee observations. (**a**) Dec 5, 2008 at 18:54: The day before the start of Hajj (Pilgrims moving towards Safa). (**b**) Dec 5, 2008 at 18:53: The day before the start of Hajj (Pilgrims moving towards Safa)

Application of Analytical Framework

The information derived from the observations made and the secondary material collected was collated using the analytical framework; namely, separated into Low-Level Actions, Procedural Factors, Physical Constraints and High-Level Outcomes (as indicated by Stage [1] in Fig. 1). This collation enabled a more detailed understanding of the underlying factors to be established, along with their relationship with the conditions produced. The presentation of a full set of derived factors is beyond this paper; therefore, a selection of the collated factors are presented in Tables 1–3. Table 1 includes the observed procedural factors deemed to have the most impact on performance. These are categorised into those relating to religious requirements and to operational and service procedures. In reality, the nature of the impact of the two sets of factors manifested in entirely different ways; however, they are represented using the same mechanisms within the simulation tool. This is achieved by initially breaking them down into their constituent parts and then representing these more basic components implicitly within the simulation tool.

Table 2 describes observations associated with individual hajji: their attributes and their actions. Both of these sets of observations are useful in configuring the agent attributes within the simulation tool; i.e. where the tool could represent the attributes and where they affected the results produced. Basic speeds ([S_LLA_1]) were derived from examining movement of the flow at several different locations around the circuit. These were collected from situations where congestion did not impinge upon free movement. These were compared with other reported movement speeds [19–21, 26, 27, 30–32, 36]. The collection of the accelerated speeds adopted by males between the green lights was more constrained [S_LLA_2], given that it only occurred in this area and inevitably involved some interaction between those moving more quickly and other hajji. It should be noted that the focus of this work is in gaining representative numerical values to configure the model use, rather than to provide a definitive numerical resource, which is beyond the scope of this paper.

Table 1 Sa'ee procedural factors

Religious requirements	Operational/service factors
[S_PrF_1] While wearing Ihram (if performing *Umrah*) and not wearing shoes, start from Safa and while facing in the *Ka'ba*, make *Dua*.	[S_PrF_9] Separate lanes for Safa to Marwah and from Marwah to Safa.
[S_PrF_2] Walk towards Marwah while reciting/reading *Duas*.	[S_PrF_10] Dedicated lanes for wheelchair users on three of the five floors.
[S_PrF_3] Area clearly marked by green lights where change of speed is required. Male hajji jog in the area bounded by the green lights.	[S_PrF_11] Water coolers at designated locations.
[S_PrF_4] Stop at Marwah, make *Dua* to complete one round.	[S_PrF_12] Small areas on either side of the lanes which can be used for resting/waiting.
[S_PrF_5] Walk towards Safa while reading/recting *duas* and jog (for male pilgrims only) between the specified area.	
[S_PrF_6] Stop at Safa, make *Dua* to complete another round.	
[S_PrF_7] Make seven rounds in total, last one ends at Marwah.	
[S_PrF_8] The Sa'ee stops at the time of five daily congregational *Salah*.	
During busy periods, this area is also used for *Salah*.	

The combined impact of the procedures, the individual attributes, the physical/environmental conditions and the low-level actions are shown in Table 3; i.e. the high-level outcomes that emerged as a result of these factors (Fig. 1). These reflect the qualitative and quantitative benchmarks against which the simulated results can be compared. In essence, the simulated results should be able to produce more reasonable quantitative and qualitative high-level results, as the low-level factors become representative of the original conditions.

As with travel speeds, population densities [S_HLO_1] were derived from a number of areas around route during peak use. These were calculated by selecting locations where the area available could be approximated, where head counts were performed. The time to complete the Sa'ee was derived from two sources: the author's direct experiences from him (and his group) completing the Sa'ee several times, and calculations derived from video footage. Both sources demonstrated that when relatively low level of congestion were experienced completion times were towards the minimum of the range (i.e. 45 min), while where a hajji consistently experienced congestion the completion times were towards the maximum of the range (75 min). These figures were moderated by individual abilities. However, given the scale of the movement and the conditions that can develop, these results should only be considered as indicative.

Table 2 Summary of Sa'ee internal hajji factors and low-level observations

Individual attributes	Gender [male/female]	Walking aid [none/stick/chair]
	Physical fitness [impaired/unimpaired]	Carrying shoes [yes/no]
	Age [yound adult – elderly]	Education level [not possible to ascertain visually]
	Wearing Ihram [yes/no]	
	Level of experience and training [inexperienced – experienced with procedures/space]	Cultural background [not possible to reliably ascertain visually]
	Encumbered [unencumbered/object/child]	Reading [yes/no]
Low-level observations	[S_LLA_1] Basic movement speed [averaging 1.0 m/s, ranging from 0.6 to 1.3 m/s, 50 observations]	
	[S_LLA_2] Speed change in the specified area with male pilgrims jogging at 2.0 m/s [ranging from 1.3 to 2.4 m/s, 30 observations] between green lights.	
	[S_LLA_3] *Duas* at both Safa and Marwah. [15–20 % of population delaying for between 1 and 3 min]	
	[S_LLA_4] Direction change at the end of each round at Safa and Marwah.	
	[S_LLA_5] Group coherence and reformation, with males typically leaving women and waiting for them after they have accelerated away in green zone. [10–20 % male population remaining in place for 10–30 s].	
	[S_LLA_6] Individuals stop to drink water and/or rest. [5 % of population remain for between 2 and 5 min, queuing and drinking at designated sites].	
	[S_LLA_7] Leave shoes in allocated areas in the mosque or keep them in a hand bag.	
	[S_LLA_8] Individuals read/recite *Duas* while walking.	
	[S_LLA_9] Movement down from the Safa hill onto the flat terrain.	
	[S_LLA_10] Some impaired individuals make use of the general purpose lanes as well. Some unimpaired pilgrims walk in the dedicated wheelchair only lanes.	
	[S_LLA_11] Individuals stop to rest/wait at non-designated areas, along the side of the path. [5 % of population remain for between 4 and 5 min]	
	[S_LLA_12] Some pilgrims are accompanied by young children/elderly relatives.	
	[S_LLA_13] Some pilgrims carry small items with them, e.g. books, handbag, etc.	
	[S_LLA_14] Given variation in travel speeds, overtaking is required.	
	[S_LLA_15] Male group members slow down after crossing the specified jog area.	
	[S_LLA_16] Between green lights those in wheelchairs pushed more quickly to approximate jogging speed.	
	[S_LLA_17] Hajji groups form and maintain during movement. This is occasionally accompanied by the grouper leader reciting prayers as they move.	
	[S_LLA_18] Sometimes hajji get hit by wheelchairs as they change direction.	
	[S_LLA_19] Staff cleaning/sweeping floors.	

Table 3 Sa'ee high level outcomes

[S_HLO_1] Population densities of between 1.0 and 3.0 p/m^2 observed locally (50 observations).
[S_HLO_2] Variation in hajji travel speeds leading to overtaking and local navigation.
[S_HLO_3] Unimpaired using the designated impaired lanes producing delays and local congestion.
[S_HLO_4] Congestion produced by hajji waiting along the edges of the path encroaching upon the movement.
[S_HLO_5] Designated water cooler areas produce congestion impeding the flow as people queue.
[S_HLO_6] Congestion developed at Safa and Marwah where pilgrims stopped to perform *Duas*. This impeded the flow of hajji and also impeded those who have completed their *Duas* from rejoining the flow.
[S_HLO_7] The flow is typically relatively smooth moving. However, it is not uniform, given the social groupings that exist within it. Strata therefore appear in the flow. The flow is further disrupted by the acceleration/deceleration of hajjis in relation to the green lights. Strata develop as hajji accelerate forward (with female hajji remaining behind) and then decelerate at the end of the green area.
[S_HLO_8] Contra-flows and cross-flows were produced given the range of hajji objectives and local activities.
[S_HLO_9] The hajji typically completes Sa'ee between 45 and 75 min, according to the (40) observations made.

As an example of the relationships drawn, an explanation of the development of congestion at key locations is presented. Congestion appeared at either end of the circuit, as pilgrims delayed their movement while performing Dua. Congestion also developed along the inner boundary of the route. This was due to hajji queuing at water coolers, and when hajji stopped at undesignated locations to rest along the perimeter of the circuit (again typically along the inside of the route). This reduced the effective width of the route. Congestion was also produced when hajji interaction increased, particularly where there were larger variations in hajji travel speeds. The acceleration of male hajji between the green lights stratified the flow into waves, given the speed variations. Initially, the flow of hajji approached the green lights. At this stage the hajji are mixed and travelling at their typical walking speed. This is constrained by the population densities present, making the movement rates more uniform than might otherwise be the case. On passing the green lights, male hajjis accelerated, jogging away from the rest of the hajji. This often required them weaving through slow moving hajji in order to gain sufficient space to accelerate and jog through the designated area. Once beyond the second set of green lights, male hajjis decelerated. Some of these continued on at normal speed (i.e. those that were not in social groups), while others waited for the rest of their party. At this stage there were three waves of hajji in the area: the male hajji who had accelerated and who were not waiting, male hajji who accelerated and were now waiting, and the other hajji who did not initially change their speed. This collapsing of the population into waves increased local population densities that then influenced the travel speeds achieved. In this manner, the identification of low-level actions and procedure factors were related to the emergence of high-level conditions.

Figure 1 outlines the steps involved in the transition from observation (see Tables 1–3) to simulation (see Sect. 4):

1. Elements observed, categorized and prioritized given their influence on performance (Stage [1]).
2. Prioritized elements broken down into their constituent parts (movement, delay, direction change, speed change) (Stage [2a]).
3. Constituent parts represented as itineraries within the simulation tool and allocated to the agent population in accordance with the observations made (Stage [2b]);
4. Simulated emergent conditions (qualitative and quantitative) compared against the original high-level conditions (Stage [3]).

The development of applicable scenarios from these constituent parts and the results produced, as part of the Stage [3] framework analysis, are now discussed.

4 Modelling Hajji Movement

Simulations were conducted using advanced features of the buildingEXODUS model [42]. This enabled the simulated impact of the low-level agent behaviours upon the high-level scenario conditions to be examined. buildingEXODUS is a simulation tool that enables the representation of pedestrian and evacuee movement, in conjunction with the procedural, structural and environmental factors present. The model represents the population at the individual level allowing the attributes, tasks and capabilities of the population to be modified and to influence (individual and high-level) performance. The model is able to produce a range of output, allowing performance to be examined at the individual, group, or population level [42]. Within the model, goals and tasks can be imposed upon the agents reflecting the action components derived from the real-world observations; i.e. an attempt to qualitatively represent the individual actions such as those shown in Fig. 2b. The high-level results simulated can then be compared to the emergent conditions noted in the original observations to determine whether they are acceptable. Three scenarios have been examined to investigate the relationship between the simulation of low-level conditions and the high-level outcomes produced:

1. (a) Using the model's default movement parameters
   (b) Using modified movement parameters to reflect the observed travel speeds
   (c) As (2), with the inclusion of a sub-set of the procedural activities observed represented as tasks inserted to reflect low-level actions.

These were conducted for a single agent (to validate the approach) and a population of 15,000 to examine the impact of agent interaction upon the conditions produced. Only the simulations involving the 15,000 agents are discussed here.

The assumptions employed in the model during the three scenarios are outlined in Table 4. In Scenario [A1], the default travel speed ranges of 1.2–1.5 m/s were

Table 4 Scenarios examined

Scenario	Travel speeds (m/s)	Mobility factor	Tasks represented
[A1] Default	1.35 [1.2–1.5]	0.7–1.0	None
[A2][a] Physical	1.0[b][0.8–1.2]	0.7–1.0	None
[A3] Physical and behavioral	Walking 1.0[b] [0.8–1.2] Jogging 2.0[c] [1.8–2.2]	0.7–1.0	1. Speed modification in area denoted by green lights −20 % probability (male only)[c] 2. Performance of Dua. [1–3 min] delay. 20 % probability.[d] 3. Stop at water dispenser. [0.75–1.5 min] delay. 5 % probability.[e] 4. Group reform post-green zone. [0.25–0.5 min] delay. 20 % probability.[f] 5. Rest at side of path. [4–5 min] delay. 5 % probability.[g]

[a] The slight discrepancy between the travel speeds observed and those imposed within the model are due to manner in which the speeds are simulated and the interaction between theses speeds and other factors, such as agent mobility

[b] Based on [S_LLA_1]

[c] Based on [S_LLA_2]

[d] Based on [S_LLA_3]/[S_LLA_4]

[e] Based on [S_LLA_6]

[f] Based on [S_LLA_5]/[S_LLA_15]/[S_LLA_17]

[g] Based on [S_LLA_11]

employed. In addition, the assumption that 10 % of the population were impaired was also employed, as indicated by Boyce [43]. Within the model this was represented as a coefficient applied to the travel speed (Mobility Factor). This was assumed to vary between 0.7 and 1.0, and produced up to a 30 % reduction factor in travel speeds for the impaired section of the population (i.e. 10 % of the population). Wheelchair users were not represented during any of the scenarios; the model is not able to simulate wheelchair movement. Therefore, the central lanes (i.e. lanes dedicated for wheelchair users and the severely impaired) were excluded in all cases. In Scenario [A2], the travel speeds imposed were broadly representative of those observed. Given the range of factors discussed previously (e.g. individual, procedural, environmental, structural, etc.), the unimpeded speeds observed averaged around 1.0 m/s. Again, those with movement impairments were also represented as in Scenario [A1]. In Scenario [A3], the movement parameters were identical to those assumed in Scenario [A2]; however, behavioural tasks were introduced to broadly represent some of the key observations made (see Tables 1 and 2). As hajji travelled between the Marwah and the Safa they had the potential for being assigned itineraries to complete. This is a functionality within the buildingEXODUS model that allows agents to be assigned tasks that take a period of time to complete; i.e. implicitly representing the actions performed by an individual. These can either be pre-determined or assigned stochastically according to the individual attributes, location and experiences during the simulation. In this case, each agent was assigned tasks when passing pre-assigned locations. The assignment of these tasks was dependent on their attributes (e.g. gender), and the percentages described in Table 4. If assigned, this meant that an agent had to perform (one or more from) Dua, rest, wait after accelerating between the green lights, and/or get water at the assigned location, for a period of time between the ranges associated with that task. In effect, the itinerary task was used to represent five behavioural tasks.

It should be recognized that the set of tasks assigned to the population only represent a small sub-set of the behaviours observed (see Table 2). Given this, the simulated differences produced by including tasks (both in terms of the quantitative differences and the qualitative emergent conditions produced) should be considered a crude estimate of those that might be generated.

A representative population of 15,000 hajji were simulated. This population size was selected as it was representative of the conditions viewed and produced the population densities typically observed. This population was randomly located about the space, typically producing population densities of between 1.0 and $3.0\,\mathrm{p/m^2}$. Each scenario was run five times. The location of the hajjis was randomized and the time for the population to complete the 3.5 circuits observed. For each scenario the minimum, maximum and average times for the 15,000 hajji to circumambulate the space was recorded, ensuring that a relatively broad and representative set of completion times was produced; i.e. this would have represented the population involved in each simulation, and the different populations used across the five simulations. This approach was adopted as there was not one representative completion time for the entire population as might normally be the case in an evacuation of a space, for instance.

5 Results

Three scenarios were simulated where the hajji population of 15,000 circumambulated the space. The numerical results from these scenarios are shown in Table 5. When the default buildingEXODUS speeds were assumed during Scenario [A1], the simulated hajji completed the route in, on average, 57 min. When the observed travel speeds were employed in Scenario [A2] the average completion time increased to 65 min. As expected, Scenario [A2] produces longer completion times than Scenario [A1] (a 14 % increase), given the reduced travel speeds assumed. It is apparent that both sets of results are 'credible' in that the fall within the observed range of completion times (45–75 min); however, the qualitative conditions produced were different from those observed in reality.

Table 5 Results produced when the full hajji population was simulated

Scenario	Completion time (s)
[A1]	3,450
Default	[3,303–3,722]
[A2]	3,912
Physical	[3,542–4,557]
[A3]	4,426
Physical and behavioural	[4,211–5,316]
Observed	[2,700–4,500]

Fig. 4 Population Density Contour of Scenario. Hajji movement is from *right* to *left*

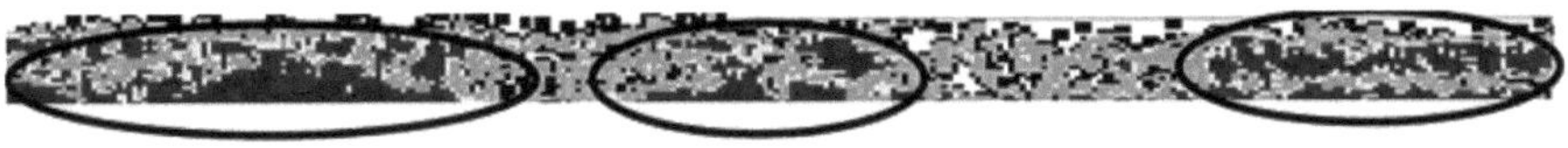

Fig. 5 Population Density Contour of Scenario. Hajji movement is from *right* to *left*. High densities are *circled*

Although there is a 14 % increase in the completion time for Scenario [A2] over Scenario [A1], the emergent (qualitative) conditions produced during the simulations were similar (and different from the actual conditions observed, see Table 3). Figure 4 shows a Population Density Contour of a section of the geometry traversed during the Sa'ee. This is a feature within the buildingEXODUS model that demonstrates the population densities (i.e. the number of people per square metre) produced at a particular time during the simulation. This is employed here as it better displays the conditions experienced given the scale of the geometry represented. In this contour, dark areas typically represent $4\,p/m^2$. It is apparent from Fig. 4 that the contour is fairly uniform and does not have a discernible pattern or include noticeable conditions specific to particular locations; for instance, congestion along the inner wall, congestion produced as people wait for others, etc., as was observed during the original event. It should also be noted that the section shown in Fig. 4 includes the area between the green lights, where significant fluctuations in the population densities were produced. This area is located in the middle third of the graphic. This example is typical of the conditions produced during Scenario [A1] and [A2] along the entire Sa'ee route.

In Scenario [A3] the average time for the simulated hajji to complete Sa'ee was 73 min. This is at the upper end of the observed completion times representing a 13 % increase over the average Scenario [A2] completion time (although the value ranges produced for the two scenarios do overlap). However, more significant is the changes in the emergent conditions that have been produced. An example of this is shown in Fig. 5. This shows the same section of the geometry that appears in Fig. 4 at a comparable time in the movement of the hajji (approximately 10 min into

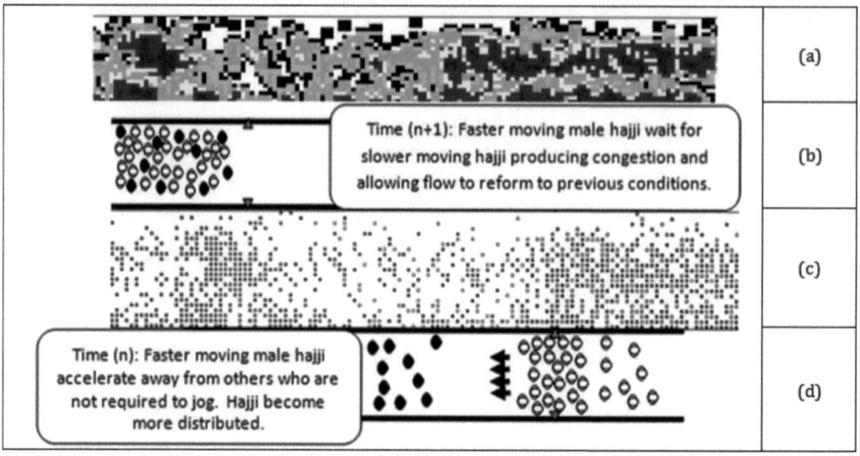

Fig. 6 Population Density Contour of Scenario [A3] – focus on area between green lights

the simulation). The flow through the section is uneven with population densities varying significantly with changes in the conditions specific to particular locations. For instance, it is apparent that there is additional congestion along some sections of the inner boundary. This is produced through agents stopping for water or resting along the inner boundary of the route and replicates both the local congestion and impedance that was observed. A similar effect was also produced at either end of the route, as a proportion of the agents delayed simulating them performing Dua.

The conditions around the green lights were more complex. In Fig. 6a, the area between the green lights shown in Fig. 5 is magnified. (This is from the centre point in the graphic to the right-hand edge of Fig. 5.) The uneven distribution of population densities is immediately apparent. The typical 2D buildingEXODUS view is shown (with individual hajji displayed, Fig. 6c), along with explanatory schematics (Fig. 6b, d). These views are presented to help identify the underlying causes producing the emergent conditions evident; i.e. by identifying individual actions and the conditions produced (see Table 3).

In Fig. 6d, the hajji population are shown approaching the green lights where some of them are expected to accelerate. In this section the movement of the hajji is complex with some heading towards the inner boundary to rest, retrieve water or moving directly forward. This is reflected in the large number of agents interacting prior to reaching the green lights – the complex movement impeding their progress leading to additional congestion (see the high densities evident at the right-hand end of Fig. 6a). The male hajji then accelerate away from the other hajji causing two obvious strata to form: one high-density group of relatively slow-moving hajji; the second, formed from a more dispersed group of male hajji accelerating away from first group. These distinct strata are apparent in the 2D buildingEXODUS image (the high density area is shown in the right-hand third of Fig. 6c, while the dispersed group is visible in the middle third). Once the accelerating hajji reach the end of the green lighted section, a portion of them delay, simulating them waiting for slower moving group members (see Fig. 6b). This leads to congestion as other hajji interact with and overtake those waiting. This in turn produced congestion at the edge of the green lighted section forming a grouping of hajji across the width of the path, as evident in Fig. 6a, c. These results are representative of the original observations made. Therefore, in this instance, not only are the quantitative results representative, but they are based on acceptable qualitative conditions.

6 Discussion

Four questions were posed in the introduction:

1. Is the simulation model employed able to represent the low-level behaviours observed?
2. Do the low-level behaviours represented have a comparable impact on the high-level conditions produced to those in the actual event?
3. Are the sets of simulated emergent conditions produced sufficiently different from each other to demonstrate the sensitivity of the model to the underlying assumptions made?
4. Is it important for the accuracy of the emergent conditions to represent low-level pedestrian behaviours, even in large-scale crowd movement?

These are now discussed.

The model is able to represent simple itineraries based on the recombination of basic behavioural tasks: movement, delay in movement associated with specific actions, change of direction, and change of speed. From these basic components, relatively complex itineraries were formed allowing procedures and behaviours to be simulated. Once the original low-level behaviours had been identified, the sub-set of those selected for representation within the model could be implicitly included. Once represented, these behaviours had an impact on the emergent conditions produced. The emergent conditions produced by the model as part of the simulated scenario were noticeably different once the low-level task itineraries were introduced. The manner in which the agents used the space changed producing more complex dynamics and different population distributions throughout the space. The fact that the inclusion of more representative agent task itineraries produced more credible high-level outcomes is encouraging. Simply producing reasonable quantitative estimates irrespective of the underlying conditions can lead to incorrect conclusions being derived. It may also indicate that a model is insensitive to the underlying conditions employed and the manner in which they interact. That was not the case here. In this instance, the model was able to cope with a large crowd movement; represent the low-level actions observed; and then produced credible emergent conditions when these low-level actions were employed that differed from when the low-level actions were excluded. As such, this model may be a useful diagnostic tool in investigating the underlying causes of emergent conditions, if applied expertly. The model was not able to produce identical conditions. This is highly unlikely in any situation, and was not expected here given that numerous low-level behaviours were not simulated and that those that were included were only represented implicitly. However, the inclusion of the sub-set of low-level behaviours certainly produced a closer approximation of the qualitative conditions originally observed than when they were not included.

It is important to recognize the limitations of computational models and of their application in investigating the relationship between low-level factors/actions and high-level conditions. Broadly speaking, there are four different relationships between low-level factors and high-level outcomes that can exist with a model. This paper presents tentative evidence that the buildingEXODUS model examined was able to generate increasingly credible emergent conditions as the low-level behaviours/factors became more representative. This applied to both qualitative and quantitative aspects of the simulation. This only relates to the case examined, but suggests that the relationship between low-level and high-level factors is reasonable within the model. This may not always be the case, in other models. Often, models are said to produce accurate quantitative outcomes with little comment made on the low-level factors/actions employed. Some models are not able to vary the low-level conditions represented and so produce high-level conditions that are insensitive to the low-level factors employed. In effect, the low-level factors assumed are only coincidental to the high-level outcome, rather than causal or influential. Similarly, other models are able to produce credible high-level outcomes irrespective of the accuracy of the low-level conditions represented. In both cases, this may lead to inaccurate conclusions being drawn concerning the relationship between low-level factors and high-level outcomes. Some models are able incorporate a representative set of low-level conditions, which do not combine to produce credible high-level outcomes. These models then have the scope to produce more complex and representative individual behaviours (i.e. low level factors), but not the sophistication to accurately represent the interaction between these individual behaviours. Finally, some models are neither capable of representing representative low-level conditions, nor generating credible high-level outcomes. The use of these models is limited, but at least predictably so.

The relationship between low-level and high-level conditions matters as it influences the credibility of any findings produced by the model, especially when used to investigate causal relationships. It is therefore important to understand the theoretical/empirical basis for these relationships and how they are represented within the model being employed prior to applying such a model to establish causal relationships as part of an analysis.

7 Conclusion

This paper presents an analysis of the pilgrim movement at the Hajj. This analysis involved the collection of data, the categorization of this data, and the investigation of the underlying dynamics of a particular event using a simulation tool.

The buildingEXODUS model proved capable of representing a set of low-level agent actions indicative of those seen during the original observations, which then combined to produce reasonable completion times and emergent high-level conditions. As such the model was able to discriminate between the impact of different low-level factors and their impact upon the outcome of an event. It was able, albeit imperfectly, to produce reasonable quantitative completion times and qualitative emergent conditions. This should enable the model to credibly produce simulated outcomes, and be used to explore potential causal factors of identifiable emergent conditions in real large-scale incidents. Given that this model (and potentially others) is able to adequately represent these relationships, the importance of understanding and simulating low-level actions in large crowd scenarios, given their impact on the conditions that emerge, seems apparent.

References

1. http://www.hajinformation.com/main/l.htm.
2. S. A. AlGadhi, "Jamarat bridge: Mathematical models, computer simulation and hajjis safety analysis," tech. rep., Ministry of Public Works and Housing - Saudi Arabia, 2003.
3. A. Johansson, "From crowd dynamics to crowd safety: a video-based analysis," *Advances in Complex Systems*, vol. 4, p. 497–527, 2008.
4. H. Klüpfel, "The simulation of crowd dynamics at very large events," in *Traffic and Granular Flow '05* (A. Schadschneider, ed.), Traffic and Granular Flow '05, Springer, 2006.
5. A. M. Shehata, "Using 3d gis to assess environmental hazards in built environments," *Journal of Al Alzhar University*.
6. Z. Zainuddin, "Simulating the circumambulation of the kaaba using simwalk," *European Journal of Scientific Research*, vol. 38, pp. 454–464, 2009.
7. N. A. Koshak, "Analyzing pedestrian movement in mataf using gps and gis to support space redesign," in *Proceedings of the Ninth International Conference on Design and Decision Support Systems in Architecture and Urban Planning*, 2008.
8. D. Clingingsmith, "Estimating the impact of the hajj: Religion and tolerance in islam's global gathering," tech. rep., Harvard University, 2008. CID Working Paper No.159.
9. A. Addelghany, "Microsimulation assignment model for multidirectional pedestrian movement in congested facilities," in *Bicycles and Pedestrians; Developing Countries 2005*, vol. 1939, pp. pp123–132, 2005.
10. H. Klüpfel, "The simulation of crowd dynamics at very large events calibration, empirical data, and validation," in *Proceedings of the 3rd International Conference on Pedestrian and Evacuation Dynamics*, 2006.
11. S.A.H. AlGadhi, "Modelling crowd behavior and movement: Application to the makkah pilgrimage," in *Transportation and Traffic Theory*, pp. 59–78, 1990.
12. N. Hussain, "Cdes: A pixel-based crowd density estimation system for masjid al-haram," *Safety Science*, 2011. (Online).
13. N. Zarboutis, "Design of formative evacuation plans using agent-based simulation," *Safety Science*, vol. 45, p. 920–940, 2007.
14. M. Moussaïda, "How simple rules determine pedestrian behaviour and crowd disasters," in *Proceedings of the National Academy of Sciences of the United States*, 2010.
15. S. Gwynne, "Simulating a building as a people movement system," *Journal of Fire Sciences*, vol. 27, pp. 343–368, 2009.

16. D. Vaughan, "The dark side of organizations: Mistake, misconduct, and disaster," in *Annual Review of Sociology*, vol. 25, pp. 271–305, 1999.
17. E. Hollnagel, "Understanding accidents - from root causes to performance variability," in *Proceedings of the 2002 IEEE 7th Conference on Human Factors and Power Plants*, pp. 1–6, 2002.
18. J. M. Lewis, *Theories Of The Crowd: Some Cross Cultural Perspectives*. Easingwold Papers, 1990. ISBN1874321043.
19. S. A. H. AlGadhi, "Simulation of crowd behavior and movement: Fundamental relations and application," in *Transportation Research Record*, vol. 1320, pp. 260–268, 1991.
20. S. AlGadhi, "Review study of crowd movement and behavior," *J. of King Saud Univ*, vol. 8, no. 1, pp. 77–108, 1996.
21. J. J. Fruin, *Engineering For Crowd Safety*, ch. The Causes And Prevention Of Crowd Disasters. 1994. 0444899200.
22. J. Drury, "Cooperation versus competition in a mass emergency evacuation: A new laboratory simulation and a new theoretical model," *Behavior Research Methods*, vol. 3, no. 41, pp. 957–970, 2009.
23. W. Grosshandler, "Draft report of the technical investigation into the station nightclub fire," Tech. Rep. NCSTAR 2, National Institute of Standards and Technology, 2005.
24. J. D. Averill, "Federal investigation of the evacuation of the world trade center on september 11, 2001," in *Proceedings 3rd International Conference on Pedestrian and Evacuation Dynamics*, 2005.
25. J. Bryan, *The SFPE, Handbook of Fire Protection Engineering(2nd Edition)*, ch. Behavioural Response To Fire And Smoke, pp. (1–241)–(1–262). National Fire Protection Association, 1996.
26. H. Nelson, *The SFPE, Handbook of Fire Protection Engineering(2nd Edition)*, ch. Emergency Movement, pp. (3–286)–(3–295). National Fire Protection Association, 1996.
27. S. A. H. AlGadhi, "A speed-concentration relation for bi-directional crowd movements with strong interaction," in *Pedestrian and Evacuation Dynamics* (S. et al, ed.), pp. 3–20, 2001.
28. C. Saloma, "Self-organized queuing and scale-free behavior in real escape panic," in *Proceedings of the National Academy of Sciences of the United States*, vol. 100, pp. 11947–11952, 2003.
29. C. E. Nicholson, *Engineering For Crowd Safety*, ch. The Investigation Of The Hillsborough Disaster By The Health And Safety Executive, pp. 361–370. Elsevier, 1994. ISBN 0444899200.
30. V. M. Predtechenskii, *Planning For Foot Traffic Flow In Buildings*. Amerind Publishing Co., 1978.
31. V. V. Kholshevnikova, "Recent developments in pedestrian flow theory and research in russia," *Fire Safety Journal*, vol. 43, p. 108–118, 2008.
32. S. M. V. Gwynne, *The SFPE Handbook of Fire Protection Engineering (4th edition)*, ch. Employing the Hydraulic Model in Assessing Emergency Movement, p. 3–355. National Fire Protection Association, 2008.
33. J. A. . Bryan, "Selected historical review of human behavior in fire," *Fire Protection Engineering*, vol. 16, pp. 4–10, 2002.
34. G. Proulx, *The SFPE Handbook of Fire Protection Engineering (4th edition)*, ch. Evacuation Time. National Fire Protection Association, 2008.
35. J. Sorenson, "Warning and evacuation: answering some basic questions," *Industrial Crisis Quarterly*, vol. 2, pp. 195–209, 1988.
36. K. Ando, "Forecasting the flow of people," *Railway Research Review*, vol. 45, pp. 8–14, 1988.
37. S. M. V. Gwynne, "Optimizing fire alarm notification for high risk groups research project," Tech. Rep., The Fire Protection Research Foundation, 2007.
38. S. M. V. Gwynne, "Conventions in the collection and use of human performance data," Tech. Rep. NIST GCR 10–928, National Institute of Standards and Technology, 2010.
39. E. Kuligowski, "Process of human behavior in fire," in *Proceedings of the Human Behaviour in Fire Symposium*, pp. 627–632, 2009.

40. E. Kuligowski, "What a user should know about selecting an evacuation model," *Fire Protection Engineering Magazine*, Human Behaviour in Fire Issue, 2005.
41. http://www.en.wikipedia.org/wiki/AlSafa_and_Al-Marwah.
42. E. D. Kuligowski, "A review of building evacuation models, 2nd edition," Tech. Rep. NIST TN - 1680, National Institute of Standards and Technology, 2010.
43. K. E. Boyce, "Towards the characterization of building occupants for fire safety engineering," *Fire Technology*, vol. 35, no. 1, 1999.

Experimental Study of Pedestrian Flow Through a T-Junction

Jun Zhang, W. Klingsch, A. Schadschneider, and A. Seyfried

Abstract In this study, series of experiments under laboratory conditions were carried out to investigate pedestrian flow through a T-junction, i.e., two branches merging into the main stream. The whole duration of the experiments was recorded by video cameras and the trajectories of each pedestrian were extracted using the software *Petrack* from these videos. The Voronoi method is used to resolve the fine structure of the fundamental diagram and spatial dependence of the measured quantities from trajectories. In our study, only the data in the stationary state are used by analyzing the time series of density and velocity. The density, velocity and specific flow profiles are obtained by refining the size of the measurement area (here 10×10 cm are adopted). With such a high resolution, the spatial distribution of density, velocity and specific flow can be obtained separately and the regions with higher value can be observed intuitively. Finally, the fundamental diagrams of T-junction flow is compared in three different locations. It is shown that the fundamental diagrams of the two branches match well. However, the velocities in front of the merging are significantly lower than that in the main stream at the same densities. After the merging, the specific flow increases with the density ρ

J. Zhang (✉) · W. Klingsch
Institute for Building Material Technology and Fire Safety Science, Wuppertal University, Pauluskirchstrasse 11, 42285 Wuppertal, Germany
e-mail: ju.zhang@fz-juelich.de; klingsch@uni-wuppertal.de

A. Schadschneider
Institut für Theoretische Physik, Universität zu Köln, 50937 Köln, Germany
e-mail: as@thp.uni-koeln.de

A. Seyfried
Computer Simulation for Fire Safety and Pedestrian Traffic, Bergische Universität Wuppertal, Pauluskirchstraße 11, 42285 Wuppertal, Germany

Jülich Supercomputing Centre, Forschungszentrum Jülich GmbH, 52425 Jülich, Germany
e-mail: seyfried@uni-wuppertal.de

V.V. Kozlov et al. (eds.), *Traffic and Granular Flow '11*, DOI 10.1007/978-3-642-39669-4_23, © Springer-Verlag Berlin Heidelberg 2013 till 2.5 m^{-2}. While in the branches, the specific flow is almost independent of the density between $\rho = 1.5$ and 3.5 m^{-2}.

1 Introduction

In recent years, research on pedestrian and traffic flow became popular and attracted a lot of attention [1–5]. A great deal of models has been proposed to simulate pedestrian movement in different situations [6, 7]. Most of these models are able to reproduce crowd phenomena such as lane formation qualitatively. On the other hand, some field studies and laboratory experiments are also carried out to obtain empirical data and improve models.

Unfortunately, the empirical database is insufficient leading to discrepancies among different handbooks. The fundamental diagram describing the empirical relation between density (ρ), velocity (v) and flow (J) (or specific flow (J_s)) is the key characteristics of pedestrian dynamics used to support the planning and design of facilities. However, there is considerable disagreement among the empirical data from different studies and handbooks even for the same type of facility or movement. The maximum flow as well as the density where the maximum flow appear has a broad range [8, 9]. These discrepancies cause big inconvenience in design and assessment. Further research is necessary to study the reasons for these differences. Carrying out laboratory experiments is a good way to do this, since secondary factors can be removed or controlled.

T-junctions are important part of most of buildings. In this kind of structure, bottleneck flow, merging flow or split flow are all possible to take place in different situations. Especially around the corners, pedestrian behavior is much more complex. Pedestrian flow characteristics in this kind of geometry are significant to study. However, except for some studies using a Cellular Automaton model [10, 11], there are few empirical studies directly considering the fundamental diagram of pedestrian flow in T-junctions.

In this study, series of well-controlled laboratory experiments are carried out to investigate the pedestrian movement in T-junctions. The Voronoi method is accepted to analyze the fundamental diagram and field profiles of density, velocity and specific flow.

2 Experiment Setup and Trajectory Extraction

Seven runs of experiments were carried out in a T-junction with corridor width $b_{\text{cor1}} = b_{\text{cor2}} = 2.4$ m. Figure 1 shows the sketch of the experiment setup. Pedestrian streams move from two branches oppositely and then merge into the main stream at the T-junction. To regulate the pedestrian density in the corridor, the width of the entrance b_{entrance} was changed in each run from 0.5 to 2.4 m, see details in

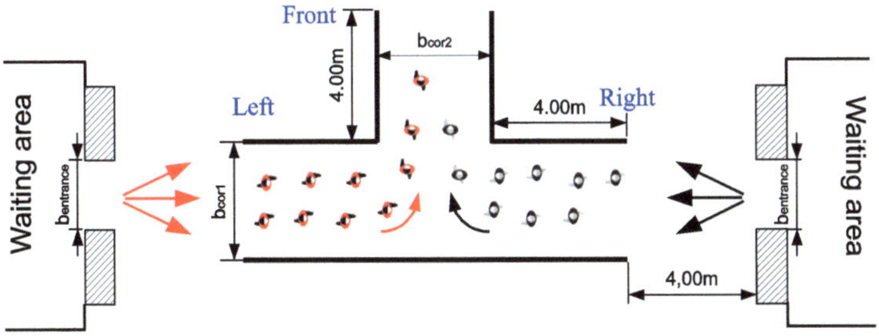

Fig. 1 Setup of T-junction experiments

Table 1 Parameters for the T-junction experiments

Runs	Name	b_{cor1} (m)	b_{entrance} (m)	b_{cor2} (m)	$N_l + N_r$
1	T–240–050–240	2.40	0.50	2.40	67+67
2	T–240–060–240	2.40	0.60	2.40	66+66
3	T–240–080–240	2.40	0.80	2.40	114+114
4	T–240–100–240	2.40	1.00	2.40	104+104
5	T–240–120–240	2.40	1.20	2.40	152+153
6	T–240–150–240	2.40	1.50	2.40	153+152
7	T–240–240–240	2.40	2.40	2.40	151+152

Table 1. For simplicity, the left and right entrances have the same width. In this way, we guarantee the symmetry of the two branch streams. More than 300 pedestrians participated in the experiments, which makes the duration of the run long enough to obtain stationary states. The average age and body height of the tested persons was 25 ± 5.7 years and 1.76 ± 0.09 m (range from 1.49 to 2.01 m), respectively. They mostly consisted of German students of both genders. The free velocity $v_0 = 1.55 \pm 0.18$ m/s was obtained by measuring 42 participants' free movement.

At the beginning, the participants were held within waiting areas in the left and right side. When the experiment starts, they pass through a 4 m passage into the corridor simultaneously and merge into the main stream at the vertical corner. The passage was used as a buffer to minimize the effect of the entrance. In this way, the pedestrian flow in the corridor was nearly homogeneous over its entire width. When a pedestrian leaves the main corridor, he or she returned to the waiting area for the next run.

The whole processes of the experiments were recorded by two synchronized stereo cameras of type Bumblebee XB3 (manufactured by Point Grey). They were mounted on the rack of the ceiling 7.84 m above the floor with the viewing direction perpendicular to the floor. The cameras have a resolution of $1{,}280 \times 960$ pixels and a frame rate of 16 fps (corresponding to 0.0625 s per frame). To cover the complete region, the left and the right part of the scenario were recorded by the two cameras separately (see Fig. 2a). The overlapping field of view of the stereo system is $\alpha = 64°$ at the average head distance of about 6 m from the cameras. With the above-mentioned height range, all pedestrians can be seen without occlusion at any time.

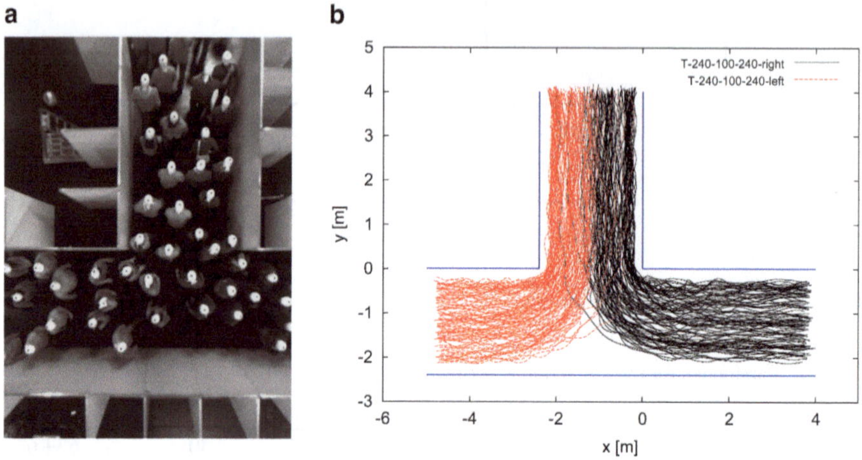

Fig. 2 Trajectories and snapshot from T-junction experiment. (**a**) Snapshot. (**b**) Pedestrian trajectories

To make a highly precise analysis, accurate pedestrian trajectories were automatically extracted from video recordings using the software *PeTrack* [12]. Lens distortion and perspective view are taken into account in this program. Figure 2b shows the pedestrian trajectories for one run of the experiments. From these trajectories, pedestrian characteristics including flow, density, velocity and individual distances at any time and position can be determined.

3 Experiment Analysis

In previous studies, different measurement methods were used to limit the comparability and fluctuation of the data. E.G. Helbing et al. proposed a Gaussian, distance-dependent weight function [13] to measure the local density and local velocity. Predtechenskii and Milinskii [14] used a dimensionless definition to consider different body sizes and Fruin introduced the "Pedestrian Area Module" [15]. This study focuses on the Voronoi method proposed in [16, 17], where the density distribution can be assigned to each pedestrian. This method permits examination on scales smaller than the pedestrians for its high spatial resolution.

3.1 Measurement Methodology

At a given time t, the Voronoi diagram can be generated from the positions of each pedestrian. It contains a set of Voronoi cells for each pedestrian i. The cell area, A_i, can be thought as the personal space belonging to each pedestrian i. Then, the density and velocity distribution over space can be defined as

$$\rho_{xy} = 1/A_i \quad and \quad v_{xy} = v_i(t) \qquad \text{if } (x, y) \in A_i \tag{1}$$

where $v_i(t)$ is the instantaneous velocity of each person (see [17]). The Voronoi density and velocity for the measurement area A_m is defined as

$$\langle \rho \rangle_v(x, y, t) = \frac{\iint \rho_{xy} dxdy}{A_m}, \tag{2}$$

$$\langle v \rangle_v(x, y, t) = \frac{\iint v_{xy} dxdy}{A_m}. \tag{3}$$

The specific flow

$$J_s(x, y, t) = \langle \rho \rangle_v(x, y, t) \cdot \langle v \rangle_v(x, y, t) \tag{4}$$

can also be calculated using the Voronoi density and velocity.

3.2 Results and Analysis

To analyze the spatial dependence of density, velocity and specific flow precisely, we use the Voronoi method to measure these quantities in areas smaller than the size of pedestrians. We calculate the Voronoi density, velocity and specific flow over small regions (10 × 10 cm) each frame. Then the spatiotemporal profiles of density ($\overline{\rho}(x, y)$), velocity ($\overline{v}(x, y)$) and specific flow ($\overline{J_s}(x, y)$) can be obtained over the stationary state separately for each run as follows:

$$\overline{\rho}(x, y) = \frac{\int_{t_1}^{t_2} \langle \rho \rangle_v(x, y, t)dt}{t_2 - t_1}, \tag{5}$$

$$\overline{v}(x, y) = \frac{\int_{t_1}^{t_2} \langle v \rangle_v(x, y, t)dt}{t_2 - t_1}, \tag{6}$$

$$\overline{J_s}(x, y) = \overline{\rho}(x, y) \cdot \overline{v}(x, y). \tag{7}$$

Figure 3 shows the profiles for two runs of the experiments under the situations of low density (T-240-050-240) and high density (T-240-240-240), respectively. These profiles provide new insights into the spatiotemporal dynamics of the motion and the sensitivity of the quantities to other potential factors. The density distribution in T-junction is not homogeneous both for low and high density

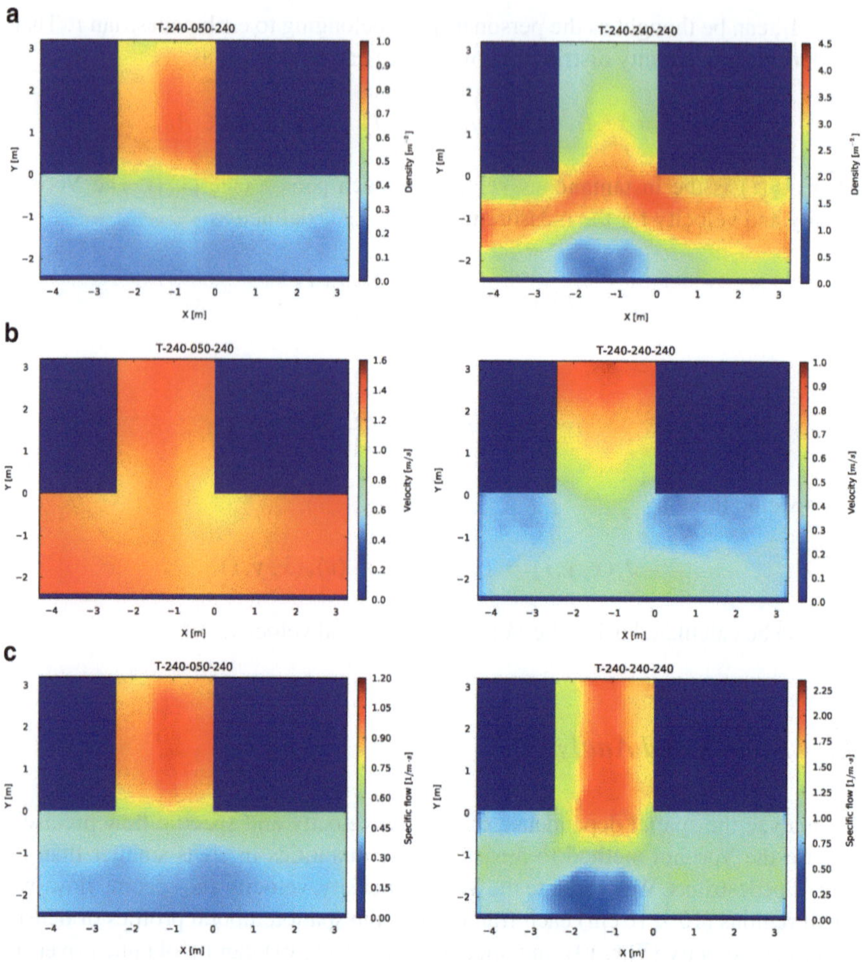

Fig. 3 The profiles of density, velocity and specific flow in T-junction obtained from Voronoi method. The resolution of the profiles is 10×10 cm. (**a**) Density profile. (**b**) Velocity profile. (**c**) Specific flow profile

situations. For the former, the higher density region locates at the main stream after merging. Whereas for the latter, the higher density region appears near the junction and the lowest density region locates at a small triangle area, where the left and right branches begin to merge. For both of the two situations, the densities in the branches are not uniform and are higher over the inner side, especially near the corners. In other words, pedestrians prefer to move along the shorter and smoother path. Moreover the density profile shows obvious boundary effects at high density situation. The spatiotemporal variation of the velocity is different. At low density situation, the velocity profile is nearly homogeneous all over the T-junction except the places near corner. Pedestrians move with free velocity and slow down round the

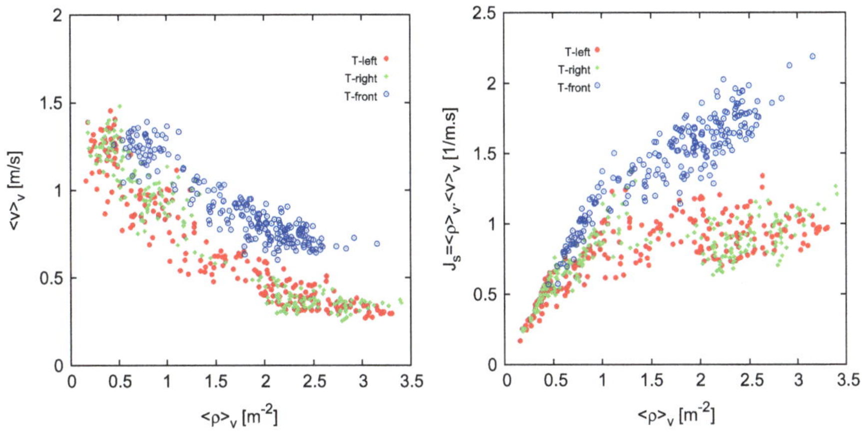

Fig. 4 Fundamental diagram of pedestrian flow at different measurement areas in T-junction

corner. While for high density condition, the velocity distribution is not uniform any more. The velocities of main stream is obviously higher than that in the branches. Boundary effect does not occur and the velocities after merging increase along the movement direction persistently. By comparison of the specific flow profiles at the two different situations, the highest flow regions are both observed at the center of the main stream after the merging. The difference is that the region extends further into the area where the two branches start to merge for the high density case, but for the other case it doesn't. This indicates that the merging process in front of the exit corridor leads to a flow restriction. Causes for the restriction of the flow must be located outside the region of highest flow.

These profiles demonstrate that density and velocity measurements are sensitive to the size and location of the measurement area. For the comparison of measurements (e.g. for model validation or calibration), it is necessary to specify precisely the size and position of the measurement area.

In Fig. 4, we compare the fundamental diagrams obtained from all T-junction experiments. The data assigned with 'T-left' and 'T-right' are measured in the areas before the streams merge, while the data assigned with 'T front' are measured in the region where the streams have already merged. The locations of these measurement areas can be seen in Table 2. For ease of comparison, we choose these measurement areas with the same size ($4.8\,\mathrm{m}^{-2}$). One finds that the fundamental diagrams of the two branches match well. However, for densities $\rho > 0.5\,\mathrm{m}^{-2}$ the velocities in the 'right' and 'left' part of the T-junction (T-left and T-right) are significantly lower than the velocities measured after the merging of the streams (T-front). This discrepancy becomes more distinct in the relation between density and specific flow. In the main stream (T-front), the specific flow increases with the density ρ till $2.5\,\mathrm{m}^{-2}$. While in the branches, the specific flow nearly remains constant for density

Table 2 The location of measurement area in T-junction

Measurement area	Range (m)
T-left	$x \in [-4.5, -2.5],\ y \in [-2.4, 0]$
T-right	$x \in [1.0, 3.0],\ y \in [-2.4, 0]$
T-front	$x \in [-2.4, 0],\ y \in [1.0, 3.0]$

ρ between 1.5 and 3.5 m^{-2}. Thus, there seems no unique fundamental diagram which describes the relation between velocity and density for the complete system.

For this difference, we can only offer assumptions regarding the causes. One is based on behavior of pedestrians. Congestion occurs at the end of the branches, where the region of maximum density appears. Pedestrians stand in a jam in front of the merging and could not perceive where the congestion disperse or whether the jam lasts after the merging. In such situation, it is questionable whether an urge or a push will lead to a benefit. Thus an optimal usage of the available space becomes unimportant. Otherwise, the situation totally changes if the location of dissolution becomes apparent. Then a certain urge or an optimal usage of the available space makes sense and could lead to a benefit. They will move in a relatively active way. That's maybe the reason why the velocities after merging are higher than that in front of merging at the same density. Whether this explanation is plausible could be answered by a comparison of these data with experimental data at a corner without the merging. This comparison is in preparation.

References

1. C. Appert-Rolland, F. Chevoir, P. Gondret, S. Lassarre, J.-P. Lebacque, and M. Schreckenberg (2009) *Traffic and Granular Flow '07* (Berlin: Springer)
2. S. Bandini, S. Manzoni, H. Umeo, and G. Vizzari (2010) *Workshop on Crowds and CA, ACRI 2010: Cellular Automata 9th International Conference on Cellular Automata for Reseach and Industry* (Berlin: Springer), Italy
3. W. W. F. Klingsch, C. Rogsch, A. Schadschneider, and M. Schreckenberg (2010) *Pedestrian and Evacuation Dynamics 2008*. Springer-Verlag Berlin Heidelberg
4. A. Schadschneider, H. Klüpfel, T. Kretz, C. Rogsch, and A. Seyfried (2009) Fundamentals of Pedestrian and Evacuation Dynamics. In: *Multi-Agent Systems for Traffic and Transportation Engineering*, pp. 124–154. IGI Global, Hershey, Pennsylvania, USA
5. A. Schadschneider and A. Seyfried (2009) Empirical Results for Pedestrian Dynamics and their Implications for Cellular Automata Models. In *Pedestrian Behavior: Data Collection and Applications*, pp. 27–43. Emerald Publishing
6. D. Helbing, I. Farkas, and T. Vicsek (2000) *Nature* 407:487–490
7. W. Song, X. Xu, B.H. Wang, and S. Ni (2006) *Physica A* 363:492–500
8. A. Schadschneider, W. Klingsch, H. Klüpfel, T. Kretz, C. Rogsch, A. Seyfried (2009) Evacuation Dynamics: Empirical Results, Modeling and Applications. In: *Encyclopedia of Complexity and System Science*, pp. 3142–3176, Springer
9. A. Seyfried, O. Passon, B. Steffen, M. Boltes, T. Rupprecht, and W. Klingsch (2009) *Transportation Science* 43(3):395–406
10. Y. Tajima and T. Nagatani (2002) *Physica A* 303:239–250
11. Y.C. Peng and C.I. Chou (2011) *Comp. Phys. Comm.* 182:205–208

12. M. Boltes, A. Seyfried, B. Steffen, and A. Schadschneider (2010) Automatic Extraction of Pedestrian Trajectories from Video Recordings. In *Pedestrian and Evacuation Dynamics 2008* (Berlin: Springer), pp. 43–54
13. D. Helbing, A. Johansson, and H. Z. Al-Abideen (2007) *Phys. Rev. E* 75:046109
14. V. M. Predtechenskii and A. I. Milinskii (1969) *Planning for Foot Traffic Flow in Buildings.* Amerind Publishing, New Dehli, 1978. Translation of: Proekttirovanie Zhdanii s Uchetom Organizatsii Dvizheniya Lyuddskikh Potokov, Stroiizdat Publishers, Moscow
15. J. J. Fruin (1971) *Pedestrian Planning and Design.* Elevator World, New York
16. B. Steffen and A. Seyfried (2010) *Physica A* 389(9):1902–1910
17. J. Zhang, W. Klingsch, A. Schadschneider, and A. Seyfried (2011) *J. Stat. Mech.* P06004.

Geodesics and Shortest Paths Approach in Pedestrian Motions

B. Nkonga, Michel Rascle, F. Decoupigny, and G. Maignant

Abstract We revisit existing ideas on the eikonal equation and combine them with a discrete Lagrangian description. Some preliminary numerical tests are reported.

1 Introduction

The first two Authors are from Applied Mathematics (M.R.) or Computational Fluid Mechanics (B.N.), whereas F.D. and G.M. are from Geography.

We present and revisit here in some detail the modeling of pedestrian flows based on the shortest path and geodesics approach, i.e. on variants of the Hamilton-Jacobi (HJ) equation, more precisely on the eikonal equation. For general mathematical references, among a huge literature, see e.g. [3, 10]. In the context of pedestrian flows [1], and many subsequent papers, e.g. [2, 14].

Our approach here was motivated for the last two Authors by previous works of Decoupigny on various problems, including discrete models of optimal paths for visitors of French national parks [12] and for Rascle by previous works on mesh generation [4], apparently far away from, but in fact closely related to this subject.

In fact, besides its historical origin on light or particle propagation, this fairly general view has been intensely used in the last decades on a huge variety of fields, e.g. by Osher and Sethian [9], Sethian [5], and many others, e.g. in the group of

B. Nkonga (✉) · M. Rascle
Laboratoire Dieudonné, UMR CNRS 6621 and University of Nice, Parc Valrose, 06108 Nice Cedex 2, France
e-mail: nkonga@unice.fr; rascle@unice.fr

F. Decoupigny · G. Maignant
Laboratoire Espace, UMR CNRS 6012 and University of Nice, BP 3209, 06204 Nice Cedex 3, France
e-mail: decoupig@unice.fr; maignant@unice.fr

V.V. Kozlov et al. (eds.), *Traffic and Granular Flow '11*,
DOI 10.1007/978-3-642-39669-4_24, © Springer-Verlag Berlin Heidelberg 2013

Markowich at Cambridge, not forgetting a lot of clever works, e.g. [7], on the numerical approximation of the viscosity solution, see e.g. [10].

This approach is thus not new in the Pedestrian experts community, see again [1], many people know it and its limitations, in particular its numerical cost. Of course, there are so many aspects in pedestrian traffic analysis that it would be naive to pretend cover all of them with this single method. Nevertheless, its beauty and the relevant global information its contains make it quite elegant and appealing.

By the way, it is perhaps worth to emphasize that, here like in (car) traffic flow, there is no physics or continuum mechanics involved, no conservation of momentum etc. The aim is simply to propose one optimization principle which might govern the choice of pedestrian trajectories in a crowded area.

It would be also quite interesting to compare it carefully with competing methods, in particular with the social forces (or acceleration) methods, see e.g. [8, 13].

In the next section, we briefly recall a few basic facts on this highly classical method, up to the notion of distance $d(x, C)$ from a curve C to any point x. Typically, say in a big hall or a plaza, each population k of pedestrians travels towards a door or a destination $C := D_k$ with a velocity determined by the gradient of $d_k(x) := d(x, D_k)$.

In Sect. 3, we present the full corresponding macroscopic model and give some details on its (expensive) approximation, before studying in Sect. 4 a (much cheaper) semi-discrete description, in which each population k of pedestrians is defined in a discrete way, but each corresponding distance d_k in a macroscopic way. We then show a few preliminary results and conclude.

2 Basic Facts on Hamilton-Jacobi Equation: Distance

We first briefly summarize the main ingredients.

2.1 Variational Problem: Distance

For any curve $\{X(t) = (X_1(t), X_2(t)), 0 \le t \le T\}$ from point $y = (y_1, y_2)$ (origin) to point $z = (z_1, z_2)$ (destination), define:

$$L_1[X] := \int_0^T L(X(t), \dot{X}(t))\, dt, \tag{1}$$

where the *cost* $L(x, v) = L(X(t), \dot{X}(t)) \ge 0$ is often called the *Lagrangian*. We set

$$d_L(y, z) = \inf\{L_1[X]\,,\ X(0) = y\,,\ X(T) = z\}. \tag{2}$$

The infimum, taken on all curves from $y = (y_1, y_2)$ to $z = (z_1, z_2)$, depends on y, z and (non locally) on the choice of L. Since $L \geq 0$, d_L is a distance, provided that $L(x, v) \equiv L(x, -v)$, which implies: $d_L(y, z) = d_L(z, y)$ for all pairs (y, z).

2.2 Euler-Lagrange Equation

If a trajectory is optimal, then a Taylor expansion and an integration by parts shows that necessarily this celebrated relation:

$$\frac{d}{dt}(\partial_v L(X, \dot{X})) = \partial_x L(X, \dot{X}), \tag{3}$$

must be satisfied. Here $v = (v_1, v_2) = \dot{X}(.)$, $x = X(.)$ and the gradients $\partial_x L$, $\partial_v L$ are $2D$ vectors.

In particular, straight lines are optimal if we choose $L(x, v) \equiv L(v) := \frac{1}{2}|v|^2$, and broken lines are optimal if $L(x, v) = \frac{1}{2c(x)}|v|^2$, with $c(x) > 0$ is piecewise-constant, say $c(x) = c_\pm$ for $\pm\, x_2 > 0$: Descartes refraction law.

2.3 Legendre Transform in v: Link with Hamilton-Jacobi Equation

For any position x and any 2D vector p, the *Hamiltonian* is defined as the Legendre transform of $L(x, v)$:

$$H(x, p) := \sup_v \{p.v - L(x, v)\}. \tag{4}$$

For convex *Lagrangian* with respect to v, extremality conditions give

$$H(x, p) = p.v - L(x, v), \quad \text{for} \quad v = \partial_p H(x, p) \tag{5}$$

In particular in the cases considered below, the *Lagrangian* is v-convex and of the following form

$$L(x, v) = \frac{1}{2c(x)}|v|^2, \tag{6}$$

where $c(x)$ is a given strictly positive function. In this case, we obtain

$$H(x, p) := \sup_v \{p.v - \frac{1}{2c(x)}|v|^2 - \frac{c(x)}{2}|p|^2\} + \frac{c(x)}{2}|p|^2 = \frac{c(x)}{2}|p|^2, \tag{7}$$

and (extremality relations), for any x, the supremum in v is reached for

$$v = c(x)p. \tag{8}$$

Link with Euler-Lagrange: Hamiltonian System

Using these extremality relations leads classically to the characterization of optimal trajectories by the celebrated Hamiltonian characteristic system:

$$\dot{X} = \frac{dX}{dt} = \partial_p H(X, p) = v, \tag{9}$$

$$\dot{p} = \frac{dp}{dt} = -\partial_x H(X, p) = \partial_x L(X, v), \tag{10}$$

which itself leads, in a highly nontrivial way, to the corresponding *eikonal* or more generally *Hamilton-Jacobi (HJ)* equation.

Hamilton-Jacobi Equation

Finally, written here in the time-dependent version:

$$\partial_t \Phi + H(x, \partial_x \Phi(x, t)) = 0, \tag{11}$$

or in its *stationary* version

$$H(x, \partial_x \varphi(x)) = f(x), \tag{12}$$

with in either case suitable *initial and/or boundary conditions*, this famous equation characterizes the optimal trajectories and therefore the solution of variational problems formulated as in Eq. (2), with the same cost L and Hamiltonian H defined in (4).

2.4 *Application: Distance from a Given Curve* (C)

In view of its sense for propagation, we should perhaps call it: distance from (C) to x. Anyway, this distance, according to a cost function L, is defined as:

$$d(C, x) := d_L(C, x) := \inf_{y \in (C)} \{d_L(y, x)\}. \tag{13}$$

In particular, we choose the cost function L given by (6) and H in (7), with $f(x) \equiv \frac{1}{2}$. In the next sections two choices are considered:

$$\text{Choice 1} : c(x) = \beta(\rho(x)), \qquad \text{Choice 2} : c(x) \equiv 1.$$

The function $\beta(\rho)$ will vanish at high local densities when $\rho(x) = \rho_\infty$, i.e. in *congested regions*. We will use $\beta(\rho(x)) = 1.0 - \frac{\rho(x)}{\rho_\infty}$. It turns out that $\varphi(x) := d(C, x)$ satisfies (12), and finally satisfies:

$$H(x, \partial_x\varphi(x)) := \frac{c(x)}{2}|\partial_x\varphi(x)|^2 = f(x), \quad \text{with} \quad \varphi(x) \equiv 0 \text{ on } (C). \tag{14}$$

Indeed, using the extremality relations (8), with choice 1 and again $f \equiv 1/2$, we see that on each optimal trajectory, parameterized by some artificial time, say $x = X(t)$:

$$v = \dot{X}(t) = c(x)\partial_x\varphi(x) \text{ and } |v| = \sqrt{c(x)}. \tag{15}$$

Therefore, as expected, *the smaller $c(x)$, the smaller the speed*. For numerous applications, see again the Osher-Sethian literature.

Moreover, $\Phi(x, t) := \varphi(x) - \frac{t}{2}$ is a solution of the evolution equation (11). Of course, similar relations would arise if we used another variant of eikonal equation:

$$\sqrt{c(x)}\,|\partial_x\varphi(x)| = 1. \tag{16}$$

Finally, we note that, see e.g. [3], even with smooth data, the solutions of (11) or (12) are not uniquely defined: *shocks* can arise, at points x which are equidistant from at least two points of curve (C). Then the theory of *viscosity solutions*, [10], provides a uniqueness criterion by selecting the first arrival "time", i.e. the trajectory coming from the nearest point on (C), and this criterion is built in efficient numerical schemes.

For instance, in some figures in Sect. 3 below, where we plot the distance $d_k(x)$ from exit door k at any point x in a plaza, these shocks appear on the graph as "passes" near a "peak" separating two different paths from the door to x.

Two Applications

As we already said, most of the material here is classical. We have been motivated by two different previous works of ours:

- One concerns mesh generation or refinement, with the same ideas as here, e.g. refine a mesh by moving at constant Riemannian speed on the graph of a previous approximation of the solution. We refer to [4] for many illustration of the results.
- The other one concerns the optimal location of parking lots in a national park in order to restrict the number of paths. The idea (Decoupigny) is to identify a

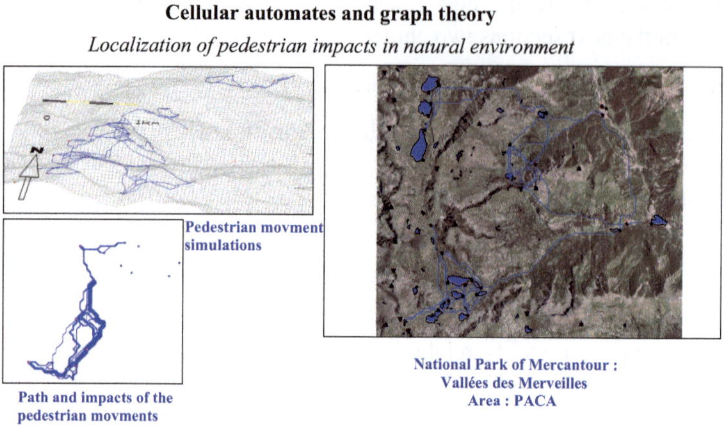

Fig. 1 Pedestrian paths in the Mercantour French National Park

cost on each path, depending on the slope and of the interest of each site, find the geodesics, compare with existing paths, make recommendations. For related ideas, see [12]. A joint work in this direction is in progress (Fig. 1).

3 Pedestrian Flow: Eulerian Description

We are going to apply the above-mentioned notion of distance depending on the density. We refer again to the literature quoted in the Introduction and so many references therein, in particular we revisit here with minor variations the approach of [1].

3.1 *Principle*

Consider, see Fig. 2 below, a plaza or a big hall with several doors $D_k, k = 1 \dots K$, all being possibly both entrance and exit doors. At each time t and point x, let $\rho_k(x,t), k = 1 \dots K$ be the local density of pedestrians moving towards door D_k. When time is discrete, set $\rho_k^n(x) := \rho_k(x, t_n)$ and let $\rho^n(x) := \sum_k \rho_k^n$ be the local *total density*. Finally, let $d_k^n(x, D_k)$ be the distance to door D_k, defined as in Sect. 2 by:

$$H(x, \partial_x d_k^n) := \frac{\beta(\rho^n(x))}{2} |\partial_x d_k^n|^2 = \frac{1}{2} \text{ in the plaza,} \tag{17}$$

$$d_k^n(x) \equiv 0 \text{ on door } D_k, \tag{18}$$

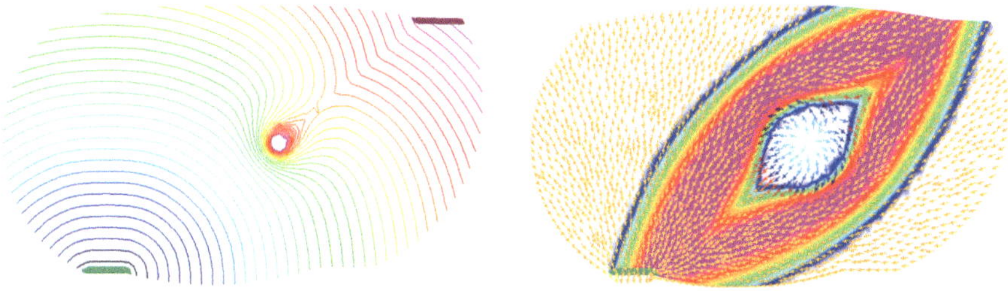

Fig. 2 Level curves of distance (*left*) from the exit door, computed for $c(x) = c_1(x)$, estimation of optimal paths (*right*) as sum of distances and local optimal trajectory direction for the exit

where the associated Lagrangian depends on the (local) total density:

$$L_k^n(x, v) = \frac{1}{2c(x)}|v|^2, \text{ with } c(x) := c_k^n(x) = \beta(\rho^n(x)), \tag{19}$$

and $f(.)$ is a given non-increasing function of ρ, vanishing for large densities.

With the background of Sect. 2 in mind, we now *postulate* that at each discrete time t_n, each population k responds to the gradient of the distance $d_k^n(x, D_k)$ by moving towards door D_k with a velocity given by Eq. (20) below.

Compare with extremality conditions (8) and note the difference of signs: here, the distance d_k^n is a nonlocal information which travels *from* door D_k, whereas the k-pedestrians travel *towards* this door. Consistently with this observation, we note that for each transport equation (21) to the unknown density ρ_k, see below, the other entrance doors $D_j, j \neq k$ only play a role in boundary data, as (given) *entering* fluxes into the plaza.

3.2 The Algorithm

Step 1: Assuming $\rho := \rho^n$ is known at time t_n, compute each *distance* $d_k := d_k^n$ (we drop the index n), by solving the above problem (17).

Step 2: Using the extremality condition, the corresponding velocity field v_k is then (note again the – sign):

$$v_k = -\frac{1}{2\,\beta(\rho(x))}\,\partial_x d_k \tag{20}$$

Step 3: For each k, refresh ρ_k from t_n to t_{n+1} by approximating the solution at time t_{n+1} to the Initial Value Problem (IVP):

$$\partial_t \rho_k + \nabla.(\rho_k\, v_k) = 0 \text{ in } \Omega \times (t_n, t_{n+1}), \tag{21}$$

$$\rho_k(x, t_n) = \rho_k^n(x) \text{ in } \Omega, \tag{22}$$

with the classical *entering boundary conditions:* the flux $\rho\, v$ at each point x of the boundary $\partial\Omega$ is imposed whenever $v_k^n \cdot \nu < 0$, $\nu(x)$ being the exterior normal vector at point x. Practically, the flux $\rho_k \cdot v_k$ is only entering in the plaza, call it Ω through the other doors $D_j, j \neq k$.

Step 4: Knowing the partial densities ρ_k^{n+1} at time t_{n+1}, refresh the total density ρ^{n+1}. Go to step 1.

4 Pedestrian Flow: Eulerian-Lagrangian Description

4.1 Principle

Consider each discrete population k, going to door D_k, assume that their individual velocities are given from a continuous velocity field.

As above, at each time t, e.g. $t = t_n$, we assume that $v_k(x) := v_k(x,t) = -c(x,t)\partial_x d_k(x, D_k)$, where $d_k(x, D_k) = \varphi_k$ is the solution to (17), and again

$$c(x) := c(x,t) = \beta(\rho(x,t)) \tag{23}$$

is a function of a total *continuous* approximation of the total discrete density $\rho = \sum_k \rho_k$ at time t, as in steps 1 and 2 of our above fully Eulerian algorithm.

Now, for each individual j in population k, the resulting semi-discrete problem at time t is, for each population k:

$$\dot{X}_j(t) = v_j(t) \quad \text{with} \quad v_j(t) = -c(X_j(t),t)\,\partial_x d_k(X_j(t), D_k). \tag{24}$$

where $c(X_j(t),t) \simeq \displaystyle\int_\Omega \sum_k \sum_{\ell \in k} \delta_\ell(x) K(x - X_j(t))dx$ with $\delta_\ell(x)$ a Dirac function centered at the position of the particle ℓ and $K(\xi)$ is a smoothing kernel.

4.2 Comments

We obviously approximate this ODE system, say by the explicit Euler scheme. At each time step, the function $c(.)$ is defined by (23), and we have to decide either to *refresh* the density (either at each time step, or more rarely, e.g. at each time of visualization), or even to keep the (obsolete) initial density.

When refreshing the density, the obvious problem is passing from the discrete positions to a continuous density. A simple choice is to count the number of particles in each numerical cell.

A crucial issue in designing pedestrian models is their ability of predicting and describing the appearance and subsequent evacuation of high concentration populations, where panic can be lethal. Of course, we do not claim to solve this challenging problem with such a simple model. Nevertheless, we see in Figures below that dense regions of space become repulsive, in a way which depends on the stiffness of function β in Eq. (20). This is related with a constrained model with a maximal ρ_{max} has been introduced and intensely studied several years ago by Maury, Venel and (on a more theoretical viewpoint) Santambrogio et al., see e.g. [11]. An obvious idea is to approximate this limit model with the above model studied here, and using in (20) a function β stiff near the maximal density ρ_{max}. In the semi-discrete approach, that requires a careful calibration in the link between discrete and continuous densities, which requires more work.

5 Numerical Tests

We present here a few preliminary tests and visualizations, in which we try to combine and/or compare some of these approaches.

Test 1: We consider a curve-shaped hall, with an (Nord-Est) entrance door and a (Sud-West) exit door. We consider that at a given time there is a dense region corresponding to $c(x) = c_1(x) = 1 - 0.99\exp(-(x - x_0)^2)$. Figure 2 shows on the left the level curves of the distance to the exit: see the shock (the "pass") in the NE corner and note that the trajectories, not shown here, are orthogonal to these level curves. On the right, we plot the sum of the two distances to (SW) and to (NE), computed on an unstructured mesh, via a numerical solver of the evolution HJ equation [6]. Even *without* any computation of the trajectories, the optimal paths from NE to SW or conversely (the purple areas) at a given time look pretty obvious, whereas running the full algorithm of Sect. 3, much more expensive, provides the velocity field, see Fig. 2, right.

Test 2: Here, a corridor or a platform with an exit door (right), with given incoming flux on the left door that is non zero for $0 < t < 120$, The geometry and positions of doors are taken from [7]. In Fig. 3 we plot the *density* at two different times. A high density front is formed near the obstacle and, by the distance update, is transversally (anisotropic) diffused. The local behavior around the obstacle is clearly different to results in [7] where high densities are avoided by adding an isotropic diffusion in the velocity (Fig. 4).

Test 3: Using the semi-discrete approach of Sect. 4 we have made a couple of comparisons between:

Choice 1: Use the above eikonal equation with a variable coefficient $c(x) = \beta(\rho(x))$, in order to prevent the density from becoming too large. Here, we have only tested it with a smooth $\beta(\rho) = 1.0 - \frac{\rho}{\rho_\infty}$ and a given ρ_∞.

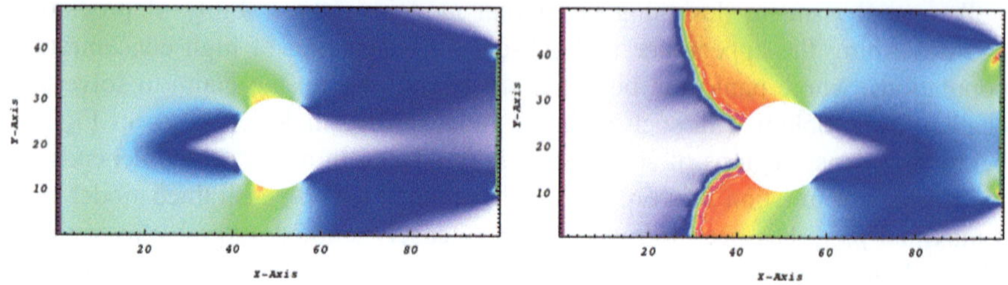

Fig. 3 Plots of the density at $t < 120$ (*left*) and $t > 120$ (*right*), using the full scheme of Sect. 3 with $\beta(\rho) = 1.0 - \frac{\rho}{\rho_\infty}$ and $\rho_\infty = 10$

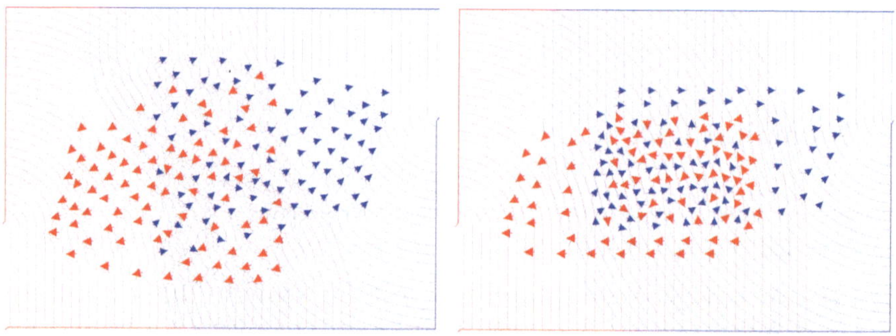

Fig. 4 Choice 1 (*left*) vs. choice 2 (*right*). Snapshots at computational midtime. For comparison, the speeds are normalized in both algorithms

Choice 2: cf. Maury, Venel (and Santambrogio et al. on more theoretical side), eikonal equation with a constant $c(x)$ and a *constraint* on the maximal (total) density with corresponding Lagrange multipliers and a numerical solution based on Uzawa algorithm. Very nice mathematically and numerically, but expensive. Its "incompressible" feature in congested regions is perhaps a good cartoon for *lane formation*.

6 Final Comments and Conclusion

As we said, we have revisited existing ideas on the eikonal equation, combined them with a discrete Lagrangian description, and made some preliminary numerical tests on some of these ideas. Clearly, by far the most tractable method is the semi-discrete algorithm used in Sect. 4.

We could also see, on numerical movies that we cannot show here, that this method can nicely describe counterintuitive motions, like moving *away* from the target in order to avoid crowded areas, which is a neat advantage on fully discrete methods. Going further requires a more refined calibration, with an ad hoc function β, in order to see how far one can go in this direction. We also mention again the importance of this notion of geodesics in geography and more generally in any field related to space organization.

References

1. R.L. Hughes, A continuum theory for the flow of pedestrians. *Transportation Research, Part B, 36, 507–535 (2002).*
2. R. L. Hughes. The flow of human crowds. *Annual Review of Fluid Mechanics,. 35:169–182, (2003).*
3. Lawrence C. Evans, Partial differential equations. *Graduate studies in mathematics, 19, American Mathematical Soc. (2010).*
4. Ph. Hoch, M. Rascle, Hamilton-Jacobi Equations on a Manifold and Applications to Grid Generation or Refinement, *SIAM J. Sc. Comp., 23, 6, (2002).*
5. J.A. Sethian, Level Set Methods and Fast Marching Methods: Evolving Interfaces in Computational Geometry, Fluid Mechanics, Computer Vision and Materials Science , *Cambridge University Press, (1999).*
6. R. Abgrall. Numerical discretization of boundary conditions for first order Hamilton–Jacobi equations. *SIAM J. Numer. Anal., 41(6):2233–2261, (2003).*
7. Y. Xia, S.C. Wong and C.-W. Shu, Dynamic continuum pedestrian flow model with memory effect , *Physical Review E, v79 (2009).*
8. D. Helbing, Traffic and related self-driven many-particle systems, *Reviews of Modern Physics 73, 1067–1141 (2001).*
9. S. Osher, J.A. Sethian, Fronts propagating with curvature-dependent speed: Algorithms based on Hamilton-Jacobi formulations, *J. Comp.. Phys. 79 (1988).*
10. M. Crandall and P. L. Lions, Viscosity solutions of Hamilton-Jacobi equations, *Trans. Amer. Math. Soc. 277 , 1–42 (1983).*
11. B. Maury, A. Roudneff-Chupin, F. Santambrogio, J. Venel, Handling congestion in crowd motion models *Networks and Heterogenous Media, 485–519 (2011).*
12. F. Decoupigny, Mobilités potentielles et émergence de structures réticulaires en région Provence-Alpes-Côte d'Azur,*L'Espace géographique 38/3 (2009).*
13. T. Kretz, C. Bönisch, and P. Vortisch, Comparison of various methods for the calculation of the distance potential field, In *Pedestrian and Evacuation Dynamics 2008*. Springer Berlin Heidelberg New York, 2009.
14. D. Hartmann. Adaptive pedestrian dynamics based on geodesics. *New Journal of Physics, 12(4): 043032, 2010.*

Modeling the Desired Direction in a Force-Based Model for Pedestrian Dynamics

Mohcine Chraibi, Martina Freialdenhoven, Andreas Schadschneider, and Armin Seyfried

Abstract We introduce an enhanced model based on the generalized centrifugal force model. Furthermore, the desired direction of pedestrians is investigated. A new approach leaning on the well-known concept of static and dynamic floor-fields in cellular automata is presented. Numerical results of the model are presented and compared with empirical data.

Force-based models try to describe the dynamics of pedestrians as reaction to forces acting on each single pedestrian. Basically two kinds of forces can be distinguished:

- *Driving forces* designed to drive pedestrians to a desired direction with a desired speed.
- *Repulsive forces* which are responsible for preserving the volume exclusion of pedestrians.

Since the introduction of force-based models [1] many works were dedicated to investigations of the repulsive forces and finding new and better forms [2–6]. These efforts for improving the form of the repulsive force is understandable, since

M. Chraibi (✉) · A. Seyfried
Jülich Supercomputing Centre, Forschungszentrum Jülich GmbH, 52425 Jülich, Germany

Computer Simulation for Fire Safety and Pedestrian Traffic, Bergische Universität Wuppertal, Pauluskirchstraße 11, D-42285 Wuppertal, Germany
e-mail: m.chraibi@fz-juelich.de; a.seyfried@fz-juelich.de

M. Freialdenhoven
Jülich Supercomputing Centre, Forschungszentrum Jülich GmbH, 52425 Jülich, Germany
e-mail: MartinaFreialdenhoven@web.de

A. Schadschneider
Institut für Theoretische Physik, Universität zu Köln, 50937 Köln, Germany
e-mail: as@thp.uni-koeln.de

V.V. Kozlov et al. (eds.), *Traffic and Granular Flow '11*,
DOI 10.1007/978-3-642-39669-4_25, © Springer-Verlag Berlin Heidelberg 2013

the interactions between pedestrians dominate the dynamics, especially at high densities. Surprisingly, not much work has been done on the influence of the specific form of the driving force which is expected to dominate the behavior at low or intermediate densities.

The standard form of the driving force is

$$\overrightarrow{F_i}^{\text{drv}} = m_i \frac{\overrightarrow{v_i^0} - \overrightarrow{v_i}}{\tau}, \tag{1}$$

with a relaxation time τ and a desired velocity $\overrightarrow{v_i^0}$. Although this expression is simple, it is not clear how to choose the desired direction

$$\overrightarrow{e_i^0} = \frac{\overrightarrow{v_i^0}}{\| \overrightarrow{v_i^0} \|} \tag{2}$$

in a given situation and only very few works were concerned with modeling the desired direction (2). In [7] an Ansatz with directing lines was introduced to steer pedestrians around 90° and 180° corners. Gloor et al. [8] used a path-oriented approach to model the desired direction of agents on given hiking paths.

In [9] Moussaïd et al. have formulated the determination of the desired direction in form of a minimization problem.

It should be mentioned that the *directing problem* we discuss here, i.e. the determination of the desired direction for each pedestrian, is conceptually different from the classical *routing problem*. In [10] an algorithm for generating automatically a navigation graph in complex buildings in combination with directing lines at corners was proposed. Another algorithm for way finding in buildings was proposed in [11]. Recently a further development of the notion of the "quickest path" using a non-iterative method to estimate the desired direction in the social force model (SFM) was introduced [12]. The main concern in this class of problems is how to define and connect intermediate targets, in order to facilitate the evacuation of pedestrians. By contrast, in the directing problem the existence of such intermediate targets is in general assumed.

In this work we introduce enhancements of the generalized centrifugal force model (GCFM) and investigate on their basis the modeling of the desired direction (2). For the sake of demonstration we test our model in two different geometries: a bottleneck and a corner.

1 The Model

In this section we give a brief overview of the GCFM and its definition. Furthermore we introduce an effective modification of the pedestrian-wall interactions that simplifies the definition of the repulsive force.

1.1 Pedestrian-Pedestrian Repulsive Interactions

Introducing the vector connecting the positions of pedestrians i and j,

$$\overrightarrow{R_{ij}} = \overrightarrow{R_j} - \overrightarrow{R_i}, \qquad \overrightarrow{e_{ij}} = \frac{\overrightarrow{R_{ij}}}{\parallel \overrightarrow{R_{ij}} \parallel}, \tag{3}$$

the repulsive force in the GCFM reads

$$\overrightarrow{F_{ij}}^{\mathrm{rep}} = -m_i k_{ij} \frac{(\eta {v_i}^0 + v_{ij})^2}{d_{ij}} \overrightarrow{e_{ij}}, \tag{4}$$

with $m_i = 1$ the mass of i and the effective distance between pedestrian i and j,

$$d_{ij} = \parallel \overrightarrow{R_{ij}} \parallel - r_i(v_i) - r_j(v_j), \tag{5}$$

and the polar radius r_i of pedestrian i.

The relative velocity v_{ij} is defined such that slower pedestrians are less affected by the presence of faster pedestrians in front of them:

$$\begin{aligned} v_{ij} &= \frac{1}{2}[(\overrightarrow{v_i} - \overrightarrow{v_j}) \cdot \overrightarrow{e_{ij}} + |(\overrightarrow{v_i} - \overrightarrow{v_j}) \cdot \overrightarrow{e_{ij}}|] \\ &= \begin{cases} (\overrightarrow{v_i} - \overrightarrow{v_j}) \cdot \overrightarrow{e_{ij}} & \text{if } (\overrightarrow{v_i} - \overrightarrow{v_j}) \cdot \overrightarrow{e_{ij}} > 0 \\ 0 & \text{otherwise.} \end{cases} \end{aligned} \tag{6}$$

The parameter

$$\begin{aligned} k_{ij} &= \frac{1}{2} \frac{\overrightarrow{v_i} \cdot \overrightarrow{e_{ij}} + \mid \overrightarrow{v_i} \cdot \overrightarrow{e_{ij}} \mid}{v_i} \\ &= \begin{cases} (\overrightarrow{v_i} \cdot \overrightarrow{e_{ij}}) / \parallel \overrightarrow{v_i} \parallel & \text{if } \overrightarrow{v_i} \cdot \overrightarrow{e_{ij}} > 0 \;\; \& \parallel \overrightarrow{v_i} \parallel \neq 0 \\ 0 & \text{otherwise,} \end{cases} \end{aligned} \tag{7}$$

reduces the effective range of the repulsive force to the angle of vision. Through the coefficient k_{ij} the strength of the repulsive force depends on the angle: it is maximal when pedestrian j is in the direction of motion of pedestrian i and minimal when the angle between j and i is bigger than 90°.

1.2 Wall-Pedestrian Repulsive Interactions

In the GCFM the interactions between pedestrians and walls are modeled by a force similar to the pedestrian-pedestrian repulsive force. A wall is represented by three

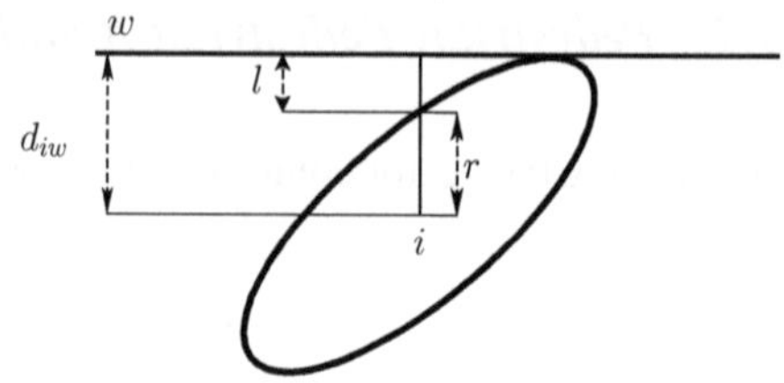

Fig. 1 Illustration of distances used in the definition of the wall-pedestrian repulsive force (8)

point masses acting on pedestrians within a certain range. From a computational point of view this analogy exhibits an overhead since the repulsive force between a pedestrian and a wall is calculated three times.

We now make use of the "distance of closest approach" as defined in [6] to formulate the repulsive force between a pedestrian i and a wall w as

$$\overrightarrow{F_{iw}}^{\text{rep}} = \eta' \parallel \overrightarrow{v_i}^0 \parallel k_{iw} b_{iw}, \tag{8}$$

with

$$b_{iw} = H\left(1 - \frac{d_{iw}}{r+l}\right) \cdot \left(1 - \frac{d_{iw}}{r+l}\right), \tag{9}$$

where l is the distance of closest approach between an ellipse and a line, r is the polar radius determined by the nearest point on the line to the center of the ellipse i (Fig. 1). $H()$ is the Heaviside step function, k_{iw} is defined in Eq. (7), $\parallel \overrightarrow{v_i}^0 \parallel$ is the desired speed of i and η' is a parameter to control the strength of the force.

The repulsive force (8) is a contact force that is different from zero if the effective distance of the center of the ellipse to the segment line is non-positive. For the simulations in this paper we set the strength of the repulsive forces as $\eta = 0.2$ and $\eta' = 5$.

2 Influence of the Desired Direction

In this section we study the effects of the desired direction on the dynamics of a system by measuring the outflow from a bottleneck with different widths. See Fig. 2 for the simulation set-up. Four different methods for setting the direction of the desired velocity are introduced and discussed. Finally, simulation results will be compared.

2.1 Strategy 1: Directing Towards the Middle of the Exit

The first strategy is probably the most obvious one. Herein, the desired direction $\overrightarrow{e_i^0}$ for pedestrian i is permanently directed towards a reference point that exactly lies

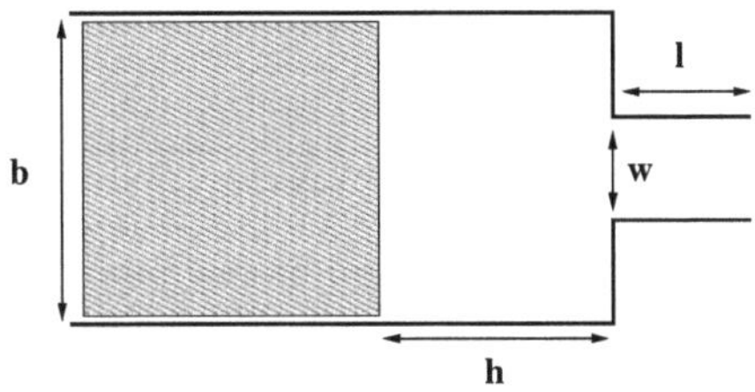

Fig. 2 Scenario set-up: Pedestrians move from a holding area (*shaded region*) through the bottleneck ($l = 2.8$ m, $h = 4.5$ m, $b = 4$ m and w variable)

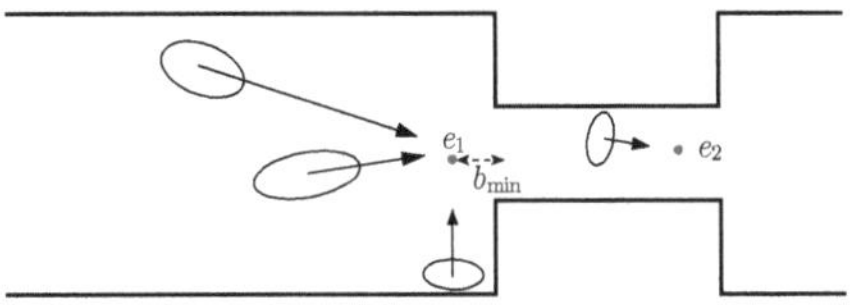

Fig. 3 Strategy 1: All pedestrians are directed exactly towards the reference points e_1 and e_2

on the middle of the exit. In some situations it happens that pedestrians cannot get to the chosen reference point without colliding with walls. To avoid this and to make sure that all pedestrians can "see" the middle of the exit the reference point e_1 is shifted by half the minimal shoulder length $b_{\mathrm{min}} = 0.2$ m (Fig. 3). Pedestrians that pass to the right of the reference point e_1 head towards e_2.

2.2 *Strategy 2: Enhanced Directing Towards the Middle of the Exit*

This is a modification of strategy 1. Pedestrian are still directed to the shifted reference point e_1. However, from a certain position pedestrians can see through the bottleneck the second reference point e_2. In this case e_1 is ignored and the desired direction is set to be parallel to the line $\overrightarrow{e_1e_2}$. Since, pedestrians that are inside the bottleneck can always see e_2 the desired direction is kept parallel to $\overrightarrow{e_1e_2}$.

Here again the reference points and the delimiting range of the bottleneck is shifted in x- and y-direction by b_{max} (Fig. 4).

2.3 *Strategy 3: Directing Towards the Nearest Point on the Exit*

Another possibility to choose the desired direction $\overrightarrow{e_i^0}$ is to define a line l in front of the exit and take at each time the nearest point from the pedestrian i to l (Fig. 5). In comparison with strategy 2, pedestrians that are not in the range where the point e_2 is not visible choose one of the end points of the line l. In strategy 2 this would be the middle of l.

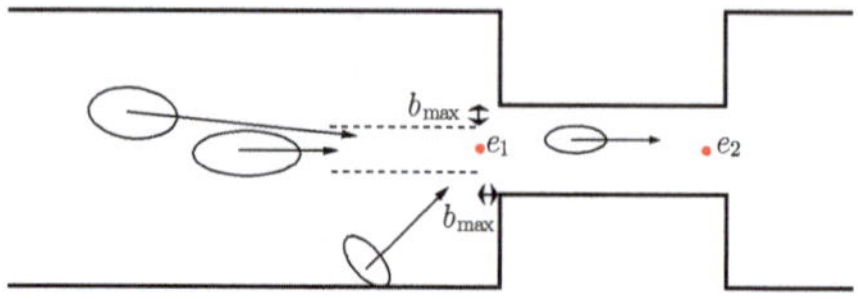

Fig. 4 Strategy 2: Depending on their position pedestrians adapt their direction. In the range where the exit of the bottleneck is visible (marked by *dashed lines*) the direction is longitudinal. Outside this area they are directed towards the middle of the bottleneck

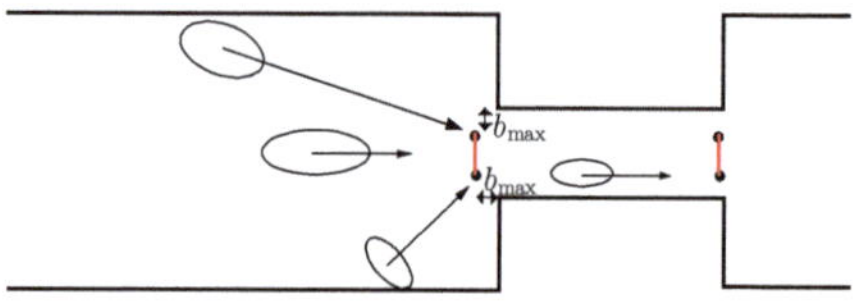

Fig. 5 Strategy 3: Directing towards the nearest point on the exit. Molnár published in [13] a very similar strategy. The only difference is the placement of the line, which is away from the corner by b_{max}

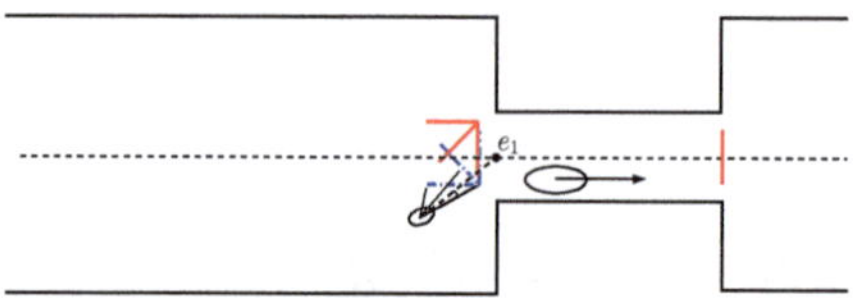

Fig. 6 Strategy 4: Guiding line segments in front of the bottleneck. For each corner a set of three line segments is generated. The length of all directing lines is equal to 3.5 m

2.4 *Strategy 4: Guiding Line Segments*

Without loss of generality we introduce the main idea of strategy 4 with help of the previous bottleneck. Unlike the previous strategies this strategy is applicable to all geometries with corners even if the exit point is not visible. We recall that in strategy 3 a line in front of the bottleneck was defined. The nearest point from each pedestrian to this line was set to define the desired direction. As a generalization we make use in strategy 4 of three different lines to "smoothen" merging in front of the bottleneck (Fig. 6). The blue line set (down the dashed line segment) is considered by pedestrians in the lower half and the red line set by pedestrians in the upper half of the bottleneck. For a pedestrian i at position p_i we define the angle

$$\theta_i = \arccos\left(\frac{\overrightarrow{p_i e_1} \cdot \overrightarrow{p_i l_{ij}}}{\parallel \overrightarrow{p_i e_1} \parallel \cdot \parallel \overrightarrow{p_i l_{ij}} \parallel}\right), \tag{10}$$

with l_{ij} the nearest point of the line j to the pedestrian i.

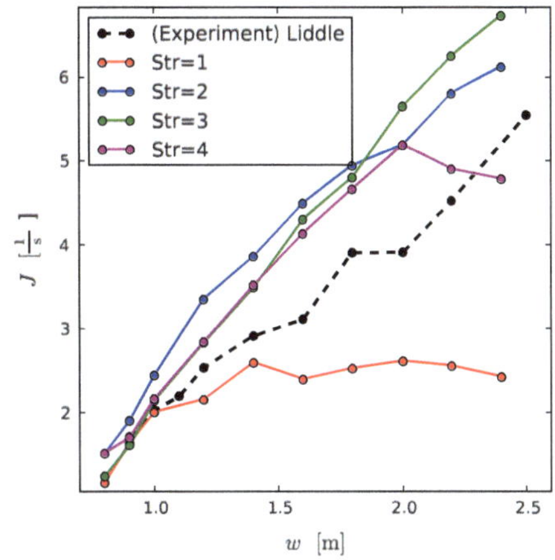

Fig. 7 Flow through a bottleneck with different widths. Simulation results with different strategies for the desired direction of pedestrians in comparison with empirical data from [14]. The experiments were conducted with 180 persons

The next direction is then chosen as

$$\overrightarrow{e_i^0} = \frac{\overrightarrow{p_i l_{ij}}}{\| \overrightarrow{p_i l_{ij}} \|} \tag{11}$$

with j such that $\theta_j = \min\{\theta_1, \theta_2, \theta_3\}$. As in strategy 3 the direction lines are shifted in x- and y-direction by $b_{\min}$.

2.5 Numerical Results

In the previous section we have proposed different methods (called strategies) for choosing the desired direction $\overrightarrow{e_i^0}$. To compare these strategies we have performed simulations for a bottleneck using the same set of parameters for the GCFM. For each strategy only the width of the bottleneck was varied from 1 to 2.4 m.

On the basis of a quantitative analysis the importance of the choice of strategy for the observed behavior can be estimated. In the following, for each strategy we measure the flow through bottlenecks of varying width w. The flow is measured directly at the entrance of the bottleneck according to

$$J = \frac{N_{\Delta t} - 1}{\Delta t}, \tag{12}$$

with $N_{\Delta t} = 60$ pedestrians and Δt the time necessary that all pedestrians pass the measurement line.

In Fig. 7 the resulting flow for all four strategies is presented.

The flow for strategy 1 saturates independently of the width. This was expected since pedestrians do not use the whole width of the bottleneck and keep indeed oriented to the middle. The picture changes for strategies 2–4, where the effective width of the bottleneck is clearly larger. Strategy 2 shows a better usage of the middle widths ($\leq$1.8 m). Here, the slight blocking near the corners, that emerges from strategy 3 is particularly disadvantageous. Strategy 4 produces higher flows for widths up to 2 m. Up 2 m the flow stagnates. The main observations are:

- The choice of the strategy for the desired direction influences considerably the outcome of the simulation.
- An inconsiderate choice of strategy, in that case strategy 1, can lead to large variations from experimental results.
- In contrast to strategy 1, strategies 2–4 show better usage of the bottleneck width and lead to higher flow values.

3 An Application: Motion Around a Corner

Basically, force-based models are functional only in areas, where the exit is constantly visible by all pedestrians. Obviously, this cannot always be guaranteed which is a problem since a proper initialization of the desired direction $\overrightarrow{e_i^0}$ for each pedestrian is not possible. In order to overcome this problem one has to introduce "virtual" exits. This was showcased previously with strategy 4.

In this section, we introduce enhancements of strategy 4 and study their impact on the movement time, i.e. the time until all pedestrians have left the simulation set up. For simplicity we consider the movement of $N = 100$ pedestrians in a 90°-corner-like corridor.

The basis of our enhancements is the following observation: Given a guiding line l, the desired direction of a pedestrian i is determined in dependence of its position and the nearest point to l. This choice neglects two important factors:

1. The perception of space: Individuals try to minimize, when possible, their path to the exit. In our example, pedestrians would take a point near the corner as goal and not the nearest point on the guiding line. Depending on the starting position of pedestrians, this can be far away from the corner and much longer than the shortest path to the exit.
2. The dynamical and collective influence of pedestrians: In the presence of other pedestrians and depending on the magnitude of the local density, the nature of the "quickest path" [15] changes dynamically and differs in most cases from the "shortest path" to the exit.

We therefore adopt a concept similar to ideas introduced in [16] which are well established and widely used in cellular automata models [17, 18].

At a time step t a pedestrian i heads towards a point on the line which minimizes the distance to the inner point of the corner $l^i(t)$. This is a natural territorial effect which leads to the shortest path to the exit. If all pedestrians try to take the shortest path, large jams will be observed right at the inner point of the corner. If, however,

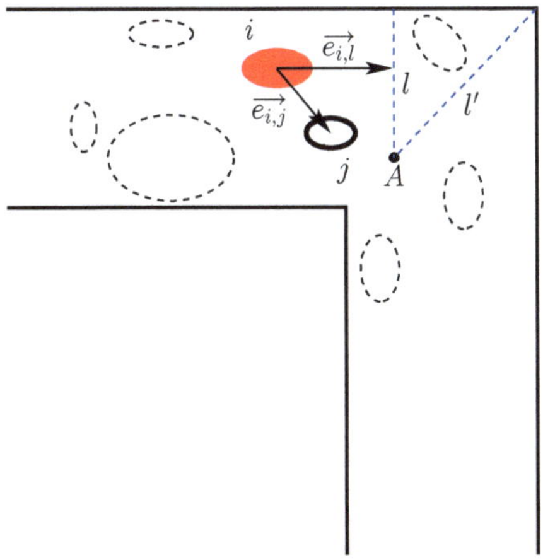

Fig. 8 How to get around the corner? Pedestrian i that is heading toward the first guiding line, considers the positions of its neighboring pedestrians as well as its initial position to decide whether or not to head closer to the edge of the corner

the collective influence of pedestrians dominates the choice of the desired direction, pedestrians will choose their desired direction to be orthogonal to the guiding lines and thus make better use of the whole directing line.

For this reason we include a dynamical factor that depends mainly on previous decisions taken by other pedestrians:

$$p^i(t) = \exp\left(-k_d \cdot \text{occ}^i_{\text{rel}}(t)\right), \tag{13}$$

where

$$\text{occ}^i_{\text{rel}}(t) = \frac{n^i}{n^i_{\max}} \tag{14}$$

is a measure of the occupation of the line. n^i is the cardinality of the set

$$A_l = \left\{l^j \mid j \in B_l \;\&\; l^j < l^i\right\}$$

and $n^i_{\max}$ is the cardinality of the set

$$B_l = \left\{j \in [1, N] \mid i \neq j \;\;\&\;\; \overrightarrow{e_{i,l}} \cdot \overrightarrow{e_{i,j}} \geq 0\right\}. \tag{15}$$

B_l is the set of all relevant neighbors of i, that influences its desired direction by means of a contribution to $\text{occ}^i_{\text{rel}}(t)$ (14). For the scenario depicted in Fig. 8 the set B_l for i (red ellipse) contains only one pedestrian j (bold ellipse).

Large values of $\text{occ}^i_{\text{rel}}(t)$ imply small values of $p^i(t)$. As a consequence pedestrians prefer not to change their desired direction closer to the edge of the corner.

Finally, the update rule of the distance $l^i(t)$ is given by:

$$l^i(t + \Delta t) = l^i(t) \cdot \left(1 - p^i(t)\right). \tag{16}$$

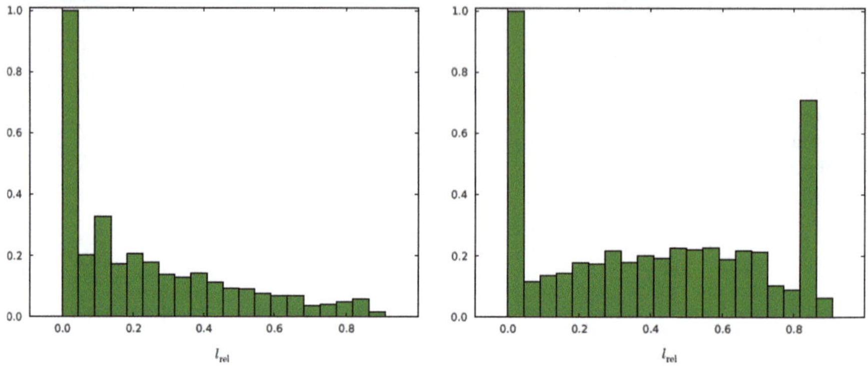

Fig. 9 Movement of $N = 100$ pedestrians around a corner with different values of k_d. The width of the corridor is $w = 3$ m. (**a**) $k_d = 0$. (**b**) $k_d = 10$

$p^i(t) \in [0, 1]$ gives the rate of change from the initial "guess" of pedestrian i. For $p^i(t) = 0$ the desired direction of i stays orthogonal to the guiding line, while $p^i(t) = 1$ displays the case where i's desired direction is directed to the edge of the corner A. In the next section we study the influence of the parameter k_d on the dynamics of pedestrians. For the second and third line we set $k_d = 0$ and vary it only for the first line.

4 Analysis of the Sensitivity Parameter

To understand the impact of the collective influence of pedestrians on the chosen target point for each pedestrian i, we study the time evolution of the relative length for different values of k_d. The relative length is defined as

$$l^i_{\text{rel}}(t) = \frac{l^i(t)}{l_{\max}} \tag{17}$$

where $l_{\max}$ is the length of the guiding line.

Figure 9a shows the probability distribution of the relative length for $k_d = 0$. Pedestrians are mainly heading towards A and the full length of the directing line is rarely used.

The situation changes considerably for $k_d = 10$. Figure 9b shows that the distribution of the length is more balanced which indicates that pedestrians make better use of the directing lines.

To showcase the impact of collective influence of pedestrians on the desired direction, we show in Fig. 10 the variation of the movement time in dependence of k_d.

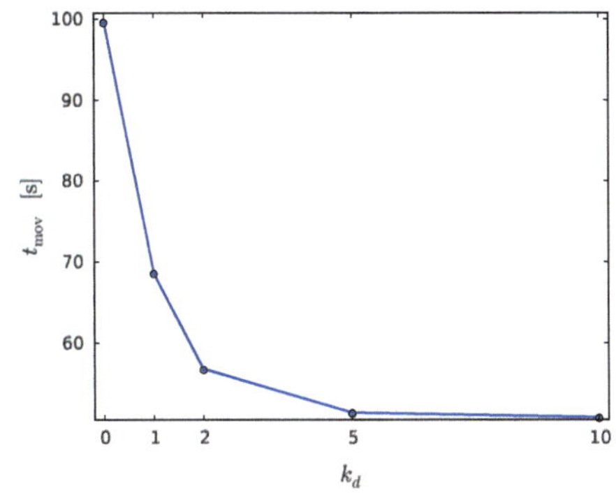

Fig. 10 Movement time for a simulation with $N = 100$ pedestrians around a corner for different values of k_d

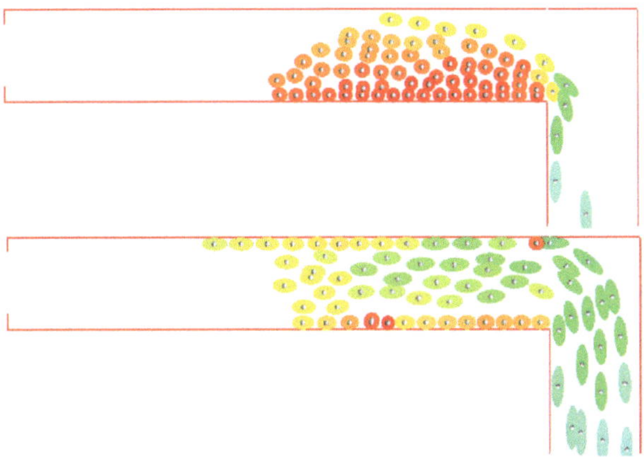

Fig. 11 Screen shot of a simulation with 100 pedestrians, $k_d = 0$ (*top*) and $k_d = 10$ (*bottom*)

A qualitative comparison shown in Fig. 11 confirms the above-mentioned quantitative analysis.

For $k_d = 0$ a jam forms immediately before the corner as indicated by the large number of slowly moving pedestrians (red ellipses). This results from a strong competition between the pedestrian to pass close to the edge A of the corner. In contrast, for $k_d = 10$ pedestrians move quicker since they make optimal use of the guiding line.

5 Summary

We have developed a strategy to determine the desired direction $\overrightarrow{e_i^0}$ for each pedestrian i. This method is rather general and can be used in each geometry characterized by the existence of corners, e.g. bottlenecks (two corners), T-Junction (two corners). In analogy to CA models we introduced and tested a factor to model the static and dynamic interactions of pedestrians with the geometry.

Our work was based on an enhanced version of the GCFM [6]. The enhancements use a considerable simplification of the repulsive forces acting on pedestrians from walls. Furthermore, we addressed an important issue in force-based models, namely the choice of the desired direction of pedestrians. Several strategies were implemented and compared with empirical data. This comparative investigation showed that the outcome of a simulation depends strongly on the chosen direction of the desired direction of pedestrians. Finally, we introduced a new mechanism to direct pedestrians in 90°-corners by means of directing lines. The main concept of this strategy base on the well-known concept of dynamical floor-field. For further works, the parameter k_d that expresses the tendency of pedestrians to take the shortest path (or not) should be varied individually as the geometrical and dynamical conceptions of pedestrians differ.

Acknowledgements This work is within the framework of two projects. The authors are grateful to the Deutsche Forschungsgemeinschaft (DFG) for funding the project under Grant-No. SE 1789/1-1 as well as the Federal Ministry of Education and Research (BMBF) for funding the project under Grant-No. 13N9952 and 13N9960.

References

1. Helbing D (1991) Behavioral Science, 36:298–310
2. Helbing D, Molnár P (1995) Phys. Rev. E, 51:4282–4286
3. Lakoba T I, Kaup D J, Finkelstein N M (2005) Simulation, 81:339–352
4. Yu W J, Chen L Y, Dong R, Dai S Q (2005) Phys. Rev. E, 72(2):026112
5. Johansson A, Helbing D, Shukla P K (2007) Adv. in Compl. Sys., 10(2):271–288
6. Chraibi M, Seyfried A, Schadschneider A (2010) Phys. Rev. E, 82:046111
7. Steffen B, Seyfried A (2009) Modelling of Pedestrian Movement around 90° and 180° Bends. In Topping B H V, Tsompanakis Y (eds) The First International Conference on Soft Computing Technology in Civil, Structural and Environmental Engineering. Civil-Comp Press, Stirlingshire, UK
8. Gloor C, Mauron L, Nagel K (2003) A pedestrian simulation for hiking in the alps. In Proceedings of the Swiss Transport Research Conference (STRC), Monte Verita, CH
9. Moussaïd M, Helbing D, Theraulaz G (2011) Proc. Nat. Acad. Sc., 108(17):6884–6888
10. Höcker M, Berkhahn V, Kneidl A, Borrmann A, Klein W (2010) Graph-based approaches for simulating pedestrian dynamics in building models. In 8th European Conference on Product & Process Modelling (ECPPM), University College Cork, Cork, Ireland
11. Kemloh Wagoum A U, Seyfried A (2010) Optimizing the evacuation time of pedestrians in a graph-based navigation. In Panda M, Chattararaj U (eds) Developments in Road Transportation, Macmillian Publishers India Ltd
12. Kretz T, Große A, Hengst S, Kautzsch L, Pohlmann A, Vortisch P (2011) Adv. in Compl. Sys., 14(5):733–759
13. Molnár P (1995) Modellierung und Simulation der Dynamik von Fußgängerströmen. PhD Thesis, Universität Stuttgart, Stuttgart

14. Liddle J, Seyfried A, Klingsch W, Rupprecht T, Schadschneider A, Winkens A (2009) An Experimental Study of Pedestrian Congestions: Influence of Bottleneck Width and Length. In Traffic and Granular Flow 2009 (arXiv:0911.4350)
15. Kretz T (2009) J. Stat Mech.: Theory and Experiment, P03012
16. Burstedde C, Klauck K, Schadschneider A, Zittartz J (2001) Physica A, 295:507–525
17. Nishinari K, Kirchner A, Namazi A, Schadschneider A (2004) IEICE Transactions, 87-D(3):726–732
18. Kretz T, Schreckenberg M (2006) The F.A.S.T.-Model. In Lect. Notes Comp. Sc., 4173:712–715, Springer Berlin/Heidelberg

[illegible]

A Simple Model for Phase Separation in Pedestrian Dynamics

Christian Eilhardt and Andreas Schadschneider

Abstract The occurrence of phase separation is a common feature observed in vehicular traffic. Experiments have shown a similar behavior for pedestrians, though the situation in pedestrian dynamics is more complicated. The two separate phases in one-dimensional "single-file" pedestrian movement are a jammed high-density phase and a phase of medium to high density with slowly moving pedestrians. Both phases consist of interacting particles (pedestrians). To understand the emergence of this kind of phase separation, we develop a simple cellular automaton model. The transition probabilities of the modeled pedestrians in general depend on their current velocities and on the occupancy of the next two cells in front of them. For inhomogeneous initial conditions the simulated pedestrian trajectories clearly feature two distinct phases analogous to the experiment. The lifetime of the decaying jammed phase is consistent with experimental results.

1 Introduction

Pedestrian dynamics is becoming a more and more important field of research for scientists of many different fields. This is understandable due to the multitude of related applications. The main areas of interest are firstly evacuation scenarios that include both forecasts of live evacuations [1] and planning of pedestrian facilities to improve safety during evacuations and avoid crowd disasters. Secondly, knowledge about the movement of pedestrian crowds can be applied to optimize the daily use of pedestrian facilities in buildings that are frequented by large numbers of pedestrians such as airports, railway stations, shopping centers etc.

In both cases an accurate understanding of the underlying principles of pedestrian motion in necessary. In particular for evacuation simulations the high density regime

C. Eilhardt (✉) · A. Schadschneider
Institut für Theoretische Physik, Universität zu Köln, 50937 Köln, Germany
e-mail: ce@thp.uni-koeln.de; as@thp.uni-koeln.de

V.V. Kozlov et al. (eds.), *Traffic and Granular Flow '11*,
DOI 10.1007/978-3-642-39669-4_26, © Springer-Verlag Berlin Heidelberg 2013

can be very important. The effect we are interested in is a separation of pedestrian traffic into two distinct phases which consist of standing pedestrians and slowly moving pedestrians, respectively. This can generally be observed at high densities, but has barely been studied systematically. To understand complex phenomena, one typically has to understand the underlying basic mechanisms first. Therefore we here only consider phase separation in one-dimensional movement without overtaking.

Typical pedestrian models such as the social force model [2] or the floor field cellular automaton model [3] can successfully describe qualitative phenomena like lane formation and are even used for quantitative predictions. However, they do not feature phase separation.

2 Phase Separation in Vehicular Traffic

The occurrence of phase separation is a common and well-understood feature in vehicular traffic [4]. There are two empirically observed, distinct phases: A jammed phase with high density and a free-flow phase with low density where cars move at their desired velocity. This can be reproduced in appropriate models such as the VDR model [5]. By using a "slow-to-start" rule the outflow of the jam is reduced compared to the maximal flow of the system. This leads to a region of non-interacting cars and thus to the free-flow phase, see Fig. 1.

The existence of two distinct phases is connected to the existence of metastable states in the fundamental diagram, see Fig. 2. For intermediate density values the flow is not a unique function of the density: the free flow branch can spontaneously break down into a congested state. This is called a capacity drop and leads to a hysteresis loop [6].

3 Phase Separation in Pedestrian Traffic

The situation in pedestrian dynamics is more complicated for several reasons. The most obvious one is the generally two-dimensional nature of pedestrian movement in contrast to the one-dimensional movement in vehicular traffic. Furthermore, collision avoidance is less of an issue for pedestrians – pedestrians bump into each other regularly, cars do not. Finally, the movement of pedestrians is at least potentially influenced by multiple other pedestrians whereas the interaction range in vehicular traffic is very small.

Consequently, experiments [7] show a similar, but somewhat different behavior for pedestrians compared to what is observed in vehicular traffic. Firstly, the pedestrian fundamental diagram does not show metastable states similar to the vehicular fundamental diagram. However, the trajectories of the one-dimensional "single-file" pedestrian movement feature two separate phases, namely a jammed high-density phase and a phase of medium to high density with slowly moving pedestrians.

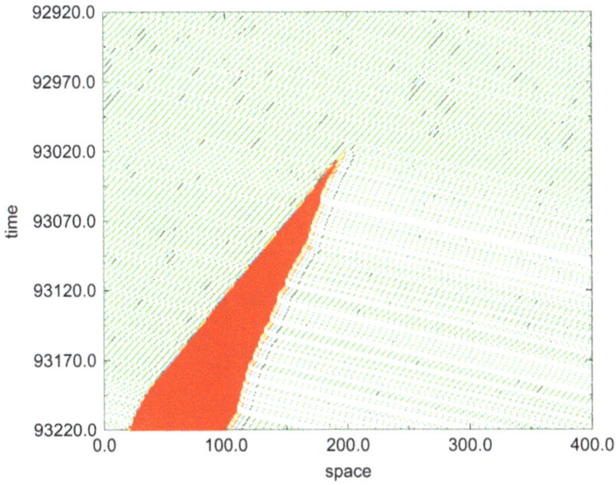

Fig. 1 Simulated car trajectories in the VDR model

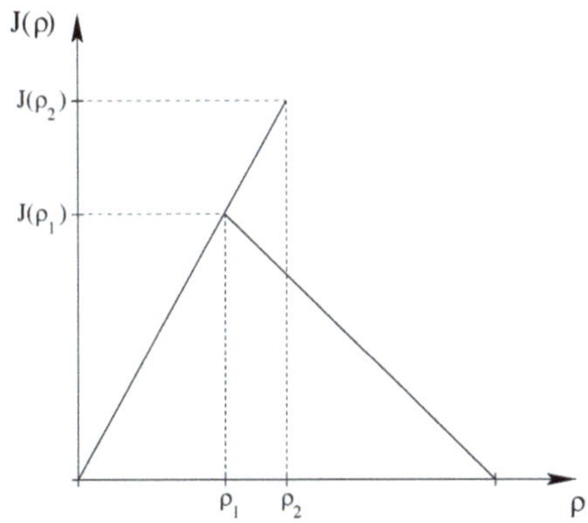

Fig. 2 Schematic fundamental diagram of the VDR model

To analyze pedestrian phase separation, one therefore has to study microscopic quantities (trajectories), it does not suffice to study the (macroscopic) fundamental diagram. In contrast to vehicular traffic, the distance between pedestrians in the moving phase is small, not allowing them to move with their desired velocity. Both phases consist of interacting particles. The mechanism creating the phase separation therefore differs from a "slow-to-start" rule.

Figure 3 shows the experimental setup of an experiment performed by Seyfried et al. [7] in 2006 with up to 70 pedestrians. The length of the experimental setup was 26 m including a 4 m long measurement section in which the pedestrian trajectories have been measured by automatically tracking the pedestrian heads. The resulting trajectories for the very high global density of 2.7 pedestrians/m are shown in Fig. 4 as a space-time plot. Note the separation into phases with slowly

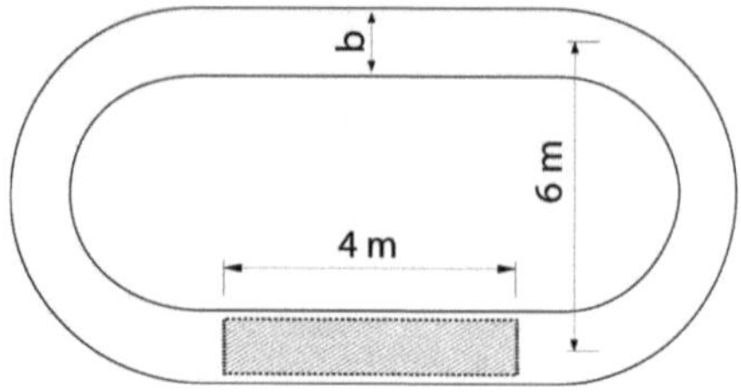

Fig. 3 Experimental setup [7]

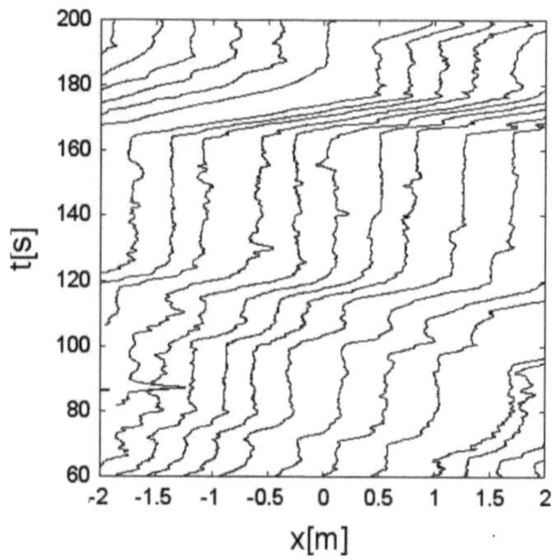

Fig. 4 Experimental trajectories [7]

moving pedestrians with velocity $v = 0.15\,\text{m/s}$ and phases in which the pedestrians do not move. Small movements in the trajectories come from the head movement of standing pedestrians.

The experimental data only extend over the small time-scale of 140 s and small spatial scales of 26 m of the whole experimental setup and only 4 m of the measurement section. This prevents judging the stability and locality of the empirical phase separation. Firstly, it is unclear whether the phases remain separate for longer timescales. Secondly, we do not know whether the observed jam in the measurement section is the only jam in the experimental setup. The existence of additional jams at the same time indicates that jams may be forming and decaying dynamically.

4 Model Definition

We develop a simple rule-based cellular automaton model that aims at reproducing the observed phase separation. This work tries to show that our model is in principle capable to create phase separation that is in qualitative agreement with the experimental data, i.e. two separate phases with slowly moving and not moving

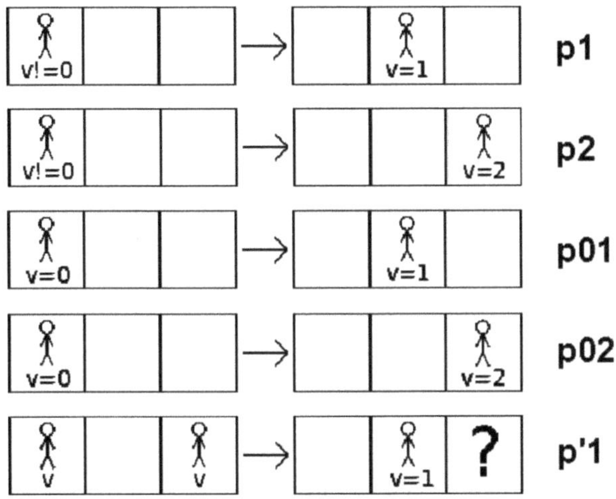

Fig. 5 Definition of the five transition probabilities used in the model

particles, respectively. The model is defined on a one-dimensional lattice containing 200 cells. As in the experiments we use periodic boundary conditions to connect the last cell to the first cell. Each cell can either be occupied by one pedestrian or empty. Densities are defined relative to the maximal density, $\rho = \frac{N_{\text{occupied}}}{N_{\text{total}}}$. The current velocity of a pedestrian is defined as the number of cells it moved in the last timestep. In every timestep the pedestrians are updated in parallel according to the following update rules which are also shown in Fig. 5.

Pedestrians that have a velocity unequal zero and at least two free cells in front of them move to the cell directly in front of them with probability p_1 and move two cells forward with probability p_2. Pedestrians with current velocity zero and two free cells in front of them are allowed to move to the cell directly in front of them and the cell two steps in front of them with probabilities p_{01} and p_{02}, respectively. Note that this requires both $p_1 + p_2 \leq 1$ and $p_{01} + p_{02} \leq 1$. Pedestrians with only one free cell in front of them move one cell forward with probability p'_1, independent of the movement of the pedestrian two cells in front of them.

The simulation results can be mapped to experimental results by identifying every cell with an area of size $0.4 \times 0.4\,\text{m}^2$. This is standard for cellular automata models such as the floor field cellular automaton [8] and corresponds to a maximum density of 6.25 pedestrians/m^2 or a maximum line density of 2.5 pedestrians/m. Note that this is a little lower than the density of 2.7 pedestrians/m used in the experiment. Identifying each cell instead with a one-dimensional length of 0.3 m leads to a maximum density of 3.33 pedestrians/m. The experimental density 2.7 pedestrians/m then corresponds to the relative density 0.81. It may not be reasonable to naively extend this to two dimensions because the resulting maximum density of 11.1 pedestrians/m^2 is very high. A timestep corresponds to about 0.35 s.

With standard parameter values the free velocity of about 1.85 cells/timestep (see Sect. 5.1) equates to

$$v_{\text{free}} = 1.85 \frac{\text{cells}}{\text{timestep}} = 1.85 \times \frac{0.3\,\text{m}}{0.35\,\text{s}} \approx 1.6\,\frac{m}{s}$$

in accordance with typical experimental values.

5 Results

5.1 Modeled Pedestrian Trajectories

Figure 6 shows pedestrian trajectories from a simulation of 140 pedestrians with inhomogeneous initial conditions. This density corresponds to a relative density of 0.7. The resulting trajectories of the pedestrians clearly feature two distinct phases, namely a completely jammed phase and a phase in which pedestrians are slowly moving. This phase separation is very slowly decaying into a single congested high density state with slowly moving pedestrians, see Sect. 5.2. However, the results are in good agreement with the experimental data when observed over a corresponding timescale.

The parameter values used here are

$$p_1 = 0.05, \qquad p_2 = 0.9, \qquad p_{01} = 0.01, \qquad p_{02} = 0.99, \qquad p_1' = 0.99.$$

The resulting free velocity can be approximated by $v_{\text{free}} = p_1 + 2 \cdot p_2 = 1.85$ because the configurations that require the use of p_{01}, p_{02} and p_1' are very unlikely to occur for free moving pedestrians.

The velocity in the congested phase is about 0.7 cells/timestep and therefore much smaller than the free-flow velocity. This qualitatively reproduces the experimental results. Small changes of the parameters generally produce only small and non-significant changes of the resulting trajectories.

Figure 7 shows two simulation runs with inhomogeneous initial conditions but higher/lower densities. The pedestrian trajectories at density 0.85 show an increased jam length compared to 0.7 density which is easily explained by the larger amount of particles in the simulation. The velocity in the congested phase is roughly the same at all three densities 0.55, 0.7 and 0.85. For relatively low densities such as 0.55 the initial jam decays within about 800 timesteps. This clearly shows the finite lifetime of the phase separation. Note that the density 0.55 is much smaller than the densities at which phase separation has been observed experimentally, even though it still is a relatively high density!

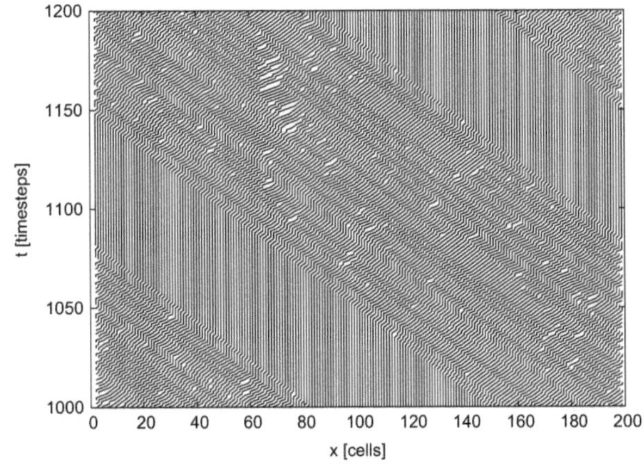

Fig. 6 Simulated pedestrian trajectories with inhomogeneous initial conditions and density 0.7. The space-time plot clearly features two separated phases

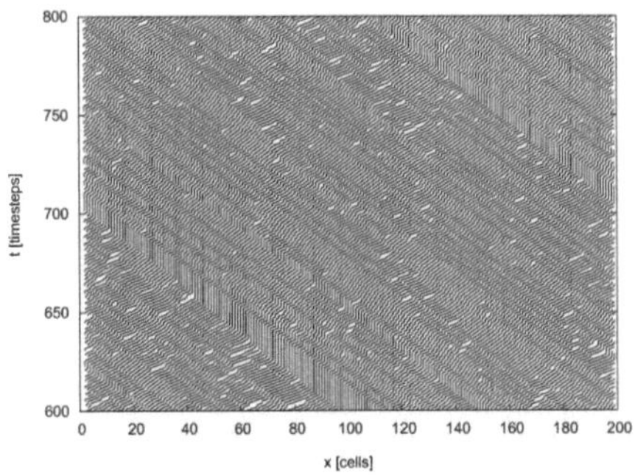

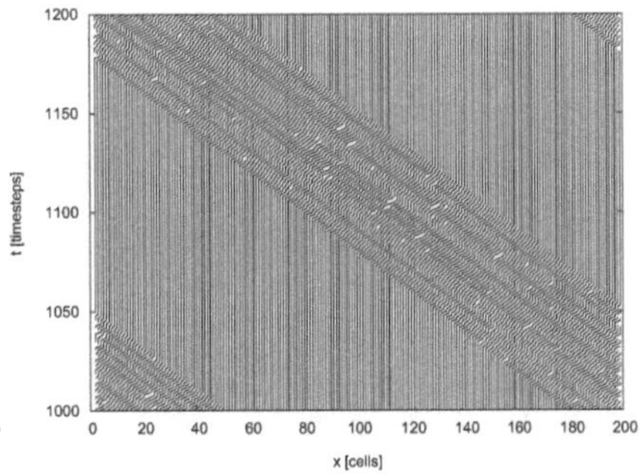

Fig. 7 Simulated pedestrian trajectories with inhomogeneous initial conditions and density 0.55 (*left*) and 0.85 (*right*)

Figure 8 shows the simulation results for homogeneous initial conditions and densities 0.7 and 0.85, Fig. 9 shows trajectories for the same densities but random initial conditions.

The trajectories of the simulation runs with homogeneous and random initial conditions and with densities 0.7 and 0.85 are barely distinguishable after 1,000 timesteps. All four space-time plots show only local phase separation. In other words: at any given time there is more than one jam in the simulation. These jams are created and die dynamically. The jams are separated by congested areas in which pedestrians can move slowly. There are however small differences in size and frequency of the jams. With random initial conditions the simulation develops less but larger jams. It is unclear whether this local phase separation is in accordance with the empirical situation because of the small measurement area in the experimental setup.

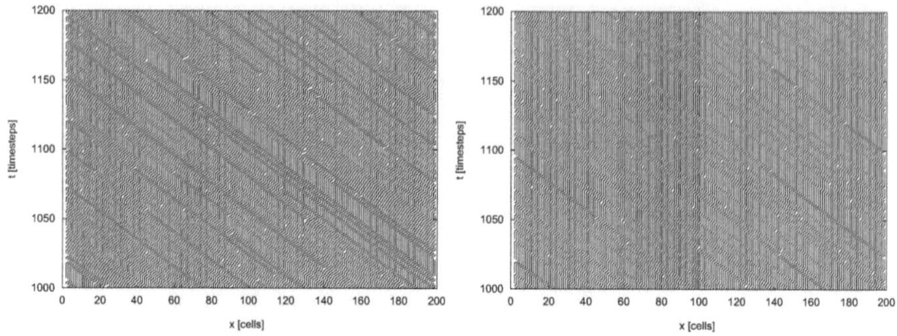

Fig. 8 Simulated pedestrian trajectories with homogeneous initial conditions and density 0.7 (*left*) and 0.85 (*right*)

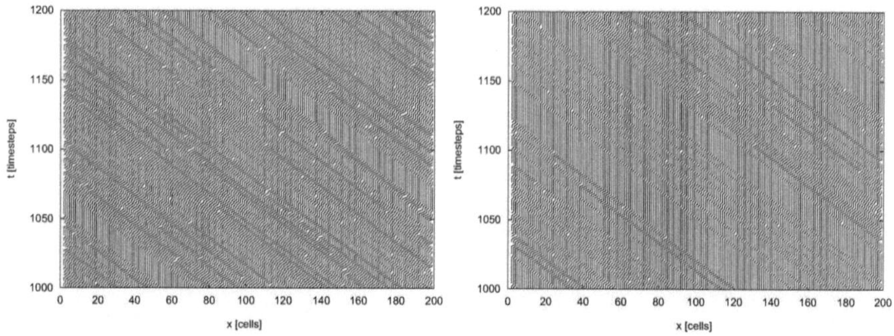

Fig. 9 Simulated pedestrian trajectories with random initial conditions and density 0.7 (*left*) and 0.85 (*right*)

5.2 *Jam Stability*

A jam is defined as a collection of multiple successive non-moving pedestrians, the length of the largest jam in the system is called jam length. For inhomogeneous initial conditions the largest jam typically is the initial jam, at least until the initial jam has decayed such that another randomly created jam can be larger.

Figure 10 shows the jam length during a typical simulation run as a function of the elapsed time for three different densities. For high to intermediate densities such as 0.85 (top left) and 0.7 (top right) the jam length does not change much over the course of the simulation. Even though the jam length is obviously decaying, its lifetime is much longer then the duration of the experiment in which phase separation has been observed. At the low density 0.55 the decay is more quickly: The initial jam has decayed completely after about 800 timesteps.

Note that this is a low density only with respect to phenomena that occur at very high densities such as phase separation. In most circumstances a density of 0.55 which corresponds to 1.83 pedestrians/meter would be regarded as a high density.

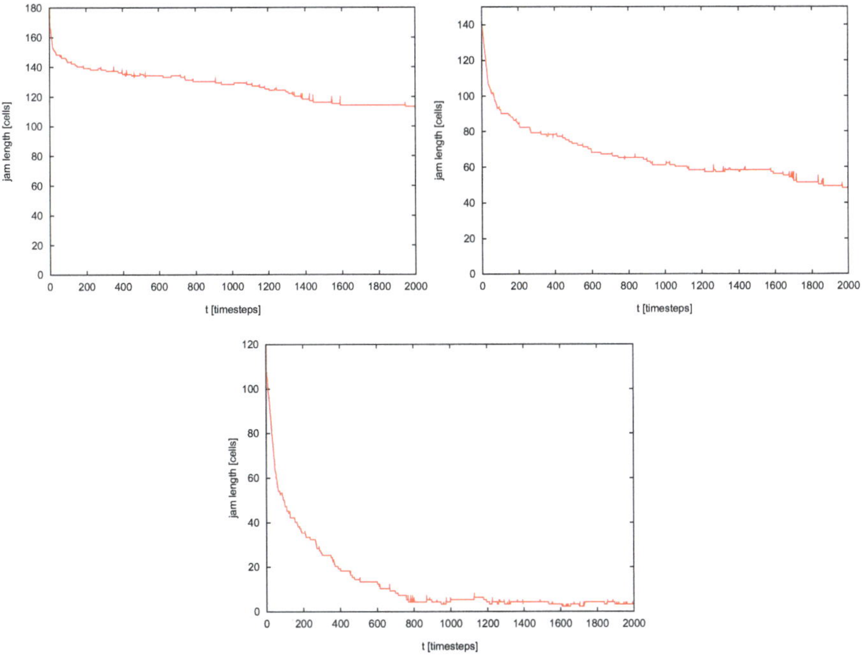

Fig. 10 Jam length for simulations with densities 0.85 (*top left*), 0.7 (*top right*) and 0.55 (*bottom*) and inhomogeneous initial conditions as a function of elapsed time

In a time frame of 140 s or about 470 timesteps corresponding to the experimental data the simulation results are consistent with the experimental results in the sense that despite the decay the model clearly features phase separation at every point in time.

Simulations with inhomogeneous initial conditions show only one large but decaying jam in the system. However the lifetime of the jam is larger than the timescale of the experimental data. In simulations with random or homogeneous initial conditions jams appear and disappear dynamically. In this case there can be several jams in the system at the same time. This is however consistent with the experimental data which only shows one jam at any given time due to the small measurement section. To prove or disprove the existence of local phase separation large scale experiments are needed.

6 Outlook

The understanding of pedestrian phase separation deepens the general understanding of pedestrian dynamics and can therefore improve the design of airports, shopping centers and other crowded areas as well as improve safety during evacuations.

Phase separation in pedestrian traffic is different from phase separation in vehicular traffic. The mechanism responsible for creating pedestrian phase separation is yet unknown. The model presented here is capable of qualitatively reproducing the experimental results. It features two distinct phases in which the pedestrians (a) move with greatly reduced velocity or (b) do not move at all. The relatively small time-scale as well as spatial scale of the actual measurement do not allow to judge the stability and locality of the phase separation.

The next step is to quantitatively reproduce the experimental results. To do that, the system size needs to be decreased. However, this increases finite size effects and leads to a few other challenges as well. Phase separation occurs in the very high density regime in which the total number of free cells in simulations of cellular automata models becomes very small. The limited spatial resolution of standard cellular automata models leads to a limited velocity resolution as well. Therefore it is extremely difficult to distinguish between standing and very slow moving pedestrians.

This could in principle be resolved by reducing the cell size to $0.15 \times 0.15\,\mathrm{m}^2$ or smaller and increasing the size of the particles to several cells as well as increasing the velocity of the particles.

Acknowledgements This work has been performed within the Hermes project funded by the German Federal Ministry of Education and Research (BMBF) as part of the "Research for Civil Security – Protecting and Saving Human Life" program. It was supported by the Bonn-Cologne Graduate School of Physics and Astronomy.

References

1. Hermes project, http://www.fz-juelich.de/jsc/hermes.
2. D. Helbing, P. Molnár. *Social force model for pedestrian dynamics*, Phys. Rev. E **51**, 4282 (1995)
3. C. Burstedde, K. Klauck, A. Schadschneider, J. Zittartz. *Simulation of pedestrian dynamics using a 2-dimensional cellular automaton*, Physica A **295**, 507 (2001)
4. A. Schadschneider, D. Chowdhury, K. Nishinari. *Stochastic Transport in Complex Systems: From Molecules to Vehicles* (Elsevier, 2010)
5. R. Barlovic, L. Santen, A. Schadschneider, M. Schreckenberg. *Metastable states in cellular automata for traffic flow*, Eur. Phys. J. B **5**, 793 (1998)
6. F.L. Hall, B.L. Allen, M.A. Gunter. *Empirical analysis of freeway flow-density relationships*, Transp. Res. A **20**, 197 (1986)
7. A. Seyfried, A. Portz, A. Schadschneider. *Phase Coexistence in Congested States of Pedestrian Dynamics*, Lect. Notes Comp. Sc. **6350**, 496 (2010)
8. A. Kirchner, A. Schadschneider. *Simulation of evacuation processes using a bionics-inspired cellular automaton model for pedestrian dynamics*, Physica A **312(1–2)**, 260–276 (2002)

HERMES: An Evacuation Assistant for Large Sports Arenas Based on Microscopic Simulations of Pedestrian Dynamics

Andreas Schadschneider, Christian Eilhardt, Stefan Nowak, Armel Ulrich Kemloh Wagoum, and Armin Seyfried

Abstract The improvement of safety at mass events has become an important issue not only due to several disasters involving large crowds. To support security services in case of emergencies we have developed an evacuation assistant which allows forecasting the emergency egress of large crowds in complex buildings. Such a forecast requires pedestrian models which produce realistic crowd dynamics and, at the same time, can be simulated faster than real-time. Here we give an overview of the project and present first results.

1 Introduction

In recent years we have seen an increasing number of mass events with sometimes more than 100,000 spectators. Unfortunately also the number of crowd disasters with many injured persons or even fatalities has increased [1]. Multifunctional building structures in combination with a wide range of large-scale public events present new challenges for the quality of security concepts [2]. Rather general legal regulations should ensure that in case of an emergency all spectators are able to leave the danger zone quickly. This is achieved by specifying e.g. the minimal width and

A. Schadschneider (✉) · C. Eilhardt · S. Nowak
Institut für Theoretische Physik, Universität zu Köln, 50937 Köln, Germany
e-mail: as@thp.uni-koeln.de; ce@thp.uni-koeln.de; sn@thp.uni-koeln.de

A.U.K. Wagoum
Jülich Supercomputing Centre, Forschungszentrum Jülich GmbH, 52425 Jülich, Germany
e-mail: u.kemloh@fz-juelich.de

A. Seyfried
Jülich Supercomputing Centre, Forschungszentrum Jülich GmbH, 52425 Jülich, Germany

Computer Simulation for Fire Safety and Pedestrian Traffic, Bergische Universität Wuppertal, Pauluskirchstraße 11, 42285 Wuppertal, Germany
e-mail: a.seyfried@fz-juelich.de

V.V. Kozlov et al. (eds.), *Traffic and Granular Flow '11*,
DOI 10.1007/978-3-642-39669-4_27, © Springer-Verlag Berlin Heidelberg 2013

maximal length of escape routes. Sometimes computer simulations are performed [3–9], but these are usually restricted to the planning stage.

In the event of an emergency, not all rescue routes might be available due to fire or structural damage to the facility. For an optimal crowd management which aims to prevent critical situations, e.g. due to dangerously high crowd densities and bottlenecks, accurate and up-to-date information about the current status is needed. In the project HERMES we have developed a new concept to increase safety in large sports arenas. An *evacuation assistant* provides valuable information for the security forces (e.g. police or firefighters) which supports them in their decision [2, 10]. In addition, a forecast is made which shows the evolution of the evacuation based on a microscopic computer simulation.

2 The Concept

The idea behind the evacuation assistent HERMES is to provide the security forces with additional information about the current situation, e.g. the number of people in the danger zone. It thus supports the decision makers to rate the actual danger, to decide on a successful evacuation strategy and to optimally employ the security staff. To test this concept the evacuation assistent has been implemented in the ESPRIT arena in Düsseldorf (Germany). This multifunctional arena with a capacity of 60,000 spectators will show how crowds of people at big events can be guided also considering the current risk situation.

Various universities, industrial partners and users cooperate in this project. For an overview we refer to [10,11]. HERMES consists of several components as shown in the schematic diagram of its functional layout (Fig. 1). In a first step information about the situation when an emergency occurs is gathered. The distribution of the spectators in the facility is obtained by automated person counting at entrances and doors using a video-based system. Information about the available escape routes blocked due to smoke, locked doors or other dangers comes from the safety and security management system of the facility.

This information is used as basis for a computer simulation of a microscopic model of pedestrian dynamics. It will provide a forecast of the evacuation process by predicting the movement of all people during the next 15 min and updates it at 1-min intervals. This forecast is used by the security personnel to identify critical situations (e.g. congestion areas, high densities) and decide about possible provisions for their avoidence. Moreover a macroscopic network model will calculate the optimal distribution of occupants on the available routes. A communication module provides the information to the emergency teams on site.

The approach is very flexible and allows to take into account the actual situation concerning the distribution of persons and availability of exits. In contrast, currently emergency plans exists only for some selected scenarios. The computer simulation uses microscopic models (floor field cellular automaton model [12], the TraffGo

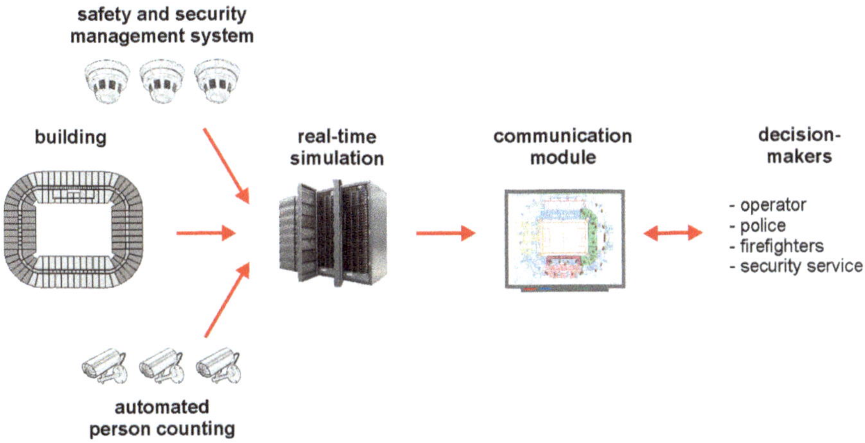

Fig. 1 Schematic diagram of the evacuation assistant

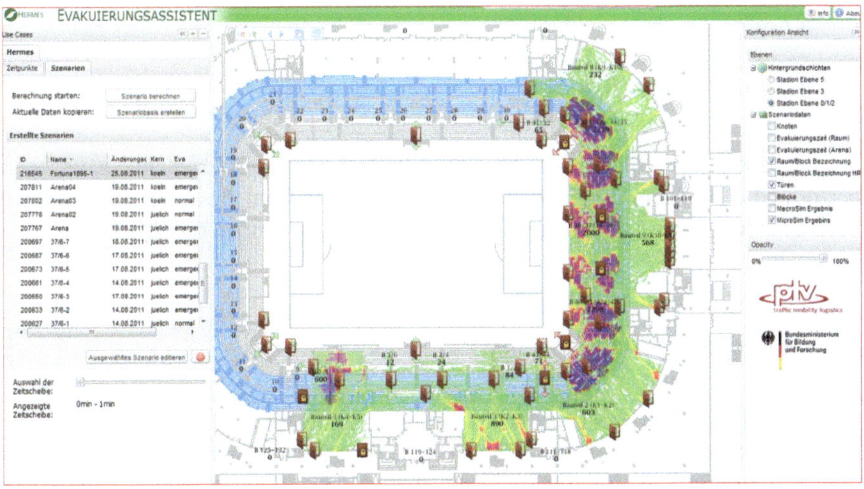

Fig. 2 The evacuation assistant showing the results of a simulation. The congestion areas are displayed

cellular automaton [4] and the generalized centrifugal force model [13, 14]) which provide rather detailed predictions about the likely development of an evacuation.

Different types of information are displayed, for instance real-time information like the spatial distribution of the pedestrians, as well as the states of the different doors and areas. The simulation results are shown using the Level of Service (LOS) classification [15]. LOS is a measure used in traffic to determine the effectiveness of elements of transportation infrastructure, for instance highways. The LOS of the areas are displayed in three colors (red, yellow and green) where red stands for high density and green for low density (Fig. 2).

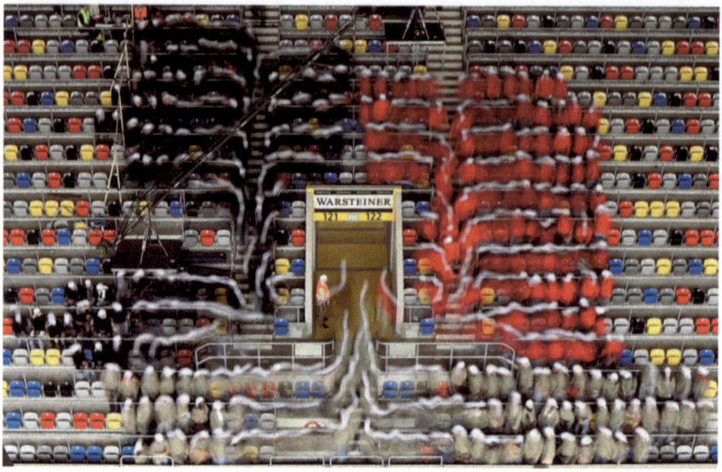

Fig. 3 Snapshot from an evacuation experiment in the arena

3 Empirical Studies

To increase the reliability of the forecast, large-scale experiments have been performed in the laboratory and the arena. The results from these experiments have been used to calibrate the models in order to make quantitative predictions.

3.1 Laboratory Experiments

Numerous laboratory experiments have been performed to study pedestrian streams in a controlled environment. The focus has been on basic situations, e.g. the motion in a corridor, the behavior at bottlenecks, the influence of counterflow etc. [16, 17]. The results for T-shaped corridor [18] are described elsewhere in these proceedings.

3.2 Arena Experiments

In addition to the laboratory experiments we have also performed experiments in the arena itself. Figure 3 shows a snapshot of one of these experiments. Here it was studied how the stands are emptied, e.g. to identify the areas of congestion. It was found that the congestions occur near the seating area and on the stairs, not at the exit itself.

4 Microscopic Models

To obtain more reliable predictions, different modeling strategies are combined. One approach is based on cellular automata models which allows to achieve an good performance and obtain results quickly. The second approach is a spatially continuous force-based model which gives a better spatial resolution. The advantages and drawbacks of these two model classes are e.g. discussed in [19]. The application of these models to security sensitive tasks within the Hermes project requires their quantitative verification and calibration. Currently it is not known how reliable forecasts based on simulations are. However large differences in the predictions of commercially available simulations even for the simplest scenarios have been found [20]. Partly this is a consequence of the insufficient and sometimes even contradictory experimental data which makes testing and calibration difficult [19, 21]. In the following we briefly describe the modeling approaches used in the Hermes project.

4.1 Cellular Automata

Cellular automata (CA) models are characterized by discrete space, time and state variables. The dynamics is usual based on simple rules which have a rather intuitive justification. Space is discretized into small cells which can be occupied by at most one pedestrian (exclusion principle). The cell size corresponds to the space requirement of a person in a dense crowd. Assuming a maximal density of 6 persons/m^2 yields a cell size of $40 \times 40\,\text{cm}^2$. Time is also discrete and the pedestrians move synchronously in each time step. The time step can naturally be identified with the reaction time of a pedestrian. Thus one time step corresponds to a few tenths of a second in real time.

In the CA approach pedestrians are represented by particles moving according to stochastic rules. Usually the motion is restricted to neighbouring cells. The transition probabilities are determined by three factors: (1) the desired direction of motion (e.g. given by origin and destination), (2) interactions with other pedestrians, and (3) interactions with the infrastructure (walls, doors, etc.). In the simplest models, the latter two factors are only taken into account through an exclusion principle, i.e. occupied or wall cells are not accessible. More sophisticated approaches like the floor field model try to model these interactions in more detail which leads to more realistic results, especially for collective effects and self-organization phenomena [21]. Figure 4 illustrates the definition of the transition probabilities p_{ij} for a particle representing a pedestrian located at (0, 0) to one of the neighbour cells (including the current position). Some model variants also allow diagonal steps.

One of the most popular models is the floor field cellular automaton (FFCA) [12, 22–25]. The transition probabilities are determined by the three contributions (1)–(3) mentioned above. The interactions with other pedestrians and with the infrastructure are incorporated via two discrete floor fields D and S. These act

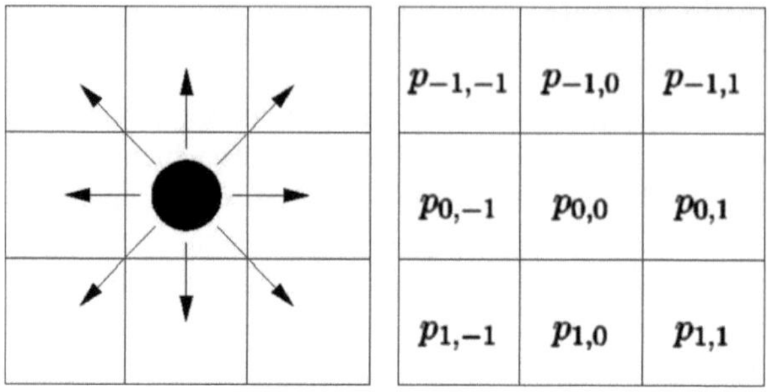

Fig. 4 Definition of the transition probabilities p_{ij} for a particle

like virtual chemotaxis by enhancing transitions in the direction of stronger fields. The static floor field S does not change in time and represents the environment (surrounding infrastructure) and can be used to encode e.g. the shortest path to the destination or exit. It decreases with increasing distance from the destination. The dynamic floor field D represents a virtual trace left by moving pedestrians. It reflects attractive interactions between pedestrians, e.g. the preference to follow moving persons, e.g. to minimize interactions with oncoming persons. Similar to the process of chemotaxis, this trace has its own dynamics (diffusion and decay). Two parameters, α and δ, control the broadening and dilution of the trace.

In each time step for each pedestrian the transition probabilities p_{ij} for a move to a neighbour cell (i, j) (see Fig. 4) is determined:

$$p_{ij} = N \exp(k_s S_{ij}) \exp(k_D D_{ij})(1 - n_{ij})\zeta_{ij} \,. \tag{1}$$

The relative influence of the static and dynamic fields S and D is controlled by sensitivity parameters k_S and k_D. The occupation number is $n_{ij} = 0$ for an empty and $n_{ij} = 1$ for an occupied cell, but $n_{00} = 0$ for the cell currently occupied by the considered particle. The obstacle number is $\zeta_{ij} = 0$ for forbidden cells, e.g. walls, and $\zeta_{ij} = 1$ otherwise, and the factor$N$ ensures the normalization $\sum_{ij} p_{ij} = 1$ of the probabilities. A more detailed description of the update rules can be found in [12, 23].

The floor field model has been applied to various situations. It is not only able to reproduce collective effects like lane formation [26] or oscillations at bottlenecks [12, 24], but gives also realistic results for evacuation scenarios [12, 22, 27, 28]. The realism of the FFCA can further be improved by taking into account various modifications. A more detailed discussion of these variants can be found in [29].

4.1.1 Generalized Centrifugal Force Model

The Generalized Centrifugal Force Model (GCFM) [13,30] is a spatially continuous which describes the movement of pedestrians and their interactions in terms of fields or forces [31–33]. It is based on modifications of the Centrifugal Force Model introduced in [34]. The acceleration of a pedestrian i is – analogy to Newtonian dynamics – determined by the superposition of all forces acting on her/him:

$$m_i \ddot{\mathbf{R}}_\mathbf{i} = \mathbf{F}_i = \mathbf{F}_i^{\mathrm{drv}} + \sum_{j \in \mathcal{N}_i} \mathbf{F}_{ij}^{\mathrm{rep}} + \sum_{w \in \mathcal{W}_i} \mathbf{F}_{iw}^{\mathrm{rep}} . \tag{2}$$

Here $\mathbf{F}_i^{\mathrm{drv}}$ is a driving force

$$\mathbf{F}_i^{\mathrm{drv}} = m_i \frac{\mathbf{v}_\mathbf{i}^\mathbf{0} - \mathbf{v}_i}{\tau}, \tag{3}$$

with the desired velocity of pedestrian i, the current velocity and τ a relaxation time. It reflects the motion towards a desired destination. $\mathbf{F}_{iw}^{\mathrm{rep}}$ is a force imposed by walls or other stationary obstacles. Pedestrians interact via a repulsive force

$$\mathbf{F}_{ij}^{\mathrm{rep}} = -m_i k_{ij} \frac{(\eta v_i^0 + v_{ij})^2}{R_{ij} - r_i(v_i) - r_j(v_j)} \cdot \frac{\mathbf{R}_{ij}}{R_{ij}} \tag{4}$$

where $\mathbf{R}_{ij} = \mathbf{R}_i - \mathbf{R}_j$ is the vector connecting the positions of the pedestrians and

$$v_{ij} = \frac{1}{2}[(\boldsymbol{v}_i - \boldsymbol{v}_j) \cdot \mathbf{e}_{ij} + |(\boldsymbol{v}_i - \boldsymbol{v}_j) \cdot \mathbf{e}_{ij}|] \text{ and } k_{ij} = \frac{1}{2} \frac{\boldsymbol{v}_i \cdot \mathbf{e}_{ij} + |\boldsymbol{v}_i \cdot \mathbf{e}_{ij}|}{\|\boldsymbol{v}_\mathbf{i}\|} . \tag{5}$$

v_{ij} is the projection of the relative velocity of pedestrian i and j onto the direction of the vector connecting their centers, see Fig. 5 (left). The coefficient k_{ij} reduces the action-field of the repulsive force to 180° in the direction of motion.

In the GCFM the pedestrians are represented by ellipses with velocity-dependent semi-axes since faster pedestrians require more space in walking directory than slower pedestrians. In addition, the lateral swaying of slower pedestrians is stronger than that of faster pedestrians. In order to obtain a more realistic behavior, in the simulations a cut-off radius r_c for the repulsive force is introduced (Fig. 5 (right)) which also allows runtime optimization by means of neighbor list methods [35]. Further details of the model are given in [13, 14] and elsewhere in these proceedings [30].

5 The Evacuation Assistant

5.1 Simulation Speed and Distribution of Evacuation Times

A very important aspect of the evacuation assistant is the time it takes to make a forecast of the evacuation. To be helpful for the decision makers, results have to be available quickly. Its speed is determined by the speed of the simulation kernels

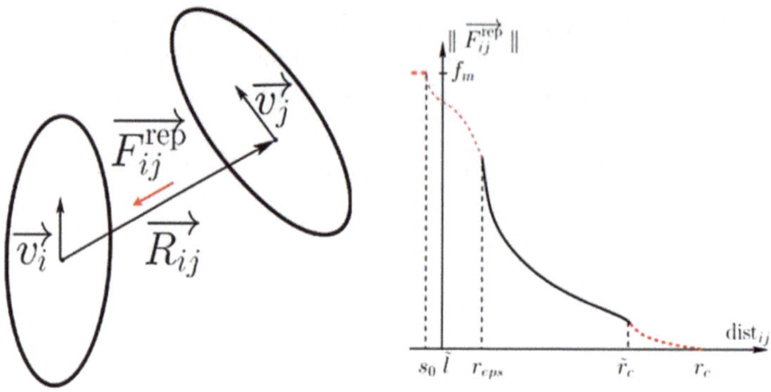

Fig. 5 *Left*: the repulsive force acting on the *center* of the ellipse representing pedestrian i. *Right*: interpolation of the repulsive force between pedestrians i and j with respect to their distance

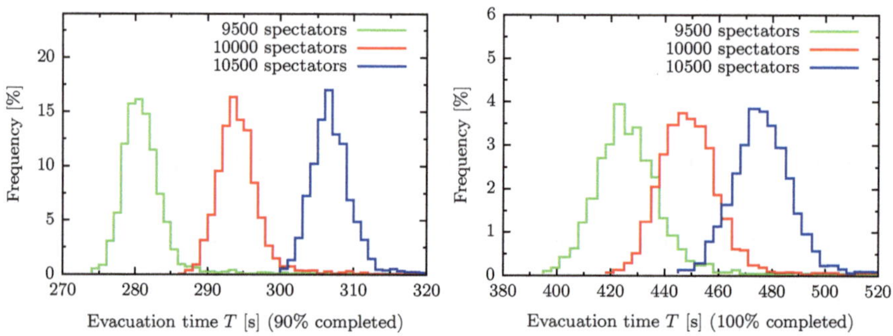

Fig. 6 Distribution of evacuation times

on the one hand and the speed of the visualization of the resulting trajectories on the other hand. The latter is mainly a question of software engineering and will be neglected here.

The deterministic GCFM is the computationally more expensive model and uses most of the computation power available for the microscopic simulations. Due to the intrinsic stochasticity of its dynamics, results of the FFCA model vary when several runs are performed. The question then is: How may runs have to be analyzed to get meaningful results?

Figure 6 shows the distribution of evacuation times for a typical scenario with 10,000 pedestrians and for two other scenarios which refer to the uncertainty of the pedestrian counting and use pedestrian numbers 5 % higher and lower, respectively. Figure 6a, b use a different definition of the evacuation time: results for the naive definition – evacuation time is the time it takes to evacuate 100 % of the pedestrians – are shown in Fig. 6b. Sometimes slightly different definitions of the evacuation time are more useful since the naive definition depends strongly on

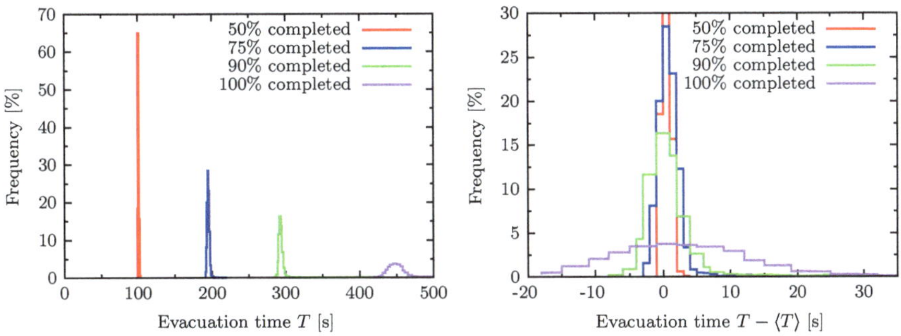

Fig. 7 Distribution of evacuation times

the last stages of the simulation when there are only a few pedestrians remaining in the simulation who might have got "lost".

A more reliable description of a typical evacuation can be gotten by defining the evacuation time as the time it takes until 90 % of the pedestrians are evacuated (Fig. 7). This is shown in Fig. 6a. For both definitions the variation of the evacuation times over 10,000 simulation runs is clearly smaller than the uncertainty related to the initial conditions. This suggests that it is sufficient to only do one simulation run and still get a sufficiently accurate evacuation time – despite the stochastic nature of the model. This is justified by the law of large numbers due to the large number of simulated pedestrians: a typical evacuation consists of more then 10^7 random decisions.

5.2 *Routing*

In addition to the uncertainty of the initial conditions there is another factor that influences the outcome of the simulation: the route choice behavior of the pedestrians. The FFCA model uses a stochastic routing. Pedestrians can leave a room only towards other rooms that are closer to an exit, they basically follow (virtual) emergency exit signs. When there are several such exits, the pedestrians choose different exits with different probabilities depending on the distances between the pedestrian and the exits. This is done such that the closest door is the most likely choice.

Routing in the GCFM is done on a navigation graph generated from the arena geometry. The graph is computed at the beginning of each simulation, to consider situations that have changed: a closed door for instance. Well established algorithms like Floyd-Warschall are used to determine the default optimal paths. The default paths are only optimal in the case there are no congestions. In that case the global shortest and the quickest path to a given destination are the same. To optimize the runtime, each node of the graph stores information such as the distances to all possible final destinations. Those destinations are pre-defined by the means of transport to the arena: train station, parking 1, parking 2. Depending on the affiliation of each pedestrian, he/she chooses either way. The graph is a directed graph. No backward movement to the previous destination point is possible.

On top of this graph, a quickest path is implemented. The idea behind this is presented in [36]. At a given point, usually when arrived in a new location, each pedestrian observes the current situation, evaluates the travel times and compares them with his/her actual default route. If the results of the evaluation are above a certain threshold, the actual route is modified. The same process is started if a pedestrian is stuck in a jam for an amount of time and can freed him/herself. This is the case if the pedestrian is not in the middle of the jam, for instance.

5.3 Runtime Optimization

An optimization of the runtime of the GCFM model is required to achieve the goals for the assistant, i.e to compute faster than real-time. The geometrical operations bounded to the GCFM together with the route choice algorithm made it very time consuming. The optimization is performed at three levels. The Message Passing Interface (MPI) is used to run the code across different computing nodes. A static domain decomposition approach has been adopted to share the work among the nodes. On each node the code is run in a multi-threaded environment using OpenMP. The threads used the linked-cells algorithm, which takes advantages of the short range character of the repulsive forces in the GCFM. More about this complexity and some preliminary results are found in [35]. A configuration for instance of 5×800 pedestrians distributed in the promenade gives a speedup of 131.44 over the brute force serial program and 12.52 over the code optimised with the linked cells and run with MPI + OpenMP [37].

6 Summary and Outlook

We have described an evacuation assistant (HERMES) for mass events. It uses information about the distribution of persons in a sports arena which is collected using video-based automated person counting at entrances and doors as a basis for computer simulations of the evacuation process. These simulations are based on two classes of pedestrian models, a cellular automaton (FFCA) and a continuous force-based approach (GCFM) which are more than 10 times faster than real time. The predictions of the simulations are than visualized using a graphical interface and provide valuable information for the security forces to support them in their decisions. HERMES was successfully tested between April and November in the ESPRIT Arena in Düsseldorf.

Acknowledgements This work has been performed within the program "Research for Civil Security" in the field "Protecting and Saving Human Life" funded by the German Government, Federal Ministry of Education and Research (BMBF). The project is supported under grant no. 13N9952 and 13N9960.

References

1. C. Rogsch, M. Schreckenberg, W. Klingsch, T. Kretz: In *Pedestrian and Evacuation Dynamics 2008*, pages 743–755. Springer, 2011, 2010.
2. A. Schadschneider: *Physics World*, pages 21–35, July 2010.
3. E. R. Galea, S. Gwynne, P. J. Lawrence, L. Filippidis, D. Blackspields, D. Cooney: *buildingEXODUS V 4.0 - User Guide and Technical Manual*, 2004. www.fseg.gre.ac.uk.
4. TraffGo HT GmbH: *Handbuch PedGo 2, PedGo Editor 2*, 2005. www.evacuation-simulation.com.
5. T. Korhonen, S. Hostikka, S. Heliövaara, H. Ehtamo: In W. Klingsch, C. Rogsch, A. Schadschneider, M. Schreckenberg, editors, *Pedestrian and Evacuation Dynamics 2008*, pages 109–120, Springer, 2008.
6. B. Raney, K. Nagel: In B. Jourquin, P. Rietveld, K. Westin, editors, *Towards Better Performing Transport Networks*, pages 305–347, Routledge, 2006.
7. P.A. Thompson, E.W. Marchant: *Fire Safety Journal*, 24:131–148, 1995.
8. I.S.T. Integrierte Sicherheits-Technik GmbH, Frankfurt/Main. *Aseri handbook, Version 4.6*, 2008.
9. R.M. Tavares, E.R. Galea: *Building and Environment*, 44(5):1005–1016, 2009.
10. S. Holl, A. Seyfried: inSiDe 7, 60 (2009)
11. Hermes project: http://www.fz-juelich.de/jsc/hermes.
12. C. Burstedde, K. Klauck, A. Schadschneider, J. Zittartz: Physica A295, 507 (2001)
13. M. Chraibi, A. Seyfried, A. Schadschneider: Phys. Rev. E82, 046111 (2010)
14. M. Chraibi, U. Kemloh, A. Seyfried, A. Schadschneider: Networks and Heterogeneous Media 6(3), 425 (2011)
15. J. J. Fruin: *Pedestrian Planning and Design*. Elevator World, New York, 1971.
16. J. Zhang, W. Klingsch, A. Schadschneider, A. Seyfried: J. Stat. Mech. (2011) P6004
17. J. Zhang, W. Klingsch, A. Schadschneider, A. Seyfried: Ordering in bidirectional pedestrian flows and its influence on the fundamental diagram. J. Stat. Mech. (2012) P02002
18. J. Zhang, W. Klingsch, A. Schadschneider, A. Seyfried: Experimental study of pedestrian flow through a T-junction. *these proceedings*, Springer, 2012.
19. A. Schadschneider, W. Klingsch, H. Klüpfel, T. Kretz, C. Rogsch, A. Seyfried: In R. A. Meyers, editor, Encyclopedia of Complexity and Systems Science, pages 3142–3176. Springer, 2009.
20. C. Rogsch, A. Seyfried, W. Klingsch: In N. Waldau, P. Gattermann, H. Knoflacher, M. Schreckenberg, editors, *Pedestrian and Evacuation Dynamics 2005*, pages 357–362. Springer, 2007.
21. A. Schadschneider, A. Seyfried: In Harry Timmermans, editor, *Pedestrian Behavior: Data Collection and Applications*, pages 27–43. Emerald Group Publishing Limited, 2009. (arXiv:1007.4058).
22. A. Kirchner, A. Schadschneider: Physica A 312, 260 (2002)
23. A. Schadschneider: In M. Schreckenberg and S. D. Sharma, editors, *Pedestrian and Evacuation Dynamics*, pages 75–86. Springer, 2002.
24. C. Burstedde, A. Kirchner, K. Klauck, A. Schadschneider, J. Zittartz.: In M. Schreckenberg and S. D. Sharma, editors, *Pedestrian and Evacuation Dynamics*, pages 87–97. Springer, 2002.
25. A. Kirchner, K. Nishinari, A. Schadschneider: Phys. Rev. E67, 056122 (2003)
26. S. Nowak, A. Schadschneider: A cellular automaton approach for lane formation in pedestrian counterflow. these proceedings, pages 149–160.

27. A. Kirchner, H. Klüpfel, K. Nishinari, A. Schadschneider, M. Schreckenberg: Physica A324, 689 (2003)
28. K. Nishinari, A. Kirchner, A. Namazi, A. Schadschneider: IEICE Trans. Inf. & Syst. E87-D, 726 (2004)
29. A. Schadschneider, D. Chowdhury, K. Nishinari: *Stochastic Transport in Complex Systems. From Molecules to Vehicles*. Elsevier, 2011.
30. M. Chraibi, A. Schadschneider, A. Seyfried: GCFM. *these proceedings*, Springer 2012.
31. K. Lewin (ed.): *Field Theory in Social Science*. Greenwood Press, 1951.
32. D. Helbing, P. Molnár: Phys. Rev. E51, 4282 (1995)
33. P. Molnár: *Modellierung und Simulation der Dynamik von Fußgängerströmen*. Dissertation, Universität Stuttgart, 1995.
34. W. J. Yu, L.Y. Chen, R. Dong, S.Q. Dai: Phys. Rev. E72, 026112 (2005)
35. A.U. Kemloh Wagoum, M. Chraibi, A. Seyfried, A. Schadschneider: Efficient and validated real-time simulation of crowds within an evacuation assistant. Comput. Animat. Virtual Worlds 23, 3–15 (2012)
36. A.U. Kemloh Wagoum, A. Seyfried, S. Holl: Modelling dynamic route choice of pedestrians to assess the criticality of building evacuation. Adv. Compl. Sys. 15, 1250029 (2011)
37. A.U. Kemloh Wagoum, B. Steffen, A. Seyfried: Lect. Notes Comp. Sc. 7203, 386 (2012)

Pedestrians and Escalators: Usage Under Normal Conditions

Christian Rogsch

Abstract For large multi-storey buildings, like shopping centers, the usage of escalators is the most comfortable way to change different levels contrary to the usage of elevators. Comparing escalators to elevators, escalators have some advantages: They are open (not a closed box like an elevator), quick to use (just enter, do not wait as in front of an elevator), groups with more than 12 pedestrians can change the building levels as one group, and in case of a failure (e.g. stopping) people can leave the escalator very quickly without extra help of a technical service team. The paper show how people use an escalator in a shopping center during their "shopping tour".

1 Introduction

For large multi-storey buildings, like shopping centers, the usage of escalators is the most comfortable way to change different levels contrary to the usage of elevators. Comparing escalators to elevators, escalators have some advantages: They are open (not a closed box like an elevator), quick to use (just enter, do not wait as in front of an elevator), groups with more than 12 pedestrians can change the building levels as one group, and in case of a failure (e.g. stopping) people can leave the escalator very quickly without extra help of a technical service team.

Based on this few general advantages of escalators against elevators it is actually unknown how pedestrians enter an escalator and how they leave it in the next upper or lower level during normal conditions (no evacuation or fire condition), like during normal opening hours of a shopping center. The paper will show how pedestrians

---

C. Rogsch (✉)
Fire Protection Engineering, Bonhoefferstr. 16, D-67435 Neustadt an der Weinstraße, Germany
e-mail: christian@rogsch.de

V.V. Kozlov et al. (eds.), *Traffic and Granular Flow '11*,
DOI 10.1007/978-3-642-39669-4_28, © Springer-Verlag Berlin Heidelberg 2013

Fig. 1 New stairs at the conference venue

- Enter an escalator, e.g. which line do they use (right, left, middle),
- How many of them use the handrail,
- Which foot enters the escalator first,
- Which foot leaves the escalator first, and
- Which "configuration" like pedestrians standing on an escalator.

To obtain reliable data a large group of pedestrians has been analyzed by using video analyses. The videos show how they enter and leave an escalator during their normal "shopping tour" in a shopping center. The data obtained from the videos should help to simulate pedestrian movement and the usage of escalators during normal conditions in a more realistic way.

2 Escalators Versus Stairs

Escalators can be called "the moving stairs". They are easy to use because on escalators the tread moves; contrary to escalators on stairs the pedestrian moves. Even if escalators need electrical power to move pedestrians one level up or down, they can be used as a "normal" stair if they have no power in a case of electrical failure. In this case case they are a normal stair only with a different size of tread. So, an escalator has a rise height with equal distance except at the beginning and the end, but stairs can be very different even if they are new as you can see in Fig. 1.

3 Data Collection

The data presented in this paper was obtained by using a video analysis from pedestrians at the Złote Tarasy Shopping Center in Warsaw, Poland. The paper is a first analysis of ca. 500 pedestrians during a normal shopping time, e.g. no rush

Fig. 2 Escalators at Złote Tarasy shopping center [1, 2]

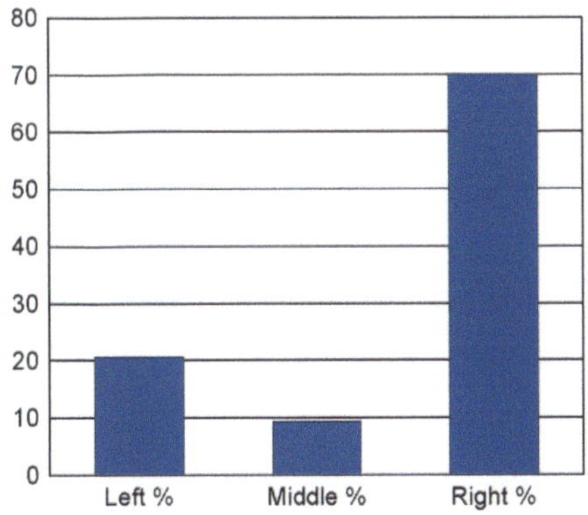

Fig. 3 Line, which is chosen by pedestrians when they enter the escalator

hour or special events. Figure 2 shows how escalators are built in this shopping center.

4 Results

In this section the first results of the video observation are shown.

4.1 Which Line?

If we look in direction of the escaltor most pedestrians use the right line (or side) if they enter the escalator. Ca. 10 % of all pedestrians change the line, e.g. move from left to right or the other way round. Figure 3 shows the results.

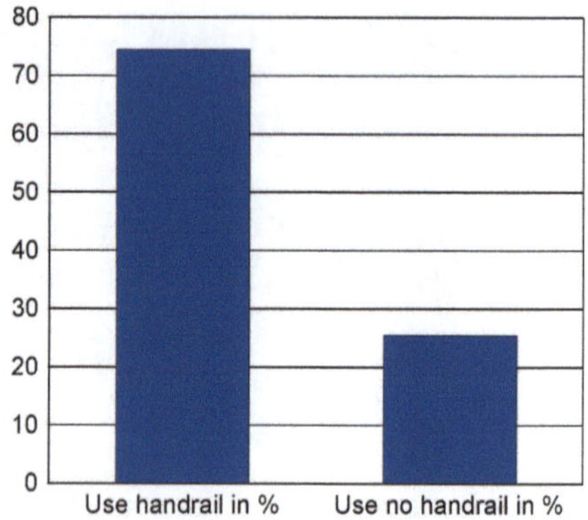

Fig. 4 Handrail usage of pedestrians

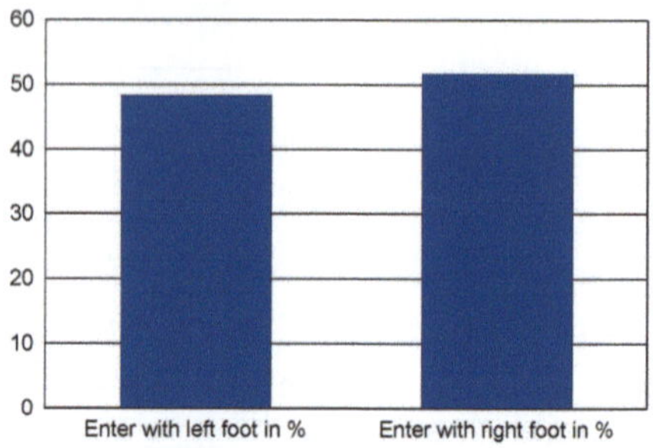

Fig. 5 Foot, which enters the escalator first

4.2 Handrail Usage

Not much interesting for modeling, but more interesting for visualisation is the question how many people use the handrail. The observation has shown that ca. 25 % of the observed pedestrians use no handrail, ca. 75 % use it. In this case it doesn't matter if pedestrians are on the right or left line, the observation has not show that e.g. people on the left line use the handrail more often than on the left line. The results are shown in Fig. 4.

4.3 Which Foot Enters First?

Like handrail usage is the question "which foot enters the escalator first?" more interesting for visualisation than for modeling. During the observation it was not possible to find a "favourite" foot which enters the escalator first. As noticed above, the selected line has no influence of the "foot choice". Results are shown in Fig. 5.

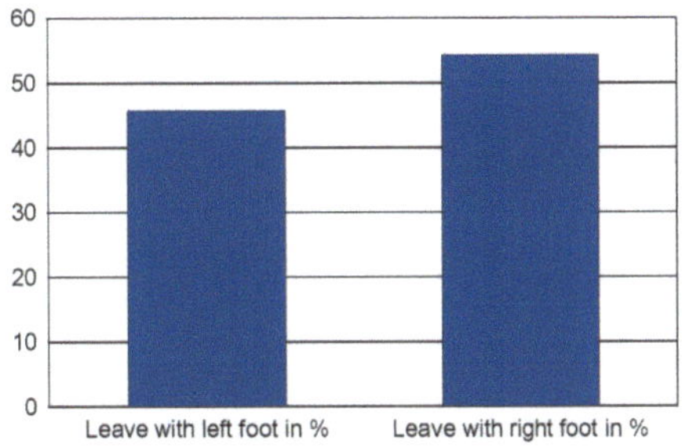

Fig. 6 Foot, which leaves the escalator first

4.4 Which Foot Leaves First?

When pedestrians enter an escalator they have to leave it also. So, the next question is: "Do they prefer any foot if they leave an escalator?" Contrary to the results shown in the case of entering an escalator, it seems that people prefer to leave the escalator with the right foot. But based on ca. 500 pedestrians, the results can change if more pedestrians are observed. Results are shown in Fig. 6.

4.5 Different "Configurations"

Pedestrians try to get as much space as possible ("personal or private space") when they enter an escalator. In the rush hour this is normally not possible, because many pedestrians use escalators thus a high density is a normal situation. This situation is also well known by all pedestrians and also accepted. But in the case of using an escalator in a shopping center pedestrians try to get their private space every time, even if they use an escalator. The observation has shown that pedestrians like to stay side-by-side or one-after-another with one stair between them, they do not like to stand directly one-after-another, because in this case their private space gets lost. Figure 7 shows a graphical sketch how people like or dislike to stand on an escalator if they have the chance to choose their position independently.

5 Conclusion

In this paper the first results are shown how escalators are used by pedestrians. In detail, the author has shown that

- Pedestrian like to stay on the right side,
- Pedestrian like to have one treat of space between them if they stay on the same side,
- Pedestrians like to use the handrail,

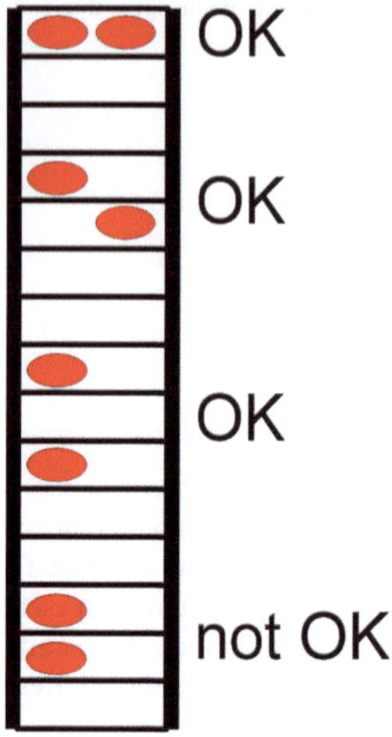

Fig. 7 Different "configurations" which people like or dislike when they stand on an escalator

- No significant foot is used by pedestrians to enter or to leave (maybe the right foot) the escalator, and
- Only 10 % of pedestrians change the line after entering the escalator.

Based on the number of pedestrians which have been observered (ca. 500 pedestrians) it is well know that these values might be changed if more pedestrians are observed. However, this investigation shows at least a basic observation about pedestrians and how they use an escalator.

References

1. http://warsawdailyphoto.wordpress.com
2. http://www.zlotetarasy.pl

Part III
Granular Flow

Model of Sand Transport by Consecutive Particle Collisions with Wind Interaction

Luc Oger, Alexandre Valance, and Madani Ammi

Abstract We study successive collisions of incident energetic beads with granular packing in the context of Aeolian saltation transport. We investigate the collision process for the case where the incident bead and those from the packing have identical mechanical properties. We analyze the features of the consecutive collision processes. We used a molecular dynamics method known as DEM (soft Discrete Element Method) with an initial static packing of 20,000 2D particles. The packing is created randomly in a box of dimension 250 × 60 disk diameters. The incident disks are launched at a given velocity ranging between 25 and 100 (normalized by $\sqrt{g.d}$) and an angle of 15° with random position on top of the flowing zone. Few disks are launched to initiate the dilation of the upper surface. This situation due to the previous collisions is responsible for a need of less input energy for maintaining the flow by ejecting new disks from the static bed. This new behavior is obtained by the presence of the wind which can accelerates the flowing disks during their ballistic trajectories.

1 Introduction

The advance of the dunes in the desert is a threat to the life of the local people. The dunes invade houses, agricultural land and perturb the circulation on the roads. It is therefore very important to understand the mechanism of sand transport in order to fight against desertification. Saltation, in which sand grains are propelled by the wind along the surface in short hops, is the primary mode of blown sand movement [1]. The saltating grains are very energetic and when impact a sand

L. Oger (✉) · A. Valance · M. Ammi
Institut de Physique de Rennes, UMR U.Rennes1-CNRS 6251, Université de Rennes1, CS 74205, 263 Avenue du Général Leclerc, 35042 Rennes CEDEX, France
e-mail: Luc.oger@univ-rennes1.fr; Alexandre.valance@univ-rennes1.fr; Madani.ammi@univ-rennes1.fr

V.V. Kozlov et al. (eds.), *Traffic and Granular Flow '11*,
DOI 10.1007/978-3-642-39669-4_29, © Springer-Verlag Berlin Heidelberg 2013

surface, they rebound and consequently eject other particles from the sand bed. The ejected grains, called reptating grains, contribute to the augmentation of the sand flux. Some of them can be promoted to the saltation motion.

The main outcomes of our numerical studies follow. Firstly, we make an short overview of the previous experimental and numerical results concerning the successive collision between an incident beads and a granular packing. Then, we describe briefly our mechanical model based on the Discrete Element Method and the special wind conditions used in this study. Finally, we present the main results available from this study.

2 State of the Art

Some experimental models of collision between an incident bead and the granular packing have been performed by different authors. They studied the effect of the impact angle [2] and the incident speed [3, 4] on the collision process. The authors have used steel beads (4 mm by Mitha et al. [2]), PVC beads (6 mm by Rioual et al. [4]) and the sand grain (850 μm by Werner [3]). The experimental results showed that the mean restitution coefficient for the impacting bead (defined as the ratio between the rebound speed and the incident speed) depends only on the impact angle. The restitution coefficient decreases with increasing impact angle. Concerning the ejection process, Mitha et al. [2] showed that the number of ejected beads does not vary significantly when the impact angle increases from 17° to 31° and that the average of vertical speed of ejection is of order of $3\sqrt{\mathbf{g}.d}$ (where $\mathbf{g}$ is the gravitational acceleration and d the bead diameter). However, the number of ejected grains increases with increasing incident speed [4].

In addition to the experiments, some simulations of the collision process have been performed [5–7]. The authors made simulation using a discrete element method (DEM). The numerical simulations allowed to extract a law for the incident and ejected beads behaviors. The numerical results confirm the experimental result, in particular:

- The dependence of the mean normal restitution coefficient for impacting bead with impact angle. This coefficient is inversely proportional to the sinus of the impact angle,
- The mean number of ejected beads increases linearly with incident speed and impact angle,
- The distribution of the velocities for the ejected beads can be fitted by an exponential law.

These simulations were limited to a single shot but the DEM can be easily extended for sequential or simultaneous shots of particles on partially immobile packings. This is the goal of our present study. We study successive collisions of incident energetic beads with granular packing in the context of Aeolian saltation transport (i.e. with the presence of lateral wind).

3 Description of the Numerical Simulation

The full description of the different steps of the numerical simulations to mimic the collision process between an incident bead and a packing of identical beads was already made by Oger et al. [7]. So we will briefly present here the two main steps of the numerical simulations. Indeed, in order to save time of computation, we have simplified the technique to create the static packings of disks prior to the active part of the simulation; we perform two consecutive and complementary steps: a geometric piling step then a dynamic one.This second step can be also separated by an initiation one and a flowing one according how the flow is maintained.

3.1 DEM Model: Soft Model Approach

Our mechanical model is similar to the two-dimensional formulation of Savage [8]. The *i*th particle is characterized by its radius $\mathbf{r_i}$, the position of its center (x_i, y_i). A "soft-particle" approach is used, where each particle can have multiple contacts that can persist for extended durations. Both normal and tangential forces develop at the contact between two particles. The normal and tangential contact forces increase as the centers of the particles approach each other.

The normal force F_n at the contact is modeled as viscoelastic. It consists of an elastic (a linear spring linked to K_n) contribution and viscous damping (a linear dashpot b_n). The force in the tangential direction is modeled also as a viscoelastic one; a linear spring and a linear dashpot are used to generate a tangential contact force. The tangential force is also limited to a maximum value which is chosen according to a Coulomb friction law when slipping can occur according to the intergrain coefficient of friction. For this mechanical model, the time step calculation remains constant.

It is convenient to cast the governing equations in a non dimensional form and perform the computations based upon these dimensionless equations [8]. It is straightforward to revert back to physical variables if desired. Hence, all lengths are nondimensionalized by D, the diameter of the largest sized particles used in the computations. Time is nondimensionalized by dividing by $\sqrt{M/K_n}$, where M is the mass of the largest particle and K_n the effective spring constant. Velocities are nondimensionalized by dividing by $D/\sqrt{M/K_n}$. Thus we introduce the following non dimensional time and spatial coordinates

$$(\tilde{t}, \tilde{x}, \tilde{y}) = (t\sqrt{\frac{K_n}{M}}, \frac{x}{D}, \frac{y}{D}). \tag{1}$$

This dimensionless approach lets us model either large bead collisions such as in Rioual et al. [4] experiments (6 mm) or sand grains (200 μm) [9–11]. For each particle, the forces acting on it are calculated; then the particle is displaced according to the resultant of all these forces. Local displacements of each particle are recorded.

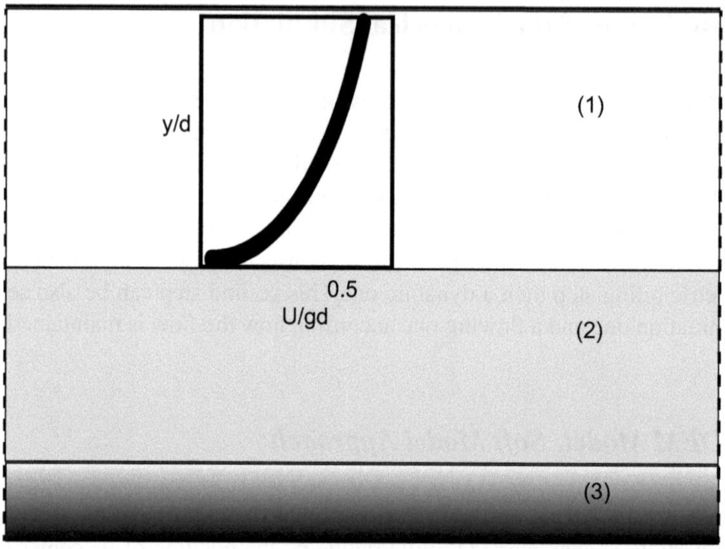

Fig. 1 Schematic representation of the disk packing (*2*). The *lower part* (*3*) is the damping zone. The lateral wind profile is represented in the *upper empty zone* (*1*)

3.2 Initial and Running Conditions

In the first step, we are doing a geometrical piling of disks in which the particles are just touching each other. In a mechanical point of view, this stage corresponds to a dense packing of perfectly rigid, immobile grains. In order to generate the 2D packings, we followed the Powell's algorithm [12]. The total number of disks used for the packing is between 20,000 and 30,000 (150–250 large by 60–100 height). We have already assigned periodic boundary conditions for the lateral wall during the piling and this periodic boundary condition will be maintained for the saltating simulation.

For a mechanical point of view, we assign some physical parameters to each disk. We choose those corresponding to classical sand grains. The disk diameter is taken equal to 200 μm and the density is 2,500 $\mathrm{kg/m^3}$. The coefficient of restitution and the friction coefficient will be selected according to the desired studies (inside a range of 0.75–0.95 and 0.2–0.4 respectively). The spring constant is chosen equal to $K_n = 10^9\,\mathrm{N\,m^{-1}}$ for normal force and for the tangential one: $K_t = 0.3\,\mathrm{K_n}$. The normal dashpot coefficient b_n is calculated according to the restitution coefficient and the tangential one is $b_t = 0.5\,\mathrm{b_n}$. The problem and its boundaries are defined as shown in the Fig. 1:

- Perfectly lateral periodic conditions;
- The lower part of the packing ((3) in the figure) is defined as a dumping surface in which the two components of the particle velocity are dumped by an amount

proportional to the particle altitude: i.e. the particles are less and less mobile as they are deeply located.

- In order to limit the possible flights of moving disks we have defined a horizontal upper limit like a perfect mirror for the moving disks (same horizontal velocity component and reverse vertical one after touching the mirror) (see Fig. 1).

After this first step, the shooting sequence is activated and the collision process proceeds. At each time step of the process, position and velocity of the particles are recorded. A bead becomes an ejecta if its center moves, from its initial coordinate, more than one diameter in the vertical direction. Two different ignitions of the flow process are studied: a full consecutive series of launched disks with wind interaction (case 1) or few initial launched disks coupled with lateral wind interaction (case 2).

4 Results

There many adjustable parameters in this study that can be tuned such as the angle, the velocity for each individual launched disk, the variable time between two consecutive shots, and so on.... So, in order to start easily this study, we have decided to keep constant the angle and the input velocity of each launched disk during one run. Only the interval time between two consecutive shots will be varied for either the full series of launched disks (case 1) and also for the the few initial launched disks (case 2). The randomness of the simulation will be assured by two variables: the random horizontal initial position of the shot (the vertical one is defined at some altitude from the initial static bed) and a fluctuating time interval around the mean value given as an input.

4.1 Case 1: Without External Wind Force

Firstly, to see the evolution of the flowing disks or by analogy the energy increase for the flowing grains, we have observed the number of moving disks versus time for different numbers of disks launched. The mean time between consecutive shots varied from $2.5\ 10^{-5}$ up to $2.0\ 10^{-3}$ s which characterize, by analogy, the fluctuation of the energy flux. We have seen that the amount of moving disks can increase according to the energy flux increase. The generation of the number of moving disks is less efficient when the input energy flux is large which can be easily demonstrated by the logarithmic relation (dashed line in the figure) between this slope and the interval time. In complementary, if we increase the energy flux by only increasing the velocities of the launched disks we can also observe a higher slope on Fig. 2.

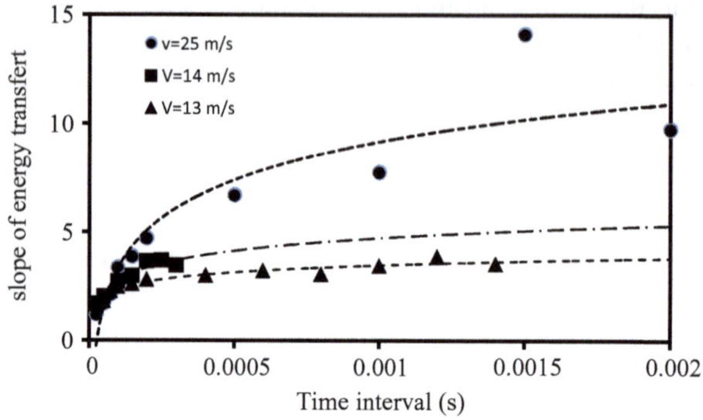

Fig. 2 Efficiency of the energy flux defined by the time interval for different velocities of the launched disks. The *dashed lines* correspond to the best logarithmic profiles obtained for each input velocity

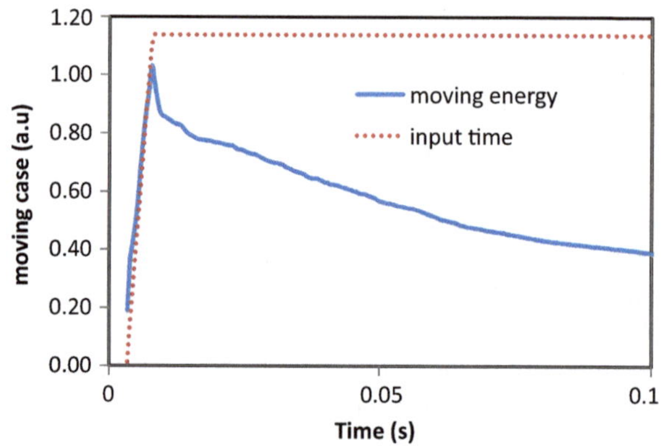

Fig. 3 The energy of the full setup of the grain packing versus time. The *dashed line* characterize the energy increase during the time in which few energetic new disks are launched. The *continuous curve* represents the energy kept by the system during the time run

4.2 Case 2: With External Wind Force

In the previous case, it was very easy to analyze the efficiency of the energy input due to the flow of launched disks as no external forces were involved. But in the real saltation, the wind is present and plays a crucial role in the acceleration of the grains. In order to mimic this effect we have developed a second case which starts exactly as the previous one (i.e.: initial dense packing then few launched disks which already interact with the wind) but continues with only wind interaction. The Fig. 3 shows the two different steps of this numerical conditions. The dashed line characterize the energy increase during the time in which few energetic new disks are launched and

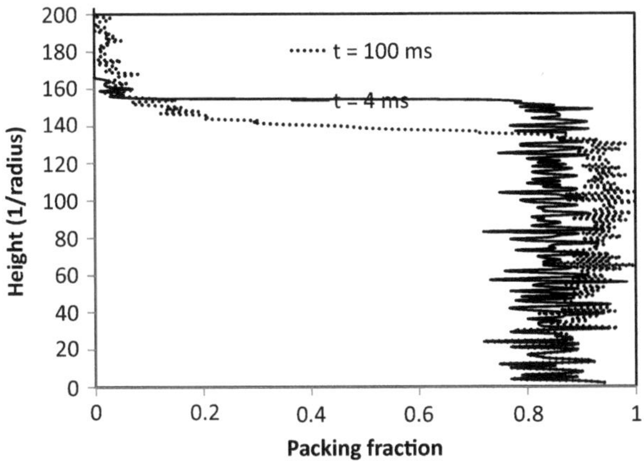

Fig. 4 Packing fraction versus height of the packing for $t = 4$ and 100 ms. The packing fraction increase just below the flowing zone in which the packing fraction decrease continuously down to zero

then the horizontal constant situation. The continuous curve represents the energy kept by the system during the time run.

At the initial stage, the packing is homogeneously packed and dense which can be seen by the perfect vertical relation between the packing fraction and the height of the packing in Fig. 4 for $t = 4$ ms. When some energy flux coming from the launched disks and also from the saltating ones starts to punch the packing surface, two effects appear: a dilation of the upper part of the packing and also a compaction at the bottom of this upper flowing zone. This can be observed for the profile at time $t = 100$ ms

In order to analyze more precisely what is the behavior close to the transition between the reptating grain areas and the saltating ones, we have looked at the other well known parameters for granular flows such as the granular temperature and the local pressure. The transition of the reptating threshold visible in the displacement of the packing fraction profile can, of course, be observed also on the granular temperature Fig. 5. We can easily seen that, close to the beginning of the experiments, few grains have been put in displacement which implies that they are more energetic Fig. 5 for $t = 13$ ms. The granular temperature is in this case very high on top of the upper surface. Then more and more grains are moving which implies a kind of reorganization of the flowing zone.

After some time of the runs, the moving reptating and dilute grains area is larger and the packing fraction is more 'uniformly' diluted from the dense close packing up to the fluidized area (i.e. packing fraction from 0.8 up to 0.0).

The Fig. 6 shows also that during the initial process the packing has a compression wave due to the initial external shots which impact the surface. This creates the appearance of a higher pressure close to the surface and also at the limit of the dumping zone. After this initial reorganization period, the packing founds some

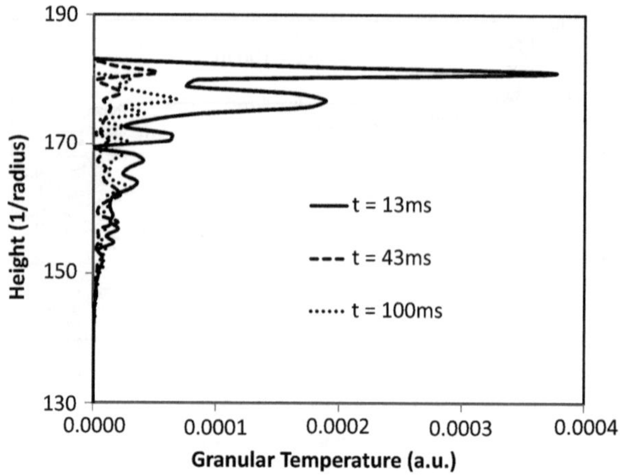

Fig. 5 Granular temperature versus height at different time during the run

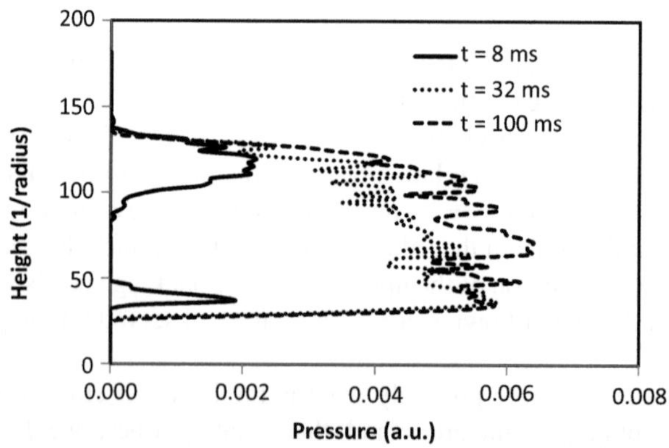

Fig. 6 Vertical pressure profile for the same packing at different time of the run. At time $t = 8$ ms only the upper surface feels the pressure due to the impact. During the rest of the run the pressure equilibrium is reached inside all the packing

kind of homogeneity behavior which can be observed by a high pressure profile just behind the reptating zone (i.e.: inside all the dense moving zone).

The Fig. 7 presents another very interesting point of view: the horizontal velocity profile for the grains is almost independent of the horizontal wind velocities. This fact can only be explained by the presence of the other adjusting parameters of the classical splash function: the number of saltating grains, higher velocity, higher moving grains. This conclusion is in conformity with the previous results obtained for the splash function when the shot velocity increases: the number of ejecta increases but not the mean value of their velocities [13].

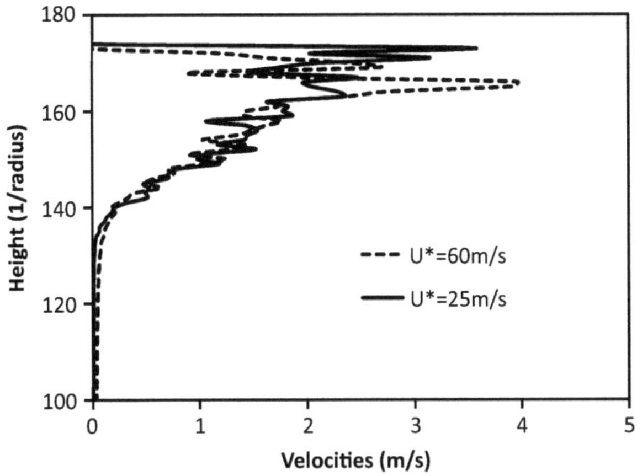

Fig. 7 Vertical velocity profiles for two different horizontal wind velocities. The two profiles are very similar in intensities

5 Conclusion

Changing the different parameters of the input energy by selecting, for a fixed angle and vertical position of the incident beads, given velocities and variable time between launches gave us the ability to determine one of the crucial parameters for the saltating splash function. We can observe that the energy kept by the bed decreases with the global energy input. In the same time, we need less and less input energy flux in order to keep the same amount of disks inside the flying zone. When the wind is present to maintain the energy input we can observe the same global behavior which can permit to keep the grain flux if enough energy is given to the medium.

References

1. R. A. Bagnold. *The Physics of blown sand and desert dunes*. Chapmann and Hall, Methuen, London, 1941.
2. S. Mitha, M.Q. Tran, B.T. Werner, and P.K Haff. The grain-bed impact in aeolian saltation. *Acta Mechanica*, 63:267, 1986.
3. B. T. Werner. A steady-state model of wind-blown sand transport. *Journal of Geology*, 98:1–17, 1990.
4. F. Rioual, A. Valance, and D. Bideau. Experimental study of the collision process of a grain on a two-dimensional granular bed. *Phys. Rev. E*, 62,2:2450–2459, 2000.
5. B.T. Werner and P.K. Haff. The impact process in aeolian saltation: two-dimensional simulations. *Sedimentology*, 35:189, 1988.
6. R. S. Anderson and P. K. Haff. Wind modification and bed response during saltation of sand in air. *Act. Mech.*, [Suppl.]1:21–52, 1991.

7. L. Oger, M. Ammi, A. Valance, and D. Beladjine. Discrete element method studies of the collision of one rapid sphere on 2d and 3d packings. *The European Physical Journal E*, 17(4):467–476, 2005.
8. S. B. Savage. Disorder, diffusion and structure formation in granular flows. In D. Bideau, editor, *Disorder and Granular Media*, pages 255–287, Amsterdam, 1993. North Holland.
9. B. R. White and J. C. Schulz. Magnus effect in saltation. *J. Fluid Mech.*, 81, 3:497–512, 1977.
10. P. Nalpanis, J. C. R. Hunt, and C. F. Barrett. Saltating particles over flat beds. *J. Fluid Mech.*, 251:661–685, 1993.
11. M. A. Rice, B. B. Willets, and I. K. McEwan. Observations of collisions of saltating grains with a granular bed from high-speed cine-film. *Sedimentology*, 43:21–31, 1996.
12. M. J. Powell. Computer-simulated random packing of spheres. *Powder Technology*, 25:45–52, 1980.
13. L. Oger, M. Ammi, A. Valance, and D. Beladjine. Study of the collision of one rapid sphere on 3d packings: Experimental and numerical results. *Computers and Mathematics with Applications*, 55:132–148, 2008.

"Faster Is Slower" Effect in Granular Flows

Paula A. Gago, Daniel R. Parisi, and Luis A. Pugnaloni

Abstract We investigate the faster is slower (FIS) effect for a granular flow system which consist of a quasi two-dimensional hopper placed on a vibrating inclined plane. Increasing the angle of the plane (θ) is similar to increasing the driving force in the social force model. We measured the distribution of the time intervals between two successive particles (*dt*) and found that for narrow exits (3 particle diameter) it displays a power-law tail with a negative exponent α, were $|\alpha| < 2$, indicating that the mean *dt* is not defined. Hence, we proposed to use α as a measure of the ability of the system not to develop long-lasting blockages and in this sense, the FIS effect was observed for the experimental granular system. This is the first experimental evidence of the FIS effect in granular flows. Moreover, our results suggest that a different approach might be necessary to quantify the evacuation time for pedestrians since its mean may not be defined for highly competitive crowd systems.

1 Introduction

The study of the dynamics of high density crowds is of fundamental importance in pedestrian evacuation. The availability of validated models that describe this state of high density would allow the design of safer pedestrian facilities, ensuring that all people can egress in the case of an emergency.

P.A. Gago (✉) · L.A. Pugnaloni
Instituto de Física de Líquidos y Sistemas Biológicos (CONICET La Plata, UNLP), Calle 59 Nro. 789, 1900 La Plata, Argentina
e-mail: paulaalejandrayo@gmail.com; luis@iflysib.unlp.edu.ar

D.R. Parisi
CONICET and Instituto Tecnológico de Buenos Aires, 25 de Mayo 444, (1002) C. A. de Buenos Aires, Argentina
e-mail: dparisi@itba.edu.ar

V.V. Kozlov et al. (eds.), *Traffic and Granular Flow '11*,
DOI 10.1007/978-3-642-39669-4_30, © Springer-Verlag Berlin Heidelberg 2013

However, an experimental characterization of this type of systems is not straightforward because, under competitive or emergency conditions, the high pressure zones caused during the blockage of an exit may lead to choking and other fatal injuries.

This is the main reason why computer simulations are the state of the art to study this type of phenomena. However, given the lack of information about how real systems behave in extreme conditions, computational and physical models are not properly validated. One should be very careful when drawing conclusions for real systems based on computational models.

A key problem in evacuation dynamics is the escape of highly competitive pedestrians through a narrow door. Using the social force model (SFM), it was predicted the *faster is slower* (FIS) effect [1]. FIS can be observed in a plot of mean evacuation time vs. desired velocity. There exists an optimum pedestrian desired velocity that yields the shortest mean evacuation time. This optimum is related to the appearance of *blocking clusters* [2] of pedestrians that intermittently clog the exit door mainly due to frictional contact forces [3].

Permanent full blockages have been observed in real catastrophes such as in the Rhode Island fire (at the disco "The Station") in USA, 2003. Such events are not a proof of the existence of FIS. Each point on the FIS curve is an average of many realizations. We do not know what is the occurrence probability of these blockages under controlled initial conditions. Thus, the FIS curve has not been measured for humans under laboratory conditions yet.

Lacking information of human evacuation, it is very helpful to study, experimentally, similar systems. One possibility is to consider the evacuation of biological agents (e.g., ants [4, 5] or mice [6]). The analogy with human crowds is based on both kind of agents having biological sensors, communication mechanisms, alarm behavior and being self-propelled.

Another system similar to the room evacuation that can be study is granular flow. Implications on pedestrian evacuation drawn from granular experiments can already be found in the literature [7]. In Ref. [7], the granular flow through an exit at the bottom of a vertically shaken box is studied. It was found that variations in the geometry of the box can increase the outflow. In particular the suitable placement of a cylindrical obstacle above the exit allows the enhancement of the mean rate of discharge up to 3.5 times the rate without obstacle.

In the case of granular flow, the analogy with pedestrian flow is based on the contact forces. It was shown in [3] that the contact forces play a key role in the FIS effect emerging in the simulation of competitive egress with the SFM. The contact force in the SFM is directly related to the contact force of grains in a granular system. The social force term of the SFM is not relevant to the existence of FIS [3]. Furthermore, it is natural to expect for a high density crowd, that social interactions (i.e., the desire to stay away from other pedestrians) are not relevant. So, the absence of remote interactions in granular systems is analogous to what happen in a packed crowd under panic conditions.

We have carried out an experiment that consists of a quasi two-dimensional silo and hopper placed on a vibrating inclined plane. Increasing the angle of the plane leads to an increase in the component of the force of gravity parallel to the incline. This is similar to increasing the driving force in the SFM. Since blocking arches are much more stable in granular flows than in self-propelled particle flows, it is necessary to add an external perturbation to the system [7]. We achieve this by applying an external vibration which is an aid to the rearrangement of grains that promote breaking of blockages and, otherwise permanent, arches.

In the present work we will study the aforementioned granular system for two exit widths of the hopper that correspond to the flowing and jammed regime. For the jammed regime, it will be shown that the experimental system do exhibit the FIS effect.

The rest of the paper is organized as follows. In Sect. 2 the experimental device is described. In Sect. 3 the results of the experiment are presented via the analysis of the times intervals, *dt*, between the output of two successive grains. Finally, in Sect. 4, the conclusions are presented.

2 Experimental Set Up

The experiment consists in a two-dimensional silo and hopper on an inclined plane (Fig. 1). The incline is made of a Plexiglas bottom sheet covered in a grounded aluminum foil to reduce electrostatic charges. The entire incline is covered with a second Plexiglas sheet supported by a frame slightly thicker than the plastic strips. The silo and hopper are built at the center of the incline by means of plastic strips 3.0 mm thick and 4 mm wide glued to the Plexiglas cover. The whole system is mounted on a pivot that allows the control of the angle, θ, that the incline makes with the horizontal. A secondary hopper allows the refilling of the silo by simply inverting the inclination angle. Attached to the plane, from below, there is a speaker that is used to vibrate the system. This vibration was a sinusoidal wave of 130 Hz with a maximum acceleration of 0.23 g in the $\hat{z}$ direction, perpendicular to the plane. In the $\hat{x}$ and $\hat{y}$ directions, parallel to the plane, there exists a small transfer of the vibration whose maximum acceleration is of 0.09 g. The vibration was measured by means of an accelerometer (ADXL321) attached to the lower Plexiglas sheet. Although mechanical modes of the structure distort the sinusoidal input vibration, the amplitude and overall shape proved to be relatively independent of our main control parameter, i.e., the inclination angle θ.

The granular material consists in 200 mono-disperse glass spheres (diameter $d =$ 3.5 mm). Two exit widths were investigated: $W = 3d = 10.5$ and $W = 5d =$ 17.5 mm. The silo itself is 4.5 cm wide and 8.5 cm long. The hopper at the bottom has an inclination with respect to the silo wall of 105°.

After filling the silo with the glass spheres at a given θ while keeping the exit closed, the vibration is started and the particles allowed to rearrange for a few seconds before opening the exit.

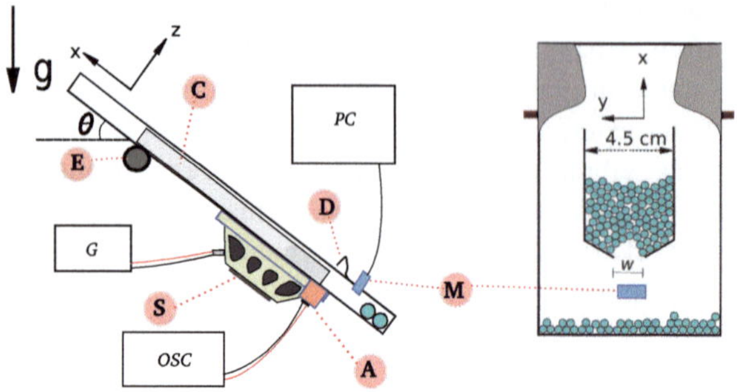

Fig. 1 Lateral and zenithal view of the experimental device: *C* is the incline, *D* is the door, *M* is the microphone, *A* is the accelerometer, *S* is the speaker, *G* is the wave generator and *E* is the rotation axis of the incline

Each discharge process was registered by a microphone placed below the exit such that particles leaving the hopper knock it. From this data, the total discharge time (T) and the time intervals between the passage of two consecutive beads (dt) can be obtained.

3 Results

3.1 Flowing Regime

In this regime, the hopper presents a smooth continuous discharge without intermittencies. The exit has 5 particle diameters ($W = 17.5$ mm). In Fig. 2 we plot the total time of discharge as a function of the angle of the incline. Thirty realization ($N_R = 30$) were carried out for each of the four angles studied. It can be seen that, in this regime, *faster is faster*: the increase of the driving force (i.e., the inclination angle) induce a decrease of the evacuation time. The mean discharge time in this regime is inverse to the flow-rate which in turn is proportional to $\sqrt{g}$ for vertical configurations [8]. In this inclined set up, the effective acceleration of gravity is $g\sin(\theta)$. Therefore, $\langle T\rangle \propto \sin(\theta)^{-1/2}$ (see fit in Fig. 2.)

This regime is characterized by very short time delays (dt) between the passage of two particles because blockages are not present. In Fig. 3 the probability distribution function (PDF) of dt is shown. It can be seen that as the angle increases the probability of having longer dt decreases. Time delays beyond 0.5 s have not been observed.

In Fig. 4 we plot the probability of finding a delay $dt > dt_0$, with dt_0 an arbitrary threshold. As we can see, for any threshold dt_0 the probability of finding a large dt falls as the inclination angle grows. This feature is hard to appreciate from Fig. 3, and it is an indication that, in this regime, a larger "driving force" does not promote long delays between the passage of consecutive grains.

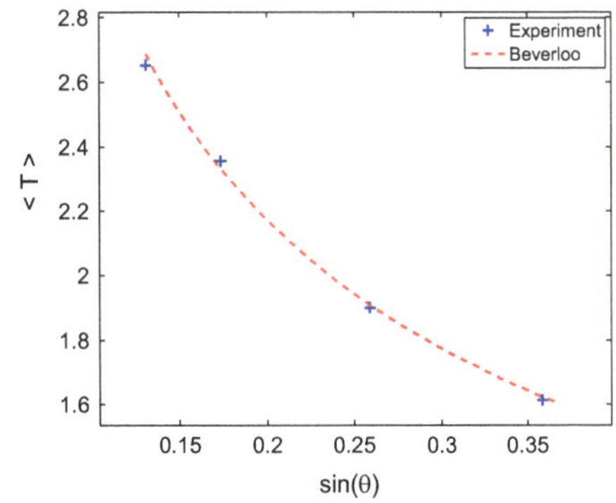

Fig. 2 Mean discharge time $< T >$ as a function of the sine of the angle of the inclined plane (θ) for an aperture $W = 5d$ (flowing regime). The *dashed line* corresponds to the fit with $\langle T \langle \propto sin(\theta)^{0.5}$

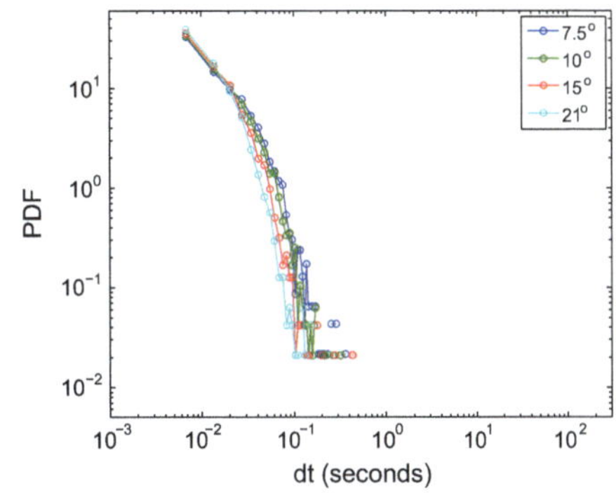

Fig. 3 Probability distribution function of *dt* for $W = 17.5$ mm (flowing regime)

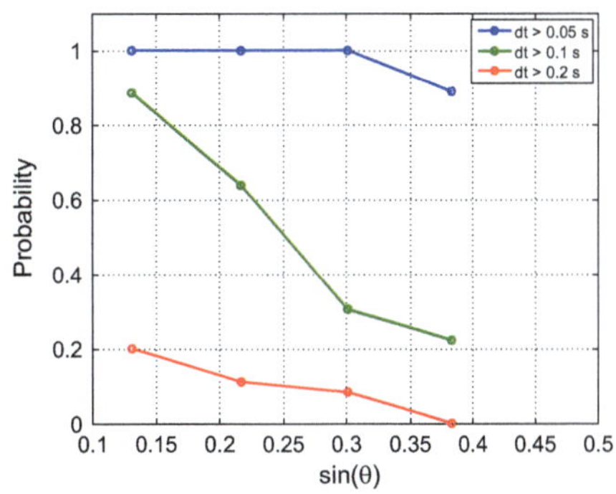

Fig. 4 Probability of having a delay *dt* longer than: 0.05 (*blue*), 0.1 (*green*) and 0.2 (*red*) s. Results for $W = 17.5$ mm (flowing regime)

3.1.1 Intermittent Regime

This regime is characterized by the occurrence of very long lasting blockages. The PDF of the interparticle delays *dt* show slowly decaying tails as can be seen in Fig. 5a. This behavior has been observed in previous studies of discharge of vibrated vertical silos [9, 10] and in the evacuation of mice [6].

The long time delays dominate the total discharge time. Looking in detail at the long tails of the distributions, for *dt* longer than 0.5 s, it is possible to

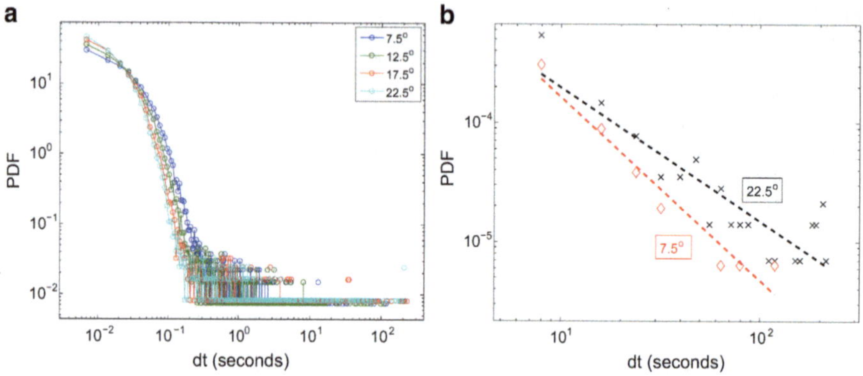

Fig. 5 (**a**) Probability distribution function of dt for $W = 3d = 10.5$ mm (intermittent regime). (**b**) A close up of the potential tails of the PDF for the extreme angles $\theta = 7.5°$ and $22.5°$

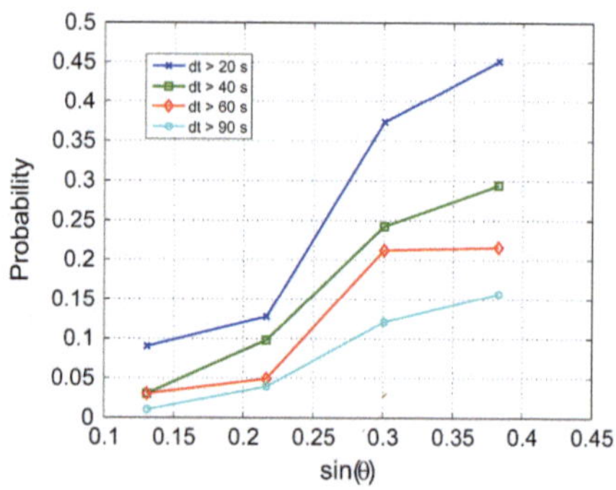

Fig. 6 Probability of having a delay dt longer than: 20 (*blue*), 40 (*green*), 60 (*red*) and 90 (*cyan*) s. Results for $W = 3d = 10.5$ mm (intermittent regime)

approximate these by a power law as shown in Fig. 5b. The exponent of the power law increases with the angle θ, indicating that longer dt are more probable for larger angles, in contrast to what we have observed in the flowing regime (see Fig. 3). Furthermore, the absolute values of the exponent are less than 2 for the present experimental conditions, indicating that the mean dt of the distribution is not defined. Distributions with absolute value of the power-law exponents below 2 have been also reported in other granular discharges under low intensity vibrations [9]. This is a crucial observation if we consider that FIS has been intensively studied in the past in terms of the mean evacuation time. If realistic conditions lead to undefined mean evacuation times, FIS should be rethought from a different perspective.

In Fig. 6, we look at the probability of having a delay dt between the passage of two particles lasting more than a certain time dt_0. Notice that in this regime longer dt_0 can be considered since long-lasting blockages appear. As it can be expected from the tendency of the exponent of the power-law tails, longer time delays are more probable for grater angles. This is the opposite situation to that of the flowing regime (see Fig. 4). Therefore, a FIS effect is observed in terms of the probability of having long-lasting jams rather than in terms of mean evacuation times, which are not defined.

4 Conclusions

We have studied a two-dimensional granular flow on a vibrated inclined plane for two hopper exits (3 and 5 particle diameter). The angle of the incline was varied in the range 7.5°–22.5°, which allows to mimic different degrees of hurry in a pedestrian system. The two exit widths generate two different states in the system: continuous flow and intermittent flow.

For the narrow exit (3 particles diameters) the distribution of dt (the delay between the passage of two consecutive grains) shows a long tail which is the main responsible for the total discharge time. This tail can be fitted by a power law [9] with a negative exponent α, were $|\alpha| < 2$, indicating that the mean dt is not defined. As a consequence, the mean discharge time depends on the number of discharges averaged. This implies that a different approach is necessary to quantify the evacuation dynamics for pedestrians since the mean evacuation time may not be defined in such systems. Hence, we suggest a different observable to characterize the experimental results in the intermittent regime.

The probability of finding a discharge with a delay (dt) between the passage of two grains that lasts longer than a prescribed cutoff seems to be a suitable observable. For the narrow exit (intermittent regime), this probability increases with the angle of the inclined indicating that longer blockages are more probable when grater forces are exerted by the particles over the exit. This is clear evidence of the FIS effect in the intermittent regime. For the wider exit, this probability decreases monotonically, indicating that *faster is faster* in the flowing regime.

Based on our results, it can be stated that, depending on the exit width, two different states of the system can be found. In the one hand, in the intermittent regime the discharge performance decreases when greater forces are exerted by particles (when the angle of the incline increases). On the other hand, for a wider exit ($W = 5d$), a continuous flow regime is found, where the efficiency of the discharge improves for increasing forces exerted by the particles. Consequently an optimum exit width should exist being the minimum width that allows a regime where the total discharge time improves for greater driving forces. Notice that this is related, but not identical to the problem of finding a critical exit width beyond which jamming will never occur in non-vibrated silos [11, 12].

Although the existence of an optimum exit door suggested by our experiments on granular flow should be taken with care when extrapolating to pedestrian dynamics, the results are encouraging and attempts should be made to pursue similar analysis regarding realistic experiments with biological agents.

Acknowledgements PAG acknowledges a scholarship from CONICET (Argentina). LAP acknowledges support from CONICET (Argentina). DRP acknowledges support from CONICET(Argentina) via grant P.I.P. 2010-0304. The authors acknowledge support from grant P.I.C.T. 2011-1238.

References

1. Helbing D, Farkas I J, Vicsek T (2000). "Simulating dynamical features of escape panic". Nature 407: 487–490
2. Parisi D R, Dorso C O (2005). "Microscopic Dynamics of Pedestrian Evacuation". Physica A: Statistical Mechanics and its Applications 354: 606–618
3. Parisi D R, Dorso C O (2007). "Morphological and Dynamical Aspects of the Room Evacuation Process". Physica A, 385: 343–355.
4. Altshuler E, Ramos O, Núñez Y, Fernández J, Batista-Leyva A J, Noda C (2005). "Symmetry breaking in escaping ants". The American Naturalist, 166(6):643–649.
5. Soria S, Josens R, Parisi D R, unpublished.
6. Saloma C, Perez G J, Tapang G, Lim M, Palmes-Saloma C (2003). "Self-organized quenching and scale-free behavior in real scape panic". PNAS 100: 11947
7. Zelinski B, Goles E, Markus M (2009). "Maximization of granular outflow by oblique exits and by obstacles". Physics of Fluids, 21(3):031701
8. Beverloo W A, Leniger H A, Van de Velde J (1961). "The flow of granular solids through orifices", Chemical Engineering Sciences 15: 260–269
9. Janda A, Maza D, Garcimartï£¡n A, Kolb E, Lanuza J, Clement E. (2009). "Unjamming a granular hopper by vibration". Europhys. Lett. 87: 24002
10. Mankoc C, Garcimartï£¡n A, Zuriguel I, Maza D, Pugnaloni L A (2009). "Role of vibrations in the jamming and unjamming of grains discharging from a silo". Phys. Rev. E 80: 011309
11. Zuriguel I, Pugnaloni L A, Garcimartín A, Maza D (2003). "Jamming during the discharge of grains from a silo described as a percolating transition". Phys. Rev. E 68: 030301(R)
12. Janda A, Zuriguel I, Garcimartín A, Pugnaloni L A, Maza D (2008). "Jamming and critical outlet size in the discharge of a two-dimensional silo". Erophys. Lett. 84: 44002

Modeling of Nanoparticles in a Gas Flow

J. Kaupužs, Sh. E. Guseynov, and J. Rimšāns

Abstract A mathematical model is developed to describe the motion of solid particles of micro- and nano-size in a gas flow. The model is represented by certain integro-differential equation with appropriate initial and boundary conditions. A probabilistic interpretation of the model is provided and its solvability is studied. We find that unique solution exists at certain sufficient condition. In this case, the Fokker–Planck equation can be obtained from the integro–differential equation.

1 Introduction

In the present work, the behavior of solid particles of micro- and nanosize in a gas flow is modeled. The motion of solid particles is considered in a coordinate system which moves along with the gas flow. We describe this as a Brownian motion, resulting from the influence of gas on the particles due to thermal fluctuations in the gas pressure. In absence of a temperature gradient (∇T) the time-averaged value of the Brownian fluctuations is zero. Otherwise, it is nonzero because of asymmetric action of the gas flow on the nanoparticles, resulting in a thermophoretic force [1,2] which shifts the mean position of a particle. The modeling of the thermophoretic force is related to the problem of making traps for particles and their agglomerates without any electro-magnetic or chemical actions on these particles. Here we treat only the simplest case of $\nabla T = 0$.

J. Kaupužs (✉) · Sh. E. Guseynov · J. Rimšāns
Institute of Mathematical Sciences and Information Technologies, University of Liepaja, 14 Liela Street, Liepaja LV-3401, Latvia
e-mail: kaupuzs@latnet.lv; sh.e.guseinov@inbox.lv; janis.rimsans@liepu.lv

V.V. Kozlov et al. (eds.), *Traffic and Granular Flow '11*,
DOI 10.1007/978-3-642-39669-4_31, © Springer-Verlag Berlin Heidelberg 2013

The following assumptions are made:

- The influence of the gas on the micro- nanoparticles is negligible;
- The influence of the micro- nanoparticles on each other is negligibly small as compared to the influence of gas on the particles;
- Micro-nanoparticles can move in the gas flow in different directions. At each time step a particle moves by certain distance, not necessarily to the next position.

These assumptions are quite natural if, e.g., the concentration of solid particles is sufficiently small, and if external forces are absent. This situation is typical in many important applications dealing with dilute particle flows. Based on these assumptions, first we build up a discrete one-dimensional model, and then we obtain a continuous model as a initial-value boundary problem for certain integro-differential equation. A probabilistic interpretation of the model is given and its solvability is studied. Finally, a generalization to the two-dimensional case is performed.

2 A Discrete One-Dimensional Model

We have build a discrete one-dimensional model based on the following assumptions:

(A) At each time moment $t = 0, \Delta t, 2\Delta t, 3\Delta t, \ldots$ each of the micro–nanoparticles can have one of the coordinates $0, \pm\Delta x, \pm 2\Delta x, \pm 3\Delta x \ldots.$

(B) If any of the particles in the 1D volume has coordinate $i\,\Delta x$ $(i = 0, \pm 1, \pm 2, \ldots)$ at time moment $n\,\Delta t$ $(n \in \{0 \cup \mathbb{N}\})$, then in the next time moment $(n+1)\Delta t$ $(n \in \{0 \cup \mathbb{N}\})$ it can have any one of the coordinates $j\,\Delta x$ $(j = 0, \pm 1, \pm 2, \ldots)$ with probability

$$p_{n;i,j} \quad (n \in \{0 \cup \mathbb{N}\}\,;\, i, j = 0, \pm\Delta x, \pm 2\Delta x, \pm 3\Delta x \ldots) \tag{1}$$

(C) The transition probabilities correspond to the Markov process.

(D) The particle-gas system is considered as almost homogeneous and it is assumed that the properties of the system do not depend on the direction of particle motion.

In the limit case, where the number of particles $U(m,n)$ at any coordinate $m\Delta x$ and time $n\Delta t$ tends to infinity, we have the following balance equation

$$U(m_1, n+1) - U(m_1, n) \approx \sum_{\substack{m_2=0,\pm 1,\pm 2,\ldots \\ m_2 \neq m_1}} U(m_2, n)\, p_{n;m_2,m_1} - U(m_1, n) \sum_{\substack{m_2=0,\pm 1,\pm 2,\ldots \\ m_2 \neq m_1}} p_{n;m_1,m_2}\,. \tag{2}$$

This relation is useful to derive an integro-differential equation in the continuum limit considered further on.

3 Continuum Model

Considering the limit, where the number of particles $U(m,n)$ is large, we introduce the particle mass density

$$\rho(x,t) = \frac{M}{\Delta x} U(m,n) . \tag{3}$$

Here M is the mass of a particle, which is assumed to be constant. In the continuum limit $\Delta x \to 0$, $\Delta t \to 0$ we have

$$\frac{\partial \rho(x,t)}{\partial t} = \lim_{\Delta x \to 0} \lim_{\Delta t \to 0} \frac{M}{\Delta x} \frac{U(m,n+1) - U(m,n)}{\Delta t} . \tag{4}$$

Dividing both sides of (2) by Δt, we obtain after certain limit procedures a continuous model as a initial-value boundary problem for the integro-differential equation

$$\frac{\partial \rho(x,t)}{\partial t} = \int_{-l}^{+l} K(t;y,x)\,\rho(y,t)dy - \rho(x,t) \int_{-l}^{+l} K(t;x,y)dy , \tag{5}$$

where

$$K(t;x,y) = \lim_{\substack{\Delta x \to 0 \\ m_i \to \pm\infty}} \lim_{\substack{\Delta t \to 0 \\ n \to \pm\infty}} \frac{p(n\Delta t; m_1 \Delta x, m_2 \Delta x)}{\Delta x \; \Delta t} \tag{6}$$

(here $i = 1,2$) is the kernel of the integro–differential equation. The used here limit procedures are such that the coordinates $x = \lim\limits_{\substack{\Delta x \to 0 \\ m_1 \to \pm\infty}} m_1 \Delta x$, $y = \lim\limits_{\substack{\Delta x \to 0 \\ m_2 \to \pm\infty}} m_2 \Delta x$ and the time $t = \lim\limits_{\substack{\Delta t \to 0 \\ n \to \pm\infty}} n\Delta t$ tend to finite values.

Equation (5) has to be solved with appropriate initial and boundary conditions, e.g.,

$$\rho(x,t)\,|_{t=0} = \rho_0(x) , \quad x \in [-l,+l] \tag{7}$$

$$\left(\alpha_1 \rho(x,t) + \beta_1 \rho'(x,t)\right)|_{x=-l} = \rho_1(t) , \quad t \in [0,T] \tag{8}$$

$$\left(\alpha_2 \rho(x,t) + \beta_2 \rho'(x,t)\right) \mid x = l = \rho_2(t) , \quad t \in [0,T] , \tag{9}$$

where $\rho_0(x)$ is the initial distribution which obeys the boundary conditions.

The kernel of the integro–differential equation $K(t; z_1, z_2)$ is equal to the probability density for moving a particle from $z_1 \in [-l, +l]$ to $z_2 \in [-l, +l]$ $(z_1 \neq z_2)$ at $t \in [0, T]$. More precisely, $K(t; z_1, z_2)dz_2dt$ is the probability that a jump from a given point z_1 to $z \in [z_2, z_2 + dz]$ occurs within a time interval dt. Due to the assumed isotropy and homogeneity of the model, we have $K(t; z_1, z_2) \equiv K(t; \mid z_1 - z_2 \mid)$.

The density $\rho(x, t)$ of micro-nanoparticles in gas flow can be interpreted as the density of mathematical expectation of particle mass. If we consider a random walk of a single particle, then $\rho(x, t)$ is the probability density for finding this particle at coordinate x at time t.

4 Solvability of the Model

We have investigated the solvability of the model and have found a sufficient condition in the form of an inequality

$$12\,(\Delta x)^2 \int_0^l K(t;z)z^2 dz \gg \int_0^l K(t;z)z^4 dz\,, \qquad \forall x \in [-l, +l],\ t \in (0, T] \tag{10}$$

for the existence and uniqueness of the solution, where Δx is the average jump interval. It implies a sufficiently fast convergence of a Taylor series expansion around $y = x$ in our integro–differential equation (5).

5 The Fokker–Planck Equation

If the stability condition (10) is satisfied, then the Fokker–Planck equation [3]

$$\frac{\partial \rho(x, t)}{\partial t} = a^2(t)\,\frac{\partial^2 \rho(x, t)}{\partial x^2} \tag{11}$$

is obtained by truncating the Taylor series, where

$$a^2(t) = \int_0^l K(t;z)\,z^2 dz > 0 \tag{12}$$

is the sensitivity or the diffusion coefficient. It can depend on time t in general.

6 The Two-Dimensional Case

Our model can be easily extended to the two-dimensional case, where we have

$$\frac{\partial\rho(x_1,x_2,t)}{\partial t} = \int_{-l_1}^{+l_1} dy_1 \int_{-l_2}^{+l_2} K(t;y_1,y_2;x_1,x_2)\,\rho(y_1,y_2,t)dy_2 - \rho(x_1,x_2,t)\int_{-l_1}^{+l_1} dy_1 \int_{-l_2}^{+l_2} K(t;x_1,x_2;y_1,y_2)dy_2\,. \quad (13)$$

Here $\rho(x_1, x_2, t)$ is the two-dimensional particle mass density (or probability density for finding a single randomly walking particle at certain position). The kernel $K(t; x_1, x_2; y_1, y_2)$ is the probability density for moving a particle from certain point (x_1, x_2) in the two-dimensional domain to vicinity of another point (y_1, y_2).

7 Conclusions

A mathematical model is developed, describing the motion of micro-nanoparticles in a gas flow by means of an integro–differential equation. Solvability of this model has been studied, showing that it has a unique solution under certain conditions. The Fokker–Planck equation can be obtained from the integro–differential equation at this condition.

References

1. S.P. Bakanov, *Thermophoresis in gases at small Knudsen numbers*, Uspekhi Fizicheskikh Nauk, 35:783–792, 1992.
2. M.X.Fernandes, J.G de la Torre, *Brownian Dynamics Simulation of Rigid Particles of Arbitrary Shape in External Fields*, J. Biophys., 83:3039–3048, 2002.
3. H. Risken, *The Fokker-Planck Equation: Methods of Solution and Applications*, Berlin, Springer-Verlag, 1989.

Multiple Shear Banding in Granular Materials

Robabeh Moosavi, M. Reza Shaebani, Maniya Maleki, János Török, and Dietrich E. Wolf

Abstract We present numerical and experimental evidences for multiple shear band formation in sheared granular materials. A modified Couette cell with a split bottom near the outer cylinder is made rough by gluing glass beads on all boundaries. The cell is filled with the same beads and sheared by slowly rotating the inner cylinder and the attached bottom disk. A wide shear band is mostly observed at the free surface of the material. However, depending on the filling height and grain size, simultaneous shear bands may form near the confining walls and in the middle of the system. By minimizing the rate of energy dissipation, we numerically find similar velocity profiles for intermediate filling heights and relatively large grain sizes.

1 Introduction

As dense granular materials are sheared, they finally yield under stress and start flowing. The resulting flow is not necessarily homogeneous like in normal fluids. At low shear rates, grains form solid like regions separated by narrow regions between them, called shear bands [1–3]. The rigid blocks move relative to each other in such a way that the strain is localized in shear bands, i.e. along the narrow interfaces between the unstrained parts. Understanding the mechanism of material failure is of great importance in industry and geophysics, and hence shear band formation is widely investigated experimentally [4–14] and numerically [15–18].

R. Moosavi · M. Maleki
Department of Physics, Institute for Advanced Studies in Basic Sciences (IASBS), Zanjan 45137-66731, Iran
e-mail: r_moosavi@iasbs.ac.ir

M.R. Shaebani (✉) · J. Török · D.E. Wolf
Department of Theoretical Physics, University of Duisburg-Essen, 47048 Duisburg, Germany
e-mail: shaebani@lusi.uni-sb.de

V.V. Kozlov et al. (eds.), *Traffic and Granular Flow '11*,
DOI 10.1007/978-3-642-39669-4_32, © Springer-Verlag Berlin Heidelberg 2013

The shear bands are dependent on size and shape of the grains [5, 8] and, most of the time, are narrow (of the order of a few grain diameters) and localized near the boundaries [1, 7, 13], which makes the theoretical description more difficult. For example, in a Couette cell geometry one always observes that a narrow shear band forms near the inner cylinder [4–6]. However, using a modified Couette geometry, Fenistein and co-workers managed to generate wide shear zones in the bulk of the material away from the confining cylinders [2, 8]. The experimental setup was a Couette cell (two co-axial cylinders) modified by splitting the bottom into two rings which were attached to the cylinders. The cell was filled with grains up to height H and one of the cylinders together with its co-moving bottom ring were rotated. In this way, they drove the system from the split bottom, which initiates a shear band with cylindrical symmetry. The shear band is pinned to the split bottom and its width W grows with height. The angular velocity profile at the surface follows an error function characterized by the width and center position of the shear zone at the surface. The center of shear zone shifts towards the inner cylinder with increasing the filling height H independent of the particle properties. The width W, however, depends on the shape and size of the grains. Using a theory based on the principle of minimum dissipation, Unger et al. described the position of the shear band with the assumption of negligible width of the shear band [15]. The model produced results in excellent agreement with the experiments with no fitting parameter. Later, a generalized version of this method proposed which was capable of reproducing the width of the shear bands [18].

In the present work, we introduce a new geometry which initiates a wide shear band in bulk and narrow ones near the confining walls simultaneously. The effective friction coefficient is different in bulk and near the rough boundaries. This difference, together with the possibility of choosing different grain size and filling height, leads to different scenarios for shear band formation. We use the least dissipation principle to describe the behavior and compare the results with experimental data.

2 Shear Band Formation as a Variational Problem

According to the experimental findings (from Refs. [2, 8]), the width of shear band W depends on the grain properties. One can make W arbitrarily small, by using a suitable choice of the size and shape of the grains. This justifies the narrow band approximation used in Ref. [15] to calculate the center position of the shear band. Applying the idea of the principle of least dissipation, they required a time-independent (steady) flow which leads to the minimum rate of energy dissipation and fulfils the external constraint conditions. Assuming a narrow shear band, the dissipation occurs throughout the interface between the two sliding rigid blocks which leads to the following variational problem

$$\int v\, p\, \mu_{\text{eff}}\, dS = \min, \tag{1}$$

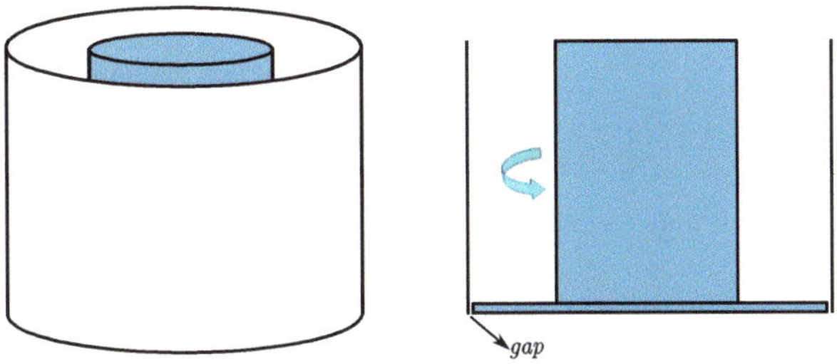

Fig. 1 (*left*) Schematic view of our modified Couette cell. (*right*) Side view

where v, p, and dS are the relative velocity of the blocks, the pressure, and the sliding surface at height h, respectively, and the dissipation is integrated over the whole sliding surface. The effective coarse-grained friction coefficient μ_{eff} can be assumed to remain constant in the bulk of the material. Then, assuming hydrostatic pressure, Eq. (1) for the modified Couette geometry leads to

$$\int r^2 (H - h) \left[1 + (dr/dh)^2\right]^{1/2} dh = \min. \tag{2}$$

Equation (2) provides a unique path for the shear band inside the bulk of the material, which is in agreement with the center position obtained from experiments [15].

In practice, the path of the shear band is not frozen due to the presence of disorder and inhomogeneities. Because of the relative displacements in the shear band, the spatial distribution of the inhomogeneities changes with time, which may change the global minimum path and produce instantaneous shear bands. If a random potential is replaced (in order to take into account the disordered nature of the system), then an ensemble average over the instantaneous shear bands, obtained via the principle of least dissipation, reproduces a wide shear band [18]. The width W obtained this way matches remarkably with the experimental data.

3 Experimental Setup

Our experimental setup is basically a modified Couette geometry with a split bottom near the outer cylinder (see Fig. 1). The gap between the bottom plate and the outer cylinder is around 400 μm. The space between the two cylinders is filled with spherical glass beads up to a height H, and the inner cylinder rotates together with the attached bottom plate. The diameter of the grains ranges from 1 to 3 mm, thus no particle can escape from the split bottom. The side walls and bottom disk made rough by gluing a layer of grains. The system is driven from the bottom by a motor. The resulting flow at the free surface is monitored from above by a fast CCD camera

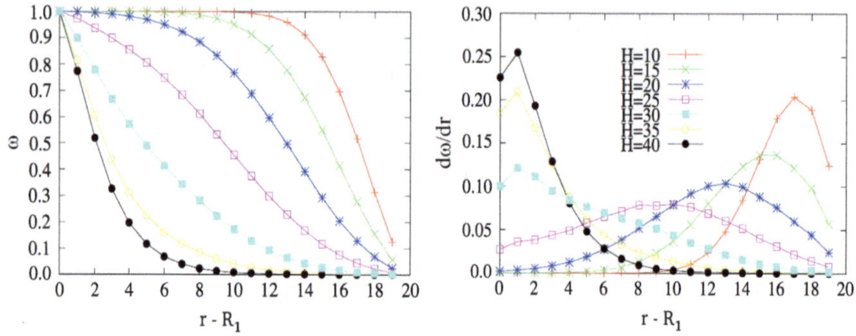

Fig. 2 (*left*) Angular velocity (ω) profile as a function of the distance $r-R_1$ from the inner cylinder for different filling heights. R_1 is the radius of the inner cylinder. (*right*) The derivative of ω with respect to r. The ratio between the effective friction coefficients in bulk and wall $\mu_{\text{bulk}}/\mu_{\text{wall}}$ is set to 1.0

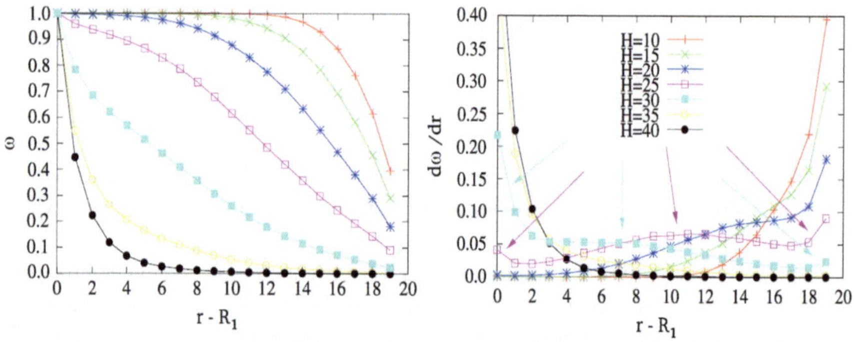

Fig. 3 The same plots as in Fig. 2 but for $\mu_{\text{wall}}/\mu_{\text{bulk}} = 0.7$. The *arrows* show the center position of the shear bands for intermediate filling heights

at a rate of 60 frames/s with pixel resolution of 100 μm. The flow rapidly reaches to a stationary state (of the order of few seconds) where the angular velocity profile is symmetric with respect to the cylinder axis.

4 Numerical Results

We first apply the numerical method of Sect. 2 to the geometry of our experimental setup. We use a random potential field and find the path which minimizes the rate of energy dissipation and fulfils the constraints. As the local structure is changed along the shear path, we change the randomness in the neighborhood of the shear band and search for the optimal path anew. This process is repeated many times, and the shear band is finally determined as the ensemble average over all realizations.

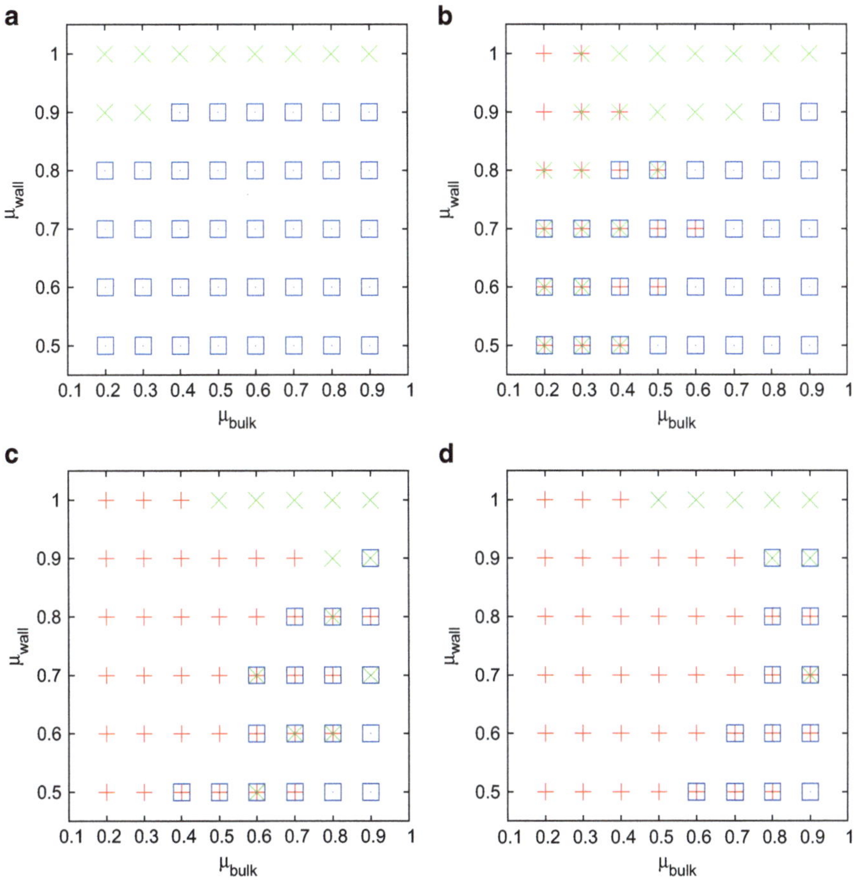

Fig. 4 Numerical results for the phase diagram of the possible shear bands in the $(\mu_{\mathrm{wall}}, \mu_{\mathrm{bulk}})$ space for different ratios between the filling height H and the width of the system L: $H/L=$ (**a**) 0.5, (**b**) 1.0, (**c**) 2.0, and (**d**) 3.0. The *blue square*, *green cross*, and *red plus* denote, respectively, a shear band near the outer cylinder, a wide shear band in the middle, and a shear band near the inner cylinder

We fix the ratio R_1/R_2 between the radius of the inner and outer cylinders and vary the ratio H/L (aspect ratio, $L = R_2 - R_1$), the lattice size (larger than the grain diameter d), and the ratio between the effective friction coefficient in the bulk and near the rough walls $\mu_{\mathrm{bulk}}/\mu_{\mathrm{wall}}$. Most of the time, the numerical results produce an angular velocity profile at the surface which can be well fitted by an error function, i.e. a wide shear band exists in the bulk of the system (see Fig. 2). However, for relatively large grains ($d/L \geq 0.04$) and in the intermediate filling heights we find different types of velocity profile, which imply the existence of simultaneous shear bands: relatively narrow shear bands near the confining walls, and a wide shear zone in the middle (Fig. 3).

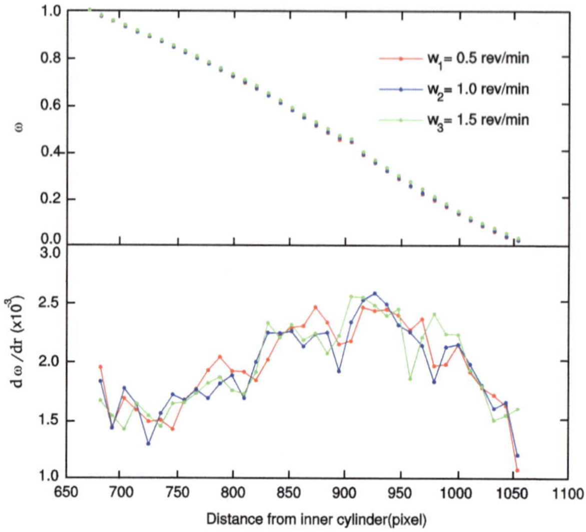

Fig. 5 Experimental data for the angular velocity profile (*top*) and its derivative (*bottom*) as a function of the distance from the inner cylinder for different shear rates

The phase behavior for different values of μ_{bulk} and μ_{wall} is shown in Fig. 4. Since the walls are made rough with the same grains as those used in the bulk, one expects that $\mu_{\text{wall}} < \mu_{\text{bulk}}$ in our setup. Therefore, Fig. 4b reveals that the chance of multiple shear band formation is remarkably larger for intermediate filling heights ($L <= H$).

5 Experimental Evidence

The experimental data show two different situations depending on the particle size. For small grains, there is always a single shear band which starts at the split bottom and reaches the surface. It is located near the outer cylinder for small filling heights, but it becomes wider and shifts towards the inner cylinder with increasing H, a behavior reported previously in the literature. Interestingly, for larger grains, there can be more than one shear band on the surface in agreement with numerical results. An example is shown in Fig. 5 belonging to 2 mm glass beads and $H/L \sim 1.2$. Besides the wide middle shear band, a narrow one also forms near the inner cylinder. The behavior is robust with respect to the choice of shear velocity. By varying the grain size and filling height, we are currently looking for cases where two shear bands form near the rough walls or all three types of shear band coexist simultaneously.

Acknowledgements We would like to thank W. Losert for fruitful discussions. We gratefully acknowledge support by the Institute for Advanced Studies in Basic Sciences (IASBS) Research Council under grant No. G2010IASBS136.

References

1. Jaeger H. M., Nagel S. R., and Behringer R. P. (1996) Rev. Mod. Phys. **68**, 1259.
2. Fenistein D. and van Hecke M. (2003) Nature **425**, 256.
3. GDR MiDi (2004) Eur. Phys. J. E **14**, 341.
4. Howell D., Behringer R. P., and Veje C. (1999) Phys. Rev. Lett. **82**, 5241.
5. Mueth D. M., Debregeas G. F., Karczmar G. S., Eng P. J., Nagel S. R., and Jaeger H. M. (2000) Nature **406**, 385.
6. Losert W., Bocquet L., Lubensky T. C., and Gollub J. P. (2000) Phys. Rev. Lett. **85**, 1428.
7. Hartley R. R. and Behringer R. P. (2003) Nature **421**, 928.
8. Fenistein D., van de Meent J. W., and van Hecke M. (2004) Phys. Rev. Lett. **92**, 094301.
9. Toiya M., Stambaugh J., and Losert W. (2004) Phys. Rev. Lett. **93**, 088001.
10. Fenistein D., van de Meent J. W., and van Hecke M. (2006) Phys. Rev. Lett. **96**, 118001.
11. Börzsönyi T., Unger T., and Szabó B. (2009) Phys. Rev. E **80**, 060302(R).
12. Knudsen H. A. and Bergli J. (2009) Phys. Rev. Lett. **103**, 108301.
13. Scott D. R. (1996) Nature **381**, 592.
14. Török J., Krishnamurthy S., Kertész J., and Roux S. (2000) Phys. Rev. Lett. **84**, 3851.
15. Unger T., Török J., Kertész J., and Wolf D. E. (2004) Phys. Rev. Lett. **92**, 214301.
16. Cheng X., Lechman J. B., Fernandez-Barbero A., Grest G. S., Jaeger H. M., Karczmar G. S., Möbius M. E., and Nagel S. R. (2006) Phys. Rev. Lett. **96**, 038001.
17. Unger T. (2007) Phys. Rev. Lett. **98**, 018301.
18. Török J., Unger T., Kertész J., and Wolf D. E. (2007) Phys. Rev. E **75**, 011305.

Anisotropic Elasticity in Sheared Packings of Frictional Disks

M. Reza Shaebani, Jens Boberski, and Dietrich E. Wolf

Abstract We study the effect of unilaterality of the interparticle interactions on the elastic response of granular materials. The anisotropy of the contact network is related to the opening of contacts during quasi-static shear deformations. As a result, local incremental constitutive relations describing the evolution of stress in terms of shear and volumetric strains are proposed, and it is shown that the macroscopic elastic behavior of anisotropic granular assemblies under biaxial deformations can be described by three independent elastic moduli: bulk, shear, and anisotropy modulus. We show that the bulk and shear moduli are independent of the imposed shear deformation if scaled by the contact density, and the magnitude of the anisotropy modulus is proportional to the ratio between shear and volumetric strain. The theoretical predictions are qualitatively in agreement with MD simulation results far from the jamming transition.

1 Introduction

Mechanical response of granular materials has been widely studied because of scientific challenges and many industrial and geophysical applications [1,2]. A static granular assembly does not behave like an ordinary elastic solid. The stress-strain relation is nonlinear and depends on the loading path and fabric. The nonlinear behavior originates from the nonlinearity of contact forces, disorder, and friction. However, unilaterality of the contact forces is another source of the nonlinear elastic response in granular media [3]. In the absence of attractive contact forces in dry granular media, the contacts open, and hence no elastic restoring or frictional force is transmitted. This renders the macroscopic elastic behavior nonlinear, even in the case of linear restoring interparticle forces. This source of nonlinearity is studied

M.R. Shaebani (✉) · J. Boberski · D.E. Wolf
Department of Theoretical Physics, University of Duisburg-Essen, 47048 Duisburg, Germany
e-mail: shaebani@lusi.uni-sb.de

V.V. Kozlov et al. (eds.), *Traffic and Granular Flow '11*,
DOI 10.1007/978-3-642-39669-4_33, © Springer-Verlag Berlin Heidelberg 2013

in the present work. We investigate the fabric anisotropy and relate it to the shear deformation. The elements of the stiffness tensor are expressed as functions of the average packing properties and the probability distribution of contact orientations. We derive analytical expressions for the elastic moduli and compare them with molecular dynamics simulation results.

2 Linear Elastic Response

The relationship between the *stress* $\boldsymbol{\sigma}$ and *strain* $\boldsymbol{\epsilon}$ tensors expresses the elastic response of solids. The relation between the incremental strain and stress for small deformations (by first order Taylor expansion of the strain increment) is given by

$$\delta\sigma_{ij} = \sum_{k,l} C_{ijkl}\, \delta\epsilon_{kl} \,, \tag{1}$$

where the fourth rank tensor C is the *stiffness* tensor. Due to symmetry considerations for the stress and strain tensors, C usually has few independent elements. Isotropic materials can be described by only two independent elements, represented by the *Lamé* coefficients. We first recall how the elements of stiffness tensor are related to the average packing properties and the probability distribution of contact orientations [4, 5]. Using a harmonic force law with normal and tangential spring constants k_n and k_t (which approximates the interaction for 2D disks for small deformations [6, 7]) and starting from a weakly deformed state, the change in the force exerted by particle B on particle A due to a small change of the branch vector $\boldsymbol{l}^c$ is

$$\delta \boldsymbol{f}^c = k_n(\delta \boldsymbol{l}^c \cdot \boldsymbol{n}^c)\boldsymbol{n}^c + k_t(\delta \boldsymbol{l}^c \cdot \boldsymbol{t}^c)\boldsymbol{t}^c \,, \tag{2}$$

and the average stress increment within the volume V can be approximated by [8, 9]

$$\delta\sigma_{ij} = \frac{1}{V}\sum_{c=1}^{N_c}\left(k_n\big(\sum_k \delta l_k^c\, n_k^c\big)n_i^c + k_t\big(\sum_k \delta l_k^c\, t_k^c\big)t_i^c\right) l_j^c \,. \tag{3}$$

If one assumes *affine* displacements of the particle centers Eq. (3) can be simplified. The affine part of the relative displacements reads

$$\delta l_k^c = \sum_l \delta\epsilon_{kl}\, l_l^c \,. \tag{4}$$

Then, using the affine assumption, the stiffness tensor elements can be obtained as [10, 11]

$$C_{ijkl} = \frac{\ell^2}{V}\sum_{c=1}^{N_c}\left(k_n n_i^c n_j^c n_k^c n_l^c + k_t t_i^c n_j^c t_k^c n_l^c\right) . \tag{5}$$

In the case of polydisperse packings, a correction factor has to be also taken into account which only depends on the moments of the particle size distribution [12]. If the sum over the contacts located inside the volume V is replaced by an integral over the contact orientation distribution $P(\alpha)$ [5], Eq. (5) becomes

$$C_{ijkl}=\frac{2}{\pi}z\phi\int_{-\pi}^{\pi}\left(k_n n_i n_j n_k n_l + k_t t_i n_j t_k n_l\right) P(\alpha)\, d\alpha\,, \tag{6}$$

where z, ϕ, and $P(\alpha)$ denote, respectively, the average coordination number, the volume fraction of the packing, and the probability distribution of the contact orientations. Therefore, the fabric properties (z, ϕ, and $P(\alpha)$) determine the elements of the stiffness tensor.

By choosing the principal axes of the strain tensor increment $\delta\boldsymbol{\epsilon}$ as coordinate system so that

$$\delta\boldsymbol{\epsilon} = \frac{1}{2}\begin{pmatrix} \delta\epsilon_v+\delta\gamma & 0 \\ 0 & \delta\epsilon_v-\delta\gamma \end{pmatrix} \quad \text{with } \delta\gamma \geq 0\,, \tag{7}$$

where $\delta\epsilon_v$ and $\delta\gamma$, denote the change in the volumetric strain and in the shear deformation respectively, Eq. (1) can be written as

$$\begin{pmatrix} \delta\sigma_{11} \\ \delta\sigma_{22} \\ \delta\sigma_{12} \\ \delta\sigma_{21} \end{pmatrix} = \begin{pmatrix} C_{1111} & C_{1122} \\ C_{2211} & C_{2222} \\ C_{1211} & C_{1222} \\ C_{2111} & C_{2122} \end{pmatrix} \begin{pmatrix} (\delta\epsilon_v+\delta\gamma)/2 \\ (\delta\epsilon_v-\delta\gamma)/2 \end{pmatrix}. \tag{8}$$

The torque balance requires that the average stress tensor remains symmetric, i.e. $\delta\sigma_{12}=\delta\sigma_{21}$, which leads to the following constraints

$$C_{1211}=C_{2111}\,, \quad C_{1222}=C_{2122}\,. \tag{9}$$

3 Isotropic Packings

For isotropic assemblies of grains, we can assume a uniform distribution of contact orientations i.e. $P(\alpha)=\frac{1}{2\pi}$. Using Eq. (6) one obtains the stress-strain relation in the principal axes of incremental strain for an isotropic packing as

$$\begin{pmatrix} \delta\sigma_{11} \\ \delta\sigma_{22} \\ \delta\sigma_{12} \\ \delta\sigma_{21} \end{pmatrix} = \begin{pmatrix} C_{1111} & C_{1122} \\ C_{2211} & C_{2222} \\ 0 & 0 \\ 0 & 0 \end{pmatrix} \begin{pmatrix} (\delta\epsilon_v+\delta\gamma)/2 \\ (\delta\epsilon_v-\delta\gamma)/2 \end{pmatrix}. \tag{10}$$

with $\mathcal{C}_{1111}=\mathcal{C}_{2222}=\frac{z\phi}{4\pi}(3k_n+k_t)$, $\mathcal{C}_{1122}=\mathcal{C}_{2211}=\frac{z\phi}{4\pi}(k_n-k_t)$. Equation (10) shows that $\delta\sigma_{12}=\delta\sigma_{21}=0$, so the principal axes of strain and stress coincide. One can write the stress tensor increment in its principal axes as

$$\delta\sigma = \begin{pmatrix} \delta P + \delta\tau & 0 \\ 0 & \delta P - \delta\tau \end{pmatrix}, \tag{11}$$

where δP and $\delta\tau$ are the incremental isotropic pressure and shear stress. The stiffness tensor elements can be related to the two Lamé coefficients, λ and μ, as $\lambda=\mathcal{C}_{1122}$ and $\lambda + 2\mu=\mathcal{C}_{1111}$. Then, Eq. (10) is further reduced to

$$\begin{pmatrix} \delta P + \delta\tau \\ \delta P - \delta\tau \end{pmatrix} = \begin{pmatrix} \lambda+2\mu & \lambda \\ \lambda & \lambda+2\mu \end{pmatrix} \begin{pmatrix} (\delta\epsilon_v + \delta\gamma)/2 \\ (\delta\epsilon_v - \delta\gamma)/2 \end{pmatrix}. \tag{12}$$

Therefore, one obtains δP and $\delta\tau$ as

$$\begin{pmatrix} \delta P \\ \delta\tau \end{pmatrix} = \begin{pmatrix} \lambda + \mu & 0 \\ 0 & \mu \end{pmatrix} \begin{pmatrix} \delta\epsilon_v \\ \delta\gamma \end{pmatrix}. \tag{13}$$

It then follows from Eq. (13) that the bulk modulus E and the shear modulus G are [11]

$$E = \frac{\delta P}{\delta\epsilon_v} = \lambda + \mu = \frac{z\phi}{2\pi} k_n,$$
$$G = \frac{\delta\tau}{\delta\gamma} = \mu = \frac{z\phi}{4\pi}(k_n + k_t). \tag{14}$$

As expected for isotropic materials, the elastic response can be described via only two independent parameters E and G that are determined by the spring coefficients k_n and k_t, and the contact density of the structure.

4 Anisotropy Induced by Unilaterality

We investigate the effect of unilaterality on the elastic response by considering the possibility of opening and closing of contacts, after a finite volumetric strain ϵ_v and shear deformation γ are imposed to the unstrained reference state with zero overlap. Assuming the affine displacements of the particles, the distance between the centers of neighboring particles changes due to the strain $\boldsymbol{\epsilon}$ as

$$\varDelta\xi_n(\alpha, \boldsymbol{\epsilon}) = -\ell \sum_{i,j} \epsilon_{ij} n_i n_j = -\ell \left(\frac{\epsilon_v}{2} + \frac{\gamma}{2} \cos 2\alpha \right). \tag{15}$$

$\Delta\xi_n$ depends on $\boldsymbol{\epsilon}$ and the direction of the branch vector, α. If the two particles touched each other, i.e. $\xi_n = 0$ in the unstrained state, a positive $\Delta\xi_n$ means that the deformation leads to an overlap, while a negative value indicates that the particles are no longer in contact in the strained state. When there is a gap between the particle surfaces in the unstrained configuration, an overlap can also form, if $\Delta\xi_n$ is larger than the gap. We, therefore, extend the notion of an overlap to include small negative values, $\xi_n < 0$, which tell the size of the gap. We introduce the probability density $Q(\xi_n, \alpha, \boldsymbol{\epsilon})$ that a particle pair has a branch vector at an angle α with respect to the principal axis of the strain tensor $\boldsymbol{\epsilon}$, belonging to the eigenvalue $\epsilon_v + \gamma$, and that the overlap respectively the negative gap has a value ξ_n. This probability density depends on the strain $\boldsymbol{\epsilon}$. For $\boldsymbol{\epsilon} = 0$ it is assumed to be isotropic, i.e. independent of α, and it fulfills $Q_0(\xi_n) = 0$ for $\xi_n > 0$, as there are no overlaps in the unstrained configuration, and $Q_0(\xi_n) \neq 0$ for $\xi_n \leq 0$, as the unstrained configuration is jammed. We assume that the probability distribution simply shifts by $\Delta\xi_n$, Eq. (15), under the influence of strain:

$$Q(\xi_n, \alpha, \boldsymbol{\epsilon}) = Q_0(\xi_n - \Delta\xi_n(\alpha, \boldsymbol{\epsilon})) \tag{16}$$

The probability that a pair of neighbor particles is actually in contact is

$$\mathcal{N} = \int_{-\pi}^{\pi} d\alpha \int_0^{\infty} d\xi_n \; Q(\xi_n, \alpha, \boldsymbol{\epsilon}) \tag{17}$$

The probability density that a contact has a certain angle α with respect to the principal axis of the strain, which belongs to the eigenvalue $\epsilon_v + \gamma$, is

$$P(\alpha, \boldsymbol{\epsilon}) = \frac{1}{\mathcal{N}} \int_0^{\infty} d\xi_n \; Q(\xi_n, \alpha, \boldsymbol{\epsilon}). \tag{18}$$

Applying the assumption (16), the integral is in first order of $\boldsymbol{\epsilon}$ given by

$$\begin{aligned} \int_0^{\infty} d\xi_n \; Q(\xi_n, \alpha, \boldsymbol{\epsilon}) &\approx Q_0(0)\Delta\xi_n(\alpha, \boldsymbol{\epsilon}) \\ &= -Q_0(0)\ell \left(\frac{\epsilon_v}{2} + \frac{\gamma}{2} \cos 2\alpha \right). \end{aligned} \tag{19}$$

Integrating this over α gives the corresponding first order approximation of $\mathcal{N}$:

$$\mathcal{N} \approx -Q_0(0)\ell\epsilon_v\pi. \tag{20}$$

Hence the properly normalized first order approximation of the probability density of contact directions is

$$P(\alpha, \boldsymbol{\epsilon}) \approx \frac{1}{2\pi} \left(1 + \frac{\gamma}{\epsilon_v} \cos 2\alpha \right). \tag{21}$$

This approximation can at most be applied for $\left|\frac{\gamma}{\epsilon_v}\right| \leq 1$, as otherwise the probability density would not be positive semi definite. $\alpha = 0$ is the direction of the principal axis belonging to the eigenvalue $\epsilon_v + \gamma$. For $\epsilon_v < 0$ (compressive strain) and $\gamma \geq 0$ (by definition) this is the direction in which the precompressed system expands during biaxial deformation. Therefore contacts preferentially open in this direction so that $P(0, \boldsymbol{\epsilon}) \approx \frac{1}{2\pi}\left(1 + \frac{\gamma}{\epsilon_v}\right) < \frac{1}{2\pi}$. One finds that the elastic moduli of a granular packing are approximately

$$E = \frac{z\phi}{2\pi}k_n, \quad G = \frac{z\phi}{4\pi}(k_n + k_t), \quad A = \frac{z\phi}{4\pi}k_n\frac{\gamma}{\epsilon_v}. \tag{22}$$

Note that due to the presence of the nonzero element A, two independent experimental tests are required to determine the elastic moduli of an anisotropic material, for example:

(I) *Incremental* $\delta\epsilon_v$ *while* $\delta\gamma$=0:

$$E = -\left.\frac{\delta P}{\delta\epsilon_v}\right|_{\delta\gamma=0}, \qquad A = -\left.\frac{\delta\tau}{\delta\epsilon_v}\right|_{\delta\gamma=0}. \tag{23}$$

(II) *Incremental* $\delta\gamma$ *while* $\delta\epsilon_v$=0:

$$G = -\left.\frac{\delta\tau}{\delta\gamma}\right|_{\delta\epsilon_v=0}, \qquad A = -\left.\frac{\delta P}{\delta\gamma}\right|_{\delta\epsilon_v=0}. \tag{24}$$

This is in contrast to the isotropic case, where a single experiment with simultaneous incremental $\delta\epsilon_v$ and $\delta\gamma$ is sufficient to measure both bulk and shear moduli. In granular media, in contrast to an isotropic elastic material, a pure shear leads to a pressure increase, $\delta P = -A\delta\gamma > 0$ as $A < 0$.

5 Simulation Results

We carry out numerical simulations to verify the theoretical predictions of Sects. 3 and 4. We first generate 2D polydisperse packings of 3,000 rigid disks with particle radii uniformly distributed between $a_{\rm min}$=0.95 and $a_{\rm max}$=1.05 to avoid crystallization. By means of contact dynamics simulations [13], the initial dilute configurations of particles with fully periodic boundary conditions are compressed to construct homogeneous packings. Each of the resulting packings then provides an initial unstrained jammed configuration for MD simulations of soft particles. By modeling the interparticle interactions with normal and tangential Hookian springs (with k_t/k_n=0.5), the system is subjected to compression and then shear with the help of the LAMMPS molecular dynamics code [14, 15].

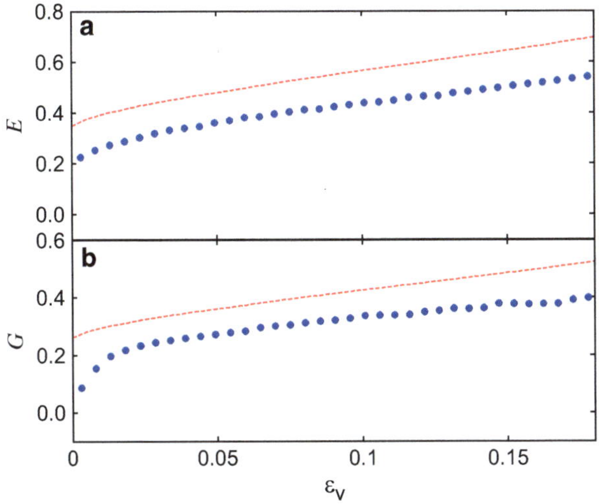

Fig. 1 (**a**) Bulk, and (**b**) shear modulus in units of k_n determined during an isotropic compression, in which the volumetric strain ϵ_v is increased in small steps and $\gamma = 0$. The moduli are calculated from fabric (method 1, *dashed lines*), or from Eq. (23) (method 2, *full circles*)

In a quasi-static compression process, the volume of the unstrained packing is gradually decreased by applying incremental volumetric strain steps and allowing the system to relax between those steps. The process is stopped when the total volumetric strain $\epsilon_v = \Delta V/V$ reaches a given value ($\epsilon_v = -0.04$ respectively $\epsilon_v = -0.09$). This precompressed packing is then sheared in a biaxial geometry, while keeping the volume of the system constant. The shear deformation is imposed via incremental steps, and the system is allowed to relax between the steps. Two different methods are used to determine the elastic moduli: (1) Using Eq. (5) the elastic constants can be computed from the fabric. This formula assumes affine deformations and that the contact network remains unchanged for incremental strains. (2) An incremental strain test is simulated by molecular dynamics and the elastic moduli are determined from Eqs. (23) and (24). This method allows for a change of the contact network and non-affine motions of the particles.

For pure compression, the values of moduli calculated from the fabric agree with the theory, whereas the ones obtained by evaluating Eqs. (23) and (24) are about 20 % smaller (see Fig. 1). Moreover, as one approaches the unstrained configuration (at the jamming transition), the elastic moduli obtained from evaluating incremental strain tests soften, whereas the ones calculated from the fabric remain unchanged. This shows that close to the jamming transition the assumption that the incremental particle movements are affine fails, in agreement with the findings of [16]. Therefore we concentrate on the regime $-\epsilon_v \geq 0.04$ in the following. Remarkably, the ratio of the two moduli, G/E is very close to the prediction of the simple theory, $G/E = (1 + k_t/k_n)/2 = 0.75$. The contact density increases during the compression process, while it decreases if a precompressed sample is sheared at

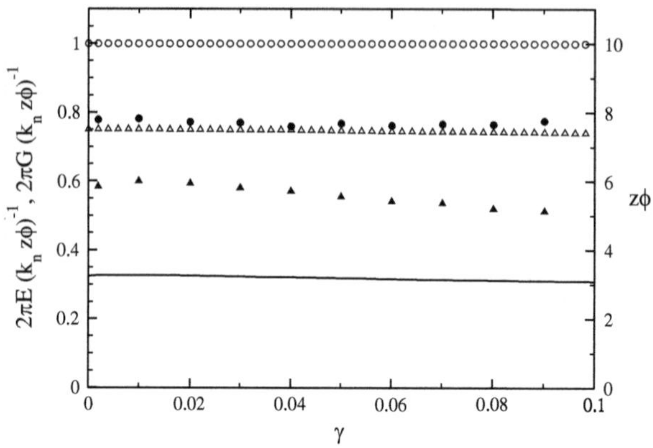

Fig. 2 Bulk (*circles*) and shear (*triangles*) modulus, as function of shear strain γ for fixed $\epsilon_v = -0.09$. Both moduli are given in units of $k_n/(2\pi)$ and divided by $z\phi$. They were calculated from fabric (method 1, *open symbols*), respectively from Eq. (23) (method 2, *full symbols*). The *full line* represents $z\phi$, referring to the scale on the right

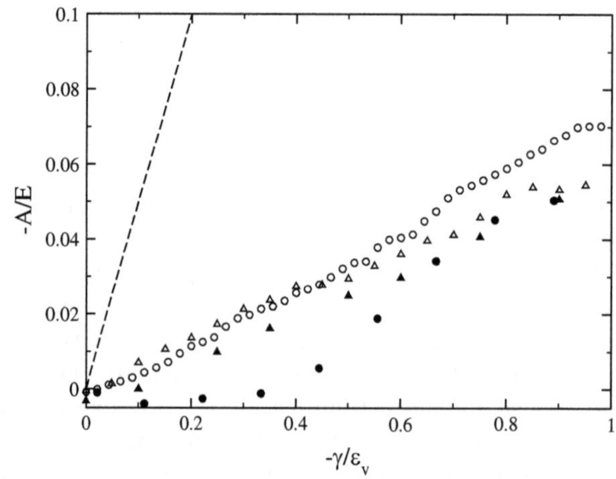

Fig. 3 Ratio of negative anisotropy modulus and bulk modulus,$-A/E$, as function of the ratio between shear strain and negative volumetric strain, $-\gamma/\epsilon_v$. Data are shown for two volumetric strains, $\epsilon_v = -0.04$ (*circles*) and $\epsilon_v = -0.09$ (*triangles*). They were calculated from fabric (method 1, *open symbols*), respectively from Eq. (23) (method 2, *full symbols*). The *dashed line* is the theoretical result, Eq. (22)

fixed volume, provided γ remains small enough. For large shear deformation, the system presumably forms a shear band and begins to flow. Then the contact density must become independent of γ.

The bulk and shear moduli, divided by the contact density, do not change when a precompressed state is sheared, in agreement with the theoretical result Eq. (22) (see Fig. 2). The model predicts, however, that the anisotropy modulus is proportional to γ/ϵ_v (dashed line in Fig. 3). Indeed, this is confirmed in the simulation. The results for A/E determined from the fabric, as well as the ones from the incremental shear tests for $\epsilon_v = -0.04$ agree with each other within the error bars and can be fitted by a linear dependence on γ/ϵ_v. Only the shear test simulation data for $\epsilon_v = -0.09$ deviate from this behavior for small shear, $-\gamma/\epsilon_v < 0.7$, although the evaluation of the fabric perfectly agrees. Further simulations are needed to clarify the origin of this peculiar behavior. The simulation data essentially confirm the linear dependence of the anisotropy modulus on the ratio γ/ϵ_v, however, the theory overestimates the slope.

Acknowledgements We are grateful to Isaac Goldhirsch and Tamás Unger for fruitful discussions. Support by the German Research Foundation (DFG) via priority program SPP 1486 "Particles in Contact" is acknowledged.

References

1. Neddermann R. M. (1992) Statics and Kinematics of Granular Materials. Cambridge University Press, Cambridge.
2. Jaeger H. M., Nagel S. R., and Behringer R. P. (1996) Rev. Mod. Phys. **68**, 1259.
3. Goldenberg C. and Goldhirsch I. (2005) Nature **435**, 188.
4. Mehrabadi M. M., Nemat-Nasser S., and Oda M. (1982) Int. J. Num. Anal. Meth. Geomech. **6**, 95.
5. Bathurst R. J. and Rothenburg L. (1988) J. Appl. Mech. **55**, 17.
6. Landau L. D. and Lifshitz E. M. (1986) Theory of Elasticity. Pergamon Press, Oxford.
7. Johnson K. L. (1987) Contact Mechanics. Cambridge University Press, Cambridge.
8. Christoffersen J., Mehrabadi M. M., and Nemat-Nasser S. (1981) J. Appl. Mech. **48**, 339.
9. Luding S. (2004) Int. J. Sol. Struct. **41**, 5821.
10. Liao C.-L. and Chang T.-C. (1997) Comput. Geotech. **20**, 345.
11. Kruyt N. P. and Rothenburg L. (1998) Int. J. Eng. Sci. **36**, 1127.
12. Shaebani M. R., Madadi M., Luding S., and Wolf D. E. (2011) submitted to Phys. Rev. E.
13. Shaebani M. R., Unger T., and Kertesz J. (2009) Int. J. Mod. Phys. C **20**, 847.
14. Plimpton S. (1995) J. Comp. Phys. **117**, 1.
15. http://lammps.sandia.gov/
16. Heussinger C. and Barrat J.-L. (2009) Phys. Rev. Lett. **102**, 218303.

Part IV
Traffic Networks and Intelligent Traffic Systems

Two-Variable Macroscopic Fundamental Diagrams for Traffic Networks

Victor L. Knoop and Serge P. Hoogendoorn

Abstract Traditionally, traffic is studied at the level of individual vehicles (microscopic), or the level of road sections (macroscopic). Recently, a higher level of study has been introduced, the level of an area. In homogeneous traffic conditions, there is a relationship between the number of vehicles in an area (accumulation) and the average flow in that area (production). This paper studies this relationship under inhomogeneity. Using simulation on a grid network it is shown that the performance of a network is a smooth function of the average network density (accumulation) and the spatial variation of the density. Hence, we introduce a two-variable macroscopic fundamental diagram (TV-MFD) which also holds in a inhomogeneous transition state, for instance in queue build up. The TV-MFD can be used for network control, for instance at ramp metering installations or at a more aggregate level perimeter control.

1 Introduction

To control traffic, its processes needs to be understood, or at least need to be predictable. Nowadays, research projects aim at collecting detailed data of driving processes, which reveals even more differences between drivers. These data will need much aggregation to come to the understanding of general traffic patterns.

As opposed to the movement of collecting more detailed data, Geroliminis and Daganzo [1] started a simplified description of traffic. The traffic state is only based on the accumulation, being the number of vehicles in an area. Drawback of this description is that it is only valid for homogeneously loaded networks, e.g. [2].

In this paper we analyse the onset of congestion in a simulated grid network. The focus for the paper is the effect of the spatial variation of density on the network

V.L. Knoop (✉) · S.P. Hoogendoorn
Department of Transport & Planning, Delft University of Technology, Delft, The Netherlands
e-mail: v.l.knoop@tudelft.nl

V.V. Kozlov et al. (eds.), *Traffic and Granular Flow '11*,
DOI 10.1007/978-3-642-39669-4_34, © Springer-Verlag Berlin Heidelberg 2013

performance. The next section summarises the recent developments in describing traffic with the macroscopic fundamental diagram. Section 3 then presents the setup of the simulation study. The 2-dimensional macroscopic fundamental diagram, performance as a function of accumulation and of spread of the accumulation, is presented in Sect. 4. Finally, Sect. 5 presents the conclusions and the further outlook.

2 Literature Overview of Macroscopic Fundamental Diagrams

In the past 5 years the theory of a macroscopic fundamental diagram (MFD) has been developed. Concepts were already proposed by Godfrey [3], but only when the concept of perimeter control was reintroduced [4], more studies started.

The best-known studies are the ones by Daganzo [4] and Geroliminis and Daganzo [1]. They show the relationship between the number of completed trips and the production function which is defined as a weighted average of the flow on all links. This means that the network production can be used as a good approximation of the utility of the users for the network, i.e., it is related to the estimated travel time of travelers. Furthermore, after some theoretical work, it was shown that MFDs work in practice [1]. With pioneering work using data from the Yokohama metropolitan area, an MFD was constructed with showed a crisp relationship between the network production and the accumulation.

Also, theoretical insights have be gained over the past years. Daganzo and Geroliminis [5] have shown that rather than to find the shape of the MFD in practice or by simulation, one can theoretically predict its shape. This gives a tool to calculate the highest production of the network, which then can be compared with the actual network production.

One of the requirements for the crisp relationship is that the congestion should be homogeneous over the network. Buisson and Ladier [6] were the first to test the how the MFDs change if the congestion is not homogeneously distributed over the network. They showed a reasonably good MFD for the French town Toulouse in normal conditions. However, 1 day there were strikes of truck drivers, driving slowly on the motorways, leading to traffic jams. The researchers concluded that leads to a serious deviation from the MFD for normal conditions. The inhomogeneous conditions were recreated in a traffic simulation of a urban motorway with several on-ramps (several kilometers) [7]. I was found that inhomogeneous congestion leads to a reduction of flow. Moreover, they advised on the control strategy to be followed, using ramp metering to create homogeneous traffic states. Cassidy et al. [2] studied the MFD for a motorway road stretch. They conclude, based on real data, that the MFD only holds in case the whole stretch is either congested or in free flow. In case there is a mix of these conditions on the studied stretch the production is lower than the production which would be predicted by the MFD.

The effect of variability is further discussed [8, 9]. Contrary to Ji et al. [7], these papers focus on urban networks. First, Mazloumian et al. [8] show with simulation that the variance of density over different locations (spatial variance) of density (or accumulation) is an important aspect to determine the total network production caused by the discrete nature of vehicles at junction, leaving gaps. So not only too many vehicles in the network in total, but also if they are located at some shorter jams at parts of the networks. The reasoning they provide is that "an inhomogeneity in the spatial distribution of car density increases the probability of spillover, which substantially decreases the network flow." This finding from simulation and reasoning is confirmed by empirical analysis [9], using the data from the Yokohama metropolitan area. The main cause for this effect is claimed to be the turning movement of the individual vehicles.

A theoretical explanation for the phenomenon of the influence of the spatial variance of the accumulation is given by Daganzo et al. [10]. They show that turning at intersections is the key reason for the drop in production with unevenly spread congestion. Dayah and Daganzo [11] then use this information by adding dynamics to the MFD. If congestion solves, it will not solve instantaneous over all locations. Rather, it will solve completely from one side of the queue. Therefore, reducing congestion will increase the spatial variance of the accumulation and thus (relatively) decrease the production. This means that the production for a system of dissolving traffic jams is under the equilibrium state, thus under the MFD. This way, there are hysteresis loops in the MFD, as also noted by Ji et al. [7]. Note that these loops are an effect by themselves and are different from for instance the capacity drop [12, 13].

This paper will again look into this phenomenon by performing a macroscopic traffic simulation, but without individual turning movements. By means of this simulation we aim to reveal the importance of the microscopic (i.e., vehicle based) turning movement.

3 Experiment Setup

This section describes the traffic simulation used for this research. The section first describes what will be simulated in terms of network and demands. Then, Sect. 3.2 describes the model used for this simulation. Section 3.3 describes the output of the simulator that is used later in the paper.

3.1 Experimental Settings

In the paper an urban network is simulated, since this is the main area where MFDs have been tested. We follow [9] and choose a Manhattan network with periodic boundary conditions. This means that the nodes are located at a regular grid,

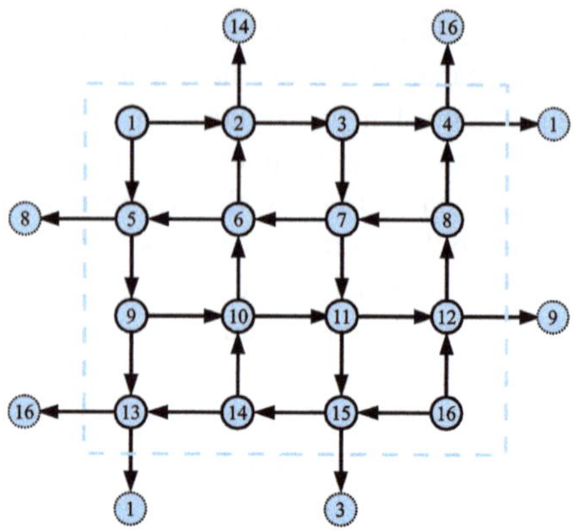

Fig. 1 Illustration of a 4 × 4 grid network with periodic boundary conditions

for which we choose a 20 × 20 size. Then, one-way links connect these nodes. The direction of the links changes from block to block, i.e. if at $x = 2$ the traffic is allowed to drive in the positive y direction, at $x = 1$ and at $x = 3$ there are one-way roads for traffic to drive in the negative y direction. We assume 2 lanes per link, a 1 km block length, a triangular fundamental diagram with a free speed of 60 km/h, a capacity of 1,500 veh/h/lane and a jam density of 150 veh/km/lane.

Furthermore, periodic boundary conditions are used, meaning that a link will not end at the edge of the network. Instead, it will continue over the edge at the other side of the network. An example of such a network is given in Fig. 1. Traffic can continue in a direct link from node 13 to node 1 or from node 5 to node 8. This way, all nodes have two incoming and to outgoing links and network boundaries have no effect.

The destinations are randomly chosen from all points in the network. In the network, there are 19 nodes chosen as destination nodes. There are no origin nodes. Instead, at the beginning of the simulation, traffic is put on the links. Vehicles are assigned to a destination, and for this distribution is equal over all destinations.

When the cars have reached their destination, they will not leave the network, but instead they are assigned a new destination. We use a macroscopic model (see Sect. 3.2), hence we can split the flow of arriving traffic equally over the 18 other destinations. The number of cars in the network is hence constant. This number will be a parameter setting for the simulations, but throughout one simulation, it is constant. The demand level is expressed as the density on all links at the start of the simulation, as fraction of the critical density. Figure 2a shows the network used under initial conditions.

3.2 *Traffic Flow Simulation*

This section describes the traffic flow model. For the traffic flow modelling we use a first order traffic model. Links are split into cells with a length of 250 m

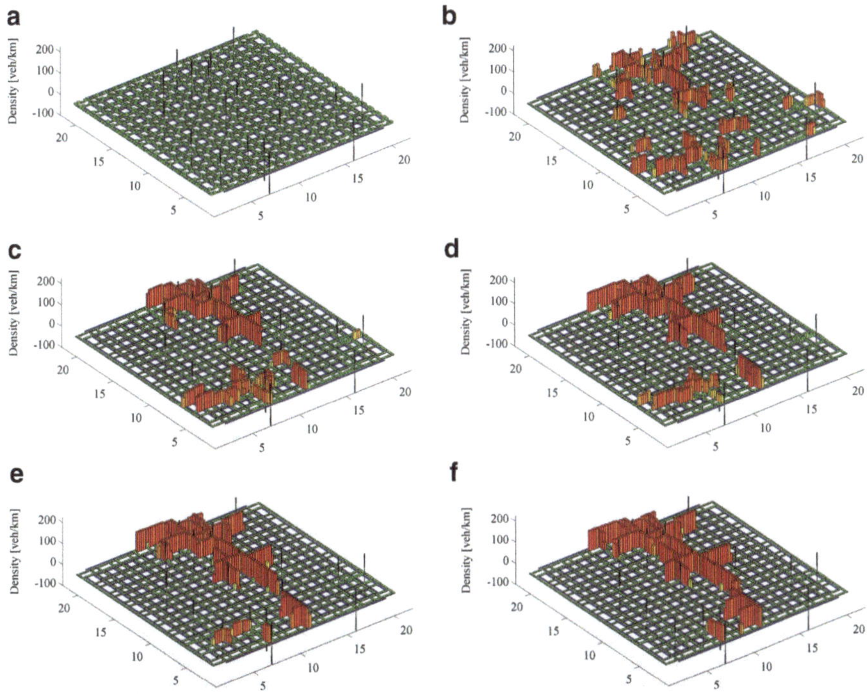

Fig. 2 Evolution of the densities (bar heights) and speeds (colours) in the network. (**a**) Start of the simulation. (**b**) 0.5 h. (**c**) 1 h. (**d**) 1.5 h. (**e**) 2 h. (**f**) 3 h

(i.e., 4 cells per link). We use the continuum LWR-model [14, 15] that we solve with a Godunov scheme [16]. Lebacque [17] showed how this is used for traffic flows, yielding a deterministic continuum traffic flow simulation model. The flux from one node to the next is basically restricted by either the demand from the upstream node (free flow) or by the supply from the downstream node (congestion):

$$\phi_{c,c+1} = \min\{D_c, S_{c+1}\}; \tag{1}$$

At a node r we have inlinks, denoted by i which lead the traffic towards node r and outlinks, denoted by j which lead the traffic away from r. At each node r, the demand D to each of the outlinks of the nodes is calculated, and all demand to one link from all inlinks is added. This is compared with the supply S of the cell in the outlink. In case this is insufficient, a factor α is calculated which shows which part of the demand can continue.

$$\alpha_r = \operatorname*{argmin}_{[j \text{ leading away from } r]} \left\{ \frac{S_j}{D_j} \right\} \tag{2}$$

This is the model developed by Jin and Zhang [18]. They propose that all demands towards the node are multiplied with the factor α, which gives the flow over the node.

This node model is slightly adapted for the case at hand here. Also the node itself can restrict the capacity. In our case, there are 2 links with a capacity of 3,000 veh/h as inlinks and 2 links with a capacity of 3,000 veh/h as outlinks. Since there are crossing flows, it is not possible to have a flow of 3,000 veh/h in one direction *and* a flow of 3,000 veh/h in the other direction. To overcome this problem, we introduce a node capacity (see also [19]). The node capacity is the maximum of the capacities of the outgoing links. This means that in our network, at maximum 3,000 veh/h can travel over a node. Again, the fraction of the traffic which can continue over node r is calculated, indicated by β:

$$\beta_r = \frac{C_r}{\sum_{\forall i \text{ to } r} D_i} \tag{3}$$

The demand factor γ is now the minimum of the demand factor calculated by the nodes and the demand factor due to the supply:

$$\gamma = \min\{\alpha_r, \beta_r, 1\} \tag{4}$$

Similar to [18], we take this as multiplicative factor for all demands to get to the flux ϕ_{ij}, i.e. the number of cars from one cell to the next over the node:

$$\phi_{ij} = \gamma D_{ij} \tag{5}$$

The path choice is static, and determined based on distance to the destination. Traffic will take the shortest path towards the destination. For intersections where both directions will give the same path length towards a destination, the split of traffic to that direction is 50–50.

3.3 Variables

In this paper, several traffic flow variables will be used. In this section we will explain them and show the way to calculate them.

Standard traffic flow variables are flow, q, being the vehicle distance covered in a unit of time, and density, k, the number of vehicles per unit road length. The network is divided into cells, which we denote by c, which have a length L_c. Flow and density in cells are denoted by q_c and k_c.

Furthermore, the accumulation N in an area X is the weighted average density:

$$N_X = \sum_{c \in X} \frac{k_c * L_c}{L_c} \tag{6}$$

Similarly, the production P in an area X is the weighted average flow:

$$P_X = \sum_{c \in X} \frac{q_c * L_c}{L_c} \tag{7}$$

Since the cell length are the same for all links in the network, the accumulation and production are average densities and flows. Recall that there is a strong relationship between the production and the number of completed trips [1].

This paper also studies the variations in densities. The standard deviation of the cell density is found by considering all cell densities for one moment in time, and calculate the standard deviation of these numbers.

4 Simulation Outcomes

This section describes the evolution of traffic over the time. First, the traffic flow phenomena are qualitatively described, then in Sect. 4.2 the performance and variation are quantified.

4.1 *Traffic Flow Phenomena*

This section first describes the traffic flow over time. Figure 2 shows the outcomes of the simulation, in snapshots of the density and speed over time. At the start of the simulation (see Fig. 2a), traffic is evenly distributed over all links, since this was the initial situation as it was regulated externally. The destinations of the network are indicated by the vertical lines.

When the traffic starts to run, various distributed bottlenecks become active. This is shown in Fig. 2b. After some time (Fig. 2d–f), traffic problems concentrate more and more around one location. The number of vehicles in the rest of the network reduces, ensuring free flow conditions there. This complete evolution can be found in Fig. 2a–f. The network has periodic boundary conditions, which means that the network edges do not have any effect. Any deviations from a symmetry are due to random effects and thus to the location of the destinations, since the traffic simulation is deterministic.

At the end, the situation seems to have stabilised. From 2.5 to 3 h (Fig. 2e–f) there have been little change, and the changes in the traffic state get smaller and smaller: an equilibrium has formed. Now, the number of vehicles passing the most restricting bottleneck equals the number of vehicles arriving at the end of the queue.

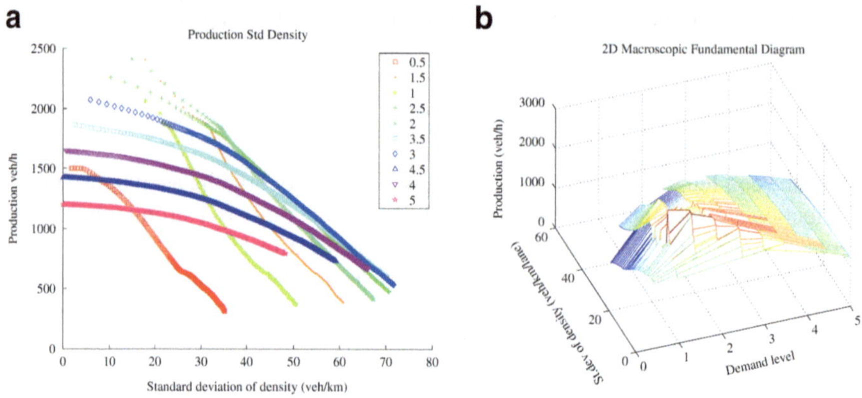

Fig. 3 The effect of demand and spatial variation on the network performance. (**a**) Different lines for different demands. (**b**) The two-dimensional macroscopic fundamental diagram

4.2 Influence of Variation in Density

Most articles describing macroscopic fundamental diagrams emphasize that the relationship is only valid as long as traffic states are similar for all links in the network (e.g. [1]). This requirement of homogeneous distributed congestion clearly does not hold for our situation (see Fig. 2f). Furthermore, a standard macroscopic fundamental diagram would relate the network performance to the network accumulation, being proportional to the number of vehicles in the network. In this case, however, the number of vehicles in the network is constant: vehicles cannot drive out of the network and at the destination they are simply given another destination. The accumulation is thus fixed, but there are various values for the performance, depending on the congestion, hence the macroscopic fundamental diagram would result in a vertical line.

Instead of linking the performance only to the accumulation, we also link it to the variation of the densities in the network. This relation is shown graphically in Fig. 3. Figure 3a shows production lines for the different network loadings. They decrease as function of the density variability, as also found by Mazloumian et al. [8]. Contrary to their setup, the simulation in our paper does not have a flow discretization in vehicles. So the effect of the decreasing production with increasing variability is not mainly caused by the individual vehicle movements since it can found in a simulation modelled with aggregated flows.

It is remarkable that the slope of the lines in Fig. 3a is different. If the average accumulation is undercritical, the demand rapidly decreases with the increase of variability. This is caused by queues, and spillover of queues. With a demand higher than the critical level (demand > 1), the decrease is less steep. In this case the temporal spillovers are less important, since vehicles will find a queue later on in the network anyway.

Figure 3b shows how these lines can be transformed into a two-dimensional macroscopic fundamental diagram. Note that the impact of the variation on the production is of a similar magnitude as the impact of the total accumulation. It is therefore essential that the variation of the density is also used in the state estimation, and thus the prediction of the production or performance. Note furthermore that we did not exclude any points from the observations. This two-dimensional macroscopic fundamental diagram can also be used if congestion is homogeneously distributed over the network.

5 Conclusions and Outlook

This paper presented a simulation study to the effect of variation in the macroscopic fundamental diagram. Even with very simple simulation tools, being a first-order traffic simulation, a proportional node model, and a node capacity, we find a clear influence of the variation of density on the network performance. In fact, the variation has an influence on the performance which is similar in magnitude to the influence of the accumulation. We therefore propose to use a two-variable macroscopic fundamental diagram (TV-MFD) instead of the one-variable macroscopic fundamental diagram used up to now.

The traffic processes themselves cause a the variation of the congestion. This depends on the OD matrix, as well as on traffic dynamics and the route choice. Future work therefore includes how the shape of the two-dimensional fundamental diagram depends on the network internal structure, and route choice. Also the effect of route choice is topic of future study.

Acknowledgements This research was sponsored by a IP-CC subsidy from ICTregie/NWO in the project *SI4MS*, Sensor Intelligence for Mobility Systems, and by the foundation Next Generation Infrastructures in the project *JAMS*, Joint Approach for Multi-level Simulation.

References

1. N. Geroliminis, C.F. Daganzo, Transportation Research Part B: Methodological **42**(9), 759 (2008)
2. M. Cassidy, K. Jang, C. Daganzo, in *Proceedings of the 90th Annual Meeting of the Transportation Research Board* (2011)
3. J. Godfrey, Traffic Engineering and Control **11**(7), 323 (1969)
4. C. Daganzo, Transportation Research Part B: Methodological **41**(1), 49 (2007)
5. C. Daganzo, N. Geroliminis, Transportation Research Part B: Methodological **42**(9), 771 (2008)
6. C. Buisson, C. Ladier, Tranportation Research Records, Journal of the Transportation Research Board (2009)
7. Y. Ji, W. Daamen, S. Hoogendoorn, S. Hoogendoorn-Lanser, X. Qian, Transportation Research Record, Journal of the Transporation Research Board **2161**, 42 (2010)

8. A. Mazloumian, N. Geroliminis, D. Helbing, Philosophical Transactions of the Royal Society A **368**, 4627 (2010)
9. N. Geroliminis, Y. Ji, in *Proceedings of 90th Annual Meeting of the Transportation Research Board* (2011)
10. C. Daganzo, V. Gayah, E. Gonzales, Transportation Research Part B: Methodological **45**(1), 278 (2011)
11. V. Gayah, C. Daganzo, Transportation Research Part B (2011)
12. F.L. Hall, K. Agyemang-Duah, Transportation Research Record: Journal of the Transportation Research Board No. 1320 pp. 91–98 (1991)
13. M. Cassidy, R. Bertini, Transportation Research Part B: Methodological **33**(1), 25 (1999)
14. M.J. Lighthill, G.B. Whitham, Proceedings of the Royal Society of London. Series A, Mathematical and Physical Sciences **229**(1178), 317 (1955)
15. P.I. Richards, Operations Research 4 **4**, 42 (1956)
16. S.K. Godunov, Math. Sb. **47**, 271 (1959)
17. J.P. Lebacque, in *Proceedings of the 13th International Symposium on Transportation and Traffic Theory* (1996)
18. W.L. Jin, H.M. Zhang, Transportation Research Part B: Methodological **37**(6), 521 (2003)
19. C. Tampère, R. Corthout, D. Cattrysse, L. Immers, Transportation Research Part B **45**, 289 (2011)

A Uzawa Algorithm for Static Traffic Assignment in an Orthotropic Network

Tibye Saumtally, Jean-Patrick Lebacque, and Habib Haj-Salem

Abstract We consider a static traffic assignment problem in a bidimensional network, viewed as a continuum of orthotropic roads. The commuters are spilled over a dense urban area, and try to reach a possible destination. There are a finite number of destinations, which are represented by a small area. The traffic flows from the city area to a determined destination are considered as one commodity. The network supply is defined by the side constraints, or road capacities. The travel demand is variable. The traffic assignment problem is built as a mathematical program inspired from the Beckmann objective function. It is solved by a Lagrangian scheme and a dual method.

1 Introduction

The assignment problem has been formulated well in the discrete network case. Nevertheless, this approach is difficult to develop to give general patterns of traffic in a dense urban area. There are some big disadvantages to this approach, due to the elevated level of details and time computing. Several authors have suggested a continuous modelling approach [1–5]. The main advantage of this approach is that it involves a small number of parameters, hence the computing time is much shorter. A variable demand has been introduced in a continuum approach in [6], but these articles didn't take the network supply into account. The network supply is the road

T. Saumtally (✉)
Université Paris Est, GRETTIA, Le Descartes 2, 2 rue de la Butte Verte, 93166 Noisy le Grand Cedex, France
e-mail: tibye.saumtally@ifsttar.fr

J.-P. Lebacque · H. Haj-Salem
IFSTTAR, GRETTIA, Le Descartes 2, 2 rue de la Butte Verte, 93166 Noisy le Grand Cedex, France
e-mail: jean-patrick.lebacque@ifsttar.fr; habib.haj-salem@ifsttar.fr

V.V. Kozlov et al. (eds.), *Traffic and Granular Flow '11*,
DOI 10.1007/978-3-642-39669-4_35, © Springer-Verlag Berlin Heidelberg 2013

capacity, or the maximum flow which can be accepted on a given road. A description of road capacity also called side constraints can be found in [7].

2 Road Traffic on a Large and Dense Urban Area

We consider a city in a two-dimensional plane. We denote the geographical surface of this city by $\mathfrak{A}$. The surface $\mathfrak{A}$ can be identified as a subset of the affine space $\mathbb{R}^2$. We suppose that all the commuters are uniformly distributed in the area $\mathfrak{A}$, and try to get to their destination. There are M possible destinations for the customers in the area. We denote them by $\mathfrak{B}_m$, $m \in \{1, \ldots, M\}$. The traffic flows from all the area to one determined destination are called a *commodity*. Hence, there are M commodities on the area $\mathfrak{A}$. The network of the area $\mathfrak{A}$ is viewed as a continuum. To illustrate it, one could imagine a big city observed from so far away that it would be impossible to distinguish separately the different roads of the network.

2.1 Distance in the Network

If the area $\mathfrak{A}$ is provided with an orthotropic network (a set of orthogonal streets), the distance between two points A and B of $\mathfrak{A}$, which coordinates are (x_A, y_A) and (x_B, y_B), is $d(A, B) = |x_A - x_B| + |y_A - y_B|$. This distance is called orthotropic as far as it defines an orthotropic scheme of paths on $\mathfrak{A}$. What has to be pointed out is that with an orthotropic metric, several paths can have the same distance between two points (e.g. the green and red paths of the Fig. 1-left). This is not the case with an isotropic metric, where there is a single shortest path between A and B: the segment $[AB]$ (Fig. 1-right). In the following, we will develop an orthotropic model.

3 Demand Function of Commuters

The traffic demand to the destination $\mathfrak{B}_m$ from a small area $\mathrm{d}x\mathrm{d}y$ around the point located by its coordinates (x, y), is the quantity we denote $\Theta_m(x, y)\mathrm{d}x\mathrm{d}y$. This quantity is positive over the area $\mathfrak{A} \setminus \cup_{m=1}^{M} \mathfrak{B}_m$, and strictly negative over $\cup_{m=1}^{M} \mathfrak{B}_m$.

3.1 Demand Function While Getting to a Destination (Generation)

We assume that the demand for each destination is elastic. This means that for every m, it exists a monotonic decreasing function of the travel cost τ from the origin (x, y) to the destination $\mathfrak{B}_m$. The travel cost τ from origin (x, y) to the

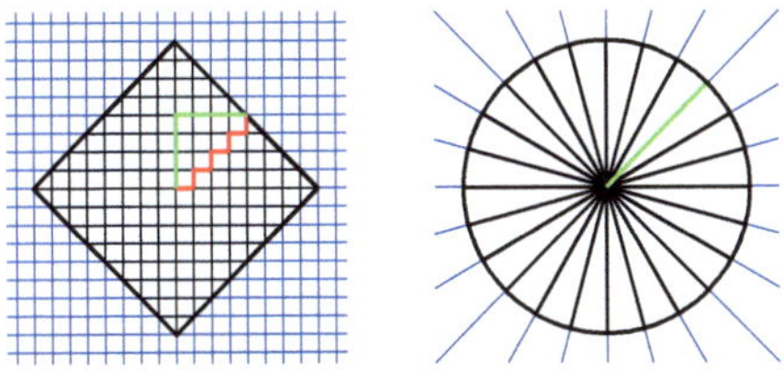

Fig. 1 Examples of short distance paths between two points: orthotropic case (*left*), isotropic case (*right*)

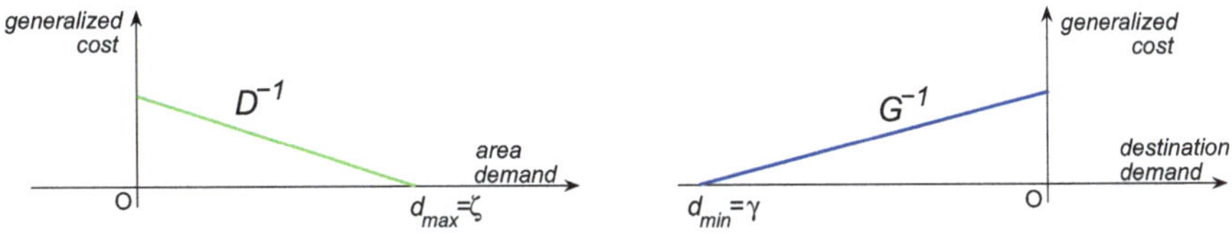

Fig. 2 Affine demand functions

destination $\mathfrak{B}_m$ only represents the time cost (we assume it does not include the transportation cost). The demand at point (x, y) is set in $[0, \zeta_m(x, y)]$. This means that the demand is generated, and it cannot exceed a maximum $\zeta_m(x, y)$, which depends on the network generation possibility (e.g. vehicle exits on (x, y)). For all m and for all $(x, y) \in \mathfrak{A} \setminus \mathfrak{B}_m$, we denote the area demand function $\mathcal{D}_{m,(x,y)} : \tau \mapsto d_m\tau + \zeta_m(x, y)$, where $d_m < 0$. We assume that for every (x, y), the demand functions are all parallel (same slope d_m), but do not have necessarily the same demand-intercept ($\zeta_m(x, y)$ may vary with (x, y)). For every m, the function $\mathcal{D}_{m,(x,y)}$ is supposed to be strictly decreasing, so $\mathcal{D}^{-1}_{m,(x,y)}$ is. If ϑ is the travel demand at a point to the destination m, the travel cost from this point to the destination m is $\mathcal{D}^{-1}_{m,(x,y)}(\vartheta)$, and the higher is $\mathcal{D}^{-1}_{m,(x,y)}(\vartheta)$, the lower is ϑ.

3.2 *Demand Function into a Destination Area (Absorption)*

While entering the area of their destination, the commuters feel satisfied. For a given destination, we can hypothesize that there will be more demand as the gain of being at its area is high. As the vehicles are absorbed by the destination, we can model the demand function as on Fig. 2-right. In this case, the cost has to be seen as the opposite of a gain, that's why we represent the demand as a strictly increasing function of the cost. The demand at point (x, y) is set in $[\gamma_m(x, y), 0]$. This means that the demand is absorbed, and it cannot underpass a minimum $\gamma_m(x, y)$, which depends on the network absorbtion possibility (e.g. vehicle entrance on (x, y)). For all m and for all $(x, y) \in \mathfrak{B}_m$, we note the destination demand function $\mathcal{G}_{m,(x,y)} : \tau \mapsto g_m\tau + \gamma_m(x, y)$, where $g_m > 0$. We assume that for every (x, y), the demand functions to the destination m are all parallel (same slope g_m), but do not have necessarily the same demand-intercept ($\gamma_m(x, y)$ may vary with (x, y)).

4 Road Traffic Physic

4.1 Flow Conservation

At each point (x, y) of the network, the vector $q_m(x, y)$ represents the flow state of vehicles that want to reach the destination m. If we represent the vector $q_m(x, y)$ by its coordinates $(q_{1m}(x, y), q_{2m}(x, y))$, each one of its coordinates is expressed in number of vehicles per length unit.

On every disc D included in $\mathfrak{A}$, the demand refered to the destination m is equal to the projection of the flow over the normal exterior vectors of D, what we can mathematically express by the relation $\iint_D \Theta_m(x, y)\mathrm{d}x\mathrm{d}y = \oint_{\partial D} \langle q_m(x, y)|\mathbf{n}\rangle \mathrm{d}\sigma$. If we suppose that the flow q is sufficiently smooth, by using the Green-Riemann theorem, we obtain the relation of conservation in the static case: $\Theta_m = \mathrm{div} q_m$.

4.2 Travel Cost

As vehicles in opposite ways do not share the same lane, a crucial difference between traffic flow and fluids is that two opposite traffic movements do not cancel each other out. To illustrate this, one can see on Fig. 3 that the flows q_a and q_b have nearly the same influence on the travel cost at point P, whereas each one of the flows q_a and q_c nearly do not disturb the other. Then, the travel cost at a point where exist two opposite flows of traffic is the sum of the cost of each flow (but not the cost of the sum of the norms of each flow).

We consider the two privileged network lines of the area. These lines are orthogonal, and give us four possible directions of moving. We explain with the simple example of the Fig. 4 how we construct the travel cost for the commuters at any point P of the area $\mathcal{A}$. Let the four vectors e_1, e_2, e_3 and e_4 be as in Fig. 4. The three flows q_A, q_B and q_C at the point P can be written as $q_A = q_{A1}e_1 + q_{A2}e_2$, $q_B = q_{B3}e_3 + q_{B2}e_2$ and $q_C = q_{C3}e_3 + q_{C4}e_4$ where each of the coordinates q_{ij} (with $(i, j) \in \{A, B, C\} \times \{1, 2, 3, 4\}$) are positive numbers. On direction i, the travel cost is equal to the cost of the sum of the projection on this direction. As only two orthogonal directions are privileged, it is just as if the travel cost at any point were the sum of the costs of the sum of each flow coordinates on every possible direction of moving: $C(P, (q_A, q_B, q_C)) = c_1(P, q_{A1}) + c_2(P, q_{A2} + q_{B2}) + c_3(P, q_{B3} + q_{C3}) + c_4(P, q_{C4})$. We have written four travel cost functions c_i, $i \in \{1, 2, 3, 4\}$ because we allow the cost to be different for each sense of one

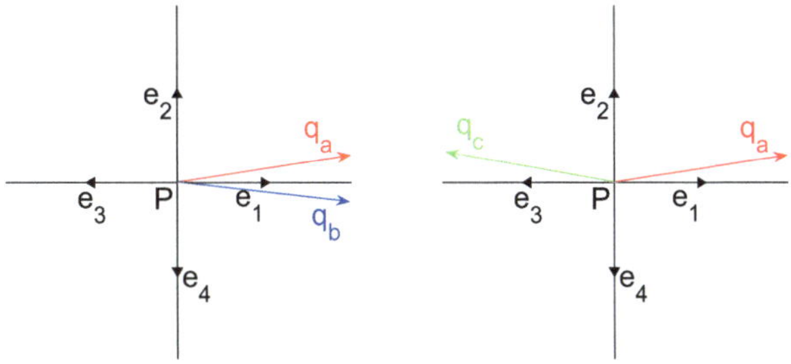

Fig. 3 Different flows of traffic at point P

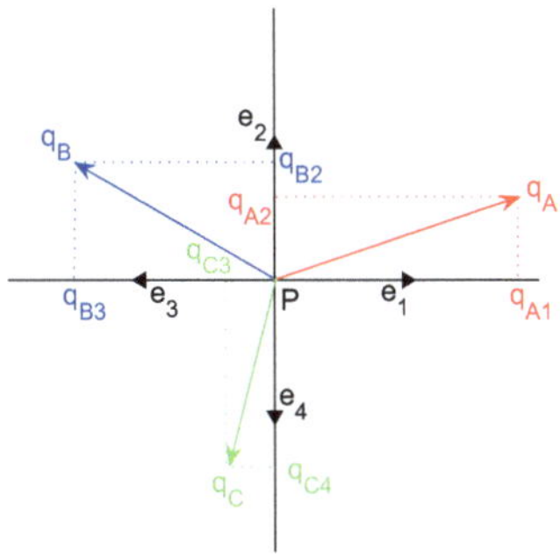

Fig. 4 Three different flows at point P

determined direction (think about two lanes in one sense and only one lane in the opposite sense).

5 The Network's Supply

The displacement scheme is the result of the confrontation between a demand and a supply. We have explained what is the demand in Sect. 3. Now let us explain what is the network's supply.

Each road has a finite limit of traffic flow, congestion point being reached while flow getting to this limit. To take this into account roads are provided with an upper bound on traffic flows, so the roads could not carry arbitrarily large volumes of traffic. This upper bound is called a side constraint [7, 8].

6 Formulation of the Equilibrium

For any real number a, we note $R(a)$ or simply a^+, the quantity $\max(0, a)$. The function R is called *Ramp function*. It is convex (but not strictly convex), it is not differentiable at 0.

6.1 Formal Approach of the Constraint Flows and Demands Set

The functions q_m and Θ_m we are looking for must verify conditions that come from the network's topological properties, the capacity constraints and the underlying traffic model.

A first constraint is the positivity (which corresponds to a generation of vehicles) and the upper boundedness (which corresponds to a maximum number of generated vehicles per unit area and per hour) of each demand function Θ_m at every point of the area which is not in destination $\mathfrak{B}_m$: $\forall m \in \{1, \ldots, M\}$, $\forall (x, y) \in \mathfrak{A} \setminus \mathfrak{B}_m$, $0 \leqslant \Theta_m(x, y) \leqslant \zeta_m(x, y)$.

A second constraint is the negativity (which corresponds to an absorption of vehicles) and the lower boundedness (which corresponds to a maximum number of absorbed vehicles per unit area) of each demand function Θ_m at every point of the area which is in destination $\mathfrak{B}_m$: $\forall m \in \{1, \ldots, M\}$, $\forall (x, y) \in \mathfrak{B}_m$, $\gamma_m(x, y) \leqslant \Theta_m(x, y) \leqslant 0$.

A third constraint is the capacity constraint on every direction, the traffic flow cannot exceed it: $\forall i \in \{1, 2, 3, 4\}$, $\forall (x, y) \in \mathfrak{A}$, $\sum_{m=1}^{M} \langle q_m(x, y) | e_i \rangle^+ \leqslant K_i(x, y)$. This constraint takes into account the network's topology (an orthotropical network) too.

A fourth constraint is the flow conservation for every destination: $\forall m \in \{1, \ldots, M\}$, $\forall (x, y) \in \mathfrak{A}$, $\mathrm{div} q_m(x, y) = \Theta_m(x, y)$.

A fifth constraint is the border conditions, the traffic flow goes along the boundary of the area and cannot cross it: $\forall m \in \{1, \ldots, M\}$, $\forall (x, y) \in \partial\mathfrak{A}$, $\langle q_m(x, y) | \mathbf{n} \rangle = 0$.

We denote by X_0 the set of couples (q, Θ) such that

$$\begin{cases} 0 \leqslant \Theta_m(x, y) \leqslant \zeta_m(x, y) & \forall m \in \{1, \ldots, M\}\ \forall (x, y) \in \mathfrak{A} \setminus \mathfrak{B}_m \quad (1.a) \\ \gamma_m(x, y) \leqslant \Theta_m(x, y) \leqslant 0 & \forall m \in \{1, \ldots, M\}\ \forall (x, y) \in \mathfrak{B}_m \quad (1.b) \\ \displaystyle\sum_{m=1}^{M} \langle q_m(x, y) | e_i \rangle^+ \leqslant K_i(x, y) & \forall i \in \{1, 2, 3, 4\} \quad \forall (x, y) \in \mathfrak{A} \quad (1.c) \\ \mathrm{div}\, q_m(x, y) = \Theta_m(x, y) & \forall m \in \{1, \ldots, M\}\ \forall (x, y) \in \mathfrak{A} \quad (1.d) \end{cases}$$

It is relevant to see that we have quit the fifth constraint from the set we call X_0. Indeed, we are going to *incorporate* this constraint directly in the functional space of the flow q, the Hilbert $(\mathbf{H}_0(\mathrm{div}))^M$.

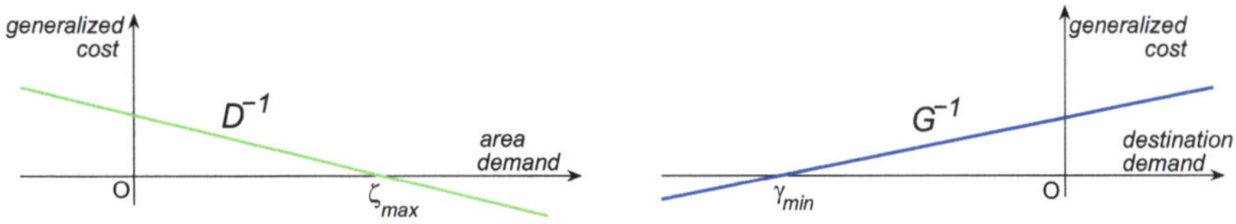

Fig. 5 Area and destination demand functions for the mathematical study

6.2 A Beckmann-Like Objective Function

A city commuter wants to reach his destination. The problem of user equilibrium in a dense and large urban area with an orthotropic network viewed as a continuum can be formulated in analogy to Beckmann's model for graphs [1]. We will prove that the obtained traffic assignment obeys Wardrop's first principle [2]. We denote $\mathscr{J}_0 : (\mathbf{H}_0(\text{div}))^M \times (\mathbf{L}^2)^M \to \mathbb{R}$ the function defined by

$$\begin{aligned}\mathscr{J}_0(q,\Theta) = &\iint\limits_{\mathfrak{A}} \sum_{i=1}^{4} \int_0^{\sum_{m=1}^{M}\langle q_m(x,y)|e_i\rangle^+} c_i(x,y,\xi)\mathrm{d}\xi \; \mathrm{d}x\mathrm{d}y \\ &- \sum_{m=1}^{M} \Bigg[\iint\limits_{\mathfrak{A}\backslash\mathfrak{B}_m} \int_0^{\Theta_m(x,y)} \mathcal{D}^{-1}_{m,(x,y)}(\vartheta)\mathrm{d}\vartheta \; \mathrm{d}x\mathrm{d}y \\ &+ \iint\limits_{\mathfrak{B}_m} \int_{\Theta_m(x,y)}^{0} \mathcal{G}^{-1}_{m,(x,y)}(\vartheta)\mathrm{d}\vartheta \; \mathrm{d}x\mathrm{d}y \Bigg]\end{aligned} \tag{1}$$

Here we can remark that the functions $\mathcal{D}_m^{-1}$ we chose in Fig. 2 do not fit. We will use in place of them the functions of the Fig. 5. Such two inverse demand functions do not have any physical sense, they are just a way to build the function $\mathscr{J}_0$ on its definition domain $(\mathbf{H}_0(\text{div}))^M \times (\mathbf{L}^2)^M$, from a mathematical point of view.

6.3 Minimization Problem

The minimization problem is $\mathcal{P}$:

$$\min_{(q,\Theta)\in X_0} \mathscr{J}_0(q,\Theta) \tag{2}$$

It is relevant at this point to explain why the minimum we are looking for exists. The function to minimize is convex. Moreover, the couples (q,Θ) are in the non-empty (it contains $(0,0)$) closed convex set X_0 defined by the constraints. The set X_0 is closed and it is bounded ($\forall (q,\Theta) \in X_0$, $\|(q,\Theta)\|^2_{(\mathbf{H}_0(\text{div}))^M \times (\mathbf{L}^2)^M} \leqslant 2\sum_{m=1}^{M}(\|\zeta_m \mathbf{1}_{\mathfrak{A}\setminus\mathfrak{B}_m} + \gamma_m \mathbf{1}_{\mathfrak{B}_m}\|^2_{L^2}) + 2M\sum_{i=1}^{4}\|K_i\|^2_{L^2}$). Hence, generic properties of separable Hilbert spaces give us the existence of the minimum.

7 Two Important Properties of the Minimizers

We note $\mathscr{J}_0(q,\Theta) = \Gamma(q) + \Upsilon(\Theta)$, with

$$\Gamma(q) = \iint\limits_{\mathfrak{A}} \sum_{i=1}^{4} \int_0^{\sum_{m=1}^{M}\langle q_m(x,y)|e_i\rangle^+} c_i(x,y,\xi)\mathrm{d}\xi \;\mathrm{d}x\mathrm{d}y$$

and

$$\Upsilon(\Theta) = -\sum_{m=1}^{M}\Bigg[\iint\limits_{\mathfrak{A}\setminus\mathfrak{B}_m}\int_0^{\Theta_m(x,y)} \mathcal{D}^{-1}_{m,(x,y)}(\vartheta)\mathrm{d}\vartheta\;\mathrm{d}x\mathrm{d}y + \iint\limits_{\mathfrak{B}_m}\int_{\Theta_m(x,y)}^{0} \mathcal{G}^{-1}_{m,(x,y)}(\vartheta)\mathrm{d}\vartheta\;\mathrm{d}x\mathrm{d}y\Bigg].$$

If f is an increasing (respectively strictly increasing) function, then $u \mapsto \int_0^u f$ is convex (resp. strictly convex). By using this fact, it is easy to see that the function Υ is strictly convex; and that the function Γ is convex (but not strictly convex as the *Ramp function* is only convex). Hence, if we have two minimizers (q^*,Θ^*) and $(\bar{q},\bar{\Theta})$, then $\Theta^* = \bar{\Theta}$ almost everywhere on $\mathfrak{A}$, and for every $i \in \{1,2,3,4\}$, $\sum_{m=1}^{M}\langle q^*_m(x,y)|e_i\rangle^+ = \sum_{m=1}^{M}\langle \bar{q}_m(x,y)|e_i\rangle^+$ almost everywhere on $\mathfrak{A}$.

8 Approximation of the Equilibrium with a Smooth Formulation

8.1 *Smooth Functions and Mere Properties*

As we have yet noticed it, the *Ramp function* is not differentiable. It only has a distributional derivative, called the *Heaviside step function*, but this type of derivative is too weak for our problem. Let's fix $\epsilon > 0$. For every $x \in \mathbb{R}$ we denote $r_\epsilon(x) = \frac{1}{2}\big(x + (x^2+\epsilon^2)^{\frac{1}{2}}\big)$, a smooth approximation of the *Ramp function*. The parameter $\epsilon > 0$ is bound to tend towards 0 (see Fig. 6). We note $h_\epsilon = r'_\epsilon$.

Fig. 6 Ramp function R and its approximation r_ϵ

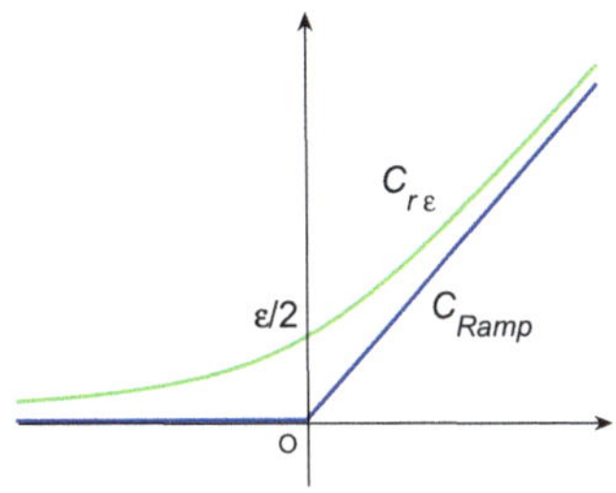

8.2 *The New Problem*

For every $\epsilon > 0$, we denote $\mathscr{J}_\epsilon : (\mathbf{H}(\mathrm{div}))^M \times (\mathbf{L}^2)^M \to \mathbb{R}$ the function defined by:

$$\begin{aligned}\mathscr{J}_\epsilon(q,\Theta) = &\iint_{\mathfrak{A}} \sum_{i=1}^{4} \int_0^{\sum_{m=1}^{M} r_\epsilon(\langle q_m(x,y)|e_i\rangle)} c_i(x,y,\xi)\mathrm{d}\xi \; \mathrm{d}x\mathrm{d}y \\ &- \sum_{m=1}^{M} \Bigg[\iint_{\mathfrak{A}\setminus\mathfrak{B}_m} \int_0^{\Theta_m(x,y)} \mathcal{D}^{-1}_{m,(x,y)}(\vartheta)\mathrm{d}\vartheta \; \mathrm{d}x\mathrm{d}y \\ &+ \iint_{\mathfrak{B}_m} \int_{\Theta_m(x,y)}^{0} \mathcal{G}^{-1}_{m,(x,y)}(\vartheta)\mathrm{d}\vartheta \; \mathrm{d}x\mathrm{d}y \Bigg] \end{aligned} \tag{3}$$

The function $\mathscr{J}_\epsilon$ is an approximation of $\mathscr{J}_0$ when ϵ is near 0. The minimization approximated problem is $\mathcal{P}_\epsilon$:

$$\min_{(q,\Theta)\in X_\epsilon} \mathscr{J}_\epsilon(q,\Theta) \tag{4}$$

where X_ϵ is the subset of $(\mathbf{H}(\mathrm{div}))^M \times (\mathbf{L}^2)^M$ defined by (only the inequation (1.c) is changed into (3.c), the others remaining the same):

$$\begin{cases} 0 \leqslant \Theta_m(x,y) \leqslant \zeta_m(x,y) & \forall m \in \{1,\dots,M\} \;\; \forall (x,y) \in \mathfrak{A}\setminus\mathfrak{B}_m \quad (3.a) \\ \gamma_m(x,y) \leqslant \Theta_m(x,y) \leqslant 0 & \forall m \in \{1,\dots,M\} \;\; \forall (x,y) \in \mathfrak{B}_m \quad (3.b) \\ \sum_{m=1}^{M} r_\epsilon(\langle q_m(x,y)|e_i\rangle) \leqslant K_i(x,y) + M\frac{\epsilon}{2} & \forall i \in \{1,2,3,4\} \quad \forall (x,y) \in \mathfrak{A} \quad (3.c) \\ \mathrm{div}\; q_m(x,y) = \Theta_m(x,y) & \forall m \in \{1,\dots,M\} \;\; \forall (x,y) \in \mathfrak{A} \quad (3.d) \end{cases}$$

Let us remark that the notations are consistent. We can take $\epsilon = 0$ in the expression of $\mathscr{J}_\epsilon$ and X_ϵ to have $\mathscr{J}_0$ and X_0. The main fact is that for all $\epsilon > 0$ fixed, the functional $\mathscr{J}_\epsilon$ is strictly convex and smooth on $(\mathbf{H}_0(\mathrm{div}))^M \times (\mathbf{L}^2)^M$. We note $(q_\epsilon, \Theta_\epsilon)$ its unique minimum on X_ϵ. It can be proved that the sequence

$(q_\epsilon, \Theta_\epsilon)$ is a minimizing sequence of $\mathcal{J}_0$ in this sense: if we fix $(q^*, \Theta^*) \in X_0$ such that $\min_{X_0} \mathcal{J}_0 = \mathcal{J}_0(q^*, \Theta^*)$, then $\lim_{\epsilon \to 0+} \mathcal{J}_0(q_\epsilon, \Theta_\epsilon) = \mathcal{J}_0(q_*, \Theta_*)$.

Let us define $\mathcal{F}_\epsilon$ as the representative constraints functional: for every $(q, \Theta) \in (\mathbf{H}(\mathrm{div}))^M \times (\mathbf{L}^2)^M$,

$$\mathcal{F}_\epsilon(q, \Theta) = (\mathcal{F}_\epsilon^1(\Theta), \mathcal{F}_\epsilon^2(\Theta), \mathcal{F}_\epsilon^3(\Theta), \mathcal{F}_\epsilon^4(\Theta), \mathcal{F}_\epsilon^5(q), \mathcal{F}_\epsilon^6(q, \Theta))$$

where

$$\mathcal{F}_\epsilon^1(\Theta) = -\Theta, \quad \mathcal{F}_\epsilon^2(\Theta) = \Theta - \zeta, \quad \mathcal{F}_\epsilon^3(\Theta) = \gamma - \Theta, \quad \mathcal{F}_\epsilon^4(\Theta) = \Theta,$$

$$\mathcal{F}_\epsilon^5(q) = (x, y) \mapsto \Big(\sum_{m=1}^{M} r_\epsilon(\langle q_m(x, y)|e_i\rangle) - K_i(x, y) - M\frac{\epsilon}{2}\Big)_i,$$

$$\mathcal{F}_\epsilon^6(q, \Theta) = \mathrm{div}\, q - \Theta.$$

9 Lagrangian Formulation

Then, for every $(q, \Theta, \phi, \varphi, \chi, \psi, \lambda, \mu) \in (\mathbf{H}_0(\mathrm{div}))^M \times (\mathbf{L}^2)^M \times \Pi_{l=1}^M \mathbf{L}^2(\mathfrak{A} \setminus \mathfrak{B}_m) \times \Pi_{l=1}^M \mathbf{L}^2(\mathfrak{A} \setminus \mathfrak{B}_m) \times \Pi_{l=1}^M \mathbf{L}^2(\mathfrak{B}_m) \times \Pi_{l=1}^M \mathbf{L}^2(\mathfrak{B}_m) \times (\mathbf{L}^2)^4 \times (\mathbf{H}^1)^M$ we consider the Lagrangian of the problem:

$$\mathcal{L}_\epsilon(q, \Theta, \phi, \varphi, \chi, \psi, \lambda, \mu) = \mathcal{J}_\epsilon(q, \Theta) + \Big\langle \phi, \varphi, \chi, \psi, \lambda, \mu \Big| \mathcal{F}_\epsilon(q, \Theta) \Big\rangle \tag{5}$$

where $\langle | \rangle$ is the scalar product in $\Pi_{l=1}^M \mathbf{L}^2(\mathfrak{A} \setminus \mathfrak{B}_m) \times \Pi_{l=1}^M \mathbf{L}^2(\mathfrak{A} \setminus \mathfrak{B}_m) \times \Pi_{l=1}^M \mathbf{L}^2(\mathfrak{B}_m) \times \Pi_{l=1}^M \mathbf{L}^2(\mathfrak{B}_m) \times (\mathbf{L}^2)^4 \times (\mathbf{L}^2)^M$.

9.1 *Saddle Point of the Lagrangian*

The nine-uplet $(q_s, \Theta_s, \phi_s, \varphi_s, \chi_s, \psi_s, \lambda_s, \mu_s)$ such that $(\phi_s, \varphi_s, \chi_s, \psi_s, \lambda_s) \geqslant 0$ is said to be a saddle point of the lagrangian $\mathcal{L}_\varepsilon$ if for all $(q, \Theta, \phi, \varphi, \chi, \psi, \lambda, \mu)$ such that $(\phi, \varphi, \chi, \psi, \lambda) \geqslant 0$:

$$\begin{aligned}\mathcal{L}_\epsilon(q_s, \Theta_s, \phi, \varphi, \chi, \psi, \lambda, \mu) &\leqslant \mathcal{L}_\epsilon(q_s, \Theta_s, \phi_s, \varphi_s, \chi_s, \psi_s, \lambda_s, \mu_s)\\ &\leqslant \mathcal{L}_\epsilon(q, \Theta, \phi_s, \varphi_s, \chi_s, \psi_s, \lambda_s, \mu_s).\end{aligned}$$

If $(q_s, \Theta_s, \phi_s, \varphi_s, \chi_s, \psi_s, \lambda_s, \mu_s)$ is a saddle point of the Lagrangian $\mathcal{L}_\epsilon$, as $(q_s, \Theta_s) \in X_\epsilon$ and is a global minimum of $\mathcal{J}_\epsilon$ on X_ϵ, then $(q_s, \Theta_s) = (q_\epsilon, \Theta_\epsilon)$ (see e.g. [9, 10]). It is easy to show that if $(q_\epsilon, \Theta_\epsilon, \phi_\epsilon, \varphi_\epsilon, \chi_\epsilon, \psi_\epsilon, \lambda_\epsilon, \mu_\epsilon)$ is a saddle point of the Lagrangian $\mathcal{L}_\epsilon$, then it confirms the four relations (3.a)–(3.d).

9.2 Relations Between the Primal and Dual Variables at a Saddle Point

As the functions which represent the contraints are convex and differentiable, we have:

$$\forall(\hat{q},\hat{\Theta}),\ D_{(q,\Theta)}\mathcal{L}_\epsilon\big((q_\epsilon,\Theta_\epsilon,\phi_\epsilon,\varphi_\epsilon,\chi_\epsilon,\psi_\epsilon,\lambda_\epsilon,\mu_\epsilon,\nu_\epsilon)\mid(\hat{q},\hat{\Theta})\big)=0,$$

with

$$D_{(q,\Theta)}\mathcal{L}_\epsilon\big((q,\Theta,\phi,\varphi,\chi,\psi,\lambda,\mu,\nu)\mid(\hat{q},\hat{\Theta})\big)=$$

$$\iint\limits_{\mathfrak{A}}\sum_{m=1}^{M}\left\langle\hat{q}_m\middle|+\sum_{i=1}^{4}\left[\lambda_i+c_i\Big(x,y,\sum_{m=1}^{M}r_\epsilon(\langle q_m|e_i\rangle)\Big)\right]h_\epsilon(\langle q_m|e_i\rangle)e_i-\nabla\mu_m\right\rangle\mathrm{d}x\mathrm{d}y$$

$$+\iint\limits_{\mathfrak{A}}\sum_{m=1}^{M}\hat{\Theta}_m\Bigg[\Big(-\mathcal{D}_m^{-1}(\Theta_m)-\phi_m+\varphi_m-\mu_m\Big)\mathbf{1}_{\mathfrak{A}\setminus\mathfrak{B}_m}$$

$$+\Big(\mathcal{G}_m^{-1}(\Theta_m)-\chi_m+\psi_m-\mu_m\Big)\mathbf{1}_{\mathfrak{B}_m}\Bigg]\mathrm{d}x\mathrm{d}y. \tag{6}$$

The existence of a saddle point is a difficult problem in the case of general Hilbert spaces with convex constraints. It is possible to find a saddle point in an approximated sense if contraints are linear [11], but the mathematical techniques involved are far away from our article, and we will suppose the existence of such a saddle point. Supposing the existence of a saddle point, its uniqueness would also be a very difficult problem. In finite dimension, the uniqueness of the Lagrange multipliers comes from properties called *constraint qualification* which are not easily adaptable to the infinite dimension. Anyway, we will content ourselves with the discretized problem and its numerical solution.

Now let us fix m. For

$$\hat{q}_m=\sum_{i=1}^{4}\left[\lambda_{\epsilon,i}+c_i\Big(x,y,\sum_{m=1}^{M}r_\epsilon(\langle q_{\epsilon,m}|e_i\rangle)\Big)\right]h_\epsilon(\langle q_{\epsilon,m}|e_i\rangle)e_i-\nabla\mu_{\epsilon,m},$$

$$\hat{q}_l=\mathbf{0}\ \forall l\neq m \text{ and } \hat{\Theta}=0$$

(to avoid the integral on the boundary $\partial\mathfrak{A}$ anyway, we can use a sequence $\hat{q}_m^{(n)}\in\mathcal{C}_c^\infty$ that converges to this $\hat{q}_m$ with the topology of the norm $||\cdot||_{\mathbf{H}^1\times\mathbf{H}^1}$) we obtain that:

$$\forall m,\ \sum_{i=1}^{4}\left[\lambda_{\epsilon,i}+c_i\Big(x,y,\sum_{m=1}^{M}r_\epsilon(\langle q_{\epsilon,m}|e_i\rangle)\Big)\right]h_\epsilon(\langle q_{\epsilon,m}|e_i\rangle)e_i=\nabla\mu_{\epsilon,m}. \tag{7}$$

With the same types of arguments, we deduce that for all m:

$$-\mathcal{D}_m^{-1}(\Theta_{\epsilon,m}) - \phi_{\epsilon,m} + \varphi_{\epsilon,m} - \mu_{\epsilon,m} = 0 \text{ on } \mathfrak{A} \setminus \mathfrak{B}_m, \tag{8}$$

$$\mathcal{G}_m^{-1}(\Theta_{\epsilon,m}) - \chi_{\epsilon,m} + \psi_{\epsilon,m} - \mu_{\epsilon,m} = 0 \text{ on } \mathfrak{B}_m. \tag{9}$$

10 Used/Unused Paths Costs

The *sign function* is defined by $\mathrm{sgn}(a) = -1$ for every $a < 0$, $\mathrm{sgn}(0) = 0$ and $\mathrm{sgn}(a) = 1$ for every $a > 0$. First, we are going to prove that all used paths between an origin O and a destination D_m have the same cost. The cost of travel on such a path is:

$$\begin{aligned}\mathrm{Cost}(\text{used } p, m) = & \int_p \Big[\lambda_{\epsilon,1} + c_1\big(x, y, \sum_{m=1}^{M} r_\epsilon(\langle q_m | e_1\rangle)\big)\Big] h_\epsilon(\langle q_{\epsilon,m} | e_1\rangle)|\mathrm{d}x| \\ & + \int_p \Big[\lambda_{\epsilon,3} + c_3\big(x, y, \sum_{m=1}^{M} r_\epsilon(\langle q_m | e_3\rangle)\big)\Big] h_\epsilon(\langle q_{\epsilon m} | e_3\rangle)|\mathrm{d}x| \\ & + \int_p \Big[\lambda_{\epsilon,2} + c_2\big(x, y, \sum_{m=1}^{M} r_\epsilon(\langle q_m | e_2\rangle)\big)\Big] h_\epsilon(\langle q_{\epsilon m} | e_2\rangle)|\mathrm{d}y| \\ & + \int_p \Big[\lambda_{\epsilon,4} + c_4\big(x, y, \sum_{m=1}^{M} r_\epsilon(\langle q_m | e_4\rangle)\big)\Big] h_\epsilon(\langle q_{\epsilon m} | e_4\rangle)|\mathrm{d}y|. \end{aligned} \tag{10}$$

In fact, with the functions r_ϵ and h_ϵ, the cost of small coordinates of the flow are underestimated. By noticing that at any (x, y) on the used path p we have $|\mathrm{d}x| = \mathrm{sgn}(\langle q_{\epsilon,m}(x, y) | e_1\rangle)\mathrm{d}x$ and $|\mathrm{d}y| = \mathrm{sgn}(\langle q_{\epsilon,m}(x, y) | e_2\rangle)\mathrm{d}y$, we can write the previous cost as:

$$\begin{aligned}\mathrm{Cost}(\text{used } p, m) = & \int_p \Big[\lambda_{\epsilon,1} + c_1\big(x, y, \sum_{m=1}^{M} r_\epsilon(\langle q_m | e_1\rangle)\big)\Big] h_\epsilon(\langle q_{\epsilon m} | e_1\rangle)\mathrm{d}x \\ & - \int_p \Big[\lambda_{\epsilon,3} + c_3\big(x, y, \sum_{m=1}^{M} r_\epsilon(\langle q_m | e_3\rangle)\big)\Big] h_\epsilon(\langle q_{\epsilon m} | e_3\rangle)\mathrm{d}x \\ & + \int_p \Big[\lambda_{\epsilon,2} + c_2\big(x, y, \sum_{m=1}^{M} r_\epsilon(\langle q_m | e_2\rangle)\big)\Big] h_\epsilon(\langle q_{\epsilon m} | e_2\rangle)\mathrm{d}y \\ & - \int_p \Big[\lambda_{\epsilon,4} + c_4\big(x, y, \sum_{m=1}^{M} r_\epsilon(\langle q_m | e_4\rangle)\big)\Big] h_\epsilon(\langle q_{\epsilon m} | e_4\rangle)\mathrm{d}y. \end{aligned} \tag{11}$$

Then,

$$\text{Cost}(\text{used } p, m) = \int_p \left\langle \nabla \mu_{\epsilon,m} \,\middle|\, \begin{matrix} \mathrm{d}x \\ \mathrm{d}y \end{matrix} \right\rangle = \mu_{\epsilon,m}(D_m) - \mu_{\epsilon,m}(O) \tag{12}$$

We have obtained that for each commodity, the cost of the used paths have the same cost for the commuters.

Second, we are going to prove that the unused paths between the origin O and the destination D_m are more expensive than the used ones. For any unused path p, since $\langle q_{\epsilon,m}(x,y)|e_1\rangle e_1$ can be equal to the zero vector, or be parallel in the opposite sense with the path p, at any (x,y) of the unused path p, we have $|\mathrm{d}x| \geqslant \mathrm{sgn}(\langle q_{\epsilon,m}(x,y)|e_1\rangle)\mathrm{d}x$, and by the same token, $|\mathrm{d}y| \geqslant \mathrm{sgn}(\langle q_{\epsilon,m}(x,y)|e_2\rangle)\mathrm{d}y$. Thus, if we repeat the last calculus, in place of the equality, we have the following inequality:

$$\text{Cost}(\text{unused } p, m) \geqslant \int_p \left\langle \nabla \mu_{\epsilon,m} \,\middle|\, \begin{matrix} \mathrm{d}x \\ \mathrm{d}y \end{matrix} \right\rangle = \mu_{\epsilon,m}(D_m) - \mu_{\epsilon,m}(O) \tag{13}$$

We conclude that the cost of any unused path is higher than or equal to the cost of the used path. This satisfies the Wardrop's first principle of user equilibrium.

11 Algorithmical Aspects

The Uzawa Algorithm 1 computes the traffic assignment problem. It uses the Newton-Raphson-Armijo Algorithm 2 at line 7. The approximated problem we studied has one major problem: the objective function is only strictly convex. This property is not strong enough to run the classical algorithm of Uzawa of non linear programming. We will show a convenient method to obtain the property of strong convexity, but first, we are going to eliminate some dual coefficients from the algorithm, their calculus being useless.

11.1 How to Eliminate Some Coefficients

We are going to *eliminate* the dual variables ϕ, φ, χ, ψ. This result has a computational interest as far as the functions ϕ, φ, χ, ψ have no physical interest. If u and v are two $\mathbf{L}^2$ functions with $u \leqslant v$, we note for every $f \in \mathbf{L}^2$:

$$\Pi_{[u,v]}(f) = u\mathbf{1}_{\{f<u\}} + f\mathbf{1}_{\{u\leqslant f\leqslant v\}} + v\mathbf{1}_{\{v<f\}}.$$

$\Pi_{[u,v]}(f)$ is the orthogonal projection of f on the closed convex subset of $\mathbf{L}^2$: $\{g \in \mathbf{L}^2 / u \leqslant g \leqslant v \text{ a.e.}\}$. Let us examine relation (8). Let us fix m and $(x, y) \in \mathfrak{A} \setminus \mathfrak{B}_m$. As $\Theta_{\epsilon,m}(x, y) \in [0, \zeta_m(x, y)]$ and as the quantities $\phi_{\epsilon,m}(x, y)$ and $\varphi_{\epsilon,m}(x, y)$ cannot be both non equal to zero, using that $\mathcal{D}_m$ is a strictly decreasing function, one can write: for all m, $\Theta_{\epsilon,m} = \Pi_{[0,\zeta_m]}\Big(\mathcal{D}_m\big(-\mu_{\epsilon,m}\big)\Big)$ on $\mathfrak{A} \setminus \mathfrak{B}_m$. As well, with relation (9), using that $\mathcal{G}_m^{-1}$ is strictly increasing, we could show that for all m, $\Theta_{\epsilon,m} = \Pi_{[\gamma_m,0]}\Big(\mathcal{G}_m\big(\mu_{\epsilon,m}\big)\Big)$, on $\mathfrak{B}_m$. So we can summarize the obtained result in the formula:

$$\Theta_{\epsilon,m} = \Pi_{[0,\zeta_m]}\Big(\mathcal{D}_m\big(-\mu_{\epsilon,m}\big)\Big)\mathbf{1}_{\mathfrak{A}\setminus\mathfrak{B}_m} + \Pi_{[\gamma_m,0]}\Big(\mathcal{G}_m\big(\mu_{\epsilon,m}\big)\Big)\mathbf{1}_{\mathfrak{B}_m}.$$

11.2 How to Get Strong Convexity?

The first crucial point in the algorithm of Uzawa is the strong convexity of the objective function. To do so, we add to $\mathscr{J}_\epsilon(q, \Theta)$ an artificial term $\kappa\|q\|_{\mathbf{H}(\mathrm{div})^M}$, where κ is a positive constant, sufficiently small in order to not disturb too much the objective function $\mathscr{J}_\epsilon$. We do not add anything in relation with Θ because the objective function is already strongly convex with regard to Θ, as a primitive of a strictly increasing affine functions. We can interpret this new approximated problem as the research of the minimizer of the objective function $\mathscr{J}_\epsilon$ which would have the smaller norm. Physically, it does not have any relevant interpretation, and this technique only has to be seen as a convenient way to have a strongly convex structure.

11.3 Lipschitz Properties of the Constraint Functions

The second crucial point in the algorithm of Uzawa is that the constraints functionals must be Lipschitz. Nevertheless, in our case, we only need the constraint functional associated to the capacity constraints to be Lipschitz. It is possible to prove that for every $i \in \{1, 2, 3, 4\}$, $(\mathscr{F}_\epsilon^5)_i$ is $\sqrt{M}$–Lip. This result proves that the four constraint functions associated to the four inequalities contained in (3.c) are $\sqrt{M}$–Lip.

We now have all the elements we need to implement the algorithm.

11.4 Uzawa Algorithm for Static Traffic Assignment

Algorithm 1 Uzawa algorithm

Require: the domain area; the M destinations (in the following, $m = 1, \dots, M$)
the capacity constraints functions on the four directions K_i $(i = 1, 2, 3, 4)$
the inverse demand functions $\mathcal{D}_m^{-1}$ and $\mathcal{G}_m^{-1}$ (which give us the functions γ_m and ζ_m)
the small positive constants ϵ, κ, error; μ_m^{Ini}, λ_m^{Ini}

1: $\rho \leftarrow 0.95 \times \frac{2\kappa}{M}$
2: $k \leftarrow 0$
3: for every m, $\mu_m^0 \leftarrow \mu_m^{\text{Ini}}$, $\lambda_m^0 \leftarrow \lambda_m^{\text{Ini}}$,
4: **while** $k = 0$ or $||\text{div}q^k - \Theta^k||_{(\mathbf{L}^2)^M} >$ error
or $||q^{k+1} - q^k||_{(\mathbf{H}(\text{div}))^M} >$ error or $||\Theta^{k+1} - \Theta^k||_{(\mathbf{L}^2)^M} >$ error **do**
5: Calculate for every m: $\mu_m^{k+1} \leftarrow \mu_m^k + \rho(\text{div}q_m^k - \Theta_m^k)$
$\Theta_m^{k+1} \leftarrow \Pi_{[0,\zeta_m]}\Big(\mathcal{D}_m\big(-\mu_m^{k+1}\big)\Big)\mathbf{1}_{\mathfrak{A}\setminus\mathfrak{B}_m} + \Pi_{[\gamma_m,0]}\Big(\mathcal{G}_m\big(\mu_m^{k+1}\big)\Big)\mathbf{1}_{\mathfrak{B}_m}$
6: Calculate for every i: $\lambda_i^{k+1} \leftarrow P_+\Big(\lambda_i^k + \rho\big(\sum_{m=1}^{M} r_\epsilon(\langle q_m | e_i\rangle) - K_i - M\frac{\epsilon}{2}\big)\Big)$
7: Solve with the algorithm of Newton-Raphson-Armijo:
$q^{k+1} = \text{Arg} \min_{q \in \mathbf{H}(\text{div})^M} \mathscr{L}_{\epsilon,\kappa}(q, \lambda^{k+1}, \mu^{k+1})$
8: $k \leftarrow k + 1$
9: **end while**
10: **return** the flux q_m^{Fin}, the demand functions Θ_m^{Fin}
the potential functions μ_m^{Fin}, the four over cost functions λ_i^{Fin}

Algorithm 2 Newton-Raphson algorithm with an Armijo step

Require: the strongly convex function J to minimize; the first step of the iteration u^{Ini}
the small positive constant: error; the proportion w in the Armijo step

1: $j \leftarrow 0$
2: $u^0 \leftarrow u^{\text{Ini}}$
3: **while** $||DJ(u^k)|| >$ error **do**
4: Solve the descent vector d^k by solving the problem with GMRes Algorithm:
$\forall v,\ D^2J(u^k)(d^k, v) + DJ(u^k)(v) = 0$
5: $\alpha \leftarrow 1$
6: **while** $J(u^k + \alpha d^k) > J(u^k) + w\alpha DJ(u^k)(d^k)$ **do**
7: $\alpha \leftarrow \frac{\alpha}{2}$
8: **end while**
9: $u^{k+1} \leftarrow u^k + \alpha d^k$
10: $k \leftarrow k + 1$
11: **end while**
12: **return** the minimizer u^{Fin}

---

12 Numerical Example

12.1 Example Values

The roads' capacity for each of the four directions is constant: $K = 2{,}000$ vehicles/length unit. The cost function at every point of coordinates (x, y) and for each one of the four directions is chosen as $c_i(x, y, \xi) = 0,04 \times (1 + \xi/K)$. The inverse demand functions on the area are chosen for example as $D_m^{-1}(x, y, \zeta) = -0,02 \times \zeta + 9$ (this one gives a maximum vehicle generation on the area equal to 450 vehicles/km^2). The inverse demand functions on the destinations are chosen for example as $G_m^{-1}(x, y, \zeta) = 0,02 \times \zeta + 10$ (this one gives a maximum absorption on the destination equal to 500 vehicles/km^2).

12.2 An Imaginary City

We propose an illustration of our model. The city we draw is an imaginary city which looks like Paris. We don't intend here to give a *truthfull* representation of traffic in Paris. We just want to explain some possibilities of our model, and how it works with mathematical entries. These mathematical entries should be replaced by engineering measures. Paris has some special interesting geometrical aspects: a ring road that allows a clear-cut frontier with 35 gates, and the Seine river that provides a natural obstacle crossed by bridges. As in the articles [6], we only consider a continuum network over the urban area. The main difference between our network and theirs is that we build an orthotropic network with horizontal and vertical roads characterized with side constraints whereas their network is isotropic without side constraint.

We represent eight possible destinations for the commuters who are spilled over the yellow area, and want to go back home in the suburbs of Paris, which are the colored triangles (see Fig. 7-left). On Fig. 7-right, one can see the meshing implemented by the software FreeFem++. The edges of the small triangles do not represent the network's roads, they are just used for the numerical program. More precisely, one has to imagine the network as an infinite collection of vertical and horizontal lines on the city area. Figure 8-left represents the solutions we obtain for a particular generation function Θ_m. Figure 8-right represents the traffic load for direction e_1 (west to east): $\sum_{m=1}^{M} \langle q_m | e_1 \rangle^+$. To obtain the traffic assignment, we will implement the mathematical program described in the previous chapter with the cost functions, the demand functions and the side constraints.

Comments on Fig. 8-Left

We have chosen to represent the north generation function. Same results take place for the seven other possible destinations. Vehicles are absorbed at north destination

Fig. 7 Chosen destinations on the Paris ring road and meshing

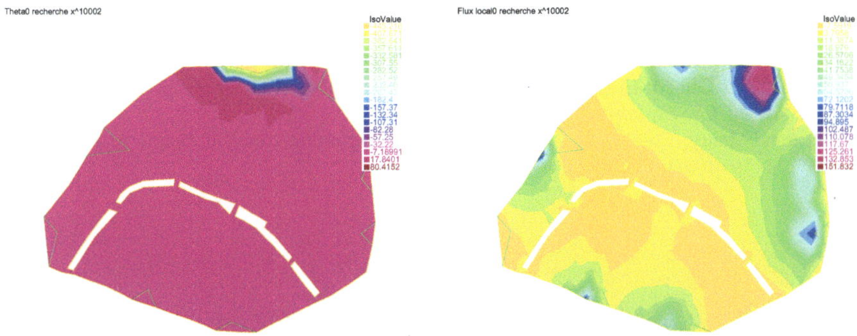

Fig. 8 Generation function for the north destination (*left*), traffic load in direction West to East (*right*)

(negative values on the destination area). They are generated on the rest of the area (positive values on the rest of the urban area). More precisely, the negative values near the destination are due to the smoothness of the demand function by the used software (FreeFEM++). The closer the vehicles are to a destination, the more they want to reach this destination.

Comments on Fig. 8-Right

We have chosen to represent the traffic load in the west to east direction. Same results take place for the three other possible directions. Figure 8-right has to be read from left to right, in the movement of the vehicles. Traffic load increases while coming close to the destinations, because vehicles pile up there, before entering the destination area.

References

1. M. Beckmann: A continuous model of transportation, *Econometrica* 20, 643 (1952)
2. John Glen Wardrop: Some theorical aspects of road traffic research, *Proceedings of the Institute of Civil Engineers*, Part II, Vol.1, 325–378 (1952)

3. Tönu Puu: Continuous economic space modelling, *The Annals of Regional Science* 43(1), 5–25 (2009)
4. Rodney James Vaughan: *Urban spatial traffic patterns* (Pion, London), 1987
5. Hai Yang, Sam Yagar and Yasunori Iida: Traffic assignment in congested discrete/continuous transportation system, *Transportation Research B* 28, 161–174 (1994)
6. S.C. Wong: Multi-commodity traffic assignement by continuum approximation of network flow with variable demand, *Transportation Research B* 32, 567–581 (1998)
7. Torbjörn Larsson and Michael Patriksson: An augmented lagrangean dual algorithm for link capacity side constrained traffic assignment problems, *Transportation Research B* 29, 433–455 (1995)
8. Azuma Taguchi and Masao Iri: Continuum approximation to dense networks and its application to the analysis of urban road networks, *Mathematical Programming Study* 20, 178–217 (1982)
9. Grégoire Allaire: *Analyse numérique et optimisation. Une introduction à la modélisation mathématique et à la simulation numérique*, Éditions de l'École Polytechnique (2005)
10. Jean Charles Gilbert: Éléments d'optimisation différentiable - Théorie et algorithmes, *Syllabus de cours à l'ENSTA, Paris* (2010)
11. Bertrand Maury: Continuous version of the Uzawa algorithm *Comptes Rendus Mathématique* 337, 31–36 (2003)

Route Advice Based on Subnetwork Properties

Victor L. Knoop, Hans van Lint, and Serge P. Hoogendoorn

Abstract Large scale routing strategies require many data, and high computational power if the basic information unit is small. This paper checks the effects of a routing strategy based on aggregated information, the average speed, of a subnetwork. In a grid-network this routing is compared with shortest-distance routing, and data-demanding shortest-time routing. Contrary to the fixed routing, the proposed average-speed routing algorithm can avoid congestion quite long time, and it still needs very few data. Furthermore, it is less sensitive to fluctuations than the control based on speeds on all links. This is a good starting point for further control based on the macroscopic fundamental diagram.

1 Introduction

Freeway traffic control have focussed on central approaches in the past decades. Only recently, more spatial spread control algorithms are proposed [1, 2]. To find optimal control in a centralized approach, one needs many data. Alternatively, one can introduce multi-level control, limiting the amount of information needed by omitting information at the lower level in the higher level optimization.

Theoretically, the information needs may be even very light: the network state can possibly be expressed in two parameters, the accumulation and production of a traffic network. Geroliminis and Daganzo [3] show that these variables even are connected to each other, by the macroscopic fundamental diagram (MFD) or network fundamental diagram. Main drawback is the homogeneity over the network which is needed. Buisson and Ladier [4] were the first to test the how the MFDs change if the congestion is not homogeneously distributed over the network. They

V.L. Knoop (✉) · H. van Lint · S.P. Hoogendoorn
Department of Transport & Planning, Delft University of Technology, Delft, The Netherlands
e-mail: v.l.knoop@tudelft.nl

V.V. Kozlov et al. (eds.), *Traffic and Granular Flow '11*,
DOI 10.1007/978-3-642-39669-4_36, © Springer-Verlag Berlin Heidelberg 2013

showed a reasonably good MFD for the French town Toulouse in normal conditions. However, 1 day strikes of truck drivers considerably impacted the traffic conditions, and the MFD changed. For more details how inhomogeneity within the network affects the MFD, see [5, 6].

The MFD can represent the traffic in an aggregated form, but up to now it is unclear whether control can be applied based on aggregated data. Therefore, in this paper we explore how alternative but still "light" information may be utilised for traffic control, and in particular the effectiveness of routing. We do so by simulating a network and applying different control strategies. The experimental set-up is given in the next section, followed by the routing strategies in Sect. 3. Section 4 presents the results and the paper concludes by the conclusions and the discussion.

2 Model

This section discusses the simulation set-up. It first describes what will be simulated in terms of network and demands. Then, Sect. 2.2 describes the model used for this simulation. Section 2.3 describes the output of the simulator that is used later in the paper. The routing is discussed separately in Sect. 3.

2.1 *Experimental Settings*

In the paper an urban network is simulated, since this is the main area where MFDs have been tested. We follow Geroliminis and Ji [7] and choose a Manhattan network with periodic boundary conditions. This means that the nodes are located at a regular grid, for which we choose a 16×16 size. Then, one-way links connect these nodes. The direction of the links changes from block to block, i.e. if at $x = 2$ the traffic is allowed to drive in the positive y direction, at $x = 1$ and at $x = 3$ there are one-way roads for traffic to drive in the negative y direction. We assume 2 lanes per link, a 1 km block length, a triangular fundamental diagram with a free speed of 60 km/h, a capacity of 1,500 veh/h/lane and a jam density of 150 veh/km/lane.

Furthermore, periodic boundary conditions are used, meaning that a link will not end at the edge of the network. Instead, it will continue over the edge at the other side of the network. An example of a smaller grid network with periodic boundary conditions is given in Fig. 1a. Traffic can continue in a direct link from node 13 to node 1 or from node 5 to node 8. This way, all nodes have two incoming and to outgoing links and network boundaries have no effect.

The destinations are randomly chosen from all points in the network. In the network, there are 15 nodes chosen as destination nodes. There are no origin nodes. Instead, at the beginning of the simulation, traffic is put on the links. Vehicles are assigned to a destination, and for this distribution is equal over all destinations.

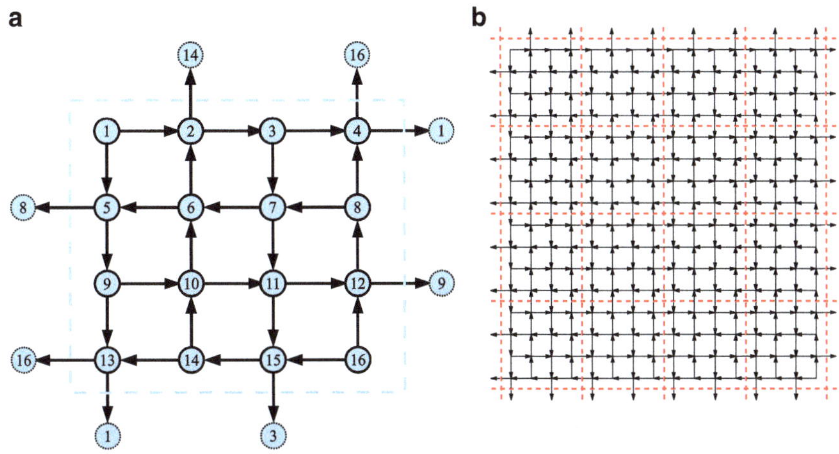

Fig. 1 A road network with periodic boundaries, and its division into subnetworks. (**a**) Illustration of a 4×4 grid network with periodic boundary conditions. (**b**) A 16×16 network with a 4-block subnetwork

When the cars have reached their destination, they will not leave the network, but instead they are assigned a new destination. We use a macroscopic model (see Sect. 2.2), hence we can split the flow of arriving traffic equally over the 14 other destinations. The number of cars in the network is hence constant. This number will be a parameter setting for the simulations, but throughout one simulation, it is constant. The demand level is expressed as the density on all links at the start of the simulation, as fraction of the critical density. Figure 2a shows the network used under initial conditions.

2.2 *Traffic Flow Simulation*

This section describes the traffic flow model. For the traffic flow modelling we use a first order traffic model. Links are split into cells with a length of 250 m (i.e., 4 cells per link). We use the continuum LWR-model [8, 9] that we solve with a Godunov scheme [10]. Lebacque [11] showed how this is used for traffic flows, yielding a deterministic continuum traffic flow simulation model. The flux from one node to the next is basically restricted by either the demand from the upstream node (free flow) or by the supply from the downstream node (congestion):

$$\phi_{c,c+1} = \min\{D_c, S_{c+1}\}; \tag{1}$$

At a node r we have inlinks, denoted by i which lead the traffic towards node r and outlinks, denoted by j which lead the traffic away from r. At each node r, the demand D to each of the outlinks of the nodes is calculated, and all demand to one link from all inlinks is added. This is compared with the supply S of the cell in the outlink. In case this is insufficient, a factor, α, is calculated which show which part of the demand can continue.

$$\alpha_r = \underset{[j \text{ leading away from } r]}{\text{argmin}} \left\{ \frac{S_j}{D_j} \right\} \tag{2}$$

This is the model developed by Jin and Zhang [12]. They propose that all demands towards the node are multiplied with the factor α, which gives the flow over the node.

This node model is slightly adapted for the case at hand here. Also the node *itself* can restrict the capacity. In our case, there are two links with a capacity of 3,000 veh/h as inlinks and two links with a capacity of 3,000 veh/h as outlinks. Since there are crossing flows, it is not possible to have a flow of 3,000 veh/h in one direction *and* a flow of 3,000 veh/h in the other direction. To overcome this problem, we introduce a node capacity (see also for instance [13]). The node capacity is the maximum of the capacities of the outgoing links. This means that in our network, at maximum 3,000 veh/h can travel over a node. Again, the fraction of the traffic which can continue over node r is calculated, indicated by β:

$$\beta_r = \frac{C_r}{\sum_{\forall i \text{ to } r} D_i} \tag{3}$$

The demand factor γ is now the minimum of the demand factor calculated by the nodes and the demand factor due to the supply:

$$\gamma = \min \{\alpha_r, \beta_r, 1\} \tag{4}$$

Similar to [12], we take this as multiplicative factor for all demands to get to the flux ϕ_{ij}, i.e. the number of cars from one cell to the next over the node:

$$\phi_{ij} = \gamma D_{ij} \tag{5}$$

2.3 Variables

In this paper, several traffic flow variables will be used. In this section we will explain them and show the way to calculate them

Standard traffic flow variables are flow, q, being the vehicle distance covered in a unit of time, and density, k, the number of vehicles per unit road length. The network is divided into cells, which we denote by c, which have a length L_c. Flow and density in cells are denoted by q_c and k_c.

Table 1 Characteristics of the routing strategies used

Characteristic	1 – Fixed	2 – Speed-based	3 – Subnetwork based
Routing type	Destination-specific, node specific split fractions		
Update frequency	fixed	15 min	15 min
Basis	Distance	Time	distance/(subnetwork speed)
Model	Analytical	Probit, 3 draws	Probit, 3 draws
Compliance	100 %	50 % per round	50 % per round

Furthermore, the accumulation N in an area X is the weighted average density:

$$N_X = \sum_{c \in X} \frac{k_c * L_c}{L_c} \tag{6}$$

Similarly, the production P in an area X is the weighted average flow:

$$P_X = \sum_{c \in X} \frac{q_c * L_c}{L_c} \tag{7}$$

Since the cell length are the same for all links in the network, the accumulation and production are average densities and flows. Recall that there is a strong relationship between the production and the number of completed trips[3].

3 Control Strategies

There are three routing scenarios considered in this paper:

1. Fixed routing
2. Speed-based routing
3. Subnetwork speed based routing

Details of these strategies follow below; a summary of the characteristics of the strategies is found in Table 1.

For the initial situation, the route choice is determined based based on distance to the destination. Traffic will take the shortest route towards the destination. For intersections where both directions will give the same path length towards a destination, the split of traffic to that direction is 50–50. Note that for the initial conditions the distances are proportional to the times, since traffic is loaded in at under-critical conditions and the traffic is at free flow speeds at the whole network.

For the case with fixed routing, the initial routes are used throughout the whole simulation period. Routing strategies 2 and 3 are adaptive strategies which vary with the travel times in the network. Each 15 min there is a route update. Both are based on a probit assignment [14, 15]. The strategies differ in the utility function which is used as basis for the probit assignment.

For strategies 2 and 3 there is dynamic information which is used for the adaptive routes. In strategy 2, the routes are determined based on the speed on the links. Strategy 3 uses the average speed in a subnetwork as representative for the speed of all links in the subnetwork. The subnetwork chosen here are 4×4 nodes, such that there are $4 \times 4 = 16$ subnetworks in total (see Fig. 1b).

The travel times which result from these interpretations are disturbed with an error of 10 % to mimic user interpretation. For each node, the shortest path (in time) to each of the destinations is determined, indicated with $\tilde{\omega}^*$. On the node, this leads to a single decision: turn or straight, indicated with $\tilde{\Psi}^*$. This process is repeated three times, which gives three decisions for the route from node r to destination s. These all give a split (Ψ) at node r, which are averaged, which is denoted by Ψ^+:

$$\Psi_j^+ = \frac{\sum_\Pi \tilde{\Psi}^*}{\pi} \tag{8}$$

Note that all routing variables are destination specific, but the destination index is omitted for reasons of notational simplicity.

Then, this average split vector is averaged with the split vector in the previous time period.

$$\Psi_t^j = (1-\kappa)\Psi_{t-1} + \kappa\Psi_j^+ \tag{9}$$

4 Results

Figure 2 shows the network state for different routing strategies at different times. The initial state is the same for all routing strategies. This situation, with the vehicles distributed evenly, is depicted in Fig. 2a. In case of no routing, the congestion clusters more and more. The reason is that the flow in these areas is low, and the flow in the lower density area is high. Vehicles from the uncongested, or less strong congested areas, can move quicker and reach the area of heavy congestion, thus increasing the area of heavy congestion – see Fig. 2b.

With routing based on speeds, routing 2, the congestion is much more spread, as is depicted in Fig. 2c, d. However, in the situation with route advice, the traffic is adaptive and the vehicles will avoid the areas with the strong congestion. Therefore, there is not such a clustered area of congestion, but there are several local bottlenecks. These are "fed" by vehicles which are routed around the strong congestion. This situation does not change considerably after this spread congestion has set in, as the evolution from 1.5 (Fig. 2c) to 3 h (Fig. 2d) shows.

This level of control is only possible with detailed information of traffic speeds at each cell of each link. However, proper traffic control is possible with much less information, namely subnetwork aggregated information, as is shown by the

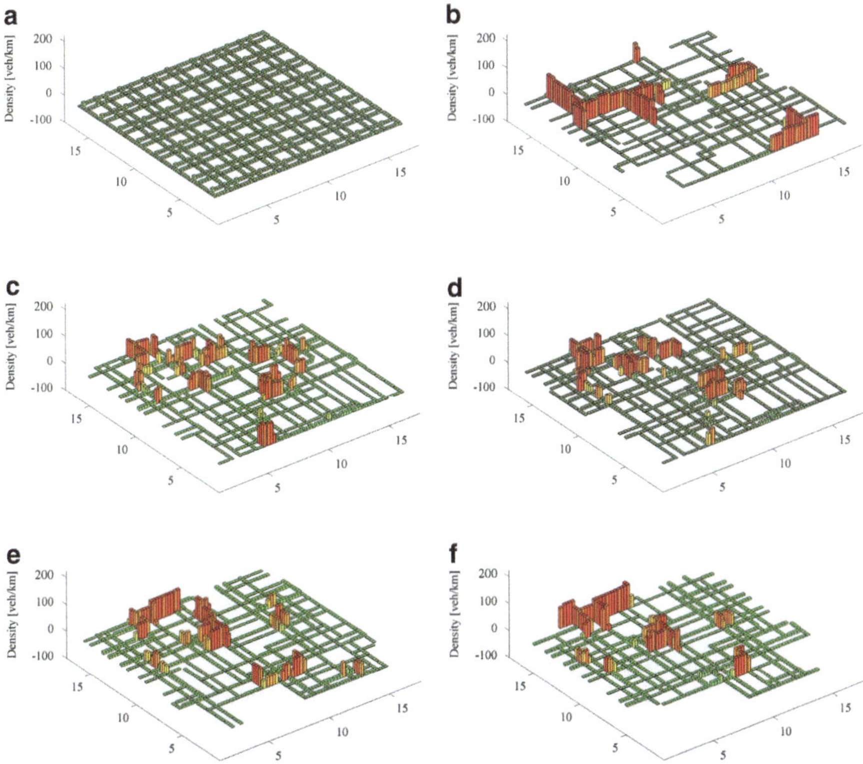

Fig. 2 Network evolutions. (**a**) Start. (**b**) End no routing. (**c**) 1.5 h speed routing. (**d**) End speed routing. (**e**) 1.5 h subnetwork routing. (**f**) End subnetwork routing

network states in Fig. 2e, f. Also here, the congestion is spread, meaning there are different bottlenecks active, and thus the production is the sum of the capacities of these bottlenecks.

Figure 3 shows the performance, i.e. the arrival rate, for the three routing strategies. It shows that the situation without routing degrades to a situation with very low performance quickly, and continously. With routing on a link level, this process is interrupted each time when a new advice is computed and communicated to the vehicles, every 15 min. This is visible in Fig. 3 by the decreasing performance for one route, followed by a sharp increase of performance every 15 min.

With routing based on subnetwork speeds, the control is almost as effective, but less sensitive to fluctuations. The performance remains on average slightly under the average level with control based on the full information, but there are less fluctuations. One reason is that the subnetwork average speeds, being the basis for control in this routing strategy, will not fluctuate quickly.

Another measure of performance might not be the average flow [3], but the arrival rate. This is what interests the end user in the end the most. There is a

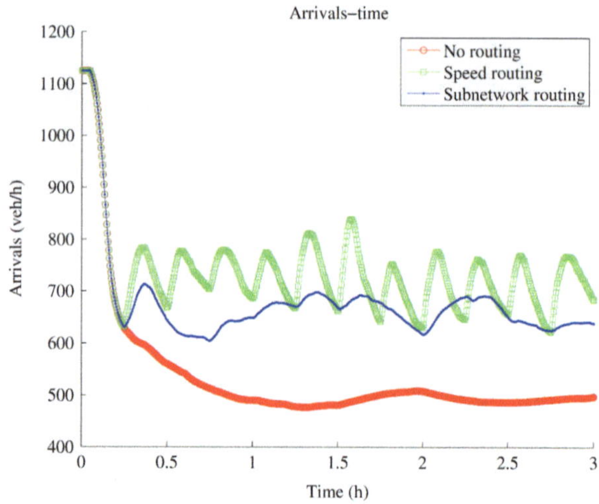

Fig. 3 Network arrivals over time

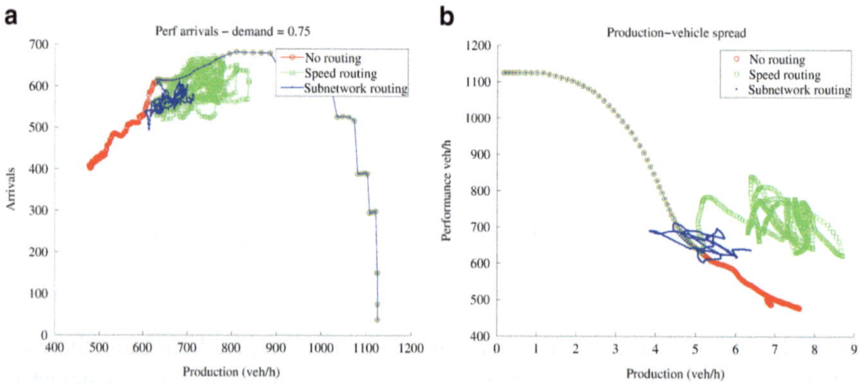

Fig. 4 Evolution of the key indicators. (**a**) Arrival rate vs production. (**b**) Standard deviation of the accumulation

good relationship between the network performance (the average flow) and the arrival rate at the destination, as is shown in Fig. 4a. The simulation starts with all users distributed over the network, at undercritical densities. In this situation, the performance is high, but the arrivals still low since drivers have to drive towards their destination. This is the line at the right hand side of the figure. This line is equal for all routing strategies since in the first time period of 15 min, the actual routing is the same. The routing strategies can be ordered as follows, in which the first gives the best result: speed routing – subnetwork routing – no routing. This results holds both for the arrivals as for the network performance, i.e. the average flow.

As shown by Mazloumian et al. [16], density variance has an impact, which has been further developed by us [17], stating that two-dimensional macroscopic fundamental diagram exists. The production depends on the accumulation *and* the spatial fluctuation of the density. Note that in the simulation we have a fixed number of vehicles, and thus have the same accumulation. We analyse the influence of variation of density again for the situation with control, which is shown in Fig. 4b. The spatial fluctuation of the density is expressed as the standard deviation of the cell density. The production reduces with an increase of spatial fluctuation of density, similar to other literature [16, 17]. However, low-level control (i.e., including cell speeds) gives a higher production *under similar spatial spread of densities*. Generally, that means that the form of the 2-dimensional network fundamental diagram changes as result of the low-level control, and its interference with traffic processes.

5 Conclusions and Discussion

In this paper the possibilities for traffic control are explored. In particular we focus at the possibilities for routing based on the speed in subnetworks. The first conclusions is that routing advice is effective and can prevent the breakdown of traffic which happens without control. It might not be as effective as routing based on actual speeds of all links, but is much more robust and the network performance varies less.

Secondly, we found that this performance increase can be found with the current speeds. Using the spatial spread of congestion, as suggested earlier [18] is not strictly needed, but might increase of the network performance even further. This is subject of further research.

Thirdly, the macroscopic fundamental diagram, even as function of two variables, accumulation and spatial distribution of accumulation, changes in shape under traffic control using link-level information.

Acknowledgements This research was sponsored by a IP-CC subsidy from ICTregie/NWO in the project *SI4MS*, Sensor Intelligence for Mobility Systems, and by the foundation Next Generation Infrastructures in the project *JAMS*, Joint Approach for Multi-level Simulation.

References

1. A. Kotsialos, M. Papageorgiou, H. Haj-Salem, S. Manfrendi, J. Van Schuppen, J. Taylor, M. Westerman, Co-ordinated Control Strategies. Tech. rep., DACCORD – Development and Application of Co-ordinated Control of Corridors (1997)
2. V. Dinopoulou, D. C., P. M., European Journal of Operational Research **175**(3), 1652 (2006)
3. N. Geroliminis, C.F. Daganzo, Transportation Research Part B: Methodological **42**(9), 759 (2008)
4. C. Buisson, C. Ladier, Tranportation Research Records, Journal of the Transportation Research Board (2009)

5. M. Cassidy, K. Jang, C. Daganzo, in *Proceedings of the 90th Annual Meeting of the Transportation Research Board* (2011)
6. V. Gayah, C. Daganzo, Transportation Research Part B (2011)
7. N. Geroliminis, Y. Ji, in *Proceedings of 90th Annual Meeting of the Transportation Research Board* (2011)
8. M.J. Lighthill, G.B. Whitham, Proceedings of the Royal Society of London. Series A, Mathematical and Physical Sciences **229**(1178), 317 (1955)
9. P.I. Richards, Operations Research 4 **4**, 42 (1956)
10. S.K. Godunov, Math. Sb. **47**, 271 (1959)
11. J.P. Lebacque, in *Proceedings of the 13th International Symposium on Transportation and Traffic Theory* (1996)
12. W.L. Jin, H.M. Zhang, Transportation Research Part B: Methodological **37**(6), 521 (2003)
13. C. Tampère, R. Corthout, D. Cattrysse, L. Immers, Transportation Research Part B **45**, 289 (2011)
14. C. Daganzo, *Multinomial Probit: The Theory and its Application to Demand Forecasting* (Academic Press, New York, United States, 1979)
15. Y. Sheffi, *Urban transportation networks: Equilibrium analysis with mathematical programming methods* (Prentice-Hall, Englewood Cliffs, New Jersey, 1985)
16. A. Mazloumian, N. Geroliminis, D. Helbing, Philosophical Transactions of the Royal Society A **368**, 4627 (2010)
17. V.L. Knoop, S.P. Hoogendoorn, *Proceedings of Traffic and Grannular Flow 2011* (in print), chap. Two-Variable Macropscopic Fundamental Diagrams for Traffic Networks
18. V.L. Knoop, J.W.C. Van Lint, S.P. Hoogendoorn, in *Proceedings of 2nd International Workshop on Traffic Data Collection & its Standardisation* (2011)

Dynamic User Optimal Route Assignment System for Traffic Flows Microsimulation

Gregory Cherniak

Abstract The occurrence of the hunting phenomenon in route assignment systems based on the Dynamic User Optimal principle is demonstrated. A method to reduce its effect on the system is developed.

1 Introduction

One of the tasks that arise during traffic flows simulation is route assignment, that is a search for the routes the vehicles follow. Described in the paper is a route assignment system for the micro-model of traffic that utilizes parameters of individual vehicles such as coordinates and instantaneous speed of an individual vehicle.

Route assignment system described in the paper is based on the Dynamic User Optimal (DUO) principle that assumes routes yielding the minimum estimated travel time for every individual vehicle [1], assigned routes being corrected on a periodic basis.

Road network model that serves as a basis for the route assignment system is comprised of links and crossroads. A link is a one-way road part that has a single entrance and a single exit. A crossroad is an element that connects two or more links. For each pair of lanes in these links a crossroad specifies whether vehicles are allowed to move from one to the other. In other words, road network is a directed graph with links as its edges and crossroads as its vertexes. Hence standard algorithms for path search in a graph are used, namely Dijkstra algorithm, reducing the task to a search for edge weights which were chosen to be estimated travel times of the corresponding links. Estimated travel time depends on static parameters of

G. Cherniak (✉)
JSC COS&HT, Dolgoprudny, Moscow Region, Russia
e-mail: grcher@cos.ru

V.V. Kozlov et al. (eds.), *Traffic and Granular Flow '11*,
DOI 10.1007/978-3-642-39669-4_37, © Springer-Verlag Berlin Heidelberg 2013

the link and vehicle distribution and their speeds along the link. Vehicular data is collected from individual vehicles and aggregated on a periodic basis.

2 Data Aggregation

In order to estimate travel time of a separate link we use data on average travel speed and queue length on the link. This approach was adopted due to effects inherent to microscopic data collection.

If one to take the simplest way to estimate travel time on a link as a ratio of the link length to the average travel speed on the link, given every vehicle on the link is in the queue and reporting zero instantaneous speed one has infinite estimated travel time whereas in reality it may prove quite low for short queues. Queueing time and free motion time should be estimated separately. So all the vehicles are separated in two groups: ones with a speed less than 5 km/h that are assumed to be queued, and other vehicles that are assumed to be in free motion. The first vehicles group provides data on the average *free* motion speed, making free motion time a ratio of the link length and the free motion speed. Free motion speed is assumed to be equal to the maximal travel speed if there are no vehicles in the first group. The second vehicles group provides data on queue position and length. In order to make travel time estimation there was developed a linear queue traversal model.

3 Queue Traversal Model

Queue traversal time depends on the nature of the queue. Described in this section is a model for a queue formed on a signal-controlled junction with two phases – red one with a duration of t_r giving zero transportation capacity and green one with a duration of t_g giving transportation capacity equal to that of adjoining link. Acceleration and deceleration processes are neglected, a vehicle is either stopped or is moving at a constant average speed v. Another assumption used is that start- and stop-waves in the queue are traveling at equal constant speed of v_q.

Shown on the Fig. 1 are the pathes that vehicles follow through the queue in the form of time dependency of vehicle position in the queue that is the distance between the vehicle and the crossroad. Green signal duration is 60 s, red signal duration is 90 s. Start- and stop-waves are shown in dashed lines.

Traversal time for the queue of the length L is an average duration of a time period from the moment vehicle entered the tail of the queue to the moment it reached the crossroad. Exact queue traversal time of a vehicle for a queue of a fixed length depends on the traffic lights phase at the moment the vehicle reached the tail of the queue. However it is difficult to predict this moment in constantly changing road situation, so it makes sense to use average queue traversal time on the traffic lights period. It is assumed that the moment above has a uniform distribution.

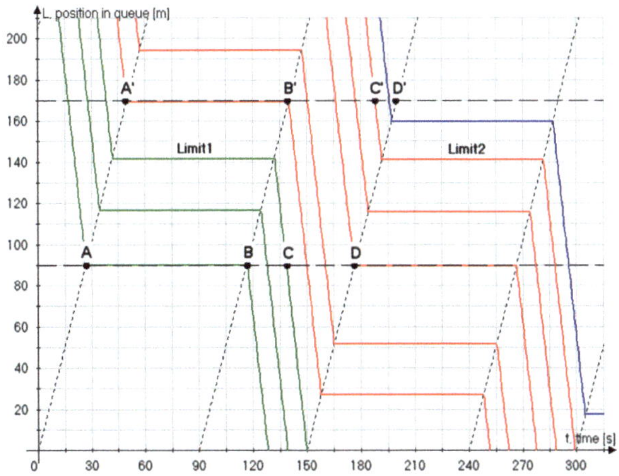

Fig. 1 Queue traversal

Consider one traffic lights period that is from 0 to 150 s. According to the Fig. 1 it may be splitted into three parts:

$$AB = t_r,\ BC = t_g - \frac{L}{v_q^*},\ CD = \frac{L}{v_q^*} \tag{1}$$

where v_q^* is a normalized queue travel speed

$$v_q^* = \left(\frac{1}{v} + \frac{1}{v_q}\right)^{-1} \tag{2}$$

Durations of BC and CD depend on queue length and are determined by the intersection point C of a constant queue length line and a limit path. Limit path is a path that have vehicle reaching the head of the queue (i.e. the crossroad) at the moment the red signal starts. Exact queue traversal time function has a break at limit pathes. We will call all the pathes between two successive limit pathes a group of pathes. There are three groups of pathes on the Fig. 1 shown in different colors, and there are two limit pathes *Limit*1 and *Limit*2.

Queue traversal time average for the queue length of L that features pathes from the first group (shown in green on Fig. 1) on AC and pathes from the second group (shown in red) on CD is

$$F^{(1)}(L) = \frac{1}{t_r + t_g}\left(I_{AB}^{(1)} + I_{BC}^{(1)} + I_{CD}^{(1)}\right) = \frac{L}{v} + \frac{t_r}{t_r + t_g}\left(\frac{t_r}{2} + \frac{L}{v_q^*}\right) \tag{3}$$

The same average for a longer queue featuring pathes from the second group on *AC* and pathes from the third group (shown in blue) on *CD* is

$$F^{(2)}(L) = F^{(1)}(L) + \frac{t_r}{t_r + t_g} t_r \tag{4}$$

So the average queue traversal time will has a break at queue lengthes of $L_{lim} = nv_q^* t_g,\ n \in N$ when point C moves from one limit path to another.

All the calculations above were made for a fixed-length queue. However in real situations it is difficult to predict or measure the exact queue length for a specific vehicle, so a linear approximation to the average queue traversal time found above may suffice:

$$T = \frac{L}{v} + a\frac{L}{v_q^*},\ a = \frac{t_r}{t_g} \tag{5}$$

4 Test Road Network

The route assignment system in hand was studied on a simple test road network of one source and one destination with two alternate pathes (legs) between them, *XPath* being shorter than *YPath*. Transportation capacity is assumed to be limited by the controlled intersection only so that there is not more than one queue per leg situated on its farther end and queues are assumed to be much shorter than the legs. Theoretical analysis of such road network was done and its results compared to the results of micromodeling. Network parameters that were used are originating flow of 1,440 veh/h, legs lengthes of 1 and 1.5 km and equal traffic signal durations of 40 s for both leg directions (Fig. 2).

Presented on the Fig. 3 are *Y Path* time dependencies of incoming flow and traffic loading that is vehicle quantity on the leg. Micromodeling results that are shown on figure in solid line demonstrate that traffic loading oscillations are present in the system. The period of oscillations for test road network is 240 s. Figures for the *X Path* leg show the same picture.

The nature of the oscillations is the following. Initially there are no vehicles on both legs. Since the X leg is shorter, all the vehicles will go to the X leg. A queue begins to form on the farther end of the X leg when it has red traffic light to it. Traffic flow would have started being equal for both legs from the moment estimated travel times for the X leg reaches that of the Y leg, but that is not the case. At that moment the X leg will have vehicles that are not in the queue but in the free motion yet. Queued vehicles are making estimated travel time of legs equal, whereas free-motion vehicles are making no effect on travel times. However after a little while, when some of the free-motion vehicles on the X leg reach the tail of the queue, the queue length on X leg will increase making its travel time larger than that of

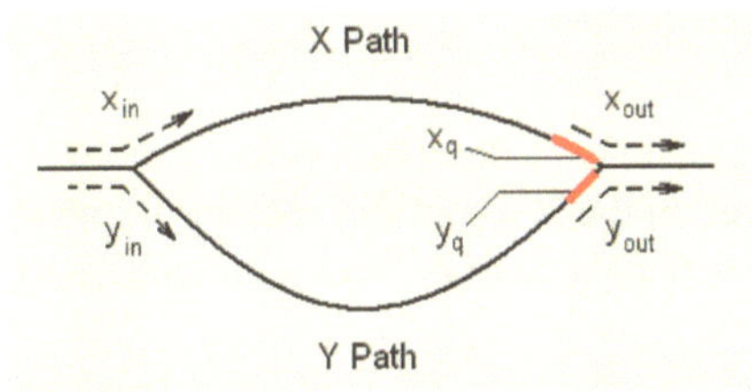

Fig. 2 Test road network

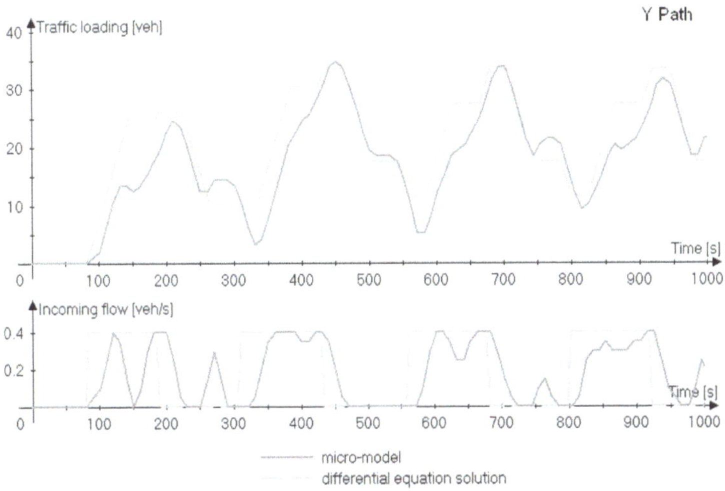

Fig. 3 System evolution

that Y leg, the legs swap their roles and the process repeats for the Y leg. So the system does not stay in the equal-travel-time state, but has oscillations instead.

This effect is called *hunting phenomenon*. It is inherent to control systems featuring an inverse feedback with a lag. Inverse feedback of the route assignment system in hand is its very principle of route assignment: when one route is overloaded the vehicles start to increase the load of the other route and vice versa. It also has the three types of lag: the first one is caused by the periodic nature of the data collection, the second one – by the periodic nature of the route updates. Decisions the system takes are based on outdated data and the routes vehicles follow are based on outdated decisions. The third type of lag, being much larger than the first two, is not a direct lag, but an effect that may be represented as a lag – unaccounted free-motion vehicles. We propose a way to substantially reduce the lag and hence the hunting phenomenon by taking these vehicles into account.

5 Hunting Phenomenon

Control systems with a lag may be described with a system of differential equations. Let x be a queue length on the X leg in the vehicles number, the distance between vehicles assumed constant and equals l. Then estimated travel time for X and Y legs is (5)

$$T_x(t) = \frac{L_x}{v} + a\frac{lx(t)}{v_q^*} \qquad T_y(t) = \frac{L_y}{v} + \frac{1}{a}\frac{ly(t)}{v_q^*} \tag{6}$$

where L_x is the length of the X leg, v_q^* is the normalized vehicle speed (2).

Let $x_{in}(t)$ be the incoming traffic flow on X leg at the moment t and $x_{out}(t)$ – the outgoing traffic flow on X leg. Initial conditions are

$$x(t)|_{t<0} = 0, \quad x_{in}(t)|_{t<0} = 0 \tag{7}$$

Instantaneous queue increase equals to the difference between vehicle flows entering and leaving the queue. Consider vehicles that entered the X leg at the moment t. Since queue length is neglected in comparison with the leg length, free motion will take

$$\Delta t_x = \frac{L_x}{v} \tag{8}$$

then vehicles will enter the queue. Instantaneous queue increase is

$$\dot{x} = x_{in}(t - \Delta t_x) - x_{out}(t) \tag{9}$$

Outgoing flow for a signal-controlled junction with two signal phases is

$$x_{out}(t) = \begin{cases} I_o, & t_{ph} \le t_g \\ 0, & t_{ph} > t_g \end{cases}, \quad t_{ph} = \left\{ \frac{t}{t_r + t_g} \right\} \tag{10}$$

where I_0 is maximum transportation capacity of the X leg.

Time dependency of incoming flow when taking into account average second type lag of $t_c/2$ (if routes are updated each t_c seconds) is

$$x_{in}(t) = f_x(t - \frac{t_c}{2})I(t - \frac{t_c}{2}) \tag{11}$$

where $f_x(t)$ is the percentage of the vehicles that are prescribed to take the X leg at the moment t, $I(t)$ – total traffic flow. On a test road network the total traffic flow is constant $I(t) = I = const$.

Decision function $f(r)$ is governed by the strategy of route assignment. In the case of the Dynamic User Optimal assignment it is

$$f_x(t) = \begin{cases} 1, & T_x(t) < T_y(t) \\ \frac{1}{2}, & T_x(t) = T_y(t) \\ 0, & T_x(t) > T_y(t) \end{cases} \tag{12}$$

So we have a system of two differential equations with a lag for the instantaneous queue increase on X and Y legs (Eq. (9) and a correspondent equation for Y leg). In order to compare its solution with the results of simulation traffic loading and the incoming flow should be determined. Incoming flow was introduced above as $x_{in}(t)$. Traffic loading that is the number of vehicles on the leg is defined as a sum of free-motion and queued vehicle numbers $x_{free} + x_{queue}$, where

$$\dot{x}_{free} = x_{in}(t) - x_{in}(t - \Delta t_x) \tag{13}$$

$$\dot{x}_{queue} = x_{in}(t - \Delta t_x) - x_{out}(t), \ x_{queue} \geq 0 \tag{14}$$

The solution of the equations system is depicted on the Fig. 3 with a dashed line. It is worth noting that developed equations describe system dynamics accurately enough to make its qualitative analysis.

6 Inferred Queue Model

The differential equation system above can be easily reduced to a single differential equation for the difference of queue length on the legs taken with phase ratio coefficient: $z = ax - y/a$

$$\dot{z}(t) = \frac{I_i}{a}\left[a^2 f_z(t - \frac{t_c}{2} - \Delta t_x) + f_z(t - \frac{t_c}{2} - \Delta t_y) - 1\right] - \frac{I_o}{a}\left[\frac{(a+1)x_{out}}{I_o}(t) - 1\right] \tag{15}$$

$$f_z(t) = \begin{cases} 1, & z(t) < z_{st} \\ 1/2, & z(t) = z_{st} \\ 0, & z(t) > z_{st} \end{cases}, \quad z_{st} = \frac{L_y - L_x}{l}\frac{v_q^*}{v} \tag{16}$$

This equation has two lags

$$\Delta t_1 = \frac{t_c}{2} + \Delta t_x, \ \Delta t_2 = \frac{t_c}{2} + \Delta t_y \tag{17}$$

each comprised of a second-type lag of $t_c/2$ and third-type lag of $\Delta t = L/v$, the third-type lag being larger than the second-type one for long enough links. One may

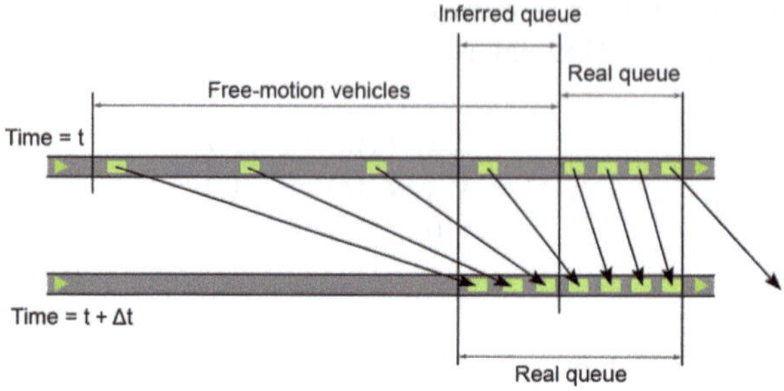

Fig. 4 Real and inferred queue

show that the period and amplitude of a hunting phenomenon are proportional to the lag. So in order to reduce hunting phenomenon one should eliminate third-type lags.

We propose that a predicted queue length should be used during travel time estimation instead of current queue length. Prediction of the queue length should be taken for the moment current vehicle reaches queue tail. In this way travel time estimation made at the moment vehicle chooses between two legs takes into account free-motion vehicles ahead and the effect that causes third-type lag is greatly reduced (Fig. 4).

It is convenient to split the predicted queue into the real queue that is the queue at current moment and the inferred queue. The inferred queue may be positive or negative, but here for simplicity we assume it to be non-negative only.

$$x_{queue}(t + \Delta t_x) = x_{queue}(t) + x_{infer}(t) \tag{18}$$

Using average transportation capacity for the traffic lights period (that is $I_0/(a+1)$ for X leg and $I_0/(a^{-1}+1)$ for Y leg for two phase traffic lights) and Eq. 14 we have an equation for the inferred queue:

$$\dot{x}_{infer}(t) = min\{0, x_{in}(t) - \bar{x}_{out}\} - min\{0, x_{in}(t - \Delta t_x) - \bar{x}_{out}\} \tag{19}$$

It is worth noting that $\dot{x}_{infer} \neq x_{in}(t) - x_{in}(t - \Delta t_x)$ since the real queue length $x_{queue} \geq 0$. The assumption that the inferred queue is not negative was used to address this issue.

Predicted queue was separated into real and inferred queue for inferred queue is quite easy to implement in a micromodel. If at the moment t, when a vehicle enters the X leg, incoming flow $x_{in}(t)$ is higher than the average transportation capacity $\bar{x}_{out}$ then the vehicle is assumed to be in the inferred queue with a probability

$$P_{infer} = \frac{x_{in}(t) - \bar{x}_{out}}{x_{in}(t)} \tag{20}$$

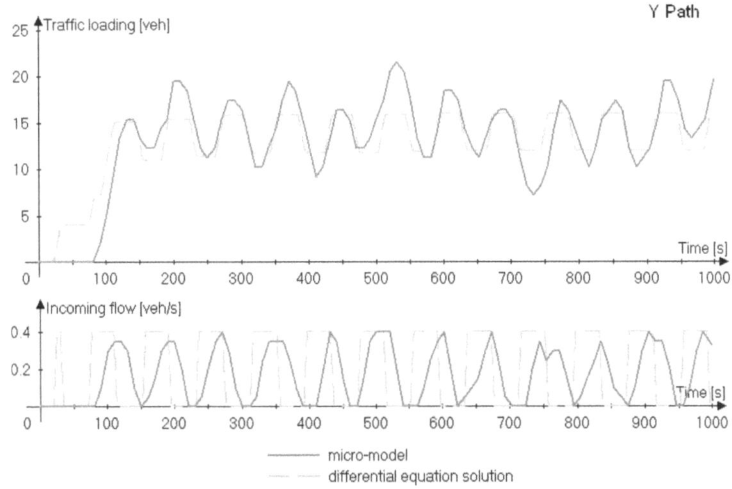

Fig. 5 System evolution with inferred queue introduced

After free motion period $\Delta t = L/v$ vehicle is removed from the inferred queue for it must have reached real queue or already left the leg.

Presented on the Fig. 5 is the evolution of the system with an inferred queue introduced. Period and amplitude of traffic loading oscillations decreased, hunting phenomenon period was reduced from 240 to 80 s, i.e. from triple total traffic lights period to a single. It should be mentioned that in both cases we observed an oscillation synchronization effect that made the hunting phenomenon period be a multiple of traffic lights period (which made it impossible to obtain periods smaller than one traffic lights period under current conditions).

Another notable result of inferred queue introduction and the consequence of hunting phenomenon reduction is the decrease of average travel time from 128 to 109 s.

7 Conclusion

Demonstrated in this paper is the occurrence of hunting phenomenon in route assignment system based on Dynamic User Optimal principle and a method to reduce its effect on the system.

Found in the research process was an effect of hunting phenomenon oscillations synchronization with a traffic lights signal. The synchronization effect was observed both in micromodeling results and in differential equations solution. It was found empirically that the period of oscillations t_{osc} depends on the full traffic lights period $t_g + t_r$ as

$$t_{osc} = m(t_g + t_r) = \left[\frac{\Delta t}{t_g + t_r} + \frac{1}{2}\right](t_g + t_r),\ m \in N \tag{21}$$

where Δt is the total lag present in the system.

Besides theoretical analysis of the developed equations on system dynamics future research areas may include a search for a method to provide further stability to the system. One of the reasonable options is to utilize a more complex decision function, for instance that of Dynamic Stochastic User Optimal. In Dynamic Stochastic User Optimal a route assigned to a vehicle is not always the one with the shorter travel time, but the shorter the route the more probability it has to be assigned to the vehicle. It is supposed that this approach may reduce feedback and decrease oscillation amplitude.

Reference

1. Kuwahara M, Horiguchi R, Yoshii T (2002) Standard Verification Process for Traffic Flow Simulation Model. Traffic Simulation Committee, Japan Society of Traffic Engineers

On Traffic Flow on the Arterial Network Model

Konstantin K. Glukharev, Nikolay M. Ulyukov, Andrey M. Valuev, and Ivan N. Kalinin

Abstract The problem of comprehensive model development for traffic flow on a city network is studied. Discrete traffic flow model with a safe distance is introduced and its properties are studied, including queue dynamics and stable modes. Flow properties in closed loops are also studied. Road intersections are decomposed onto primitive elements (uniform lane segments and elementary junctions) and the discrete flow model approach is applied, namely initial and boundary conditions are constructed and model parameters are defined. An arterial network model and topological network types are introduced. Routing is studied for certain network topological types. Computational model for traffic network is created; some simulation results and qualitative conclusions are presented.

1 On the General Problem "Traffic Jams on a City Road Network": Discrete Flow Theory – Network Navigation

Traffic jams are caused by a set of factors including traffic flows oversaturation and consequent instability that can lead traffic to a complete stop. Flow stabilization and control requires various network-scale feedback mechanisms including those based on satellite navigation systems (GPS/GLONASS/etc.). In effect, constructive solution of the problem in question suggests the development of theories and models

K.K. Glukharev (✉) · N.M. Ulyukov · A.M. Valuev · I.N. Kalinin
Institute of Machines Science named after A.A. Blagonravov of RAS, Bardina str. 4, Moscow, Russia

Moscow Institute of Physics and Technology (State University), Institutskii per. 9, Dolgoprudny, Moscow Region, Russia
e-mail: cafedravmtp@yandex.ru; weedjy@mail.ru; valuev.online@gmail.com; pupssman@gmail.com

V.V. Kozlov et al. (eds.), *Traffic and Granular Flow '11*,
DOI 10.1007/978-3-642-39669-4_38, © Springer-Verlag Berlin Heidelberg 2013

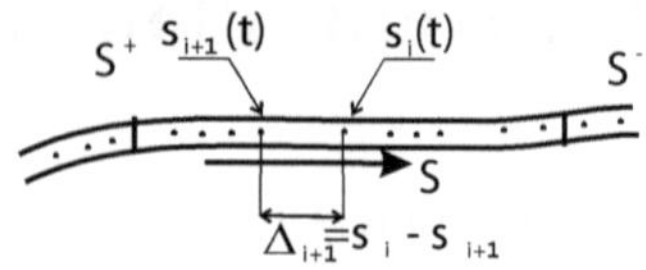

Fig. 1 Discrete flow in a uniform lane

for a new object – road network flow navigation. In this paper three types of models are presented and discussed:

- Single channel traffic flow;
- Traffic flow through crossroads;
- Traffic flow on the arterial network.

These models create a foundation for understanding flow blocking mechanisms and effects.

2 Single Channel Traffic Flow

We have developed a traffic flow microscopic model that treats a single car as a particle of a flow and describes the flow of these particles in the form of differential-functional equations [1–3]. Its simplest implementation is a model for traffic flow on a single isolated lane. We consider a part of lane in $[S^+, S^-]$, with S^+ and S^- being entry and exit sections respectively (see Fig. 1).

Lane is filled with cars that follow each other effectively forming what is called a *discrete flow*. We assume the following properties for this flow:

- Every particle's speed is either equal to zero or directed along the coordinate axis;
- Particle interaction is given with the law of safe distance [4] as follows:

$$\dot{s}_i(t) = \begin{cases} max\{\dot{s}_i(t-0), \dot{s}_{i-1}(t)\}, \Delta_i > \phi(\dot{s}_{i-1}(t)), \\ \dot{s}_{i-1}(t), \Delta_i = \phi(\dot{s}_{i-1}(t)), \\ min\{\dot{s}_i(t-0), \dot{s}_{i-1}(t)\}, \Delta_i < \phi(\dot{s}_{i-1}(t)) \end{cases}$$

$$\Delta_i = s_{i-1}(t) - s_{i-1}(t),$$

$$i = 1, 2, 3, \dots$$

$$\dot{s}_0(t) = f(t, s_0),$$

with $s_i(t)$ being particles' coordinates, $f(s)$ being the leader motion function and $\psi(\nu)$ being the safe distance function.

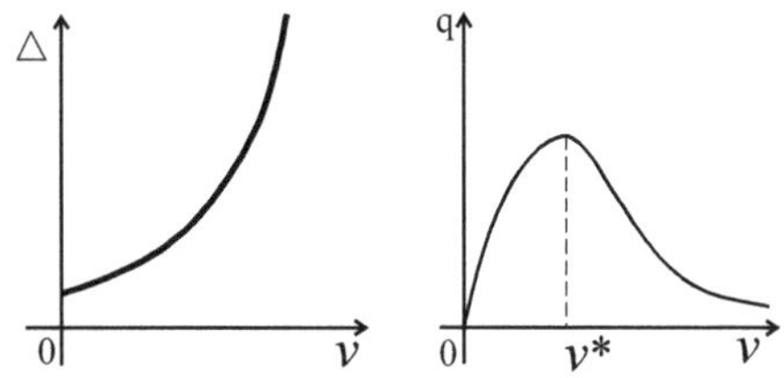

Fig. 2 Safe distance law and Fundamental diagram

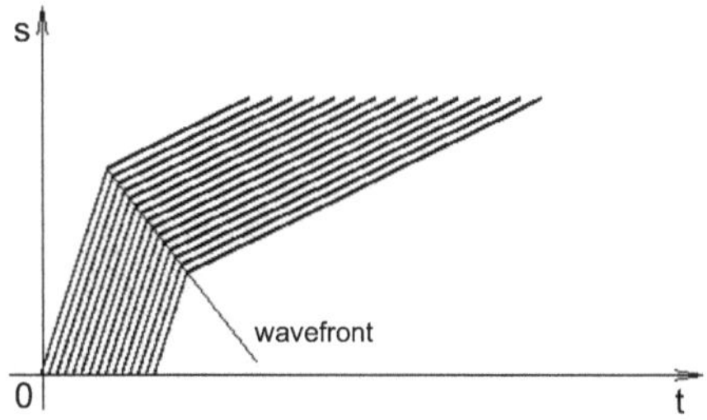

Fig. 3 Leader speed change wave front

The latter means that if particles in the flow have equal speed then the distance between any two subsequent particles is equal to $\psi(\nu)$, ν being the leader speed (see Fig. 2).

Dependencies $\psi(\nu)$ and $q(\nu) = \frac{\nu}{(\psi(\nu))}$ correspond well to the experimental data. Δ is the distance between neighbouring particles, $\nu = \frac{\delta}{\delta t}$ is the particle speed in stationary uniform flow, $q = \frac{1}{\delta t}$ is the flow intensity and $\nu^* \leq 50$ km/h is the maximum flux speed corresponding to the maximum flux $q^* \leq 2{,}000$ cars/h. Relation between intensity and speed is called the Fundamental diagram.

Car dynamics is very complex because of presence of the great number of degrees of freedom and, therefore, in a full detail doesn't suit well for the traffic flow modelling. Because of that we use a greatly simplified model for car dynamics that allows instantaneous changes of speed. This simplification is justified by the fact that in the case of high traffic flow intensity it is impossible for a particle to change it's speed in a wide range of values and acceleration/deceleration time is greatly smaller than the time it takes a particle to move from one crossroad to another. Moreover, simplified dynamics allows analytical analysis of different modes of motion.

To study various flow effects discrete flow is defined as a solution to the mixed type problems described with initial conditions (speed and position for every car in a lane) as well as boundary conditions given in the terminal sections of the lane segment. Modelling results are conveniently presented in graphical form as a set of particle trajectories.

For example, Fig. 3 presents the results for modelling of stationary flow reconfiguration due to a leader speed change in assumption of free exit conditions. The figure shows that there is so-called reverse wave that propagates in the direction opposite to the flow.

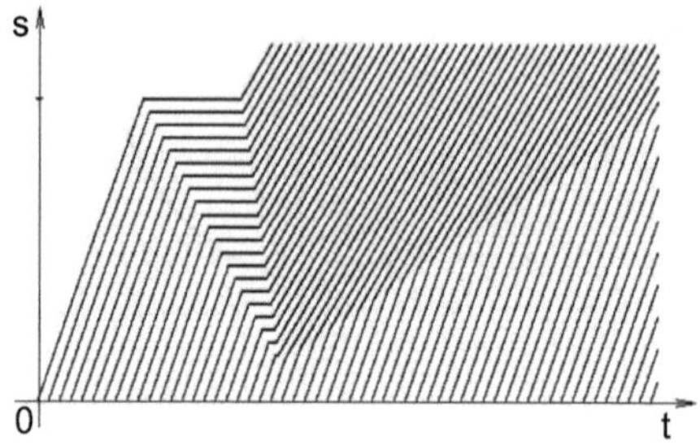

Fig. 4 Discrete flow in the uniform lane

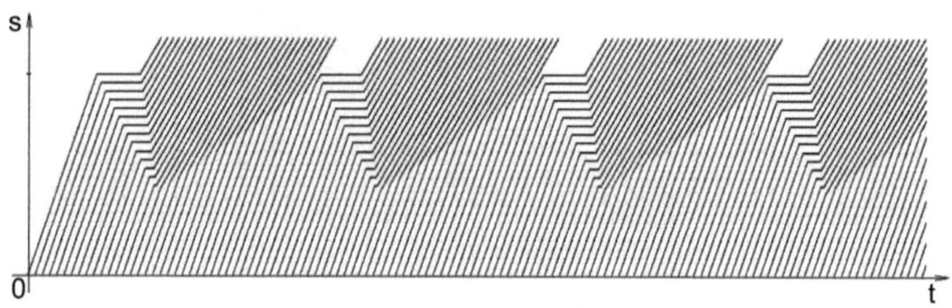

Fig. 5 Stressed periodic mode

Traffic light presence results in a boundary condition of a different kind – it prohibits leader motion through the exit section of a lane during red light intervals. It leads to queue accumulation before the traffic light, queue dynamics depending on the traffic light mode and allowed velocities before and after the traffic light.

Figure 4 depicts queue dynamics for a single-shot red light and shows a wedge filled with horizontal lines corresponding to cars with zero speed. Wedge height is given as follows:

$$h = \tau_{red}\frac{|\nu_1||\nu_2|}{|\nu_1| - |\nu_2|}, \nu_2 < \nu_1 < 0,$$

with τ_{red} being red light duration, ν_1 – left slope speed and ν_2 – right slope speed.

There are conditions that may lead to the stabilized queue dynamics (Fig. 5) and growing queue (Fig. 6). In the latter case queue is growing indefinitely and, finally, reaches previous crossroad and blocks it. However, in both cases there is a mode when a queue propagates in a direction opposite to the flow.

In the periodic queue mode it is possible to define flux count function (number of particles passed through exit section) as a sum of linear and periodic components, the former being equal to the average flux – see Fig. 7.

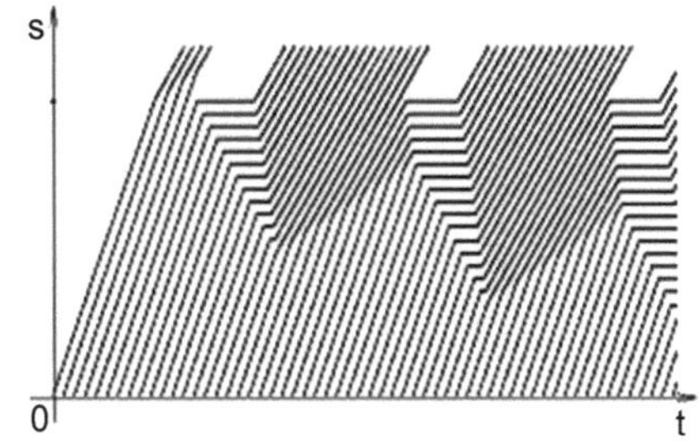

Fig. 6 Queue buildup mode

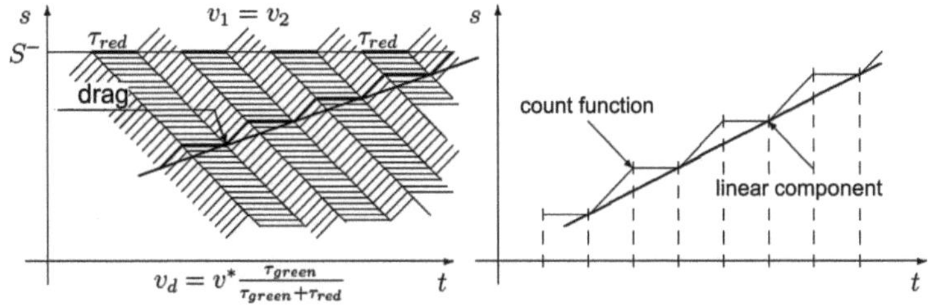

Fig. 7 Queue dynamics in periodic mode

In addition to a lane segment (isomorphic to a straight line segment) lane loops are studied. Modelling shows that the flow breaks apart to the set of uniform clusters, particles in clusters having equal speed and distance to the local leader, clusters being separated by vacant lane segments (see Fig. 8).

Considering city traffic, it's biggest part goes through the arterial network. In this network all the crossroads are either traffic light controlled or level-separated (via bridges and overpasses). In both cases any time period is separated into subperiods for which there are no traffic flow intersections on any crossroad in the arterial network. For these subperiods arterial network is divided into the set of non-intersecting uniform segments and loop lanes [12].

3 Crossroad Traffic Flow

Although the above problems do provide understanding of the traffic flow interactions in various parts of a network, description of crossroads is still incomplete and may be inadequate for the crossroads lacking traffic light control. To clarify this and to create a detailed model of traffic flow on crossroads we have conducted detailed simulation.

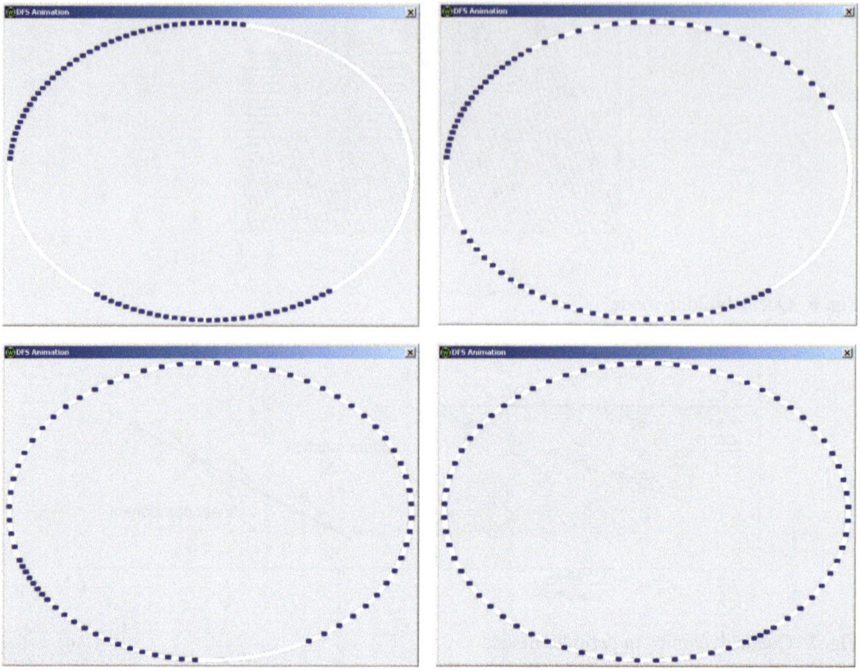

Fig. 8 Clusters on a lane loop

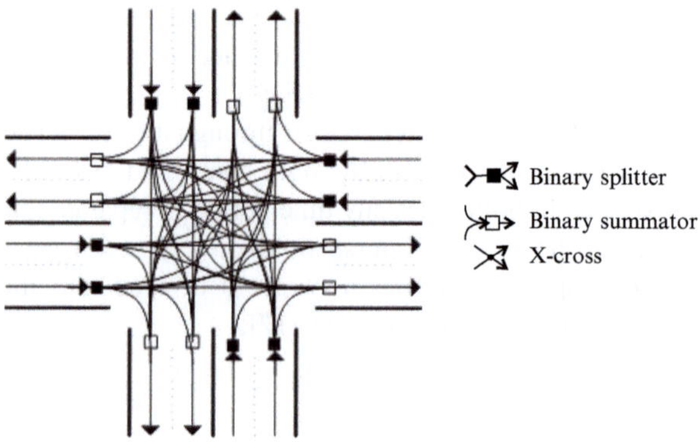

Fig. 9 Sample crossroad trajectories web set

Crossroad area is covered with web-like allowed trajectories set (see Fig. 9). Joins of this web are divided into three classes:

- Splitters
- Summators
- x-crosses

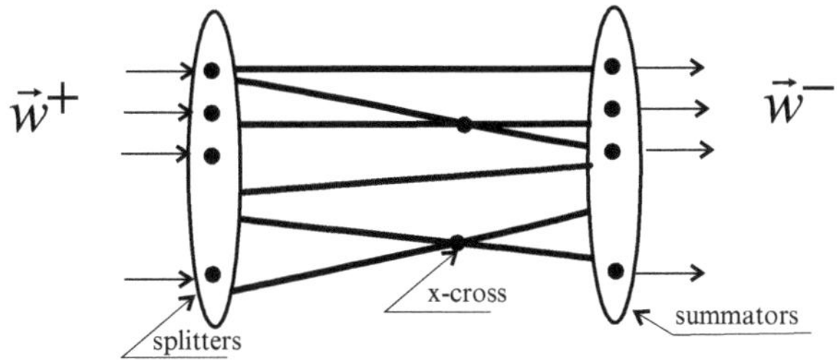

Fig. 10 Crossroad bipartite graph

In the case of x-crosses absence the web can be reduced to a bipartite graph (see Fig. 10). Combinatory schemes for crossroad construction from binary splitters and summators have been developed in [10].

The problem of maximizing the traffic flux through the crossroad [13] is studied with the following assumptions. Maximum flux is considered in terms of the linear component given previously and is defined as the mode having all the fluxes through incoming and outcoming lanes equal to the maximum flux. Flux separation matrix $\|k_{ij}\|$ is introduced for the linear component, connecting linear component intensities for pairs of enters (first index) and exits (second index) of a crossroad as follows:

$$\mathbf{w}^{-} = \|k_{ij}\|' \mathbf{w}^{+},$$

w^{-} and w^{+} being linear component intensity vectors for incoming and outcoming flows respectively.

Sufficient condition for achievement of the maximum flow through a crossroad without x-crosses is given as follows:

- Number of incoming and outcoming lanes is equal
- $\|k_{ij}\|$ matrix is bistochastic
- $l_{ij} > h_{ij}$, l_{ij} being trajectory length for $i - j$ branch and h_{ij} being queue wedge height
- $\frac{\tau_{ij}}{\Delta t}$ is an integer
- k_{ij} are rational numbers.

4 Arterial network flow

Complex and developed city road network may be practically reduced to its arterial (or highway) segment that bears most of the traffic. Concept of highway is inherently clear and highways are generally outlined in maps and atlases. Arterial network consists of highways and interconnecting crossroads, parking and stop being prohibited on the whole arterial network. This road network backbone separates and unites so-called P-zones, i.e. other parts of road network that have different rules, allowing stopping and parking. Figure 11 depicts an example of a regular arterial network illustrating the latter principles [9].

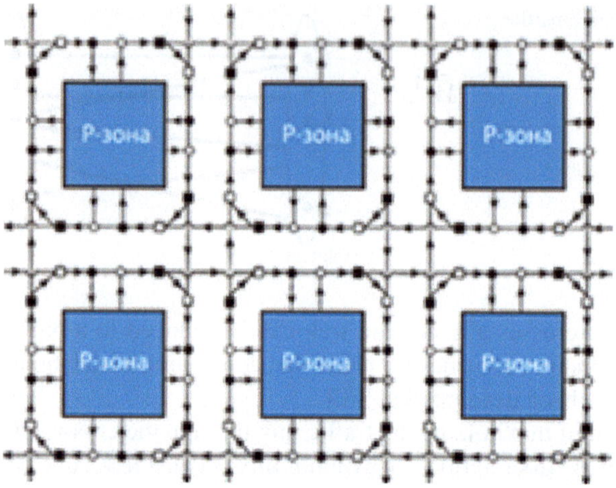

Fig. 11 Sample regular arterial network

Arterial network is covered with route lines connecting all P-zone pairs, route lines being considered as allowed particle trajectories. Shortest route connecting given pair of P-zones can be found unambiguously for some kinds of regular networks, rectangular network for example. However, route density over the rectangular network is strongly non-uniform [11]. Figure 12 depicts the number of shortest routes through given network segment as a function of the segment direction and position relatively to the boundaries of the network. It is clear that the route density increases rapidly towards the network centre.

Due to existence of such effects, leading to the inevitable overutilization of a network's central areas and underutilization of a network's peripheral areas in the case of regular rectangular network the problem of the rational arterial network topology is considered. We note that change in topology does not always require road construction, but can be done via road markings and signs change – for example, changing streets to one-way traffic or prohibiting left turn on some crossroads (see Fig. 13).

On the microscopic level we consider the single crossroad structure and traffic light control mode. Based on the introduced principles reference crossroad structure models are developed (see Fig. 14).

So-called Design equations are developed, relating the crossroad structure and traffic light mode across the network as follows:

$$\mathbf{n} = diag\left\{\xi_i^{-1}\right\} \|k_{ij}\|' diag\left\{\xi_i\right\} \mathbf{n},$$

$$\sum\nolimits_{ie\{i\}_l^+} \xi_i = 1, 0 \leq \xi_i \leq 1,$$

$$\sum\nolimits_{j\in\{i\}_l^-} k_{ij} = 1,\ i \in \{i\}_l^+,\ i \in \overline{1, I},\ l \in \overline{1, L},\ I = \sum\nolimits_{l=1}^{L} \left|\{i\}_l^+\right|,$$

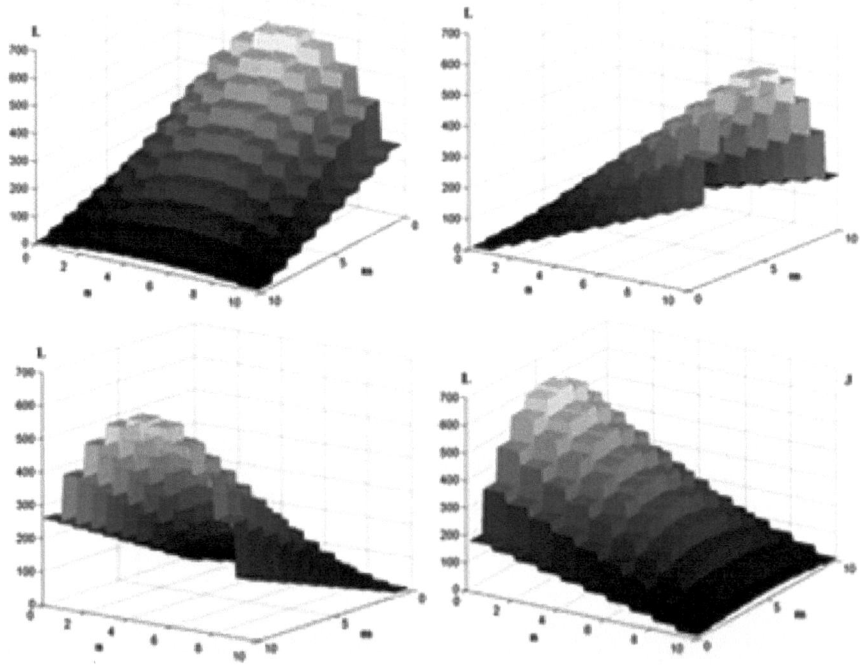

Fig. 12 Rectangular network route density

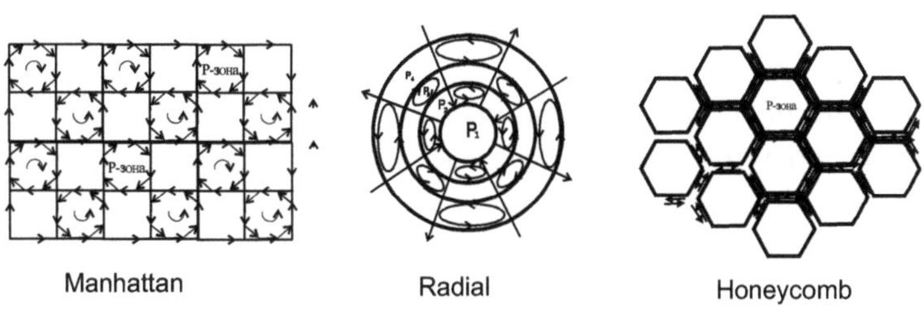

Fig. 13 Regular network topologies

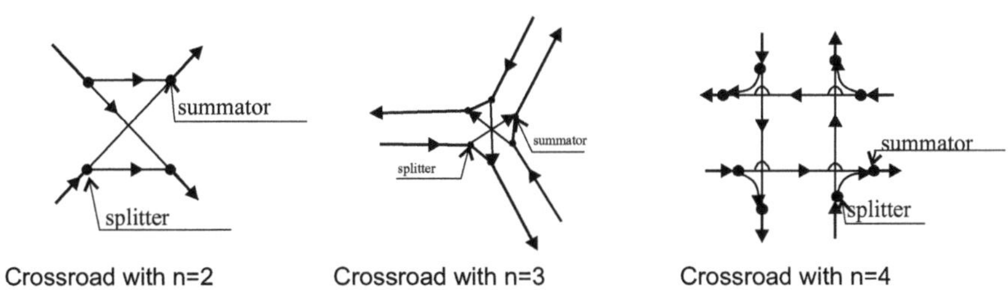

Fig. 14 Crossroads for regular networks

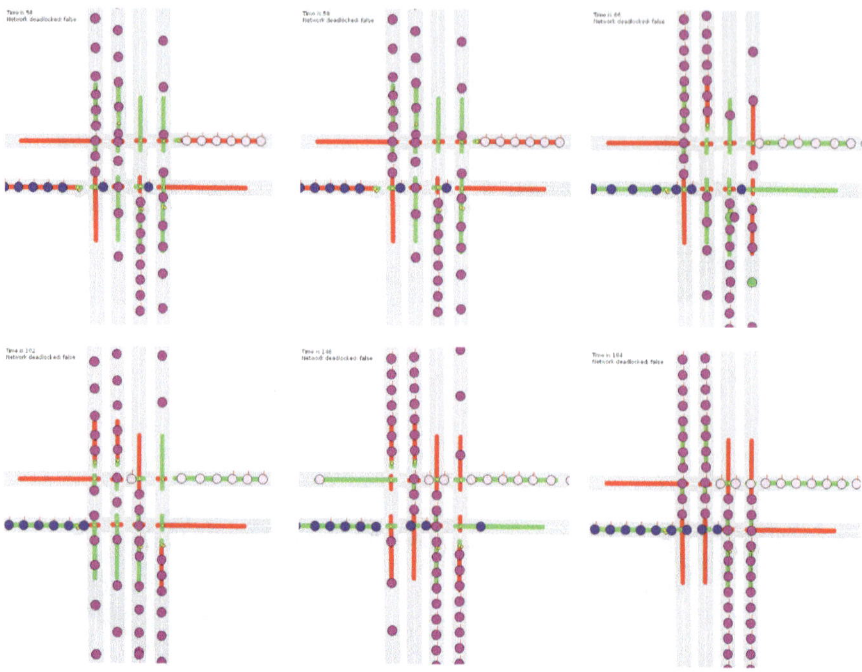

Fig. 15 Cyclic block on a sample crossroad

with **n** being vector of highway lane numbers, ξ_i being i-th traffic light period related to traffic light period τ_i, l is in $1..L$, L being the total number of crossroads in the network, $\|k_{ij}\|$ being doubly stochastic flux separation matrix, $i, j = 1..I$, I being the total number of highway segments in the network, i_l^- being the set of indices of highways outgoing from l-th crossroad and i_l^+ being the set of highway segments ingoing to the l-th crossroad.

The problem of symmetric loop design is considered [8]. It is shown that Design equations are reduced as follows:

$$n^+ - n^+ n^- + n^- = 0,$$

n^+ being the number of incoming lanes in every junction and n^- the number of outcoming ones. It is taken for granted that the equations have the only non-zero solution of $n^+ = n^- = 2$.

Studies via traffic simulations were carried out together with the analytical analysis [7].

Figure 15 depicts snapshots from traffic simulation of flows through crossing of 2-lane and 4-lane roads controlled by traffic lights. In the frames *a,b,c* one can observe the initial queue buildup and evolution and cyclic blocking in the frames *d,e,f*. This kind of simulation experiments can provide data needed for analysis and

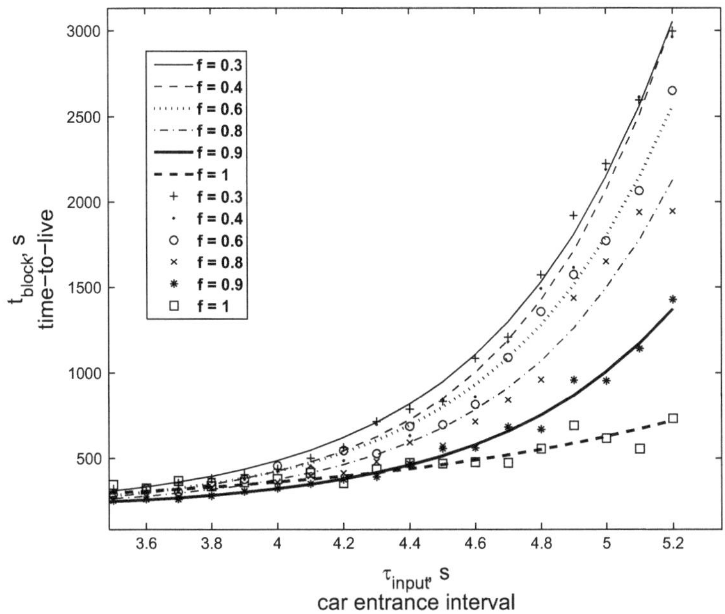

Fig. 16 Time-to-live for an uncontrolled crossroad

optimization of particular network segments performance via traffic light mode or traffic organization change. It may also be a useful tool for evaluating road network projects for newly developed territories.

Simulation can also provide means to obtain high-scale integral characteristics of network segments, like average passage time, channel capacity per road width unit etc. as a function of traffic load structure.

Figure 16 depicts simulation results for simple non-controlled crossroad as a graph of average time-to-live (TTL – time from initial moment to complete self-blocking) against inversed flux (average time between particle entry to the crossroad) with different curves corresponding to different flux separation coefficient f – part of particles that does a left turn [14]. As the figure suggests, TTL is infinite with $\tau > 5.5$ s and constant with $\tau < 4.5$ s, which means that $\tau \sim 5.5$ s corresponds to the maximum non-blocking flux through the crossroad (over 600 cars/h). Also, TTL is lower for higher f, what means that left-turning particles do cause crossroad self-blocking.

Methods for rational routing with respect to the traffic dynamics and current road situation are also developed [5, 6].

With the aid of software for simulation which development is based on the above principles different routing schemes are examined, particularly schemes for routing inside Moscow Sadovoye Ring (see Fig. 17).

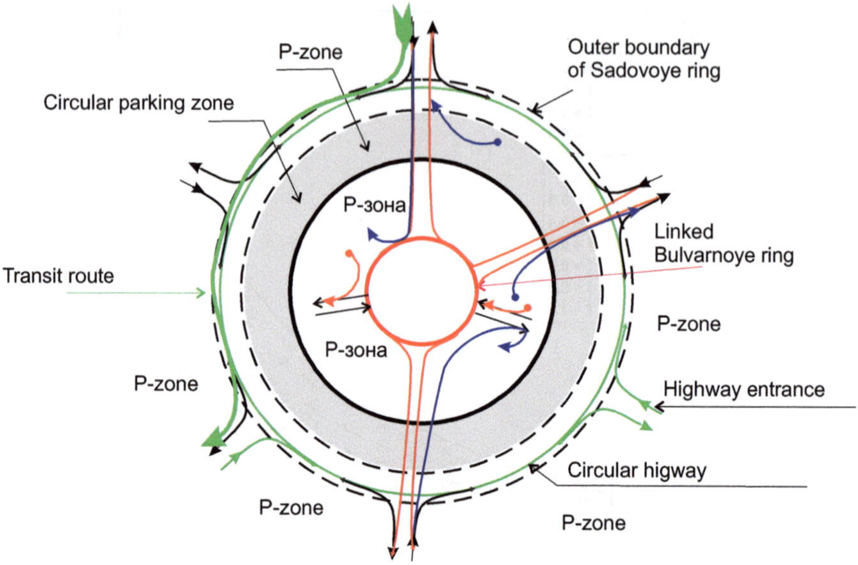

Fig. 17 Proposed routing scheme for the Sadovoye ring in Moscow

5 Conclusion

Traffic models development leads to the following problems consideration:

- Control of commuters demand;
- Quality standards and the problem of the etalon network;
- Flow observability and navigation;
- Network in congestion conditions;
- On the levels of control;
- Particular applications:
  - Conditions for buildup and propagation of queues;
  - Control of queues;
  - Recommended configurations for simple crossroads;
  - Instability of stationary flows;
  - Feedbacks in traffic processes;
  - Control of exits from P-zones;
  - Principles for flow separation.

References

1. Glukharev K, Ulyukov N (2008) Problems of mechanical engineering and machine reliability 4:29–38.
2. Glukharev K, Ulyukov N (2009) Problems of mechanical engineering and machine reliability 1:83–93.

3. Glukharev K, Ulyukov N (2010) Proceedings of Moscow Institute of Physics and Technology 2:58–66.
4. Inose H, Hamada T (1983) Traffic control. Transport, Moscow
5. Valuev A (2009) Proceeding of third international conference "Large-scale system evolution control" 2:43–43
6. Valuev A (2010) Problem of suboptimal route calculation for a network and it application to logistics. Informatisation and control. ISBN: 0236-1493
7. Glukharev K, Kalinin I (2010) Proceedings of MMSED-2010, 67–69
8. Orlov I (2006) Proceedings of 49-th scientific conference of Moscow Institute of Physics and Technology, faculty of Aerophysics and space research 87–88
9. Glukharev K, Marking I (2007) Proceedings of 50-th scientific conference of Moscow Institute of Physics and Technology. Part III: Aerophysics and space research 2:87–88. ISBN 978-5-7417-0296-3
10. Kalinin I, Glukharev K (2008) Proceedings of 51-th scientific conference of Moscow Institute of Physics and Technology. Part III: Aerophysics and space research 2:118–121. ISBN 978-5-7417-0277-2
11. Ginzburgsky K, Glukharev K (2009) Proceedings of 52-th scientific conference of Moscow Institute of Physics and Technology. Part III: Aerophysics and space research 2:142–144. ISBN 978-5-7417-0296-3
12. Glukharev K, Kalinin I (2009) Proceedings of 52-th scientific conference of Moscow Institute of Physics and Technology. Part III: Aerophysics and space research 2:145–148. ISBN 978-5-7417-0296-3
13. Glukharev K, Kalinin I (2010) Proceedings of 53-th scientific conference of Moscow Institute of Physics and Technology. Part III: Aerophysics and space research 2:72–74. ISBN 574170405-0
14. Glukharev K, Kalinin I (2010) Proceedings of 53-th scientific conference of Moscow Institute of Physics and Technology. Part III: Aerophysics and space research 2:82–84. ISBN 574170405-0

Quasi-stationary Approach in Mathematical Modeling of Traffic Flows Dynamics in a City Road Network

Andrey M. Valuev

Abstract The paper proposes a new approach for traffic flows modeling in a city road network for typical conditions of heavy traffic. The model is created according to the assumption of flow uniformity in space (but not in time) for each road segment between junctions. In format, it resembles quasi-stationary flow models for technological network that proved their usefulness for control problems. Some evidence for the model adequacy is obtained from computational experiments based on simulation of car motion ruled by the safe distance law. With a certain hypothesis of path choice by informed drivers the model gives the possibility to predict traffic dynamics. The paper expresses Bellman's type relationships for conditionally optimum route choice that may be combined with general balance equations expressing the quasi-stationary approach, which yields a closed predictive model.

1 Motivation for a Quasi-stationary Continuous Flow Model Introduction

Despite the fact that traffic flow is in fact a discrete flow of cars and other transportation units, macroscopic models were developed. In these models traffic flow is treated as analogous to a gas or fluid flow. These models may be very useful for a single road, especially a highway, and somewhat applied to practical problems of road networks study and control [1, 2]. They probably may be applied to small networks but seem to be too detailed for computations for city road networks where

A.M. Valuev (✉)
Institute of Machines Science named after A.A. Blagonravov of RAS, Bardina str. 4, Moscow, Russia

Moscow Institute of Physics and Technology (State University), Institutskii per. 9, Dolgoprudny, Moscow Region, Russia
e-mail: valuev.online@gmail.com

V.V. Kozlov et al. (eds.), *Traffic and Granular Flow '11*,
DOI 10.1007/978-3-642-39669-4_39, © Springer-Verlag Berlin Heidelberg 2013

the number of road segments between crossroads may be counted in thousands. Other obstacle for these models application is not a computational but principal question: in fact even the flow in a single segment is a mix of flows oriented to different destination points; but this situation is not reflected by known macroscopic models.

Discrete flow models representing the traffic flow on a road lane realistically as a queue of moving cars or, in a more conventional way, as a cellular automaton cannot cope with complexity of the problem of the entire network traffic flow representation as well. It results from the huge dimension of models of these kinds. So solution of any control problem for big city traffic on the basis of such models is practically impossible.

The study of a simple discrete flow models reveals a specific feature for car chains with constant input intensity periodically stopped with red traffic lights. Traffic flow as a whole may be represented as a sum of uniform and periodical components, the latter with zero average value [3]. Certainly, regular trends of traffic flow in different parts of the city road network determine its behavior much more than these periodical or stochastic fluctuations from it and must be taken into account in control problems.

To generalize the above particular results the approach is developed that happened to be very useful for technological network control (e.g., [4,5]). It is so-called quasi-stationary approach where flow in every uniform network arc is treated as having the same density and speed, which change, however, in time. The basic part of the model is Kirchhoff equations of two types, second type equation expression depending on specific features of the network [6].

The paper presents an experience of constructing a model of quasi-stationary traffic flows in a city road network realizing a known demand for transportation. Flow propagation in the model is described analogously to technical networks (e.g., mine ventilation networks), but restrictions on car velocity and the law of safe distance are taken into account as well. The model admits different ways of route choice by car drivers.

2 Substantiation of Quasi-stationary Continuous Flow Model from a Simple Hypothesis of Car Motion Ruled by the Safe Distance Law

Due to the proposed approach me treat traffic flow at any uniform road segment without junctions as a succession of cars moving with the same speed with equal distances between them. If such a flow takes place in reality then there is dependence between speed and distance known as the safe distance law [7]. This fact is confirmed with numerous natural observations and experiments. In fact, the dependence was established for really uniform flows and the question exists whether it approximately expresses average characteristics of traffic flows when the uniformity does not take place. So numerical experiments were fulfilled in which it was supposed that all cars have the same length and each driver tries to hold the safe distance law between it and the preceding car but uniformity of speeds is disturbed in some ways. Speed limitation was taken in account too.

We used for the safe distance law the formula proposed by Tanaka in 1963 [7], namely

$$\phi(v) = L_0 + T_R v + v^2/(2b), \tag{1}$$

where L_0 is an (average) car length, T_R is an average driver reaction time and b an average deceleration. For a given distance Δs between leader car and the follower we assume that the speed of the latter corresponds to (1), so

$$v_{SAFE}(\Delta s) = b(-T_R + \sqrt{(T_R)^2 + 2(\Delta s - L_0)/b}). \tag{2}$$

Within a certain oriented network segment cars are enumerated in order of their appearance at its beginning point. Position of the i-th car is characterized by distance $s_i(t)$ from segment beginning. The dynamics of the entire chain of cars on a fixed segment is described with a hybrid system in which $L(t)$ $M(t)$ are numbers of the first and the last car staying within the segment at the moment t:

$$\dot{s}_{M(t)} = \begin{cases} V_{MAX}(s_{M(t)}), & \text{if } s_{M(t)} < S'' \text{ or } t \in T_{GREEN}, \\ 0, & \text{otherwise}, \end{cases} \tag{3}$$

$$\dot{s}_{i+1}(t) = \begin{cases} v_{SAFE}(\Delta_{i+1}(t)), & v_{SAFE}(\Delta_{i+1}(t)) \leq V_{MAX}(s_{i+1}), \\ V_{MAX}(s_{i+1}), & v_{SAFE}(\Delta_{i+1}(t)) > V_{MAX}(s_{i+1}), \end{cases} \tag{4}$$

$$i = L(t), \ldots, M(t) - 1.$$

where change of $L(t)$ is determined by boundary conditions at the beginning of the segment ($s = S'$) and $M(t)$ from boundary conditions at the segment end

$$M(t+0) = M(t) + 1, \quad \text{if } s_{M(t)} = S'' + 0. \tag{5}$$

Equations (3) admit instantaneous change of the queue leader that is physically impossible. This feature of the model does not affect sufficiently the general picture of traffic presented by the model, since speed jumps for other cars due to (4) are impossible. The model (2)–(4) evolves the approach presented in [3].

Our study revealed conditions in which the dependence between mean values of distance between cars and car speed on the segment with constant $V_{MAX}(s)$ and formula (2). A series of computational experiments was fulfilled with different

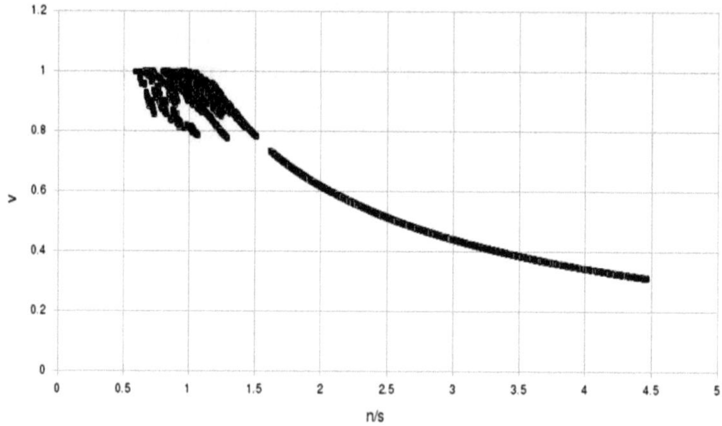

Fig. 1 Dependencies between average speed and density obtained in three computational experiments with probability values 0.1, 0.4 and 0.7 for car appearance at the beginning point

conditions of car input in the beginning point and connection between adjacent segments (with/without traffic lights regulation). Input flow was represented mainly as stochastic one. It was established that in the absence of regulation points representing instantaneous mean values of v and density $(1/\Delta s)$ practically lie on the theoretical curve, except its beginning (see Fig. 1).

In the case of regulation some degree of correspondence between the theoretical curve and points obtained from computation representing average values for the regulation cycle. However results of these computational experiments need additional checkup.

3 Model Formulation

City road network may be represented as an oriented graph which nodes represent junctions and arcs correspond to road segments between them, the j-th arc being the set of collinear lanes having the following parameters: length L_j, the number of lanes N_{ROWj}, maximum allowed speed V_{maxj}. The number of arcs is denoted as n, the numbers of initial and final node of the j-th arc are denoted as $BEG(j)$, $END(j)$. The network topology is defined with sets of adjacent arcs entering its initial point (J_{Inj}, J_{OUTj}) and leaving its end.

The aim of an individual car is to reach a definite j-th arc starting from an i-th arc. The function $Q_{ij}(t)$ expresses average amount of such cars starts per time for the time interval $[t - \Delta t, t + \Delta t]$. The set of $Q_{ij}(t)$ for all i,j represents the entire demand for transportation. Another principal item of the research is the safe distance law. It is assumed that a car moves with the maximum allowed speed V until the distance between it and the preceding car diminishes to the safe distance

$D_{min}(V)$, afterward its speed diminishes to the speed of the preceding car V_1 and a distance to $D_{min}(V_1)$. Therefore, the maximum speed for a uniform flow in the j-th arc is reached when the number of cars in it satisfies

$$N_j \leq N_{\max j}(N_{ROWj}L_j, V_{\max j}) = N_{ROWj}L_j/(D_{\min}(V_{\max j})).$$

We define the quasi-stationary speed as $V_0(N_j/(N_{ROWj}L_j), V_{maxj})$, where

$$V_0 = V_{\max j} \quad \text{for} \quad N_j \leq N_{\max j}(N_{ROWj}L_j, V_{\max j}). \tag{6}$$

Otherwise, V_0 is the root of the equation

$$N_j/(N_{ROWj}L_j) = D_{min}(V_0). \tag{7}$$

To develop the equation set for traffic flow dynamics we introduce Boolean variables $I_{lkj}(t)$ indicating whether cars moving in the l-th arc and aimed at reaching the k-th one choose the j-th arc as the next arc in their route. These variables may be treated in a model either as constants or as functions of flow intensities. With values representing the number of cars at the instant t moving in the jth arc and aimed at reaching the k-th one we obtain the balance equations for quasi-stationary flows (assumed to be uniform in every arc)

$$\begin{gathered} dN_{lk}/dt = \\ \sum_{l \in J_{INj}} \left(I_{lkj}(t) \cdot V_{0j}(\sum_{r=1}^{n} N_{lr}(t)/(N_{ROWl}L_l), V_{\max j}) \cdot N_{lk}(t)/L_l) \right) + \\ Q_{lk}(t) - V_{0j}(\sum_{r=1}^{n} N_{jr}(t)/(N_{ROWj}L_j), V_{\max j}) \cdot N_{jk}(t)/L_j, \\ k = 1, \ldots, n, \end{gathered} \tag{8}$$

$$N_{jj}(t) = 0, \quad j = 1, \ldots, n. \tag{9}$$

4 Using the Model with Different Routing Principles

With the Eqs. (6)–(9) we may compare effects of different routing principles. If all routes are chosen as optimal ones for network with minimum load, then $I_{lkj}(t)$ represent the matrix of optimum routes and are constants. The other principle is the choice among suboptimum routes (which are calculated in advance for every pair of arcs), the chosen route yielding the least time for current or prognosed traffic situation, flow density being assessed by interpolation. In the last case, $I_{lkj}(t)$ functionally depend on $N_{lk}(t)$ or even $N_{jk}(\tau), \tau \in [0, t]$ that may lead to a great amount of calculations but does not principally complicate integration of (3.3).

And the last way is to calculate routes that are (conditionally) optimal with respect to instantaneous values of flow speed in the network arcs. Values of $I_{lkj}(t)$ are expressed via Bellman functions $W_{jk}(t)$ that express conditionally minimal time of reaching the j-th arc from the beginning of the k-th one. Optimality conditions for Bellman function determination are

$$W_{BEG(j)k}(t) = \min\{W_{BEG(i)k}(t) + L_i / V_{0i}(t) \mid i \in J_{INj}\},$$
$$W_{BEG(k)k}(t) = 0, \; k = 1, \ldots, n. \tag{10}$$

Then

$$I_{jkl}(t) = \begin{cases} 1, \text{ if } W_{END(l)k} = W_{END(j)k} + L_l / V_{0l}(t), \\ 0, \text{ otherwise,} \end{cases} \quad l \in J_{OUTj}. \tag{11}$$

Equations (10)–(11) may be used altogether with (6)–(9) in simulation of traffic.

In fact, Eq. (8) are written under the assumption that does not form car queues before them. Taking into account queues before junctions with traffic lights, we must write analogous equations to different segments of arcs.

5 Conclusions

The proposed model treats somewhat ideal situations with no delays on commutation points (it means that no flows intersect). To incorporate in the model crossroads, both regulated and non-regulated, it is necessary to introduce other types of elements, namely queues before crossroads and any intersection sites. It seems that there is no principal obstacles to develop such generalization, but expression dynamics of interacting queues needs detailed study of passing intersection sites by particular cars.

References

1. Kurzhanskiy AA, Kurzhanski AB, Varaiya P (2010) Trudy MFTI 2(4): 100–118 (In Russian)
2. Kurzhanskiy AA (2007) Modeling and software tools for freeway operational planning. PhD thesis, University of California, Berkeley
3. Glukharev KK, Ulyukov NM (2010) Trudy MFTI 2(2): 58–66 (In Russian)
4. Wang YJ (1999) Journal Mine Ventilation Soc South Africa 52(2): 39–43
5. Valuev AM (2011) On Formulation and Solution of a Control Problem for Mine Ventilation Networks. In: 5th International Conference on Physics and Control (PhysCon 2011). IPACS Open Access Electronic Library, http://lib.physcon.ru/doc?id=c5b3d3c32871
6. Abramov FA, Tyan RB, Potemkin VYa (1978). Calculation of mine ventilation networks. Nedra, Moscow (In Russian)
7. Inose H, Hamada T (1975). Road Traffic Control. University of Tokyo Press, Tokyo

Some Problems of Rational Route Choice for a City Road Network

Alexander A. Yakukhnov and Andrey M. Valuev

Abstract The paper treats some problems of rational route choice for a city road network with changing traffic load. Determination of the optimum route in the strict sense as a program acting from the start time of a certain trip is impossible, the cause consisting in its dependence on future behavior of other drivers. Two approaches to the problem setup and solution are proposed. The first consists in definition of the entire set of suboptimum acyclic routes between start and destination points. For the problem solution the generalization of method by Shier for determining the K shortest paths in a network is proposed. On this basis the most reasonable route may be chosen as the suboptimum route with the minimum prognosis duration. The problem in question is that of the optimum route choice for a network with a known traffic dynamics: its implementation may be based on regularity of traffic dynamics trends taking place for a certain day of a week. The problem setup and Dijkstra algorithm modification is proposed. Some aspects of its computational complexity are studied both theoretically and by means of computational experiments. The latter are fulfilled with our computer program aimed for routes optimization and analysis of route method efficiency that may be studied on real networks as well as randomly generated ones. In addition, as an alternative to dynamic programming methods variants of branch-and-bound method are proposed for both problems.

A.A. Yakukhnov (✉)
Moscow Institute of Physics and Technology (State University), Institutskii per. 9, Dolgoprudny, Moscow Region, Russia
e-mail: yakuhnov@mail.ru

A.M. Valuev
Moscow Institute of Physics and Technology (State University), Institutskii per. 9, Dolgoprudny, Moscow Region, Russia

Institute of Machines Science named after A.A. Blagonravov of RAS, Bardina str. 4, Moscow, Russia
e-mail: valuev.online@gmail.com

V.V. Kozlov et al. (eds.), *Traffic and Granular Flow '11*,
DOI 10.1007/978-3-642-39669-4_40, © Springer-Verlag Berlin Heidelberg 2013

1 The General Problem of Rational Route Choice for a City Road Network

The problem of a rational route choice, ideally the choice of the quickest one, arises for any driver in the beginning of its route; acquisition of new information may result in several changes of the route at arbitrary moments of the route. Even for the case when the start and destination points as well as the start time stay the same for a particular driver many times, as for routes from home to the job place, changes in traffic situation may result in change of the optimum route. In fact the term "optimum route" has a conditional meaning. Realizing that in the case of heavy traffic the result of a definite route choice by any driver depends on the same choice by another drivers (including those starting later) we must come to the conclusion that at the start moment and during the entire route a driver principally has no sufficient information for the best route choice even if he would have data from exhaustive monitoring.

To be more exact, the present traffic situation prognosis, even if we take into account its faults, may indicate that a certain route stay better that any alternative routes, if difference between this optimum route and the next suboptimum route exceeds effect of possible faults. On the contrary, if the difference between optimum and suboptimum routes is small, preference of a certain route as the best one is doubtful. However, in the latter case the cost of preference of an arbitrary suboptimum route is low. Thus the suboptimum routes set definition is more useful that attempt to determine the only best route.

It must be noticed that for a certain range of traffic situations and their dynamics (this range may be wide enough) the genuine best route belongs to the set of suboptimum routes established for a certain representative of this range. To recommend a driver a rational route in the present situation it is sufficient to be able to solve two problems:

- To define the suboptimum routes sets for "typical" traffic situations; only routes containing no loops ("acyclic routes") must be taken into consideration.
- To assess present traffic situation and elaborate a reliable prognosis of its dynamics.

The more exact is prognosis, the smaller "typical" traffic situations set is sufficient to cover the best route choice in possible situations with suboptimum route sets for "typical" situations. The problem of prognosis having many ways of solution based on various traffic flow models and prognosis methods is not considered, however, in the paper.

The second approach to optimum route choice is based on knowledge of regular trend of traffic situation within a day (trends for week days and weekends must be treated separately). For long enough routes the optimum path with respect to deterministically changing traffic situation may be other than for initial, final or any intermediate traffic situation. For this type of route choice conditions (they are very typical for large cities) the generalization of the quickest path problem is needed. The proposed problem setup and the method of its solution are proposed below.

2 The Problem of Definition of the Entire set of Suboptimum Acyclic Routes

The double search method was introduced that solve the following problem of K "shortest" (suboptimum) routes [1]. A network is given with known length of each arc and a marked node; a number K is given too. For each node including the marked node (e.g., the 0-th node) one should establish the first K values in the row of length of paths connecting the marked node and this node arranged in the ascending order. Note that the first length is the length of the shortest path (or paths) between nodes, the second is the length of path (or paths) having the next length value (first suboptimum path) etc. To guarantee the problem solvability one must admit paths having loops; for paths from the marked node to itself it goes without saying. Path containing loops are, however, useless as routes for cars in the road network.

The proposed modified method enables simple restoration of routes itself, not only their length, from calculated matrices of preceding nodes; second, it excludes from the resulting paths sets those containing cycles.

In general case the problem is solved for an oriented graph. To determine it, for each j-th node two sets of adjacent nodes are determined, I_{INj} for nodes arcs from which enter the j-th node and I_{OUTj}, for nodes entered by arcs initiating in the j-th node. Generalized Bellman function W_j is a vector with components $W_{j1},\ldots,$ W_{jK} expressing the above mentioned row of suboptimum paths length. For an arbitrary set $\{W^{(1)},\ldots,W^{(m)}\}$ of K-dimensional vectors we determine the operation $\text{MAX}_K\{W^{(1)},\ldots,W^{(M)}\}$ as the operation of fetching K initial values from the succession of values of $\{W^{(1)},\ldots,W^{(m)}\}$ components taken in the ascending order. The operation $\oplus$ of adding a scalar value to all vector components is introduced too. Via these operators Bellman vector equations is represented as

$$W_{01} = 0,\ W_j = \text{MAX}_K\{W_k \oplus d_{kj} \mid k \in I_{INj}\},\ j = 0, 1, \ldots, m-1. \quad (1)$$

where m is the total number of network nodes. In iteration process of (1) solution the value ∞ for components is admitted with obvious rules of its use. Modified algorithm determines simultaneously with W_j a cortege of K sets of nodes $J_{j1},\ldots,$ J_{jK} as subsets of I_{INj} where begin the last arc of paths from 0-th node to the j-th one having respectively lengths $W_{j1},\ldots, W_{jK}$. The set J_{jk} is empty iff $W_{jk} = \infty$. For J_{jk} determination the operation $\text{ARGMAX}_K\{j^{(1)}, W^{(1)},\ldots,j^{(M)}, W^{(M)}\}$ is used which result is the cortege of K sets, the k-th one consisting of all $j^{(l)}$ for which some component of $W^{(l)}$ is equal to the k-th component of $\text{MAX}_K\{W^{(1)},\ldots,W^{(M)}\}$.

With the sets J_{jk} the problem of restoration of the entire acyclic suboptimal paths is solved with a simple algorithm.

3 The Problem of the Optimum Route Choice a for Network with a Given Dynamics of Traffic Flow Intensities

We consider here formulation and properties of the optimum route problem for a network with given dynamics. In networks with a heavy traffic all cars move in a certain uniform road practically with the same speed and pass it in the same time. This speed depends on flow intensity and changes in time. So we introduce the value $t_{ij}(t)$ denoting the time of passing the arc between adjacent nodes i and j provided that motion begins at the moment t. It looks very natural that the moment of reaching the j-th node from the i-th node leaving the latter at the moment t is an increasing function of t; it is defined from the relationship $T_{ij}(t) = t + t_{ij}(t)$.

We solve the problem of optimum routes between the 0-th node and all other ones provided that motion begins at the moment t and denote the sought values $T_j(t)$, $j = 1,\ldots, m$. For our suppositions the following relationships take place that generalize optimality conditions for a static network

$$T_0(t) = t, \tag{2}$$

$$T_j(t) = \min\{T_i(t) + t_{ij}(T_i(t)) \mid i \in J_{INj}\},\ j = 1,\ldots,m-1,\ j \neq 0. \tag{3}$$

It is shown [2] that the solution of the problem in question for a given value of t is given with the following modification of Dijkstra algorithm

Step 1. Let $T_1 = t$, $K = \emptyset$, $K' = \{1\}$.
Step 2. If $K' = \emptyset$, halt. Otherwise determine the node $i \in K'$ on which the minimum of $T_j(t)$ for $j \in K'$ is achieved and pass it from K' to K.
Step 3. Add to K' the set of nodes $J_{2i} \setminus (K \cup K')$ and determine for them

$$T_j(t) = T_i(t) + t_{ij}(T_i(t)). \tag{4}$$

For other nodes from $K' \cap J_{2i}$ let

$$T_j(t) = \min\{T_j(t),\ T_i(t) + t_{ij}(T_i(t))\}. \tag{5}$$

Step 4. Return to Step 2.

Computing alongside with $T_j(t)$ the number $k_j(t)$ of the preceding node on the optimum route between the 0-th and j-th node gives a simple way of the entire optimum route restoration.

Importance of the computational efficiency problem for the modified Dijkstra algorithm depends on context of its solution. As for a particular driver, the problem is solved once (at most, several times) for a route and does not need a great amount of computations exceeding possibilities of a typical PC. The serious difficulty exists in providing him with needed data as for any arc we must have vast arrays of data for determination of dependence $t_{ij}(t)$, so for independent solution of a particular problem great data transfer is required from city traffic control center (CTCC) to a driver's PC. In fact, using the optimization algorithm based on (2) it is enough to know only values of $t_{ij}(T_i(t))$. But in the latter case queries for these values arise during computational process and delays for data transfer may make it too slow.

Another way of the problems solution is their permanent solution for current t and different initial and final nodes in the CTCC and transfer of their solution on query. In this case data transfer problem has less significance but the problem of computational efficiency exists in two aspects: first, to compute a particular problem with a fixed initial node and starting time with minimum computation, second, to use results of solution for previous values of t for computation of optimum routes for a new given starting time.

3.1 Computational Complexity of Dijkstra Algorithm and Possibilities to Diminish It

Complexity of the very problem is determined by the number of nodes m and the number of arcs n. Typical conditions for a city street network are:

1. For network nodes both the number $\deg_{IN}(v)$ of adjacent nodes arcs from each enter the node v and the number $deg_{OUT}(v)$ of adjacent nodes entering with arcs from v are bounded on the entire set of nodes;
2. The following relationship takes place: $2m \le n \le 5m$;
3. There are positive lower and upper bounds for distances between nodes on the entire set of arcs.

For these conditions it is reasonable to assume that dimension of K' in the entire process of algorithm fulfillment does not exceed $a\sqrt{n}$ and for oblong cities it is bounded with a constant value not depending on n.

Main computations for Dijkstra algorithm are

1. Determination of the minimum of $T_j(t)$ (Step 2)
2. Calculations of $T_j(t)$ by formulas (4) and (5) which total amount not exceeds $O(n)$.

If array of pairs $\{j, T_j(t)\}$, $j \in K$ is not ordered, then computational complexity is mainly determined with the first type of computations which amount is assessed for an iteration with $O(m^{1/2})$ and for the entire algorithm with $O(m^{3/2} + n)$.

For Gribov modification or some similar approaches [3–5] the optimum number of ranges of $\{T_j(t), j \in K'\}$ values is $O(n^{1/4})$ and the entire time estimate is $O(m^{5/4} + n)$. But is a mean value and for worse cases there is no preferences with respect to the original method.

Best results are obtained if the set $\{T_j(t), j \in K\prime\}$ is organized as a binary heap (B-tree). Then the guaranteed estimate is $O(m\log_2 m + n)$.

The difference between the above estimates is important for largest cities like Moscow where the number of road network nodes is counted in thousands. Our computational experiments results correspond to these valuations, typical B-tree height for $m = 100 \div 625$ ranges from 5 to 10. Other possible way to achieve high computational efficiency may consist in application of Levit algorithm [6] that does not require a minimization operation at all. Alternative way to methods based on dynamic programming may be application of branch-and-bound type method similar to [7]. It is considered below.

3.2 *Use of Interrelations Between Optimum Paths for Different Start Time*

The idea of using of results of computations for some values of t to solve problems for other t values underlies a principal fact: possible values of t form a continuum set, since possible paths between two nodes only a finite one. Dependencies $t_{ij}(t)$ are mostly unimodal or bimodal, so switches in time between routes serving as the optimum ones are rare; the closer initial and final nodes are situated the less switches exist. So to establish the sequence of optimum routes for all t it suffices to compute them for relatively short sequence of t. To find the optimum route for value t between two values of sequence with different optimum paths it is sufficient to compare times for both paths and choose the best one.

Implementing Dijkstra algorithm it is possible to store integer values

$$k_j(t) = \arg\min\{T_i(t) + t_{ij}(T_i(t)) \mid i \in J_{1j}\},\ j = 1, \ldots, m,\ j \neq 1.$$

If for a given t for each algorithm step the demanded index i is unique (the general case), then for small values of Δt it will be the same for $t + \Delta t$. Then the correspondence between vertices of B-tree and nodes network nodes will stay the same as for t.

Assertion 1. If all functions $T_{ij}(t)$ are monotonous and continuous on the right piecewise linear with a finite number of linearity segments, then for each pair of nodes there is a finite number of intervals on which the optimum routes set are constant.

Validity of the assertion results from the following facts:

1. The number of acyclic routes between each pair of nodes is finite;
2. On intersections of linearity segments of $T_{ij}(t)$ for all arcs for each route its total time is a linear function of time;
3. Minimum of a finite set of linear functions is a piecewise-linear function with a finite number of linearity segments.

Remark 1. We do not say "the optimum route" but "the optimum routes set" because for some time intervals existence of more than one optimum route is possible. It is obvious for regular square grid of roads with the same speed limitations.

Corollary. *For each route P on each time interval of $T_{ij}(t)$ continuity for all its arcs its duration $T_P(t)$ is Lipschitz continuous, constant L being limited with value not depending on P and intervals of t.*

Remark 2. The above substantiation of Assertion 1 results in a definite method for determination of bounds of the optimum path sets. To its usage for a marked initial moment t' it is necessary to determine alongside with which the optimum route the set St' of all suboptimum acyclic routes P, for which for a given value of ΔT motion time satisfies $T_P(t') \leq T_{OPT}(t') + \Delta T$. Then for each $t \in [t', t' + \Delta t]$ where $\Delta t = \Delta T/(2L)$ the optimum route belongs to $S(t')$.

For constancy of $k_j(t)$ values the following necessary and sufficient conditions are valid:

- Each new node added to the B-tree as its next node for $t' > t$ has in it the same position as for t.

Then the B-tree conserves its structure on each iteration (as for t) both when a new vertex is added (Step 3) or deleted (Step 2; then it is the leftmost B-tree leaf).

To get the proposed computation decrease it is necessary to fix the B-tree structure after each iteration. Thus B-trees may be constructed not from beginning.

It is very likely that if the algorithm fulfillment succession is the same for t and $t' > t$ then it will be the same for the entire interval $[t,\ t\prime]$. A simpler test for constancy of the algorithm fulfillment succession is that:

The next value computed as $T_j(t\prime) = T_{kj(t)}(t\prime) + t_{ij}(T_{kj(t)}(t\prime))$ is greater than the preceding one.

Other idea consists in determination of the optimum route to a certain node via optimum routes to adjacent nodes. More detailed results of our computational experiments will be given in the paper presentation on the conference.

The entire set of optimum routes for $t' > t$ is the same that for t if and only if $k_j(t') = k_j(t)$. In that case mostly we have $k_j(t'') = k_j(t)$ for all $t'', t' > t'' > t$. For each node the finite value of $T_i(t')$ is determined when the node is passed to the set K. To be sure that the succession $\{i_l,\ l = 1, \dots, n\}$ expressing the order in which nodes enter the set K stays for $t' > t$ the same that for t the following check must be fulfilled. It must be established whether for each member $j = i_l$ of the above succession the condition is valid

$$k_j(t) = \arg\min\{T_i(t') + t_{ij}(T_i(t')) \mid i \in J_{1j} \cap \{i_l,\ l = 1, \dots, l-1\}\}. \quad (6)$$

If the formula (6) is valid, then $T_j(t')$ may be determined by the formula

$$T_j(t') = T_i(t') + t_{ij}(T_i(t')),\ i = k_j(t).$$

From the above arguments we conclude that for determination of the optimum routes system for all t it is sufficient to determine it on a discrete succession. The

Table 1 Statistical characteristics of computational experiments

The number of nodes m	Database capacity (Kbytes)	Calculation time (s)	Amount of data (Mbytes)	Average tree height	Maximum tree height
100	548	13.4	47.4	5.49	10
100	586	13.9	48.9	5.62	13
100	838	14.1	52.7	6.13	11
100	791	14	51.6	6.02	13
625	3594	51.0	256.3	9.64	20
625	4352	55.7	312.6	10.3	22
1024	8153	120.9	568.1	18.45	40
1024	7946	115.3	549.2	18.63	38
1024	8394	131.8	598.2	18.7	39

problem of determination of t values for which the optimum routes system changes is similar to the problem of roots determination for a scalar argument function and may be solved with the similar methods.

3.3 A Computer Program for Routes Optimization and Some Results of Computational Experiments

The program developed by A. Yakukhnov is intended for solution of the following problems:

1. To create interactive interface for transportation network display and editing:
2. To implement generation of a random network;
3. To calculate the commutation table, i.e., the table of costs of optimum routes between pairs of nodes.

Statistical characteristics of computational experiments fulfilled on randomly generated networks is presented in the Table 1.

4 Variants of Branch-and-Bound Method for Both Problems

A branch-and-bound method aimed at calculation of the optimum flight plan (route) for a civil aircraft was proposed in [7]. Its main characteristic feature is that it finds that route in a dynamic medium, because dangerous weather phenomena and other aircrafts that serve as forbidden areas for free flight move rapidly. When representing these areas as moving polyhedra we obtain for horizontal flight forbidden areas (horizontal sections of spatial areas) as moving polygons. The optimum route consists of arcs linking vertices of these polygons with orthodromy intervals and all these arcs form a network changing in time. So the problem of the optimum flight plan is a particular form of optimum routing problem for a variable network.

Unlike dynamical programming the proposed method solves the only problem of finding the optimum route (or a set of suboptimum routes) between a certain pair of nodes (e.g., the I-th J-th ones) trying not to take into account all nodes that lie beside the path in question. Moreover, when using the method we reduce the network at each method iteration to its subset very slowly enlarging this subset from one iteration to the next one. The possibility of finding the optimum route not taking into account majority of nodes and arcs is based on the fact that road network is planar or almost planar and so lower valuations of route cost beginning from a certain node may be obtained from geometrical calculations. Considering a path cost as the time of passing it and supposing the uniform speed limitation we obtain that:

for an arbitrary network the minimum $T_{IJ} \geq D_{2IJ}/V_{\mathrm{MAX}}$ and
for a rectangular road grid $T_{IJ} \geq D_{1IJ}/V_{\mathrm{MAX}}$,

where D_{2IJ} is a Euclidian distance and D_{1IJ} a "city block distance".

Other way of the method application for dynamical networks is a priori determination of fastest paths on static networks for static networks with minimum or fixed traffic loads and usage of these paths to determine bounds for a dynamic network with exceeding traffic loads in all arcs. It must be emphasized that these static optimum routes are calculated once and dynamically optimum routes many times.

In general, for determination of the optimum path (or a family of suboptimum paths) each node tested as possible intermediate node on the path from the I-th node to the J-th one is assessed with a lower bound of the cost of routes passing it calculated with the formula

$$T_{FACT\ IK}(t_{START}) + T_{MIN\ KJ} \tag{7}$$

Valuation (7) excludes from the solution tree vertices corresponding to nodes that lie beyond the domain of possible optimum (suboptimum) paths except nodes in the vicinity of the initial node. So it may be said that this branch-and-bound method permanently sees the target but dynamic programming based method see possible routes to the target but not the target itself.

When seeking the only optimum path each node is represented with no more than one vertex of the solution tree. On the contrary, when finding the set of suboptimum paths with R minimum costs each node may be represented with up to R vertices, every for some different way of reaching it.

5 Conclusions

Choice of the optimum path in the city road network is not a problem for particular drivers only; it is the problem of efficiency of the road network performance. Recommendations for drivers elaborated in the CTCC in the case of their massive use may result in more rational paths and gradual establishment of dynamic equilibrium of traffic flows that is the best reachable situation. To attain it, permanent monitoring of traffic flows and processing of its results must be combined with efficient ways of particular rational route problems solution.

References

1. Shier DA (1976) Networks (6):205–230.
2. Dijkstra EW (1959) Numerische Mathematik 1:269–271.
3. Anikeich AA., Gribov AB., Surin SS (1976) Computer-aided planning of truck jobs for a shift and a day. Transport, Moscow (In Russian).
4. Denardo EV, Fox BI. (1979) Operations Research 27:161–186.
5. Cherkassky BV, Goldberg AV, Silverstein C (1999) SIAM Journal Comput 28:1326–1346.
6. Levit BYu., Livshits VN (1972). Nonlinear transportation problems on networks. Transport, Moscow (In Russian).
7. Valuev AM, Velichenko VV (2002) Journal Computer Syst Sci Internat 41:979–987.

Weighted Value Feedback Strategy in Intelligent Two-Route Traffic Systems with a Bottleneck

Bokui Chen, Wei Tong, Wenyao Zhang, and Binghong Wang

Abstract Information feedback strategies are attracting keen attention recently as the central part of intelligent traffic systems. Various strategies have been put forward by previous researchers and have been applied in the model of symmetrical two-route scenario. In this letter, a novel strategy for scenario with speed-limited bottlenecks is raised, which is called weighted value feedback strategy (WVFS). Combined with three former strategies, we simulated these four into a two-route scenario with a speed-limited bottleneck. The results show that our strategy wins over the other three in effectively enhancing and balancing the vehicle numbers, as well as increasing the average flux on both routes.

1 Introduction

With the rapid development of global economy, traffic jams in urban areas have been more and more serious, which have aroused the interest of scientists in different fields to solve the puzzle with their expertise. Physics scientists are ones who have made contribution, many concepts and methods in physics have been applied in it [1–3]. During the past half century, a series of theoretical models to solve the phenomenon in traffic flow have been put forward, such as car-following model [4], kinetic model [5, 6] and particle hopping model [7]. Recently, the research on

B. Chen (✉) · W. Tong · W. Zhang
Department of Modern Physics, University of Science and Technology of China, Hefei 230026, China
e-mail: chenssx@mail.ustc.edu.cn

B. Wang
Department of Modern Physics, University of Science and Technology of China, Hefei 230026, China

The Research Center for Complex System Science, University of Shanghai for Science and Technology and Shanghai Academy of System Science, Shanghai 200093, China

V.V. Kozlov et al. (eds.), *Traffic and Granular Flow '11*,
DOI 10.1007/978-3-642-39669-4_41, © Springer-Verlag Berlin Heidelberg 2013

intelligent traffic systems has received more and more attention and to fabricate a reasonable information feedback strategy has been the most important part to solve the traffic congestion. Physics have already proposed many strategies and put them into the application of symmetrical two-route scenario. In 2000, Wahle et al. first put forward travel time feedback strategy (TTFS) [8]. Subsequently, Lee et al. raised mean velocity feedback strategy (MVFS) [9]. Then Wang et al. proposed the third one, called congestion coefficient feedback strategy (CCFS) [10]. All of them have been simulated in the symmetrical two-route scenario with two exits and the result shows that CCFS is better than others. The others three strategies have been put forward recently by Dong et al., named vehicle number feedback strategy (VNFS) [11], weighted congestion coefficient feedback strategy (WCCFS) [12] and corresponding angle feedback strategy (CAFS) [13]. All of these have been applied into a symmetrical two-route scenario with one exit and they have proved that these three strategies are better than TTFS, MVFS and CCFS [11–13].

Accidents always lead to the change of the condition of the routes in our life, such as the traffic accident, or terrible weather like dense fog and blizzard which block the roads and limit the driving speed. Based on such situation, in this letter a novel information feedback strategy is put forward in order to deal with the two-route traffic systems with a speed-limited bottleneck, called weighted information feedback strategy (WVFS). We will compare our new strategy with VNFS, WCCFS and CAFS, and finally come to the conclusion of optimal strategy according to the simulation results.

2 Related Definitions

2.1 *Traffic Flux*

The road conditions can be characterized by flux of two routes, and flux is defined as follows:

$$F = V_{mean}\rho = V_{mean}\frac{N}{L}, \tag{1}$$

where V_{mean} represents the mean velocity of all the vehicles on one of the roads. N denotes the vehicle number on each road and L is the length of two routes.

2.2 *NS Mechanism*

Vehicles on routes move according to Nagel-Schreckenberg (NS) rules [14] as follows:

1. Acceleration: $v_i(t) \to v_i(t + 1/3) = \min\{v_i(t) + 1, v_{max}\}$;
2. Deceleration: $v_i(t + 1/3) \to v_i(t + 2/3) = \min\{v_i(t + 1/3, d_i(t))\}$;

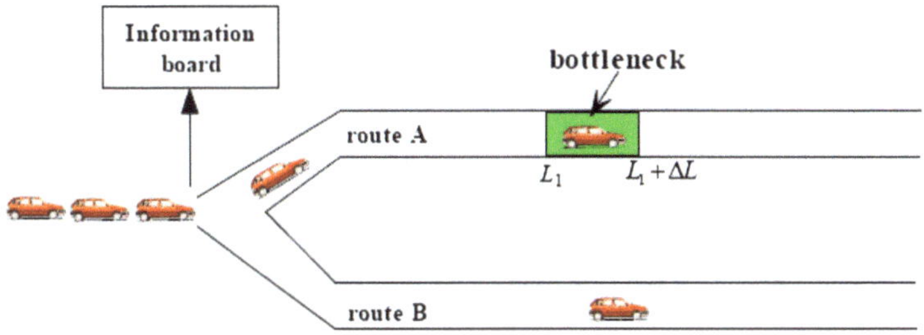

Fig. 1 Sketch of two-route scenario with a speed-limited bottleneck

3. Randomization with probability $p : v_i(t + 2/3) \rightarrow v_i(t + 1) = \max\{0, v_i(t + 2/3) - 1\}$;
4. Vehicle motion: $x_i(t + 1) = x_i(t) + v_i(t + 1)$.

The v_{max} above is the maximal velocity of a car and $d_i(t)$ is the number of empty cells in front of car i. In their simulations, the road is divided into cells of length 8 m. Each cell can either be empty or occupied by just one vehicle at a certain time. Meanwhile, the time is divided into time steps of 1 s. Thus, a velocity of n means $n \times 8$ m/s.

2.3 Two-Route Scenario with a Speed-Limited Bottleneck

In previous research, the model was symmetric two-route, that is obviously far from the real situation. To better handle the complicated road condition, symmetrical two-route scenario with a speed-limited bottleneck is introduced in this letter (see Fig. 1). At every time step, a new vehicle is generated at the entrance of two routes and one route will be selected. If a vehicle enters one of two routes, the motion of it will follow the dynamic of the NS mechanism. When the vehicle arrives at the exit, it will be deleted.

In this model there are two types of vehicles: dynamic and static vehicles. Suppose a driver is a dynamic one, he will make a choice in light of the information feedback. On the other hand, a static one selects routes to enter at random ignoring any advice. The ratio of dynamic and static travelers are S_{dyn} and $1 - S_{dyn}$, respectively.

2.4 Feedback Strategies

Therefore Dong et al. have proved that VNFS, WCCFS and CAFS are better than TTFS, MVFS and CCFS, therefore we only do the comparison between our new strategy with VNFS, WCCFS and CAFS.

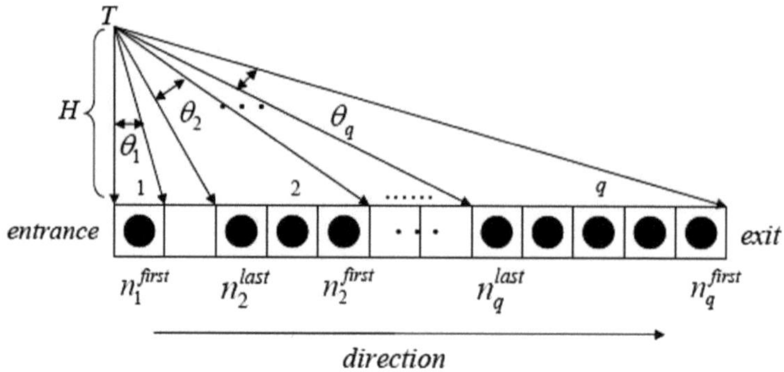

Fig. 2 Angles corresponding to each congestion cluster on the lane

VNFS: Every time step, the traffic control center will receive data of the total vehicle number of the first 500 cells of each route, then shows them on the board. Dynamic drivers at the entrance will choose the route with smaller vehicle number [11].

WCCFS: At each time step, the traffic center collects the information of the vehicles on the routes, then calculates the congestion coefficient according to a weighted function and displays the results on the board. Dynamic drivers choose the route with the smaller congestion coefficient. The weighted congestion coefficient is defined as:

$$C_w = \sum_{i=1}^{m} F(n_{ij}) n_i^w = \sum_{i=1}^{m} \left(k \times \frac{n_{ij}}{2{,}000} + 2.0 \right) \times n_i^w \tag{2}$$

where n_i stands for vehicle number of the i-th congestion cluster.. n_{ij} stands for the middle position of the ith congestion cluster. Here $w = 2$ [12]. k is a factor to be determined. According to the simulation results, we found out that for the two-route scenario with a speed-limited bottleneck, the average flux reaches maximum at $k = -1.7$.

CAFS: Every time step, the traffic control center will receive data from the navigation system (GPS). It is its function that the traffic control center computes the corresponding angle of each congestion cluster (see Fig. 2) on the route, then sums square of each corresponding angle up and displays it on the board. Dynamic drivers at the entrance will choose one road with smaller corresponding angle coefficient. The corresponding angle coefficient is defined as:

$$C_\theta = \sum_{i=1}^{m} \theta_i^2 = \sum_{i=1}^{m} \left(\arctan\left(\frac{n_i^{first}}{H} \right) - \arctan\left(\frac{n_i^{last} - 1}{H} \right) \right)^2 \tag{3}$$

where n_i^{first} and n_i^{last} stand for the positions of the first and last vehicle in the ith congestion cluster. θ_i stands for the weight (corresponding angle) of the ith congestion cluster. H denotes the vertical distance from the point T to the lane, and in this letter, we set $H = 100$ [13].

WVFS: At every time step, the traffic control center gets the information from the GPS and then calculates the weighted value of two routes according to the following:

$$Wl = \begin{cases} \frac{1}{l_A} + \frac{n_a}{L_a^2}, & \text{route } A \\ \frac{1}{l_B}, & \text{route } B \end{cases} \tag{4}$$

where l_A and l_B stands for distances of the last vehicles on route A and route B to the entrances, respectively. L_a stands for the middle position of the bottleneck. n_a is the number of vehicles in the traffic bottleneck. Then the information will be displayed on the board. Dynamic drivers at the entrance will choose the route with the smaller weighted value.

3 Simulation Results

In the simulation, the length of the routes is set to be $L_A = L_B = 2{,}000$ cells and the speed-limited bottleneck appears on road A between 300th and 500th cells. In the bottleneck region, maximum speed of vehicles is set to be $v_{max} = 1$, otherwise $v_{max} = 3$. The ratio of dynamic drivers is $S_{dyn} = 0.85$ and $p = 0.25$ is the random break probability in NS mechanism [11]. The simulation results of vehicle number, speed and flux shown here are obtained by 100,000 iterations excluding the initial 95,000 time steps. When it comes to the dependence of average flux on dynamic travelers, we obtain the simulations based on 10 times average over 100,000 iterations in order to show more precise results.

Figure 3 displays how the vehicle number changes as time passes when adopting four different strategies. From this figure we can see that the number of vehicles on route B is about 400 or so for all of the strategies. But for the route A which has a speed-limited bottleneck, the vehicle number is remarkably smaller than the route B by using VNFS, WCCFS and CAFS. On the contrary, when we adopt WVFS, the numbers of vehicles on route A and route B are almost equal. Therefore WVFS can better balance the vehicle number on two routes. Since we presume that the two routes are of the same length in our model and the density of the vehicles is defined as the vehicle number per unit length, larger vehicle number means greater vehicle density. Someone may be worried about that the high density will lead to traffic jams. We find that the vehicle number on each route is about 400 when adopting WVFS, but the length of each route is 2,000, which means each vehicle occupies 5 cells and there is 3 or 4 cells' space between neighboring vehicles. In this situation, the probability of traffic jams is low. However, high density of vehicles can easily

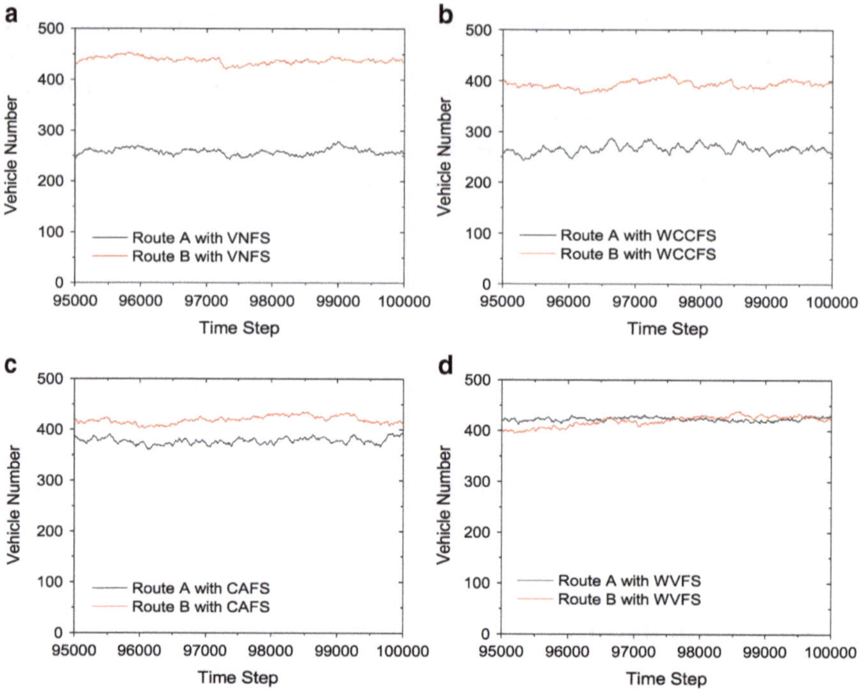

Fig. 3 Vehicle number of each route with (**a**) VNFS, (**b**) WCCFS, (**c**) CAFS, (**d**) WVFS. The parameters are $L_A = L_B = 2{,}000$, $p = 0.25$, $S_{dyn} = 0.85$, $k = -1.7$ in WCCFS

lead to jams. Therefore the stability of vehicle number is very important for the traffic. In Fig. 3, we can see that compared with other strategies, the vehicle number is more stable for WVFS. So in terms of the ability to accommodate more vehicles of a route, WVFS is the optimal strategy.

Nevertheless, it is not reasonable enough to come to the conclusion of which one is the best only according to the vehicle number. More vehicles on the road can lead to the decrease of average velocity. Fortunately, two parts should be taken into consideration about the flux–the density of vehicles multiply the average velocity. Combining these two factors can we describe the capacity of the routes more objectively. So we do further research about the flux. Figure 4 shows the result of flux versus time by applying four strategies, from which we can see that both the value and the stability are almost the same. In this way, we have to take advantage of average flux for judgment.

The average flux means the following quantity when the value of S_{dyn} is given:

$$F_{ave} = \frac{\sum_{i=1}^{n} \sum_{j=1}^{t} f_{ij}}{t \times n} \tag{5}$$

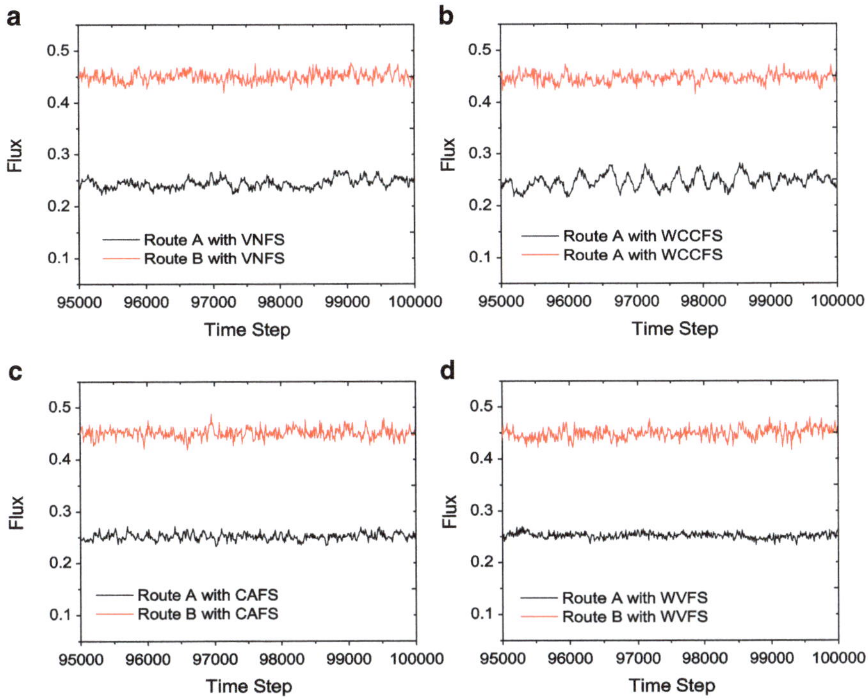

Fig. 4 Flux of each route with (**a**) VNFS, (**b**) WCCFS, (**c**) CAFS, (**d**) WVFS. The parameters are set the same as in Fig. 3

where f_{ij} is the flux of the ith route at time j, t is the total time and n is the number of route. From this definition, we find that the average flux is the reflection of flux based on both the time and route average, thus it can reflect the merits of the strategies more precisely. Figure 5 shows how the average flux changes along with the ratio of dynamic drivers when adopting four different strategies, all of which are among 0.336 and 0.350. We find that when using CAFS, the average flux nearly remains the same as S_{dyn} goes up. Therefore no matter how many drivers prefer to accept the advice on board, the average flux changes little and so does the capacity of the road. We also find that, as S_{dyn} runs from 0 to 0.5, the tendencies of the average flux are almost the same when the VNFS, WCCFS and WVFS are adopted. If S_{dyn} runs from 0.5 to 1, the average flux decreases by using the VNFS and WCCFS, but the average flux keeps stable when WVFS is adopted. Thus we can conclude that WVFS is the best choice whatever the value of S_{dyn}.

We also investigate the influence of bottleneck length and bottleneck location. Figure 6a reflects the affect of the bottleneck length on the average flux when the WVFS is adopted and the head of bottleneck is set at $L = 300$. When the length of bottleneck increases from 100 to 200, the average flux decreases obviously. But when the bottleneck length goes up from 200 to 1,500, the decrease of average flux

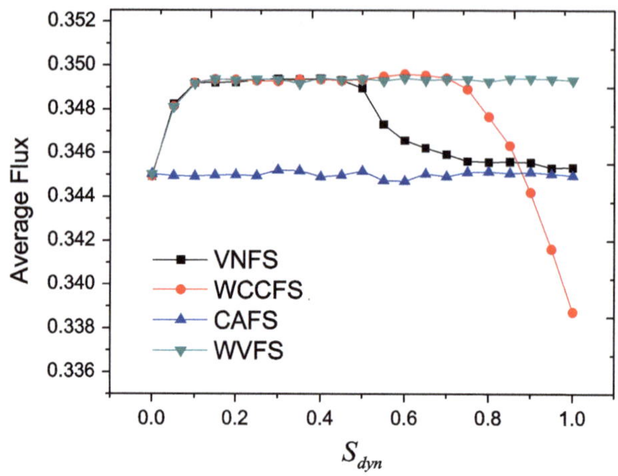

Fig. 5 Average flux by performing different strategy versus S_{dyn}

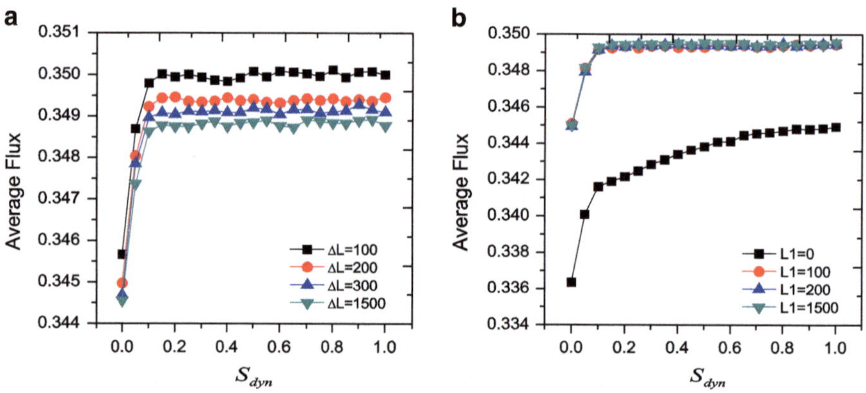

Fig. 6 (**a**) Average flux vs the length of bottleneck when adopting WVFS. (**b**) The influence of bottleneck location to the average flux when adopting WVFS

is not comparatively obvious. That is to say, when the head of the bottleneck is settled, the average flux will decrease dramatically and then keeps stable as the length increases. The traffic situation of the preceding part of the route affects greater on the capacity of the whole scenario. Figure 6b can also serve as evidence. It displays the influence of bottleneck location to the average flux when the WVFS is adopted and the bottleneck length is set as $\triangle L = 200$. If the bottleneck is decided at the entrance, the flux is low. The average flux does not increase obviously as the position of the head of bottleneck changes from 100 to 1,500. Thus we deduce that the greatest negative influence appears when the head of bottleneck is set at the entrance. As the head of bottleneck moves to the exit, the influence decreases.

4 Conclusion

In this letter, we put forward a novel information feedback strategy called weighted value feedback strategy for a two-route scenario with a speed-limited bottleneck. Combined with previous work, we compare WVFS with VNFS, WCCFS and CAFS, and apply all of them on the model for simulation. The relation of the vehicle number and flux versus time is obtained, as well as the average flux versus proportion of dynamic drivers. From the simulation results, WVFS cannot only effectively enhance and balance the vehicle number on both routes, but increase the average flux on the routes as well. The influence of length and location of bottlenecks on the scenario is also studied carefully, which shows that when the bottleneck is at the entrance, the negative impact to the whole routes is the greatest. The negative impact decreases and finally falls stable with the location of a bottleneck leaving the entrance.

Acknowledgements This work was funded by the National Important Research Project (Grant No. 91024026), the National Natural Science Foundation of China (No. 10975126), and the Specialized Research Fund for the Doctoral Program of Higher Education of China (No. 20093402110032).

References

1. Lehmann H (1996) Phys. Rev. E 54: 6058–6064
2. Wagner C, Hoffmann C, Sollacher R, Wagenhuber J, and Schürmann B (1996) Phys. Rev. E 54:5073–5085
3. Nagatani T (2002) Rep. Prog. Phys 65:1331–1386
4. Jiang R, Wu Q S, and Zhu Z J (2001) Phys. Rev. E 64: 017101
5. Helbing D (1996) Phys. Rev. E 53:2366–2381
6. Helbing D and Treiber M (1998) Phys. Rev. Lett 81:3042–3045
7. Biham O, Middleton A A, and Levine D (1992) Phys. Rev. A 46:R6124–R6127
8. Wahle J, Bazzan A L C, Klügl F, and Schreckenberg M (2000) Physica A 287:669–681
9. Lee K, Hui P M, Wang B H, and Johnson N F (2001) J. Phys. Soc. Jpn 70:3507–3510
10. Wang W X, Wang B H, Zheng W C, Yin C Y, and Zhou T (2005) Phys. Rev. E 72:066702
11. Dong C F, Ma X, and Wang B H (2010) Science China-Information Sciences 53:1–7
12. Dong C F, Ma X, and Wang B H (2010) Physics Letters A 374:1326–1331
13. Dong C F and Ma X (2010) Physics Letters A 374:2417–2423
14. Nagel K and Schreckenberg M (1992) J. Phys. I France 2:2221–2229

On Traffic Control Means Recognition in Intelligent Monitoring and Traffic Safety

M.V. Yashina and A.A. Vinogradov

Abstract Methods of automatic recognition of traffic and pedestrian control means are in high demand today. Many problems in this field are not solved yet, or solved with significant restrictions in their work. Our approaches and methods help to advance these complex problems closer to practically important goals. Here we describe our approaches to smartphone-based recognition of traffic lights, traffic signs and "zebra" pedestrian crossings. These are all needed and very important for ensuring road safety and traffic regulation.

1 Introduction

Methods of automatic recognition of traffic and pedestrian control means, such as road signs, traffic lights, pedestrian crossings, etc. in intelligent monitoring problems are very important for both organizations in charge of the placement and maintenance of those means, and drivers directly in movement. These methods are in high demand on the actively developing market of navigational computer software and systems, for analysis of traffic flow characteristics and for ensuring safety in saturated traffic and pedestrian flows.

Our recognition problems are solved using modern consumer video cameras, smartphones and computers. Developed applications can be used for driver escort and assistance as well as for driver warning in dangerous situations.

Some problems in this field can be considered as almost solved. For example, *the problem of vehicle registration number recognition* is deeply researched and is widely used in existing intelligent systems, i.e. in [1]. However, many more problems are not solved yet, or solved with significant restrictions in their work. One of such problems is *the automatic recognition of parking rules violations*, the

M.V. Yashina (✉) · A.A. Vinogradov
Moscow Technical University of Communications and Informatics (MTUCI), Moscow, Russia
e-mail: yash-marina@yandex.ru

V.V. Kozlov et al. (eds.), *Traffic and Granular Flow '11*,
DOI 10.1007/978-3-642-39669-4_42, © Springer-Verlag Berlin Heidelberg 2013

Fig. 1 MSRR mobile laboratory

automatic (or even semi-automatic) solutions of which are not yet implemented. Our approaches and methods, presented further, help to advance these complex problems closer to practically important goals.

Currently, there is a rapid increase in the degree of integration of electronic vision devices into everyday life. The improvement of technical capabilities of devices for video capture, information receiving and recording, data exchange, their raising compatibility with various computing devices allow monitoring of required objects, visual and statistical information processing, development of real-time mobile systems for objects recognition. The relevance of capture and automatic processing of data in real-time for the legal issues arising from events on the road network can also be noted.

We aim to explore capabilities of medium level consumer devices and computers that can be connected in a system, for which we can create algorithms for solving problems of recognition for ensuring driver safety, as well as for data capture, including cases where a legally relevant visual confirmation is needed.

In the Scientific-Educational Center "Intelligent Monitoring, Communication and Transport Control" (SEC IMCTC) the technologies of mobile video monitoring of flows is developed under the supervision of Prof. A. P. Buslaev since the beginning of the 2000s [2–5].

In 2005 a photo camera-based monitoring system was built for solving the problem of road sign recognition. Such technologies are being tested on mobile laboratories, on which special equipment is mounted. One of these laboratories – the Mobile Street Road Receptor ("MSRR") – is shown in Fig. 1.

It has to be noted that similar systems were then being actively developed around the world. For example, an algorithm using neural networks was demonstrated in [6]. Such approaches undoubtedly have educational value, but are not fit for usage in realtime systems due to high computational complexity of the algorithm.

Now means of video capture have reached an acceptable quality to solve the challenges of object recognition in real time while capturing the information directly from the device's video stream. In the next section we describe the recognition of traffic lights, which is very different from the problem of recognition of road signs, as described in Sect. 3. The difficulty in recognizing traffic lights on an image or a video stream is that they glow. Whereas road signs contain distinct monochrome colors and brightly stand out of natural objects, traffic lights glare and gleam, which leads to non-uniform distribution of color on the object that changes depending on the angle of view. Due to the fact that the traffic lights and road signs are in the field of vision of the camera for a short time, their real-time recognition algorithms should be as simple and fast as they can be. It will be taken into account below that the problem of object recognition can be divided into two stages: localization (A) and object feature detection (B).

2 Recognition of Traffic Light Objects

2.1 Problem Description

Input data for the problem algorithm is an image or frame of video stream in the RGB color format, from a camera mounted on the dashboard of a moving vehicle (Fig. 2.). This perspective allows us to analyze the same information flow that the driver receives while driving. We need to find the object on the image, specify the type of traffic light and calculate its characteristics.

(A) The localization process is to select the input image or video frame rectangular area containing a traffic light;
(B) The recognition process is to process the areas obtained in the localization process.

The problem of traffic light recognition is largely determined by the traffic light image capture method type:

1. Capture with a stationary camera;
2. Capture from a moving camera, mounted on the vehicle dashboard.

For the problem of recognizing a traffic light with a stationary camera, we need to restore traffic light mode function of $\eta(t)$, which is defined by the five possible values depending on the time t (frame number):

$$\eta(t) = \begin{cases} 0, \text{if at the moment t no light is active (black);} \\ 1, \text{if at the moment t red light is active;} \\ 2, \text{if at the moment t red and yellow lights are active;} \\ 3, \text{if at the moment t yellow light is active;} \\ 4, \text{if at the moment t yellow and green lights are active;} \\ 5, \text{if at the moment t green light is active.} \end{cases}$$

In the case of the recognition of traffic lights with a moving video source we only want to determine the momentarily active traffic light mode.

Also to meet the challenges of mapping the distribution of required objects and evaluation of the effectiveness of their location, a link of objects with GPS coordinates is necessary. The coordinates of the found object are transferred to the server side, where a map of recognized objects is stored.

As mentioned above, the detection of traffic lights on video sequence is complicated by a high degree of dispersion of the color signal in the image (Figs. 3 and 4).

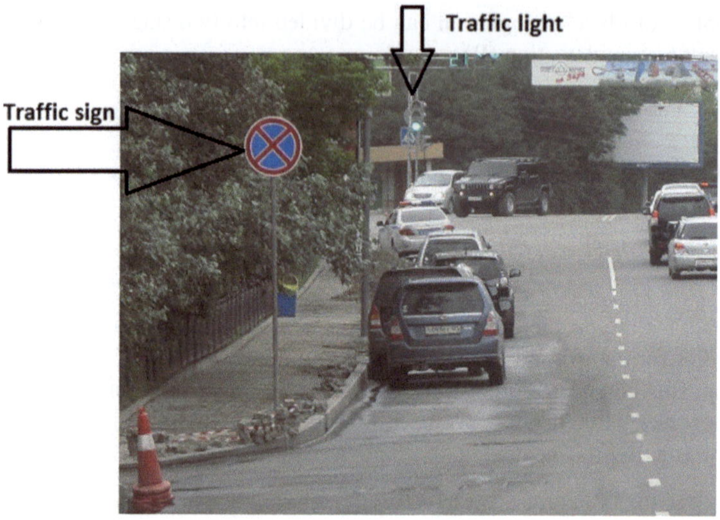

Fig. 2 Russian mentality features

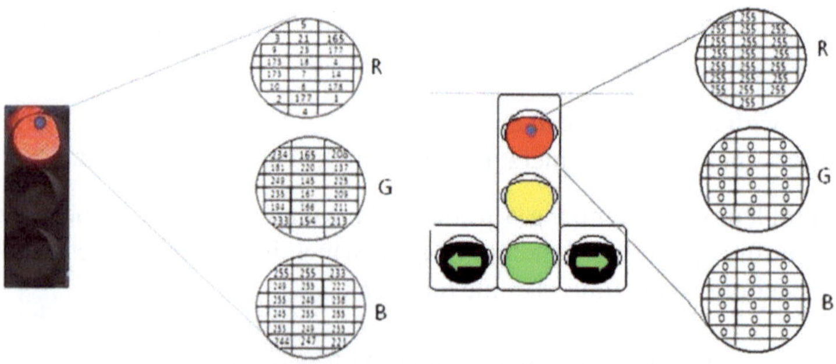

Fig. 3 *Red* signal has a high degree of dispersion in the RGB color matrix

2.2 *Traffic Light Recognition Analysis from a Stationary Video Source*

This separate problem is of great practical importance. To implement the methods of obtaining information from monitoring systems as well as information about traffic management systems, an external automatic independent control over their work is needed. For the accuracy of the information a long processing of the parameters of the system running is needed. This will allow to collect statistical data, which will help evaluate and address the shortcomings of the existing system or its current settings.

Fig. 4 Traffic light color change depending on environmental conditions

Initial localization, or finding a traffic light on the video stream, can be done manually or automatically. Traffic light is determined by the color and geometric properties: we search for colored spots that are located in a straight vertical line close to each other and having the similar size. Color clusters of primary colors (red, yellow and green) are determined. We also calculate the diameter and the center of mass of spots. If the center points of the color spots have coordinates

$$(x_r, y_r), (x_y, y_y), (x_g, y_g),$$

then the traffic light is determined by the following condition:

$$abs\begin{vmatrix} x_r - x_y & y_r - y_y \\ x_g - x_y & y_g - y_y \end{vmatrix} \le \varepsilon.$$

Thus, after a full cycle of the traffic light, we obtain a rectangular region that describes the desired object (Fig. 5).

The following requirements for video source position and angle necessary for the correct use of the developed applications were established:

Fig. 5 Automatic traffic light object localization as implemented on a smartphone

- *Stationary capture angle.* The object of analysis must be in the same location in the frame throughout the capture and must not move in the frame. To provide the necessary camera immobility a stand or bracket can be used to secure the capture device;
- *When using manual localization: conformity of the recognition area with the object of research.* The user selected area should cover the analyzed object and exclude analysis of the image background. Fulfillment of this requirement directly affects the accuracy of the analysis and the information obtained. Sufficient application accuracy is achieved with an accuracy of selection (the presence of the required object in the selected area) not less than 90 %;
- *Absence of overlapping objects that prevent recording.* Moving objects that enter the processing area frame and interfere with the analysis of the object considerably reduce the accuracy and make it impossible to obtain the correct information;
- *Weather and light conditions.* The application can operate with sufficient accuracy under different lighting conditions and weather. However, in special cases, such as heavy rain, snowstorm or night capture, manual adjustment of application parameters may be needed for proper operation.

Applications for stationary PCs and smartphones running Google Android were developed. Portable implementation functionally copies the implementation for the PC, with the exception of that the automatic localization takes longer and there is no output of calculated parameters charts and diagrams.

After finding the area of traffic light object its processing and the evaluation of numerical parameters takes place. We find numerical parameters such as:

- The current active mode of each traffic light;
- Durations of previous modes of each traffic light;
- The percentage of the activity of each mode in the current cycle;
- The mean percentage of the activity of each mode for all cycles;
- Pie charts of traffic light modes in the current cycle and in all cycles of operation (Fig. 6);
- Graphs of active modes and color channels distribution (Figs. 7 and 8).

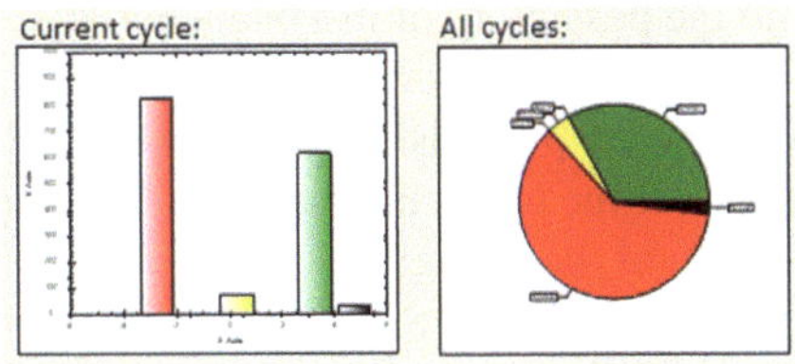

Fig. 6 Analysis of the function $\eta(t)$

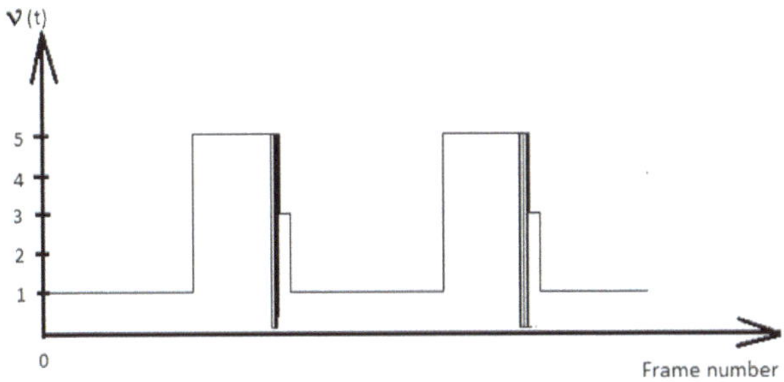

Fig. 7 $\eta(t)$ function plot

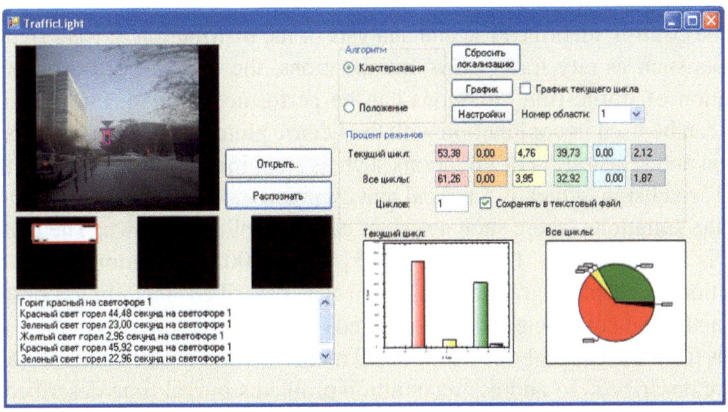

Fig. 8 TrafficLightRec PC application interface

2.3 *Traffic Light Recognition with a Moving Video Source*

To create various systems for ensuring road safety, such as "black box" vehicle systems, a solution of the problem of online object recognition is required. For example, the system can advise the driver not to go past a red traffic light signal while passing an intersection. This allows fast processing of incoming information and its verification, with the possibility of instantaneous signaling to the driver and transmitting information to the server for further processing and storage. To solve these problems the use of the simplest and most lightweight algorithms are required [7, 10]. For example, to detect the traffic light active mode a geometric criterion can be used, that is, the position of the changing segment in the area of the traffic light. This algorithm, in some cases, has a large error, but sometimes it is more convenient and faster than a color recognizing algorithm.

The usage of smartphones as client-side devices can reduce the overall system cost and increase its scalability – at the moment there are many different models of mobile devices that are suitable for the solution of problems of recognition and monitoring. Almost any of them can be used to create computer vision systems. Specific technologies and means of implementation of the recognition problem depend on the choice of smartphones and their operating systems. In particular, for smartphones running Google Android the OpenCV software library exists, that contains a large number of methods of computer vision adapted to mobile platforms. Its use greatly facilitates the creation of software for recognition systems and saves time and device performance load.

3 Traffic Sign Recognition Problem

Using the control, identification and analysis of the distribution of road signs in the city tasks such as city traffic flow optimizations, the removal of traffic jams and registration of traffic rule violations can be performed (Fig. 9). For example, the system can be used in conjunction with the license plate recognition systems for the detection and registration of violations such as stopping and parking under the "No Parking" road sign, which is a typical and common violation in Moscow. In Fig. 10 one of the situations where such a system can be applied is shown. The plan of the system is shown in Figs. 11 and 12. In [2] in the 2005, a solution for traffic sign recognition using photographic equipment was described, including a description of the method of color detection of localized road signs (color vectorization). These methods form the basis of our work. In [2] methods for detecting all types of traffic signs are described. To address recognition problems in real time described in this article, only some road signs are important. Among those particularly important to recognize the signs "No Stopping", "Speed limit," "No passing," " Pedestrian crossing" can be distinguished. To be able to use the application in the real world conditions, we want to automate the process of finding the sign on the frame (localization) and the recognition of the sign, reducing the user actions required to prepare the system and for use in the operation of the system to a minimum. The localization process is conducted based on two kinds of features: color and geometric. Many signs that are important for us have a clear red borderline on the edges of the sign. Therefore, the first step is the normalization of the red color in the image, that is, bringing different shades of red to a perfect red color

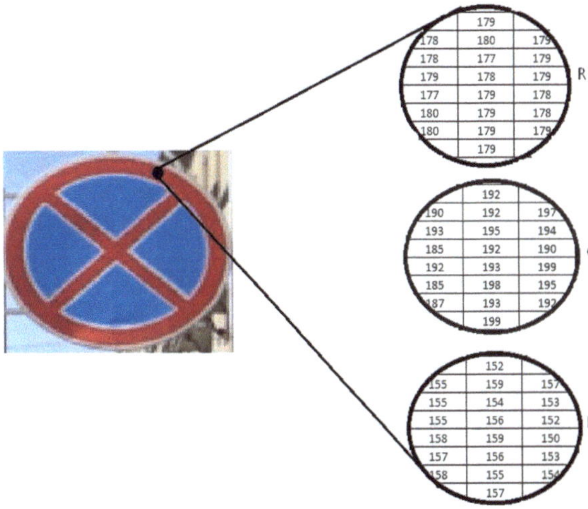

Fig. 9 *Red* color dispersion on a traffic sign

Fig. 10 A violation of the road sign 3.27 No Stopping

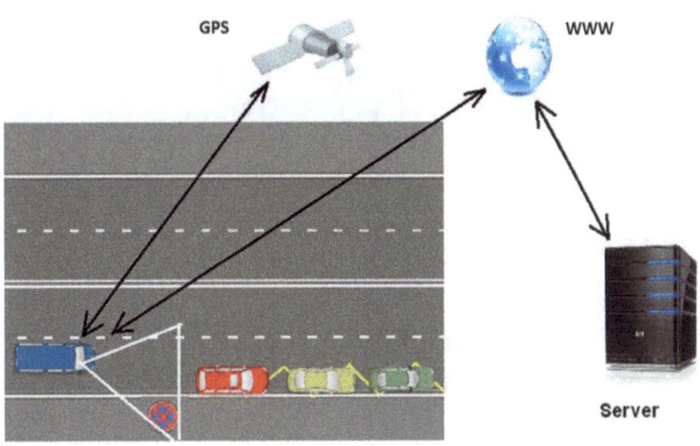

Fig. 11 Parking rules violation recognition system concept scheme

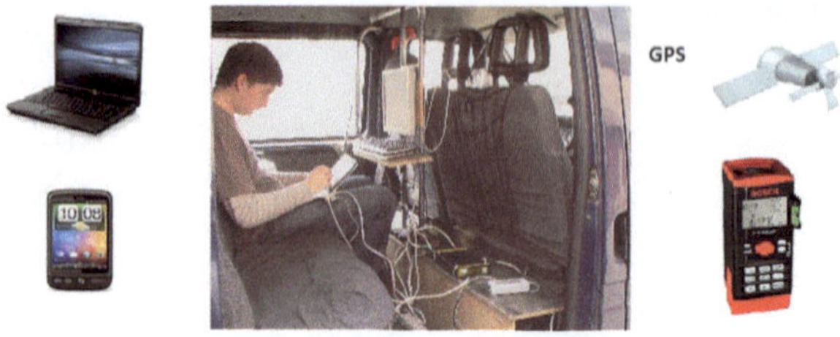

Fig. 12 An existing traffic situation recognition system (MUDRETS) [4]

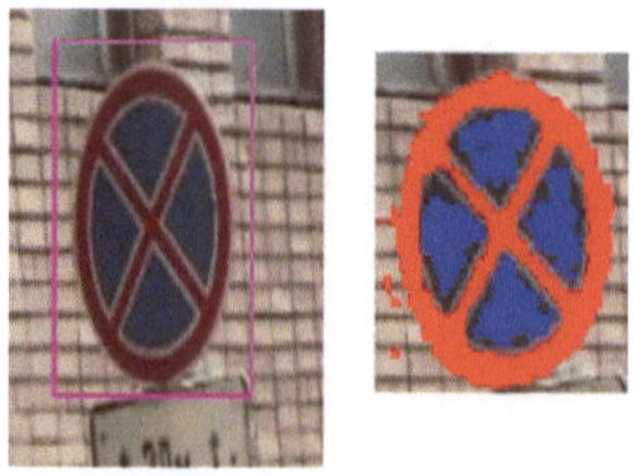

Fig. 13 *Red* color normalization

Fig. 14 Road sign localization

(RGB: 255, 0, 0), and the search of closed borders in red (Fig. 13). On the second step we evaluate the form of the found borders. If an ellipse or a triangle is recognized, then we say that the sign is localized in the found area (Fig. 14). The process of recognition of the sign is performed on a rectangular area drawn around the localized area (elliptic or triangular). The area is divided into 16 equal parts

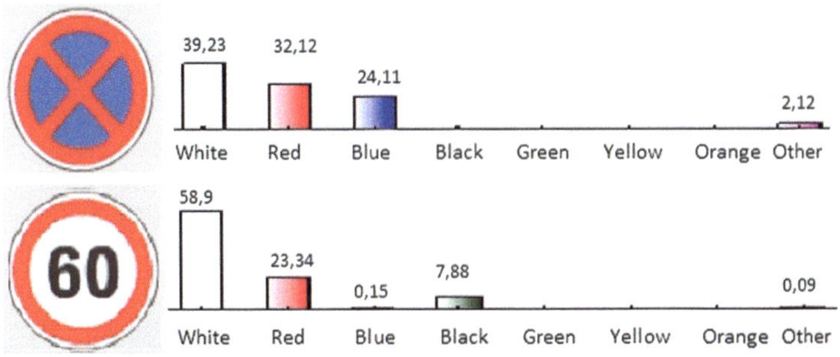

Fig. 15 Color vectorization

and color vectorization is performed for each of the parts (Fig. 15). A comparison of found color vectors with color vectors from the road sign database is then performed. If the distance between the current road sign color vector and the color vector of the database sign is less than the threshold value, the sign is considered recognized and the user is alerted.

4 Recognition of a Pedestrian Crossing

To ensure safety on the road the problem of pedestrian crossing real-time recognition on a video stream with recognition of its occupation is also very important. Moreover, the developed methods can be used for detecting violations by drivers as well as violations by pedestrians on the crossing. The task is to find a pedestrian crossing using a mobile video source, mounted on a car dashboard, [8]. According to the traffic rules of the Russian Federation, pedestrian crossing can be indicated with markings ("zebra"), and road signs (Fig. 16). This task is of high practical importance, especially in Russia. The relatively recent (2009) changes in the traffic rules introduced fines for drivers who do not stop before a pedestrian, who stepped onto the "zebra" road crossing. Unfortunately, because of this many pedestrians think they have a right to cross the road when and where they want, and literally jump under the wheels of passing cars, even during active prohibiting pedestrian traffic light signal.

Methods of road signs recognition have been described in Sect. 3. In the case of a pedestrian crossing sign, detecting the red edge of the prohibitory sign may be replaced by a blue border detection of the road crossing sign. Otherwise the recognition method remains the same. For detecting crosswalk road markings an image area binarization and evaluation of the graph of the distribution of white pixels in the search area algorithm is used. Binarization is performed for the image converted to black and white. Binarization threshold (intensity of gray) is 128 (in range from 0 to 255).

Fig. 16 Pedestrian crossing indication

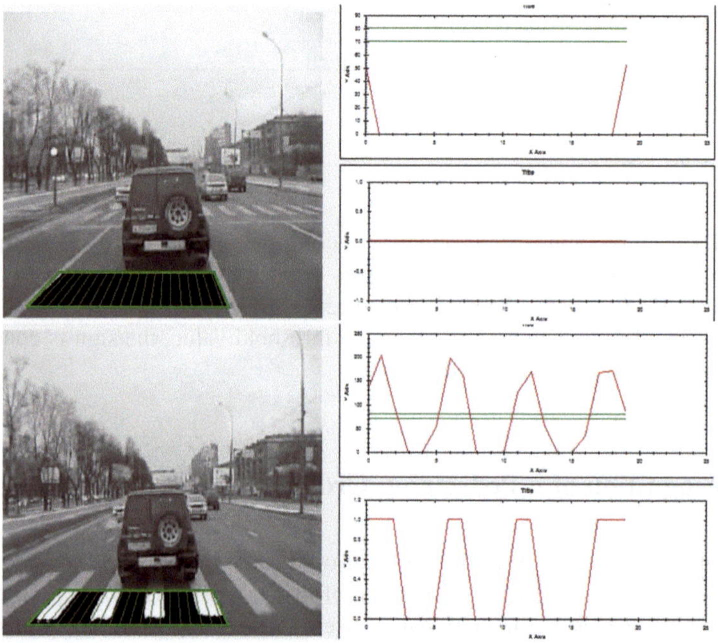

Fig. 17 Binarization of the distribution function

There are two ways of processing the selected search area, [9]:

- Wide viewing angle (DVR): Vertical rectangular search area processing. In this case, the X axis corresponds to the column number of the rectangular search area, the Y axis – to the number of white pixels in the column;
- Narrow viewing angle (video camera).

Processing is made by trapeze parts. X-axis corresponds to the number of trapezoids, Axis Y – to the mean intensity of color in the current part. Before the analysis of the distribution function of white pixels in the search area its binarization is preformed. If the Y coordinate of the specified point is greater than the upper binarization threshold value the coordinate value is set to 1. If the coordinate Y is below the lower binarization threshold value it is set to 0. If the value is between binarization thresholds it is rounded to the nearest integer value. In this case, a warning of a possible detection error is displayed (Fig. 17). If the chart contains three peaks or more, the area is considered to contain a crosswalk. In this case, an alert is

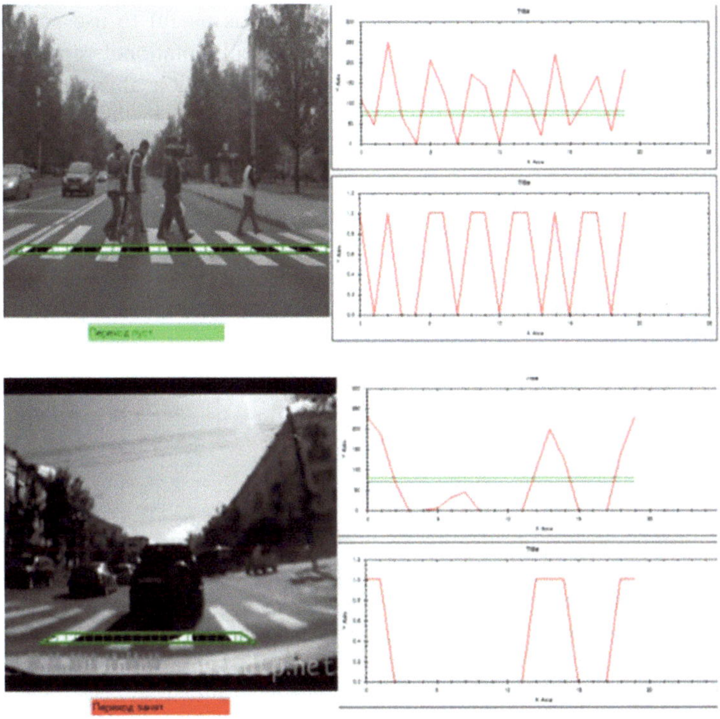

Fig. 18 Crossing occupation determination

displayed in the application window. To determine if the crossing is occupied the uniformity of the area distribution graph is estimated. We compare all the peaks and the distances between the peaks of the distribution function by their length. If there is no difference or small difference the crossing is considered empty. If there is an average difference a message of possible crossing occupation is displayed. If we see a large difference indication of crossing occupation is displayed (Fig. 18).

Functionally, the road crossing recognition algorithm consists of three steps:

1. Recognition of the crossing road sign;
2. Recognition of the zebra crossing;
3. Recognition of moving objects on the crossing (optional).

5 Conclusion

To map the distribution of traffic signs and evaluate the placement of traffic control elements a link to the current GPS-coordinates is important for the GPS-track creation. The coordinates can be sent to the server for tracking in real-time and long-term storage. To implement these methods it is useful to use smartphones as client devices. Although they have much lower computational capabilities and camera quality than the video camera + laptop connection, they are much less demanding on power, take up little space and are easy to install. The described methods have a portable implementation and can be used on smartphones running Google Android OS. An analysis of the algorithm complexity and devices' computational capabilities has been conducted, and an effective system architecture was determined and created. Thus, there are several options for creating monitoring systems using the elements of the described methods. Each option has its advantages and disadvantages, and should be selected based on the specific requirements and objectives.

References

1. ISS. Intelligent Security Systems, *Vehicle license plate recognition system Autoinspector* , http://www.iss.ru/ (In Russian).
2. Buslaev A.P., Dorgan V.V., Krulov E.A., Prikhodko V.M., Travkin V.Ur., Yashina M.V., *Automatic recognition of road sign control and evaluation of its effectiveness*, Journal of science and technology on transport, No 2, (2005) (In Russian).
3. Buslaev A.P., Dorgan V.V., Kuzmin D.M., Prikhodko V.M., Travkin V.Ur., Yashina M.V., *Image recognition and monitoring of road pavement, traffic flows and movement safety factors*, Vestnik MADI(STU), (2005) 102–109 (In Russian).
4. Buslaev A.P., Kuzmin D.M., Yashina M.V., *Computer methods of information processing. Mobile Road Laboratory.* MADI(STU) (2008) (In Russian).
5. Buslaev A.P., Novikov, V.M. Prikhodko, A.G. Tatashev, Yashina M.V., *Stochastic and simulation approach to traffic*, Moscow, Mir (2003) - p. 286.
6. Auranuch Lorsakul, Jackrit Suthakorn, *Traffic Sign Recognition Using Neural Network on OpenCV: Toward Intelligent Vehicle/Driver Assistance System*, Intelligent Service Robotics, (2009), 1–19.
7. Buslaev A., Yashina M., Kotovich I., *On problems of intelligent monitoring for traffic*. Logic Journal of IGPL: doi:10.1093 /jigpal/ jzq032 Oxford University Press (2010).
8. Buslaev A.P., Guo J.M., Wang N.J., Yashina M.V., *On recovery of plane object shape by projections*. Proc. of WorldComp-09, Conference on Image Processing, Computer Vision and Vizualization, (2009).
9. Buslaev A.P., Kuzmin D.M., Yashina M.V., *Computer methods of information processing and images recognition in problems of traffic and communication. Part 1–4.* MTUCI (2008) (In Russian).
10. Buslaev A.P., Yashina M.V., *Virtual detectors, transformation and application to pattern recognition problems*, Proc. of CMMSE-09 (2009).

Interaction of Cars and Bicycles on a One-Way Road Intersection: A Network CA-Based Model

Jelena Vasic and Heather J. Ruskin

Abstract Modelling of heterogeneous traffic including non-motorised modalities is a topic of increased interest, as 'greening' becomes an integral part of transportation science. The variation of form among these heterogeneous flows means that models developed to represent them are just as diverse. The particular case of interest here is that of lane-sharing between bicycles and motorised vehicles, with positional discipline. A cellular automata-based model is developed and applied for the study of this kind of mixed traffic.

1 Introduction

In the push towards 'greening' of urban transport, encouragement and facilitation of alternative modalities features prominently, owing to the associated benefits that span environmental, health and social domains. Heterogeneous traffic flows including non-motorised modes, especially bicycles, have attracted less extensive modelling efforts to date. Our work offers a contribution in this area by way of a model for the type of heterogeneity formed through lane sharing with 'positional discipline', which is characteristic of Dublin and other cities where dedicated bicycle infrastructure is scarce and streets relatively narrow.

The added complexity involved in modelling heterogeneous traffic, as compared to that of the mode-homogeneous case, has two components: one that stems from differences in vehicle properties and driver/rider behaviour among different mode types and another arising from unique interactions that occur between specific pairs of modalities. In terms of these, the inclusion of the bicycle in a traffic model requires a representation of cyclists' behaviour and the allowance for interactions between bicycles and motorised traffic.

J. Vasic (✉) · H.J. Ruskin
Dublin City University, Dublin 9, Ireland
e-mail: jvasic@computing.dcu.ie; hruskin@computing.dcu.ie

V.V. Kozlov et al. (eds.), *Traffic and Granular Flow '11*,
DOI 10.1007/978-3-642-39669-4_43, © Springer-Verlag Berlin Heidelberg 2013

All bicycle-focused modelling work must, indeed, include some manner of the former. However, characteristics of bicycle-only flow and related road capacities, including the bicycle-only fundamental diagram and levels of service, have been the central topic of a number of publications. These are reviewed in detail in [1], where the authors also present a cellular automaton bicycle flow model of their own, representing 'two abreast' bicycle flows. Another bicycle-only flow model using cellular automata, where multiple bicycles can occupy a single cell and distinction is made between slow and fast cyclists, is presented in [2].

Interactions between bicycles and motorised vehicles are *implicit* in models of broadly heterogeneous traffic, found, e.g., in many Asian countries, where any type of vehicle can occupy any lateral position on the road. A number of models aimed at representing this form of heterogeneity, all based on space-continuous simulation, are reviewed in [3]. A lane-based scenario including bicycles is modelled using car-following rules in [4]. Cellular automaton (CA) models described in [5] and [6] represent similar scenarios and were validated using real data and existing simulation models. While applied to heterogeneous motorised-only traffic, these CA models allow for differently-sized vehicles, including motorcycles, through multiple-cell occupancy and could easily be applied to a pedal-bicycle inclusive case.

Separately identifiable interactions between bicycles and motorised vehicles can be classified into *lateral interference* and *cross-flow*. The former occurs where bicycles and motor vehicles are moving side-by-side and interfere with each other's motion, mostly causing deceleration of the other vehicle type. The latter are interactions arising from intersections of bicycle flows with motorised ones, often in circumstances created exclusively by the presence of bicycles in traffic. For example, if bicycles and cars are sharing a lane by *positional discipline*, which means that bicycles keep to the left[1] and motorised vehicles to the right of the shared lane, cars turning left are in conflict with the bicycle flow and the two types of vehicle affect each other's movements. In the cellular automaton model presented in [7], lateral interference between a car lane and the adjacent multi-lane bicycle stream is represented through a higher probability for cars to slow down in circumstances of 'friction' or 'blockage' caused by bicycles. In [8], the lateral interference type of interaction is introduced into an optimal velocity model as a friction component accounting for the effect of pedestrians on cyclists and cyclists on motorised vehicles, with the different types of vehicles moving in designated but spatially adjacent lanes. A slightly different question, of how the general network flows are affected by different vehicle types moving side-by-side on individual links in the network, is posed in [9], where interactions are accounted for in a combined forecasting model through link impedance functions. Here, higher bicycle flows increase the impedance of motorised flows and vice versa. Cross-flow interactions are the subject of work described in [10], wherein a logit model is proposed for

[1]Our model assumes left-hand side driving, that in effect in Ireland, UK etc., without loss of generality.

the relation between the flow of turning cars and the straight-moving bicycle arrival rate, based on bicycle flow properties derived from empirical data. In [11] the same scenario is modelled building on the cellular automaton (CA) flow representation from [2], where a single cell can be occupied simultaneously by multiple bicycles. Here, general CA rule application takes place synchronously, with an exception made in instances of conflict between car and bicycle flows. These conflicts are resolved through stochastic designation of update sequence among the flows in conflict, on a case-to-case basis. Finally, interactions between left turning bicycles and straight moving cars on a two-way street are studied in [12] using a two-dimensional optimal velocity model.

Ours is a general cellular automata (CA) simulation model, primarily suited to representing lane-based traffic with positional discipline in case of lane sharing, is applicable to any vehicle type mix and accommodates both lateral interference and cross-flow types of interaction. The application here is to traffic including bicycles and cars on the intersection of two one-way streets. The scenario is used singly and as the building block of a 16-node network, in both cases under periodic boundary conditions, for a study of the relationship between traffic densities and flows. The model, which has been presented in some detail in [13], is summarised in Sect. 2, while the simulation scenario and results are presented in Sect. 3.

2 Model

The **spatial aspect** of the model uses the one-dimensional cellular automata (CA) space, or a *track*, as a building block. A track consists of cells of equal size, each occupied by a single vehicle or empty. Vehicle positions in the track are updated iteratively according to some rules with the aim of reproducing traffic flow dynamics. The iterations represent changes in the system during successive fixed time intervals or *time steps*. Any route that may be taken in the simulation scenario by any type of vehicle, as it navigates the model space, is represented as a track, resulting in a space consisting of tracks, each with cells of a size appropriate for the type of vehicle it accommodates and intersecting with other tracks, as dictated by the simulation scenario. Tracks may connect so as to form longer tracks, intersect, diverge, converge (the two latter cases corresponding to pairs of tracks that are identical up to, or starting with, a certain cell) or be adjacent to each other, forming spatial features that must be handled by the update rules, in addition to basic movement along a track.

The spatial elements of the simulated scenarios, each built from a number of tracks, are shown in Fig. 1. This figure is also intended to serve as an illustration of the general spatial modelling method, in which the model is 'extracted' from a sketch of the modelled space, hence the hand-drawn pictures. Figure 1a, b show the tracks modelling an intersection of two one-way streets and a one-way road stretch, respectively. In Fig. 1a the bicycle cells are marked in some detail. The car track

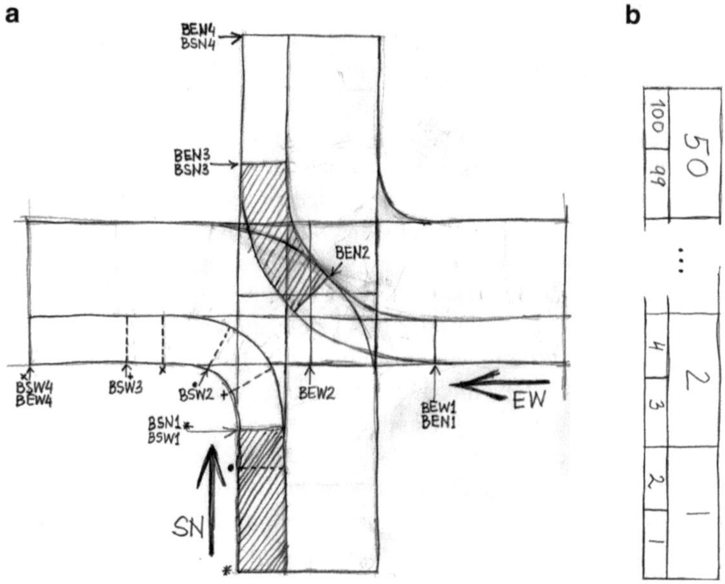

Fig. 1 Sketch for model of an intersection of two one-way streets (**a**) and for a straight road stretch (**b**), with mixed car and bicycle traffic sharing a road with positional discipline (bicycles stay to the *left* and cars to the *right*, with reference to the direction of movement). The four movement directions in the intersection (south-north [SN], south-west [SW], east-west [EW] and east-north [EN]) and the road stretch are each represented by two tracks – one for cars and one for bicycles, positioned side by side. The bicycle cell front lines are marked in (**a**) using the cell names, BSN1 for the first cell in the bicycle SN CA space, etc. A matching pair of symbols, such as * or + are used to mark the beginning and end lines of cells in the BSW CA space, indicating overlap between cells, which is used to model the slowing of vehicles caused by the turn. The shaded areas are examples of bicycle cells. The *arrows* indicate the travel direction. The car and bicycle tracks constituting the straight road stretch in (**b**) consist of 50 and 100 cells, respectively

cells are not shown to avoid cluttering the picture, however, they consist of: track CSN 2 cells, CSW 4 overlapping cells, CEW 2 cells and CEN 4 overlapping cells.

The actual spatial model used by the update rules is based on information extracted from sketches such as those in Fig. 1. This information consists of the item types described in Table 1 together with actual information for the elemental models in Fig. 1.

The space in which the update rules operate must be completed with a conflict resolution method, which can be considered a **control component** of the model. To use the elemental model in Fig. 1a as an unsignalised intersection, priority is assigned to either the south-north or the east-west direction, resulting in the left-hand-side (LHS) and the right-hand-side (RHS) rule, respectively. In the case of cross-flows on the same road, the straight moving flow always has priority, i.e., track BSN has priority over CSW and track CEW over BEN.

Table 1 Spatial model information item types and information item values extracted from the pictures in Fig. 1

Information item type	Value for inter-section model in Fig. 1a	Value for road stretch model in Fig. 1b
Description of tracks in terms of cell size, cell count, direction of cell numbering (corresponding to that of vehicle movement) and cells at which a turn (left or right) starts in the track	Car cells correspond to real lengths of 7.5 m, while bicycle cells correspond to half that length, i.e., 3.75 m; tracks BSW, CSW, BEN and CEN each have a turn, starting at cell 1	
Track connections (where one track extends another one so that the first cell of the extending track follows the last cell of the extended one)	The first cell of any track in the intersection and road stretch models may follow the last cell of another track and vice versa, if the followed and following cell are the same size	
Cell overlap instances	Shown in Table 2	None
Conflicts and divergences	Shown in Table 3	None
Indication as to whether any two tracks are geometrically positioned so as to cause inter-track interaction (other than that at conflicts) between vehicles and if yes, what kind of interaction	All adjacent car-bicycle track pairs imply interaction between bicycles and cars on those two tracks, by virtue of the tracks' proximity; cars decelerate in the presence of bicycles	

The **update rules** are based on those defined for traffic on a single-lane road by Nagel and Schreckenberg in [14]. These rules can be formulated using a *combined limit* value as follows:

0. Determine the combined limit value: $v_{\text{CLi}} = \min(v_{\text{MAX}}, d_i)$
1. Acceleration: if $v_i < v_{\text{CL}}$, $v_i \rightarrow v_i + 1$
2. Slowing: if $v_{\text{CLi}} < v_i$, $v_i \rightarrow d_i$
3. Randomisation: with probability p_{R}, $v_i \rightarrow v_i - 1$
4. Vehicle motion: each vehicle is advanced v_i cells

where v_{CLi} is the combined limit value for the ith vehicle, v_{MAX} is the maximal velocity for the vehicle type, d_i is the number of free cells to the nearest other vehicle ahead of vehicle i and v_i is the velocity of vehicle i. The variables are dimensionless: distance is measured in cells and velocity in cells per time step.

The update rules used herein modify the Nagel-Schreckenberg rules by (i) substituting d_i with $d_{\text{U}i}$, which is the number of *unimpinged* cells ahead of the ith vehicle (a cell is impinged if an overlapping cell, including itself, is occupied) and (ii) including another three limiting factors in the combined limit value:

$$v_{\text{CL}} = \min(v_{\text{MAX}}, d_{\text{U}i}, v_{\text{LT}}(d_{\text{T}i}), v_{\text{LC}}(d_{\text{C}i}), v_{\text{LB}}(d_{\text{B}i})) \quad (1)$$

where $v_{\text{LT}}(d_{\text{T}i})$ is the velocity limit imposed by the proximity of a turn, as a function of the distance to the turn, $d_{\text{T}i}$, and $v_{\text{LC}}(d_{\text{C}i})$ and $v_{\text{LB}}(d_{\text{B}i})$ are analogous values relating to unresolved conflicts ahead of vehicles and bicycles ahead of cars on

Table 2 Cell overlap table for the two one-way street intersection spatial model in Fig. 1a. Each table entry indicates whether the cell associated with its row and the cell associated with its column overlap (1) or not (0). The matrix corresponding to the table is, as expected, symmetrical

	CEN4	CEN3	CEN2	CEN1	BEN4	BEN3	BEN2	BEN1	CEW2	CEW1	BEW4	BEW3	BEW2	BEW1	CSW4	CSW3	CSW2	CSW1	BSW4	BSW3	BSW2	BSW1	CSN2	CSN1	BSN4	BSN3	BSN2	BSN1
CEN4	1	1	1	1	0	0	0	0	0	1	0	0	0	0	0	0	0	0	0	0	0	0	1	1	0	0	0	0
CEN3	1	1	1	1	0	0	0	0	0	1	0	0	0	0	0	0	0	0	0	0	0	0	1	1	0	0	0	0
CEN2	1	1	1	1	0	0	0	0	0	1	0	0	0	0	0	0	0	0	0	0	0	0	1	1	0	0	0	0
CEN1	1	1	1	1	0	0	0	0	0	1	0	0	0	0	0	0	0	0	0	0	0	0	1	1	0	0	0	0
BEN4	0	0	0	0	1	0	0	0	0	0	0	0	0	0	0	0	0	0	0	0	0	0	0	0	1	0	0	0
BEN3	0	0	0	0	0	1	0	0	1	1	0	0	0	0	1	1	1	0	0	0	0	0	1	0	0	1	0	0
BEN2	0	0	0	0	0	0	1	0	0	1	0	0	1	0	0	1	1	1	0	0	0	0	0	1	0	0	0	0
BEN1	0	0	0	0	0	0	0	1	0	0	0	0	0	1	0	0	0	0	0	0	0	0	0	0	0	0	0	0
CEW2	0	0	0	0	0	1	0	0	1	0	0	0	0	0	1	1	1	0	0	0	0	0	1	1	0	1	1	0
CEW1	1	1	1	1	0	1	1	0	0	1	0	0	0	0	0	1	1	1	0	0	0	0	1	1	0	0	0	0
BEW4	0	0	0	0	0	0	0	0	0	0	1	0	0	0	0	0	0	0	1	1	0	0	0	0	0	0	0	0
BEW3	0	0	0	0	0	0	0	0	0	0	0	1	0	0	1	1	1	1	0	1	1	0	0	1	0	0	1	0
BEW2	0	0	0	0	0	0	1	0	0	0	0	0	1	0	1	1	1	1	0	0	0	0	0	1	0	0	0	0
BEW1	0	0	0	0	0	0	0	1	0	0	0	0	0	1	0	0	0	0	0	0	0	0	0	0	0	0	0	0
CSW4	0	0	0	0	0	1	0	0	1	0	0	1	1	0	1	1	1	0	0	0	0	0	1	0	0	1	1	0
CSW3	0	0	0	0	0	1	1	0	1	1	0	1	1	0	1	1	1	1	0	0	0	0	1	1	0	1	1	0
CSW2	0	0	0	0	0	1	1	0	1	1	0	1	1	0	1	1	1	1	0	0	0	0	1	1	0	1	1	0
CSW1	0	0	0	0	0	0	1	0	0	1	0	1	1	0	0	1	1	1	0	0	0	0	0	1	0	0	1	0
BSW4	0	0	0	0	0	0	0	0	0	0	1	0	0	0	0	0	0	0	1	1	0	0	0	0	0	0	0	0
BSW3	0	0	0	0	0	0	0	0	0	0	1	1	0	0	0	0	0	0	1	1	1	0	0	0	0	0	1	0
BSW2	0	0	0	0	0	0	0	0	0	0	0	1	0	0	0	0	0	0	0	1	1	1	0	0	0	0	1	1
BSW1	0	0	0	0	0	0	0	0	0	0	0	0	0	0	0	0	0	0	0	0	1	1	0	0	0	0	0	1
CSN2	1	1	1	1	0	1	0	0	1	1	0	0	0	0	1	1	1	0	0	0	0	0	1	0	0	0	0	0
CSN1	1	1	1	1	0	0	1	0	1	1	0	1	1	0	0	1	1	1	0	0	0	0	0	1	0	0	0	0
BSN4	0	0	0	0	1	0	0	0	0	0	0	0	0	0	0	0	0	0	0	0	0	0	0	0	1	0	0	0
BSN3	0	0	0	0	0	1	0	0	1	0	0	0	0	0	1	1	1	0	0	0	0	0	0	0	0	1	0	0
BSN2	0	0	0	0	0	0	0	0	1	0	0	1	0	0	1	1	1	1	0	1	1	0	0	0	0	0	1	0
BSN1	0	0	0	0	0	0	0	0	0	0	0	0	0	0	0	0	0	0	0	0	1	1	0	0	0	0	0	1

Table 3 Conflicts and divergences for model in Fig. 1a. A conflict is defined with the first cell in conflict of each the left and the right involved track. For example, a conflict between tracks BSN and CSW starts at cell 2 of the former and cell 1 of the latter. A divergence is, similarly, defined with the first divergent cell of each the left and the right involved track

Conflict or divergence id, prefixed C or D, respectively	First cell of conflict/divergence in left track	First cell of conflict/divergence in right track
D1	BSW1	BSN1
D2	CSW1	CSN1
D3	BEW1	BEN1
D4	CEW1	CEN1
C1	BSN2	CSW1
C2	BSN2	BEW3
C3	BSN2	CEW2
C4	BSN3	BEN3
C5	CSN1	BEW2
C6	CSN1	CEW1
C7	CSN1	BEN2
C8	CSN1	CEN1
C9	BSW2	BEW3
C10	CSW1	BEW2
C11	CSW1	CEW1
C12	CSW1	BEN2
C13	BEN2	CEW1

adjacent tracks, respectively. The three velocity limit functions have been chosen so as to allow vehicles to reach a turn, conflict or bicycle at velocity 1, 0 and 1, respectively (only in the case of an unresolved conflict does a vehicle actually have to stop), while decelerating by, at most, 1 at any time step. The update is performed in parallel, which means that in each time-step the velocity update rules (1–3) are applied to all vehicles in the simulation, then the position update rule (4) is applied to all the vehicles. The rules are identical for bicycles and cars but the value of v_{MAX} is decided separately for the two types of vehicle.

Navigation of the multi-track space is handled using two system-wide parameters: probability of turning right, p_{TR}, and probability of turning left, p_{TL}.

3 Simulation Results

The two scenario spaces, shown in Fig. 2, were built using the elemental models from Fig. 1. The following parameters and initial conditions apply to both: maximal velocity, v_{MAX}, is 3 for cars and 2 for bicycles; the randomisation parameter is

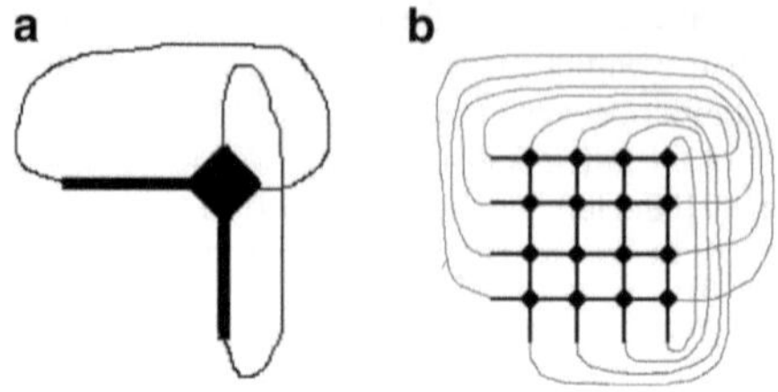

Fig. 2 Schematic representation of simulation scenario space for intersection of two one-way streets (**a**) and a 4×4 node network (**b**), both with closed boundary conditions. The *diamond shapes* each represent the intersection from Fig. 1a, while the *straight lines* between nodes represent the element in Fig. 1b. The pair of points at the ends of each free-form line are directly connected in the model, to form closed boundaries

$p_R = 0.1$; simulation length is 10^5 timesteps, each corresponding to 1s; the turning probabilities, p_{TL} and p_{TR} and vehicle densities are varied; initial vehicle spatial distribution is homogeneous among and within tracks, initial velocity for all vehicles 0. Flows are measured at the cells preceding all the southern- and east-most intersection entry points.

The results of some of simulation instances are shown in Figs. 3 and 4. While the priority bicycle flows for the LHS rule take the form of a standard fundamental diagram (Fig. 3a), the car flows are reduced at high bicycle densities (b), because of the slowing effect built into the model for adjacent bicycle and car tracks. The non-priority flows are low, as expected (c, d). The same conditions on each of the nodes in the network scenario produce similar, but fairly 'noisy' priority-flow diagrams for the case with 50-cell edges (e, f) and ones almost identical to those for the isolated intersection in the case of 200-cell edges (g, h). The diagram in Fig. 4a shows a case of a scaled fundamental diagram, owing to increased densities on two perpendicular tracks. This is reproduced in the network case (b). Another effect that can be seen is that of gridlock causing sudden flow failure, due to increase of other vehicle type density cross-flows (c). The diagrams are rather random in the network case (d, e), but the effect is still visible, particularly for the scenario with 200-cell edges (e). A simulation case that exhibits both the scaled fundamental diagram and gridlock is shown in Fig. 4f. The last two diagrams (g, h) show a case where the priority east-west bicycle flow is 'protected' from diagram scaling, due to high densities in the adjacent car track, which do not allow bicycles turning from the south-north track easy entry into the east-west one.

The application of this method of modelling is envisaged as a useful and systematic approach for the investigation of networks that accommodate heterogeneous traffic of the type encountered in old city centres, such as Dublin's, but also for other types of networks and other vehicle type mixes.

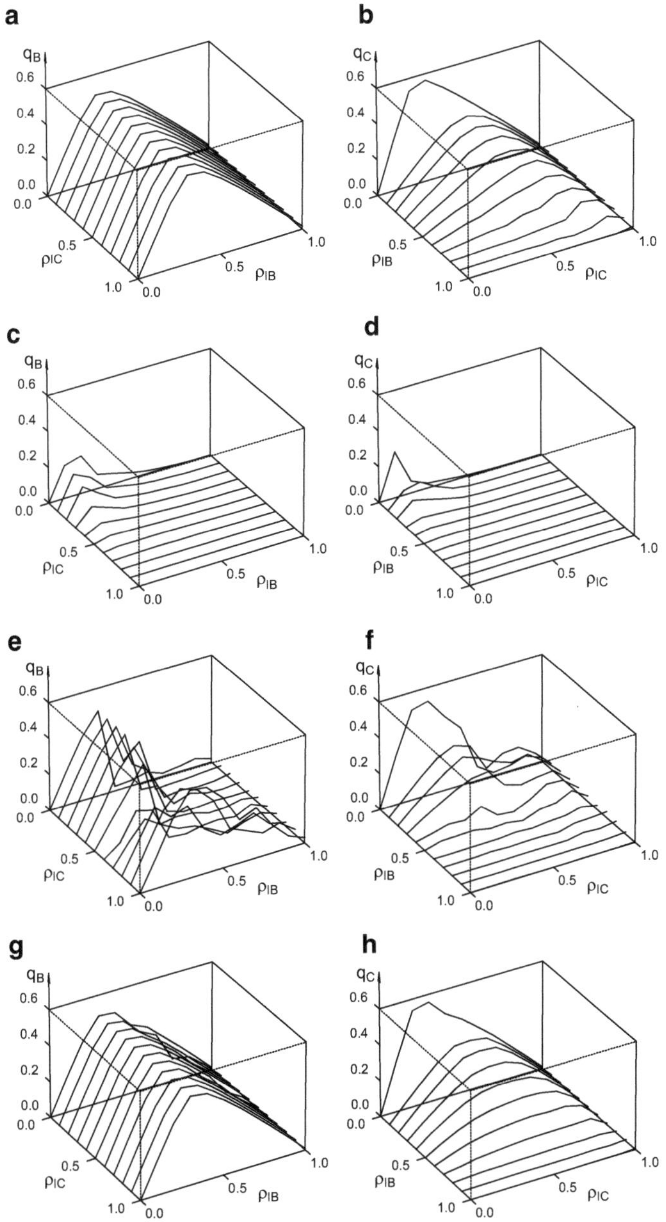

Fig. 3 Average flow for bicycles or cars, as a function of overall bicycle and car density. Each scenario instance is specified using (i) scenario name (intersection or network with length of edge in brackets), (ii) conflict resolution rule (LHS or RHS), (iii) vehicle type (B for bicycle or C for car) and (iv) direction of flow measurement (south-north or east-west). All turning probabilities are 0. (**a**) Intersection, LHS, B-SN. (**b**) Intersection, LHS, C-SN. (**c**) Intersection, LHS, B-EW. (**d**) Intersection, LHS, C-EW. (**e**) Network(50), LHS, B-SN. (**f**) Network(50), LHS, C-SN. (**g**) Network(200), LHS, B-SN. (**h**) Network(200), LHS, C-SN

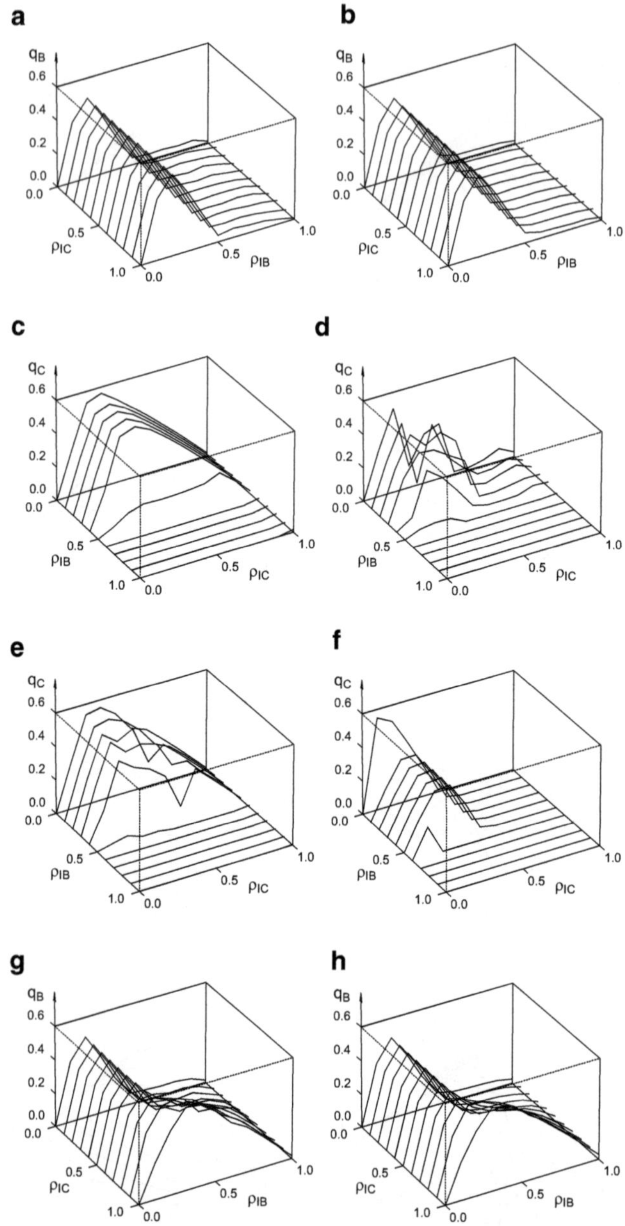

Fig. 4 Average flow for bicycles or cars, as a function of overall bicycle and car density, continued. Each scenario instance is specified using (i) scenario name (intersection or network with length of edge in brackets), (ii) conflict resolution rule (LHS or RHS), (iii) vehicle type (B for bicycle or C for car), (iv) direction of flow measurement (south-north or east-west) and (v) any turning probabilities that are not equal to 0. (**a**) Intersection, RHS, B-EW. (**b**) Network(50), RHS, B-EW. (**c**) Intersection, LHS, C-SN. (**d**) Network(50), LHS, C-SN. (**e**) Network(200), LHS, C-SN. (**f**) Intersection, LHS, C-EW. (**g**) Intersection, RHS, B-EW. (**h**) Network(50), RHS, B-EW

Acknowledgements This work is funded by the Irish Research Council for Science, Engineering and Technology (IRCSET), through an 'Embark Initiative' postgraduate scholarship.

References

1. Gould G, Karner A (2009) Modeling bicycle facility operation. Transp Res Rec 2140:157–164
2. Jia B, Li X, Jiang R, Gao Z (2007) Multi-value cellular automata model for mixed bicycle flow. Eur Phys J B 56:247–252
3. Khan S, Maini P (1999) Modeling heterogeneous traffic flow. Transport Res Rec 1678:234–241
4. Faghri A, Egyháziová E (1999) Development of a computer simulation model of mixed motor vehicle and bicycle traffic on an urban road network. Transport Res Rec 1674:86–93
5. Gundaliya P, Mathew T, Dhingra S (2008) Heterogeneous traffic flow modelling for an arterial using grid based approach. J Adv Transport 42:467–491
6. Mallikarjuna C, Rao K (2009) Cellular automata model for heterogeneous traffic. J Adv Transport 43:321–345
7. Cheng S, Yao D, Zhang Y, Su Y, Xu W (2008) A ca model for intrusion conflicts simulation in vehicles-bicycles laminar traffic flow. In: Proceedings of the 11th International IEEE Conference on Intelligent Transportation Systems Beijing, China, October 12–15, 2008
8. Tang T, Huang H, Shang H (2010) A dynamic model for the heterogeneous traffic flow consisting of car, bicycle and pedestrian. Int J Mod Phys C 21:159–176
9. Si B, Long J, Gao Z (2008) Optimization model and algorithm for mixed traffic of urban road network with flow interference. Sci China Ser E 51:2223–2232
10. Xue Z, Ming G, Kuan Y (2007) A model of potential capacity of right turn movement at signalized intersections under mixed traffic conditions. In: Proceedings of the 3rd Urban Street Symposium, June 24–27, 2007, Seattle, WA
11. Li X, Gao Z, Jia B, Zhao X (2009) Modeling the interaction between motorized vehicle and bicycle by using cellular automata model. Int J Mod Phys C 20:209–222
12. Xie D, Gao Z, Zhao X, Li K (2009) Characteristics of mixed traffic flow with non-motorized vehicles and motorized vehicles at an unsignalized intersection. Physica A 388:2041–2050
13. Vasic J, Ruskin H (2012) Cellular automata simulation of traffic including cars and bicycles. Physica A 391: 2720–2729
14. Nagel K, Schreckenberg M (1992) A cellular automaton model for freeway traffic. J Phys I 2:2221–2229

About New Dynamical Interpretations of Entropic Model of Correspondence Matrix Calculation and Nash-Wardrop's Equilibrium in Beckmann's Traffic Flow Distribution Model

E.V. Gasnikova and T.A. Nagapetyan

1 Correspondence Matrix Calculation Model

Assume that in some town there are n districts (regions), $L_i > 0$ is the number of residents living at the district i, and $W_j > 0$ is the number of residents working at the district j.[1] By $x_{ij}(t) \geq 0$ we will denote the number of residents living at the district i and working at the district j at the moment of time $t \geq 0$. Over course of the time numbered residents (whose quantity doesn't alter, and is equal to $N = \sum_{i=1}^{n} L_i = \sum_{j=1}^{n} W_j$) change their apartments (homes). And we suppsose, that changes may happen only by the means of the exchange of the apartments, i.e.

$$x_{ij}(t) \geq 0, \quad \sum_{j=1}^{n} x_{ij}(t) \equiv L_i, \quad \sum_{i=1}^{n} x_{ij}(t) \equiv W_j, \quad i, j = 1, \ldots, n. \tag{1}$$

Suppose, that at the time $t \geq 0$ resident r lives in the district k and works in the district m, and the resident l lives in the district p and works in the district q. Then $p^{L}_{k,m;\,p,q}(t)\,\Delta t + o(\Delta t)$ - is the probability for the residents with numbers r and l $(1 \leq r < l \leq N)$ to exchange their apartments in a period of time $(t, t + \Delta t)$. It's natural to consider that probability (in the unit time) of exchanging apartments depends only on the location of the workplaces and homes, which are exchanged. For instance, it may be considered that the "distance" between district i and district j is $c_{ij} \geq 0$, and

[1] This section is based on the works [1–11].

E.V. Gasnikova (✉) · T.A. Nagapetyan
Faculty of Applied Mathematics and Control, Moscow Institute of Physics and Technology (State University), Moskovskaya oblast, Russia
e-mail: egasnikova@yandex.ru; nagapetyan@gmail.com

V.V. Kozlov et al. (eds.), *Traffic and Granular Flow '11*,
DOI 10.1007/978-3-642-39669-4_44, © Springer-Verlag Berlin Heidelberg 2013

$$p^L_{k,m;\,p,q}(t) \equiv p^L \exp\Big(\underbrace{(c_{km} + c_{pq})}_{\substack{\text{sum of the distances}\\ \text{before exchange}}} - \underbrace{(c_{pm} + c_{kq})}_{\substack{\text{sum of the distances}\\ \text{after exchange}}} \Big) > 0.$$

Then, by the virtue of the ergodic theorem for discrete homogeneous Markov process with finite number of states, for all $\{x_{ij}\}_{i=1,j=1}^{n,n} \in (1)$ we have that

$$\lim_{t\to\infty} P\left(x_{ij}(t) = x_{ij}, i, j = 1, \ldots, n\right) =$$
$$= Z^{-1} \prod_{i=1,j=1}^{n,\,n} \exp\left(-2c_{ij}x_{ij}\right) \cdot \left(x_{ij}!\right)^{-1} \stackrel{def}{=} p\left(\{x_{ij}\}_{i=1,\,j=1}^{n,\,n}\right),$$

where Z is the normalizing multiplier.

Here we have the case where the final distribution, which is also a stationary distribution, satisfies detailed balance condition[2]:

$$(x_{km}+1)\left(x_{pq}+1\right) \cdot \hat{p} \cdot p^L_{k,m;\,p,q} = x_{pm}x_{kq}\,p\left(\{x_{ij}\}_{i=1,\,j=1}^{n,\,n}\right) p^L_{p,m;\,k,q},$$

where

$$\hat{p} = p\left(\{x_{11}, \ldots, x_{km}+1, \ldots, x_{pq}+1, \ldots, x_{pm}-1, \ldots, x_{kq}-1, \ldots, x_{nn}\}\right).$$

Distribution $p\left(\{x_{ij}\}_{i=1,\,j=1}^{n,\,n}\right)$ on a set (1) is concentrated with $N \gg 1$ (see below) in a neighborhood of the most probable value $\left\{x^*_{ij}\right\}_{i=1,\,j=1}^{n,\,n}$, which is determined as a solution of the following entropic –linear programming problem:

$$\sum_{i=1,\,j=1}^{n,\,n} x_{ij}\ln x_{ij} + 2\sum_{i=1,\,j=1}^{n,\,n} c_{ij}x_{ij} \to \min_{\{x_{ij}\}_{i=1,\,j=1}^{n,\,n} \in (1)}. \tag{2}$$

The solution of this problem might be presented as

$$x_{ij} = \exp\left(-1 - \lambda^L_i - \lambda^W_j - 2c_{ij}\right),$$

[2] Multipliers before probabilities, for example, in the state $\{x_{ij}\}_{i=1,\,j=1}^{n,\,n}$, arise because of the number of the ways to choose the resident, living in the district p and working in the district m, is x_{pm}, and independently the number of ways to choose the resident, living in the district k and working in the district q, is x_{kq}.

where Lagrange multipliers (dual variables) $\{\lambda_i^L\}_{i=1}^n$ and $\left\{\lambda_j^W\right\}_{j=1}^n$ are determined[3] from the system of equations (1). In practice we usually have some information about $\{L_i, W_i\}_{i=1}^n$ and $\{c_{ij}\}_{i=1,j=1}^{n,n}$. So, when we solve (2), we find

$$x_{km}\left(\{L_i, W_i\}_{i=1}^n ; \{c_{ij}\}_{i=1,j=1}^{n,n}\right)$$

If $N \sim nm, m \gg 1 \; \forall i, j = 1, \ldots, n \to L_i, W_j \sim m, \, c_{ij} = c > 0$, then the distribution of the probabilities $p\left(\{x_{ij}\}_{i=1,\, j=1}^{n,n}\right)$ on the set (1) is concentrated in $O\left(\sqrt{m}\right)$ neighborhood of the most probable value $x_{ij}^* \approx L_i W_j / N \sim m/n, \; i, j = 1, \ldots, n$. More precisely:

$$\exists \lambda > 0\colon \lim_{t\to\infty} P\left(\forall\, i, j = 1, \ldots, n \to \left| x_{ij}(t) \big/ x_{ij}^* - 1 \right| \le \lambda / \sqrt{m}\right) \ge 0.999$$

Indeed, let us note, that

$$\forall \; \{x_{ij}\}_{i=1,\, j=1}^{n,n} \in (1) \to \sum_{i=1,\, j=1}^{n,n} \frac{\partial \ln p\left(\left\{x_{ij}^*\right\}_{i=1,\, j=1}^{n,n}\right)}{\partial x_{ij}} \cdot \left(x_{ij} - x_{ij}^*\right) \le 0$$

Thus, $\forall \; \{x_{ij}\}_{i=1,\, j=1}^{n,n} \in (1) \; \exists \; \theta \in [0, 1]$:

$$\ln p\left(\{x_{ij}\}_{i=1,\, j=1}^{n,n}\right) \le \ln p\left(\left\{x_{ij}^*\right\}_{i=1,\, j=1}^{n,n}\right) +$$
$$+ \sum_{i=1,\, j=1}^{n,n} \frac{\partial^2 \ln p\left(\left\{x_{ij}^* \theta + x_{ij} \cdot (1-\theta)\right\}_{i=1,\, j=1}^{n,n}\right)}{\partial x_{ij}^2} \cdot \frac{\left(x_{ij} - x_{ij}^*\right)^2}{2}$$

Since

$$\frac{\partial^2 \ln p\left(\{x_{ij}\}_{i=1,\, j=1}^{n,n}\right)}{\partial x_{ij}^2} = \frac{\partial^2 \left(-\sum_{i=1,\, j=1}^{n,n} x_{ij} \ln x_{ij}\right)}{\partial x_{ij}^2} = -\frac{1}{x_{ij}}$$

---

[3] This can be done in different ways. For example, by Bregman's balancing method or by Newton's method [5]. The other way is to solve the dual problem for the entropy programming problem (2). There are a lot of different algorithms with the first order oracle (MART, GISM, etc. [4,5]). It can be shown that most of these methods (including Bregman's) are just barrier-multiplicative antigradient descending methods [11]. At the end (when the iteration process is achieving a sufficient small vicinity of the global minimum) it is worth to use the second order interior-point method, like Nesterov–Nemirovskii polynomial algorithm [12] (for so-called "separable" tasks).

we have "inequality of measure concentration":

$$\forall\ M > 0, \quad \forall\ \left\{x_{ij}\right\}_{i=1,\ j=1}^{n,\ n} \in (1): \quad \sum_{i=1,\ j=1}^{n,\ n} \frac{\left(x_{ij} - x_{ij}^{*}\right)^2}{2 \max\left\{x_{ij},\ x_{ij}^{*}\right\}} \geq M$$

$$p\left(\left\{x_{ij}\right\}_{i=1,\ j=1}^{n,\ n}\right) \leq e^{-M} p\left(\left\{x_{ij}^{*}\right\}_{i=1,\ j=1}^{n,\ n}\right)$$

2 Beckmann Traffic Flow Distribution Model

Let us consider the oriented graph $\Gamma = (V, E)$, which stands for transportation route in some town (V – nodes (vertices), $E \subset V \times V$ – arc of the network (edges)).[4] Let $W = \{w = (i, j) : i, j \in V\}$ be a set of pairs inlet-outlet; $p = \{v_1, v_2, \ldots, v_m\}$ – a route from v_1 to v_m, if $(v_k, v_{k+1}) \in E$, $k = 1, \ldots, m-1$ (it will be shown later (see example by V.I. Shvetsov) that, to specify the path it may not be enough to indicate only the set of vertices. In general, one must also specify exactly which edge, connecting the specified vertices, is chosen); P_w – the set of routes in correspondence $w \in W$; $P = \bigcup_{w \in W} P_w$ – the collection of all routes in the network Γ; x_p – flux on the way p, $\mathbf{x} = \left\{x_p : \ p \in P\right\}$; $G_p(\mathbf{x})$ – specific costs of travel on the road p, $\mathbf{G}(\mathbf{x}) = \left\{G_p(\mathbf{x}) : \ p \in P\right\}$; y_e – flux on the arc e: $\mathbf{y} = \Theta \mathbf{x}$, where $\Theta = \left\{\delta_{ep}\right\}_{e \in E, p \in P}$ ($\delta_{ep} = \{1,\ e \in p;\ 0,\ e \notin p\}$); $\tau_e(y_e)$ – specific costs of travel on the arc e (generally increasing, convex, smooth functions), it is natural to assume, that $\mathbf{G}(\mathbf{x}) = \Theta^T \boldsymbol{\tau}(\mathbf{y})$. Let flows on correspondences d_w, $w \in W$ to be known. Then $\mathbf{x}$, which describes flow distribution, must lie in the set:

$$X = \left\{\mathbf{x} \geq 0 \colon \sum_{p \in P_w} x_p = d_w,\ w \in W\right\}.$$

Consider a game in which each element $w \in W$ corresponds to a considerably big ($d_w \gg 1$) set of players of the same type. The set of pure strategies of such player is P_w, and profit (minus losses) is defined by the formula $-G_p(\mathbf{x})$ (a player chooses a strategy $p \in P_w$ and neglects the fact, that $|P_w|$ components of the vector $\mathbf{x}$ and hence the profit depends slightly on his choice). One can show, that Nash equilibrium is equivalent to complementarity problem, being equivalent to a solution of variation inequality, which, in its turn, is equivalent to a solution of convex optimization problem.

[4]This section is based on the works [13–24].

$$\forall w \in W,\ p \in P_w \to x_p^* \cdot \left(G_p\left(\mathbf{x}^*\right) - \min_{q \in P_w} G_q\left(\mathbf{x}^*\right) \right) = 0$$

$$\Updownarrow$$

$$\forall \mathbf{x} \in X \to \left\langle \mathbf{G}\left(\mathbf{x}^*\right), \mathbf{x} - \mathbf{x}^* \right\rangle \geq 0$$

$$\Updownarrow$$

$$\Psi\left(\mathbf{x}\right) = \sum_{e \in E} \int\limits_0^{\sum_{p \in P} x_p \delta_{pe}} \tau_e\left(z\right) dz \to \min_{\mathbf{x} \in X}. \tag{3}$$

It is easy to show, that in the case $\mathbf{G}\left(\mathbf{x}\right)$ being strictly monotonic transformation, the Nash-Wardrop equilibrium $\mathbf{x}^*$ is unique. If $\tau_e\left(y_e\right)$ are increasing functions then $\mathbf{y}^* = \Theta \mathbf{x}^*$ is unique, although, as we will see later, $\mathbf{x}^*$ isn't necessarily unique.

The route at a step $(n+1)$ player at correspondence w, choose independently according the mixed strategy with probability

$$\text{Prob}_p^w\left(n+1\right) = \gamma_n \cdot \max\left\{x_p\left(n\right), n^{-1}\right\} \exp\left(-G_p\left(\mathbf{x}\left(n\right)\right)/T\right)/Z_n^w, \quad w \in W,$$

to choose path $p \in P_w$ $(0 < \gamma_n \leq 1)$, and with probability $1 - \gamma_n$ to choose the same strategy as at the *n-th* step. Here $x_p\left(n\right)$ – the number of players at w, who have chosen at the *n-th* step strategy $p \in P_w$, and Z_n^w can be found from the normalization condition. Multiplier $\max\left\{x_p\left(n\right), n^{-1}\right\}$ describes the will to imitate and, also, the reliability of using this strategy. This multiplier notices specifics of the problem (without it there could be convergence to something different from the Nash-Wardrop equilibrium). Parameter γ describes "the conservatism", while "the temperature" T stands for "the risk appetite".

Theorem 1. *Let* $T > 0$ *be sufficiently small,* $\sum_{n=1}^{\infty} \gamma_n = \infty, \sum_{n=1}^{\infty} \left(\gamma_n\right)^2 < \infty$. *Then* $\Psi\left(\mathbf{x}\left(n\right)\right) \xrightarrow[n \to \infty]{a.s.} \Psi\left(\mathbf{x}^*\right)$, *where* $\mathbf{x}^*$ *is the minimizer from (3). Moreover, if the equilibrium is unique, then* $\mathbf{x}\left(n\right) \xrightarrow[n \to \infty]{a.s.} \mathbf{x}^*$.

In the experiments, conducted at the Laboratory of Experimental Economics in the Faculty of Applied Mathematics and Control, MIPT, in which students of the 5th course were involved, we observed the convergence to equilibrium and "vibrations" around it. Fluctuations should be explained, apparently, by the fact that in experiments the number of players was small and the hypothesis of a competitive market was not performed. We also observed, that for students $\gamma_n \equiv \gamma > 0$ it is more likely, than $\gamma_n \sim 1/n$. As a result there will be convergence not to the equilibrium point, but to its neighborhood. Size of the neighborhood depends on T, $\gamma > 0$ and the number of players.

Example 1 (Braess paradox, 1968 [15]). Let correspondence $x_{14} = 6$ thousand cars/hour (see graphs on Figs. 1 and 2). Weight of the edges is time delay (in

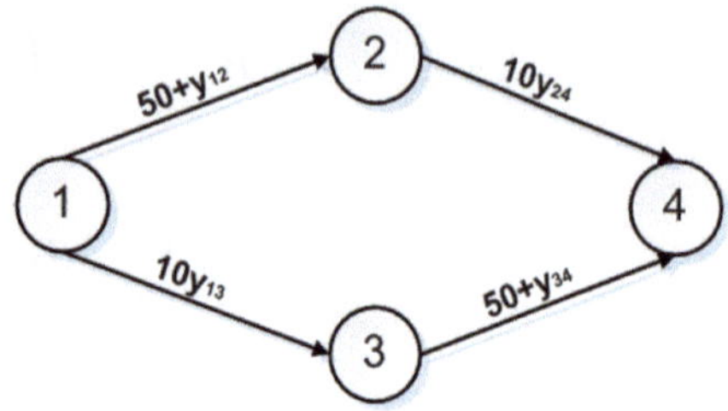

Fig. 1 Case 1. $x_{124} = x_{134} = 3$. Total time for each path is $T = 83$ min

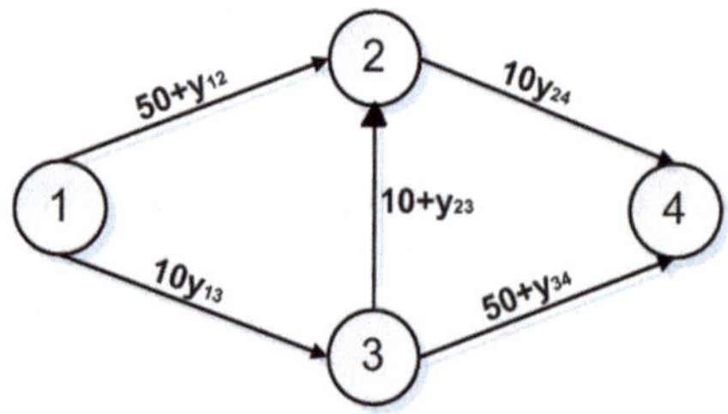

Fig. 2 Case 2. $x_{124} = x_{1,234} = x_{134} = 2$. Total time for each path is $T = 92$ min

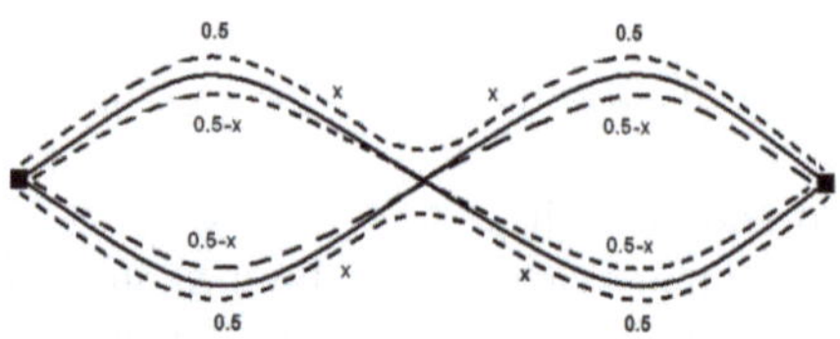

Fig. 3 Nonuniqueness of the equilibrium

minutes) when the flow on the edge is y_{ij} (1,000 cars/h). For example, in case 2 (see Fig. 2): $y_{24} = x_{1,324} + x_{124}$. It is natural that time delay (at each of the edge) is a growth function of flow.

The following example shows, that under the very natural conditions vector-function of cost of the travel $\mathbf{G}(\mathbf{x})$ can't be strictly monotone:

$$\exists \mathbf{x}, \mathbf{y} \in X \ \ (\mathbf{x} \neq \mathbf{y}) : \mathbf{G}(\mathbf{x}) = \mathbf{G}(\mathbf{y}) \Rightarrow \langle \mathbf{G}(\mathbf{x}) - \mathbf{G}(\mathbf{y}), \ \mathbf{x} - \mathbf{y} \rangle = 0.$$

This, for example, can be because of

$$\mathbf{G}(\mathbf{x}) = \Theta^T \boldsymbol{\tau}(\mathbf{y}), \quad \mathbf{y} = \Theta \mathbf{x},$$

where $\mathbf{y} = \{y_e\}_{e \in E}$ describes the loading of edges (arcs) of a graph of the transport network, $\boldsymbol{\tau}(\mathbf{y}) = \{\tau_e(y_e)\}_{e \in E}$ – vector-function of cost of the travel on the edges of transport network, Θ – incidence matrix of edges and paths, and different vectors of flow distributions $\mathbf{x}$ may correspond to the same vector $\mathbf{y} = \Theta \mathbf{x}$.

Example 2 (Nonuniqueness of the equilibrium; Shvetsov, 2010). On Fig. 3 the equilibrium flow distribution is shown, for all $x \in [0, 0.5]$.

Theorem 2. *Let $T > 0$ be sufficiently small, $\sum_{n=1}^{\infty} \gamma_n = \infty$, $\sum_{n=1}^{\infty} (\gamma_n)^2 < \infty$. Then $\Psi(\mathbf{x}(n)) \xrightarrow[n\to\infty]{a.s.} \Psi(\mathbf{x}^*)$ and $\mathbf{x}(n) \xrightarrow[n\to\infty]{a.s.} \mathbf{x}^*(\mathbf{x}(0))$. Note most of the elements of $\mathbf{x}^*(\mathbf{x}(0))$ can be equal to zero.*

We should notice, that Theorem 2 is a refutation (in case of the considered dynamics) of the hypothesis [16]. It states that in the case of non-unique Nash-Wardrop equilibrium, the equilibrium is more likely to realize, and it is a solution of the following linear-enthropy programming problem

$$\sum_{w\in W}\sum_{p\in P_w}\left(x_p \ln\left(x_p/|P_w|\right) - x_p\right) \to \min_{\mathbf{x}\in X,\ \Theta\mathbf{x}=\mathbf{y}^*},$$

where $\mathbf{y}^*$ – is the unique solution of $V(\mathbf{y}) = \sum_{e\in E}\int_0^{y_e} \tau_e(z)\,dz \to \min_{\mathbf{y}=\Theta\mathbf{x},\ \mathbf{x}\in X}$.

3 Sketch of the Proof of the Theorem 1

Lemma 1. *Let $f(w) = -T\ln w$, where $w \in (0, 1)$. $\alpha_i > 0$ — are some randomly chosen, but fixed parameters; $w_i \in (0, 1)$. Let us consider functions*

$$F_0(\mathbf{w}) = \frac{\sum_i \alpha_i w_i f(w_i)}{\sum_i \alpha_i w_i}, and F_1(\mathbf{w}) = \frac{\sum_i \alpha_i f(w_i)}{\sum_i \alpha_i}.$$

Then $F_0(\mathbf{w}) \le F_1(\mathbf{w})$, and the equality is attained only when

$$w_1 = w_2 = \ldots = w^*.$$

Proof. The proof is based on the consequent usage of the inequality between harmonic mean and geometric mean and then Cauchy inequality.

Lemma 2. *For any $\mathbf{x}(n) \in X : \mathbf{x}(n) \ne \mathbf{x}^*$ holds the following inequality*

$$\langle grad\,\Psi(\mathbf{x}(n)), E[\mathbf{x}(n+1) - \mathbf{x}(n)|\,\mathbf{x}(n)]\rangle$$

$$= \langle \mathbf{G}(\mathbf{x}(n)), E[\mathbf{x}(n+1) - \mathbf{x}(n)|\,\mathbf{x}(n)]\rangle < 0.$$

Proof. Without restricting the generality, we can assume that $\forall\, p \in P \to x_p(n) \ge 1$. Then

$$\langle \mathbf{G}(\mathbf{x}(n)),\ E[\mathbf{x}(n+1) - \mathbf{x}(n)|\,\mathbf{x}(n)]\rangle =$$

$$= \gamma_n \sum_{w \in W} d_w \left[\frac{\sum_{p \in P_w} x_p(n) G_p(\mathbf{x}(n)) \exp\left(-\frac{G_p(\mathbf{x}(n))}{T}\right)}{\sum_{p \in P_w} x_p(n) \exp\left(-\frac{G_p(\mathbf{x}(n))}{T}\right)} - \frac{\sum_{p \in P_w} G_p(\mathbf{x}(n)) x_p(n)}{\sum_{p \in P_w} x_p(n)} \right].$$

From Lemma 1 it follows that

$$\langle \mathbf{G}(\mathbf{x}(n)), E[\mathbf{x}(n+1) - \mathbf{x}(n) | \mathbf{x}(n)] \rangle \leq 0.$$

The equality can be attained only on the equilibrium vector $\mathbf{x}^*$, which can not be under considered hypothesis.

Remark 1. Lemma 2 can be more specified. At some neighborhood of the equilibrium there exists $l > 0$, such that

$$\langle \mathbf{G}(\mathbf{x}(n)),\ E[\mathbf{x}(n+1) - \mathbf{x}(n) | \mathbf{x}(n)] \rangle \leq -l\gamma_n \cdot \left(\Psi(\mathbf{x}(n)) - \Psi(\mathbf{x}^*) \right).$$

Also, with some reserves, we can change constraint $\mathbf{x}(n) \in X$ to $\mathbf{x}(n) \geq \mathbf{0}$.

Lemma 3 ([25], Chap. 2.2). *Let*

$$\sum_{n=1}^{\infty} \gamma_n = \infty, \quad \sum_{n=1}^{\infty} (\gamma_n)^2 < \infty.$$

Then

$$\Psi(\mathbf{x}(n)) \xrightarrow[n \to \infty]{a.s.} \Psi(\mathbf{x}^*)$$

and if the equilibrium is unique, then also

$$\mathbf{x}(n) \xrightarrow[n \to \infty]{a.s.} \mathbf{x}^*.$$

Proof. Lemma 2 and Theorem 1 from Chap. 2.2 of [25] allow us to consider only the situation, when $\mathbf{x}(n)$ is close to $\mathbf{x}^*$. Then by Taylor formula we have:

$$E[\Psi(\mathbf{x}(n+1)) | \mathbf{x}(n)] = \Psi(\mathbf{x}(n)) +$$

$$+ \langle \operatorname{grad} \Psi(\mathbf{x}(n)),\ E[\mathbf{x}(n+1) - \mathbf{x}(n) | \mathbf{x}(n)] \rangle + \mathrm{O}\left((\gamma_n)^2 \right).$$

If we take mathematical expectations from both sides of this equality, we will get that there exists sufficiently large $C > 0$, such that

$$E(\Psi(\mathbf{x}(n+1))) - \Psi(\mathbf{x}^*) \leq (1 - l\gamma_n) \cdot \left(E(\Psi(\mathbf{x}(n))) - \Psi(\mathbf{x}^*) \right) + C \cdot (\gamma_n)^2.$$

From more general statement from [25] we get that

$$E\left(\Psi\left(\mathbf{x}\left(n\right)\right)\right)-\Psi\left(\mathbf{x}^{*}\right)\underset{n\to\infty}{\longrightarrow}0,$$

if

$$\sum_{n=1}^{\infty}\gamma_{n}=\infty,\quad\sum_{n=1}^{\infty}\left(\gamma_{n}\right)^{2}<\infty.$$

From Kolmogorov inequality follows

$$P\left(\forall\ n\geq n_{0}\to\Psi\left(\mathbf{x}\left(n\right)\right)-\Psi\left(\mathbf{x}^{*}\right)\leq\varepsilon\right)\geq$$

$$\geq 1-\varepsilon^{-1}\cdot\left(E\left(\Psi\left(\mathbf{x}\left(n_{0}\right)\right)-\Psi\left(\mathbf{x}^{*}\right)\right)+\sum_{k=n_{0}}^{\infty}\left(\gamma_{k}\right)^{2}\right),$$

which concludes the proof.

In the end we will formulate a known result, which is in high correlation with the proved one.

Theorem 3. *Let*[5] $T>0$*. Then there* $\exists\, C,\alpha>0\colon\forall\ N\in \mathrm{N}$

$$P\left(\Psi\left(\frac{1}{N}\sum_{n=1}^{N}\mathbf{x}\left(n\right)\right)-\Psi_{\min}\geq\frac{\Omega}{\sqrt{N}}\right)\leq 2\exp\left(-C\cdot\Omega\right),$$

where $\gamma=\frac{\alpha}{\sqrt{N}}$

Acknowledgements The authors thank S.A. Avvakumov, I.B. Gnedkov, Y.V. Dorn, I.S. Menshikov, E.A. Nurminskiy, A.A. Shananin, V.I. Shvetsov and especially A.V. Gasnikov.

The work was supported by RFBR 10-01-00321-a, 11-01-00494-a 11-07-00162-a. The second author is partially supported by the Laboratory for Structural Methods of Data Analysis in Predictive Modeling, MIPT, RF government grant, ag. 11.G34.31.0073.

References

1. *Introduction to the mathematical modeling of traffic flow*. Gasnikov A.V., Klenov S.L., Nurminskiy E.A., Kholodov Y.A., Shamray N.B.; with appendices M.L. Blank, E.V. Gasnikova, A.A. Zamyatin and V.A. Malyshev, A.V. Kolesnikov, A.M. Raigorodskiy, ed. A.V. Gasnikov. M.: MIPT, 2010. 361p. (in Russian) http://zoneos.com/traffic/

[5]See formula (2.32), statement 2.2 and example on the page 1586 in [26]

2. Wilson A.G. (1970) *Entropy in urban and regional modeling. Pion ltd., London.*
3. Jaynes E.T. (1989) *Papers on probability, statistics and statistical physics*. Dordrecht: Kluwer Academic Publisher.
4. Popkov Yu.S. (1995) *Macrosystems. Theory and its applications*. Springer-Verlag.
5. Fang S.-C., Rajasekera J.R., Tsao H.-S. (1997) *Entropy optimization and mathematical programming*. Kluwer Academic Publishers.
6. Ledoux M.(2001) *Concentration of measure phenomenon*. Providence, RI, Amer. Math. Soc., (Math. Surveys Monogr. Vol. 89).
7. Weidlich W.(2005) *Sociodynamics. A systematic approach to mathematical modeling in social sciences*. Harwood Academic Publishers.
8. Castellano C., Fortunato S., Loreto V.(2009) *Statistical physics of social behavior*. Review of Modern Physics, 81:591–646. `arXiv:0710.3256v2`
9. Batishcheva Ya.G., Vedenyapin V.V. (2005) *The second law of thermodynamics for chemical kinetics*. Mat. Model., 17:106–110. (in Russian)
10. Malyshev V.A., Pirogov S.A. (2008) *Reversibility and irreversibility in stochastic chemical kinetics*. Uspekhi Mat. Nauk, 63:3–36.
11. Gasnikova E.V. (2009) *Dual multiplicative algorithms for a problem of entropy-linear programming*. Zh. Vychisl. Mat. Mat. Fiz., 49:453–464.
12. Nesterov Yu., Nemirovskii A. (1994) *Interior-point polynomial algorithms in convex programming*. Philadelphia: SIAM.
13. Sheffi Y.(1985) *Urban transportation networks: Equilibrium analysis with mathematical programming methods*. N.J.: Prentice–Hall Inc., Englewood Cliffs.
14. Patriksson M. (1994) *The traffic assignment problem. Models and methods*. Utrecht Netherlands: VSP, 1994.
15. Steenbrink P.A. (1974) *Optimization of transport networks*. John Wiley & Sons.
16. Bar-Gera H. (1999) *Origin-based algorithms for transportation network modeling*. Univ. of Illinois at Chicago.
17. Foster D., Young P. (1990) *Stochastic evolutionary game dynamics*. Theoretical Population Biology, 38:2.
18. Cressman R. (2003) *Evolutionary game theory and extensive form games*. Cambridge, Mass.: MIT Press.
19. Hofbauer J., Sigmund K. (2003) *Evolutionary game dynamics*. Bulletin of the AMS, 40:479–519.
20. McKelvey R. D., Palfrey T. R. (1998) *Quantal response equilibria for extensive form games*. Experimental Economics, 1:9–41.
21. Marsili M. (2001) *Toy models of markets with heterogeneous interacting agents*. e-print *www.unifr.ch/econophysics/*.
22. Fogel D.B. (2000) *Evolutionary Computation: Towards a New Philosophy of Machine Intelligence*. New York: IEEE Press.
23. Como G., Salva K., Acemoglu D., Dahleh M.A., Frazzoli E. (2011) *Stability analysis of transportation networks with multiscale driver decisions*. e-print *arXiv:1101.2220v1*.
24. Nesterov Yu. De Palma A. (2003) *Stationary dynamic solutions in congested transportation networks: summary and perspectives*. Networks and Spatial Economics, 3:371–395.
25. Polyak B.T. (1987) *Introduction to Optimization*. Optimization Software Inc.
26. Juditsky A., Lan G., Nemirovskii A., Shapiro A. (2009) *Stochastic approximation approach to stochastic programming*. SIAM Journal on Optimization, 19:1574–1609. http://www2.isye.gatech.edu/~nemirovs/

Feed Forward Back Propagation Neural Networks to Classify Freeway Traffic Flow State

Onur Deniz and Hilmi Berk Celikoglu

Abstract Analyses on traffic flow require accurate time-varying local traffic density information in order to effectively determine inflows to and outflows from freeway segments in several aspects of network traffic control. It is essential to specify the flow state, equivalently to derive density variable with the corresponding flow-rate measure, from primary traffic variables in order to provide accurate input to real-time traffic management strategies. In this paper, a study on the flow state specification process that employs feed forward back propagation neural network method to map sectional lane based density measure with raw traffic data collected from successive remote microwave sensor units mounted along a segment existing on the freeway network of Istanbul, is summarized. Classification of traffic flow states and matching the corresponding real-time flow state is obtained dynamically inputting raw flow measures simultaneously to neural density mapping and traffic flow modeling processes. The approach is promising in capturing instantaneous changes on flow states and may be utilized within intelligent management strategies such as incident control.

1 Introduction

Freeway traffic data is often available in the form of occupancy and volume measurements collected generally from loop detectors embedded in the pavement or from remote microwave sensor units. In conjunction with effective length data specific to detected portions of vehicle classes, these measurements can be converted into fundamental macroscopic flow variables such as density and speed. Raw data sets obtained by neither loop detectors nor microwave sensors provide full

O. Deniz (✉) · H.B. Celikoglu
Civil Engineering Department, Istanbul Technical University, Ayazaga Campus 34469, Istanbul, Turkey
e-mail: denizon@itu.edu.tr; celikoglu@itu.edu.tr

V.V. Kozlov et al. (eds.), *Traffic and Granular Flow '11*,
DOI 10.1007/978-3-642-39669-4_45, © Springer-Verlag Berlin Heidelberg 2013

satisfactory inputs to flow based traffic modeling studies since data sets are often incomplete or contain inaccurate samples. However, analysis on conserved traffic flow, such as path based network assignment procedures and freeway ramp control strategies, require accurate time-varying local traffic density information in order to effectively determine inflows to and outflows from freeway segments respectively at merges and diverges. Therefore, it is essential to derive density variable from primary traffic variables as well as to reconstruct missing traffic measurements. In this paper, an approach to specify flow state variations in real-time is proposed mainly concentrating on the incorporation of neural network theory to provide derivation for traffic density. Raw traffic measures collected by successive remote microwave sensor units existing on a stretch of freeway network in Istanbul are input simultaneously to both density mapping with functional approximation component and traffic flow modeling component of the overall state specification process. Classification of flow states and matching the corresponding real-time flow state is obtained dynamically. Performances of employed neural approximation methods on density mapping are evaluated considering the flow conservation principle. In the following, a brief review on the relevant literature is provided in Sect. 2. The overall process proposed for flow state specification is explained in Sect. 3. Formulation of density mapping problem and sought solutions with selected methods is summarized in Sect. 4. The final section concludes the paper with findings, evaluation, and possible future directions.

2 Brief Literature Review

The existing literature on the mapping of traffic data that is collected on a regular basis can be roughly classified into two that are for time interval based series prediction problems and for flow variable based modeling problems. Studies in the former class including missing data replacement and short-term variable prediction generally contain the simple applications of stochastic and heuristic methods and are relatively easier to be formulated when considered with the ones in the latter class. Solutions to flow variable based prediction problem formulations ranging from simple link performance prediction to network wide assignment require the satisfaction of desired mathematical properties, such as uniqueness and existence, besides an appropriate trade-off between accuracy and the computational burden. While the former class comprises prediction problems on fundamental traffic flow variables, i.e., speed, occupancy, volume, and etc., problems belonging to the latter class target the estimates of link performances, i.e., link-path traveling time, delay, queuing, and etc. Studies on missing data recovery covered a wide range of statistical method applications as well as the applications of stochastic methods and neural theory. In order to compare the effectiveness of widely used approaches and provide alternative for predicting missing data at traffic measurements factor, genetic, neural, and regression models are developed in [1]. A method based on principal component analysis and support vector regression for a short-term prediction of network traffic volume is proposed and its performance is evaluated in comparison to neural network and auto-regressive models in [2]. A new reliable method called probabilistic principal component analysis to impute the missing flow volume data based on historical data mining is proposed in[3]. Examining the statistical properties of traffic flow volume time series the fluctuations of traffic flow are found to be Gaussian type and principal component analysis is found to be an alternative to be used to retrieve the features of traffic flow in [3].

Within a number of studies, the fundamental traffic flow variables, especially the speed variable, are tried to be estimated with the input of data obtained from inductive loop detectors. An algorithm for estimating mean traffic speed using volume and occupancy data obtained from a single inductance loop is proposed in [4]. The PeMS algorithms for the accurate, adaptive, real-time estimation of the effective vehicle length and vehicle speeds from single-loop detector data via validating the estimates by comparison with independent, direct measurements of the g-factor and vehicle speeds from 20 double-loop detectors over a 3-month period is presented in [5]. The estimates on real-time speed and travel time derived from single-loop detector data assuming a common effective vehicle length for all detectors existing on a network are declared to possibly be in error by 50 %, and use of the PeMS algorithm is suggested to reduce those errors in [5]. Unlike in the works preceding an improved algorithm for estimating speed from single loop detector data, which is simple enough to be implemented using existing controller hardware, is proposed in [6]. A hybrid model for speed estimation by neglecting vehicle types and only considering the variational trend of lane occupancy and volume without the influence of difference of mean effective vehicle length was later on proposed in [7]. A rare method for speed estimation from traffic count and occupancy data by applying a Markov chain Monte Carlo approach with assumption of a simple random walk model for successive vehicle speeds, in which missing vehicle lengths are sampled from an exogenous data set was proposed in [8]. Unlike prior studies on speed estimation, measurement error in occupancy data is explicitly modeled and the efficacy of the estimation scheme is examined by comparing the estimates with independently collected vehicle speed data in [8].

A few numbers of studies integrating a mathematical model of traffic flow to validate fundamental variable data obtained by sensors or to derive sectional link performances in a time interval based series prediction frame is encountered in the literature. A derivation of a macroscopic traffic flow model from the cell transmission model and application to the estimation of traffic densities at unmonitored locations along a highway was presented in [9]. Both the model in [9] and a density-based version of the cell transmission model have been simulated using several days of loop detector data collected during the morning rush-hour period. Evolving the switching model in [9], a cell transmission model based switching state-space model to estimate vehicle densities and congested flow states at unmeasured locations on a highway section was proposed in [10]. A method for estimating density in a freeway lane between detector stations and measuring the net number of vehicles to enter (or leave) the lane, i.e., the lane inflow was also proposed in [11]. Vehicle arrivals at each detector station and information from a sparse vehicle re-identification system are used in [11].

Considering the works summarized above, the novelty in proposed state specification approach is two-fold: the neural network approximator to map sectional density from vehicle count and speed measures at successive cross-sections, and the simultaneous functioning of the neural approximator with the traffic flow model that simulates actual traffic dynamics.

3 Proposed Flow Specification Sub-process

The proposed process for flow state specification involves three sub-processes, two of which are simultaneous, i.e., the density derivation sub-process and the flow modelling sub-process. The third sub-process is the classification of mapped density variable with the corresponding flow-rate measure considering a dynamic segmentation on flow states over the fundamental diagram of flow and density. The overall flow state specification process is shown in Fig. 1.

Two simultaneous sub-processes run with the input of raw traffic flow measures that are collected by microwave sensors mounted on a freeway stretch. The neural network method incorporated model maps the dynamic variation of density measure along the section bounded by successive sensors, while the flow model accounts for obtaining the fundamental diagram of flow-density within a macroscopic approach to be partitioned to provide a base for the final sub-process. The critical measures of density variable, including optimum and jamming densities, with the corresponding flow-rates are used to derive a classification on the fundamental diagram. The current flow state is specified by the match-up of neural network mapping with the appropriate fundamental diagram segment.

Each sub-process is coded by C language and run with a batch file in order to simulate the overall process in real-time fashion. In the following, each sub-process is explained following the information on field data.

3.1 Information on Study Area and Data Set

In this study the remote traffic microwave sensor (RTMS) data is processed to derive the macroscopic features of traffic flow and to have a comparison between the model prediction values with the corresponding observations. RTMS unit is a traffic sensor which uses microwave signals by transmitting frequency modulated continuous waves (FMCW) to detect vehicles. These sensors transmit an FMCW waveform in which the transmitted frequency is constantly changing with respect to time in a fixed fan-shaped beam, as shown in Fig. 2 [12].

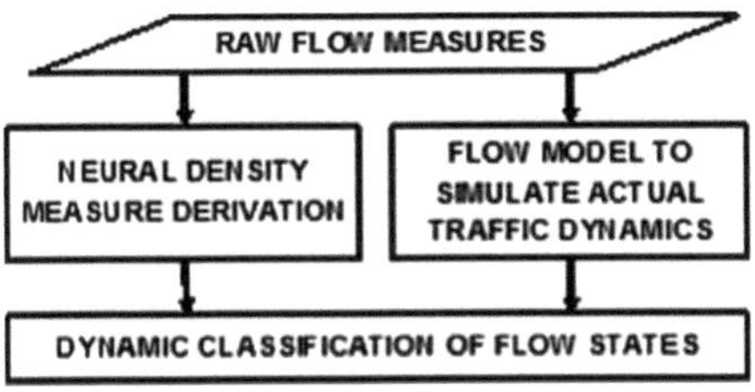

Fig. 1 Overall flow state specification process

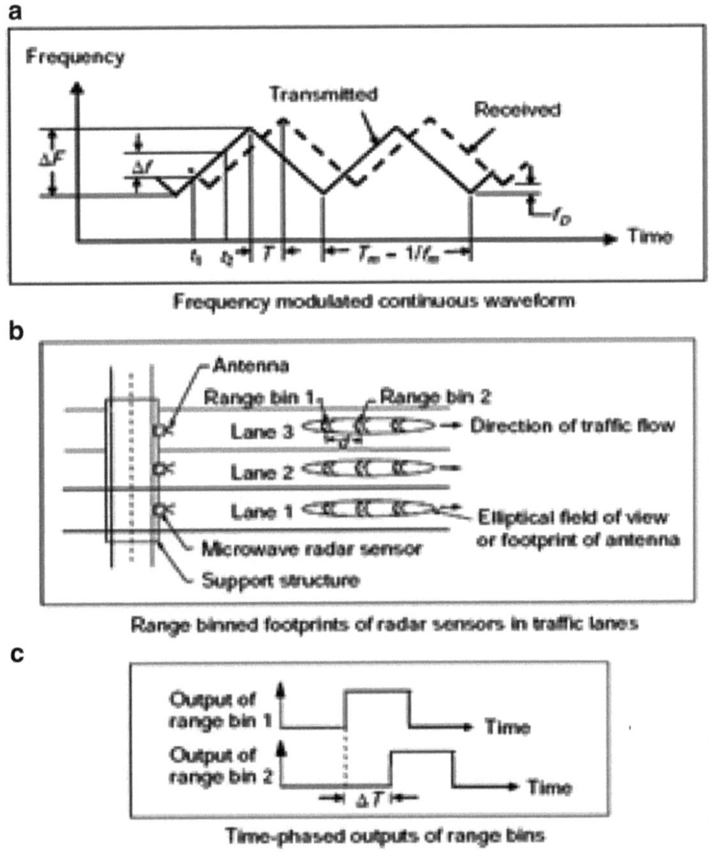

Fig. 2 FMCW signal and radar processing as utilized to measure vehicle presence and speed [13]

The FMCW radar continuously modulates the transmitted frequency and can therefore detect stationary objects. The RTMS can be used to count vehicles and their speed in either a side-fired or forward-looking setup.

Field data is obtained from two successive RTMS units both located in side-fired position on a multi-lane, 4 × 4, approach of a Bosphorus strait crossing, the Fatih

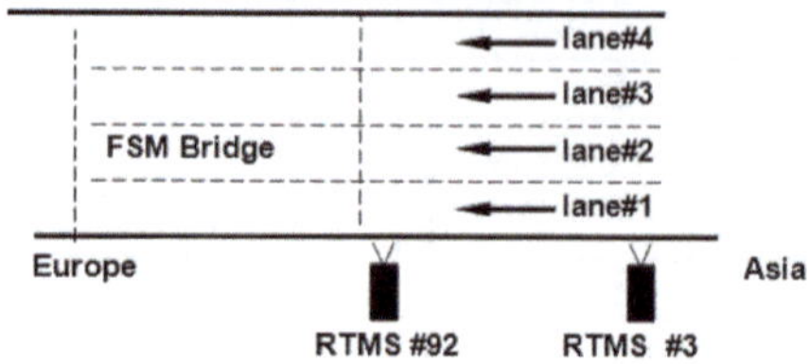

Fig. 3 Location of successive RTMS units on schematic freeway approach to strait crossing

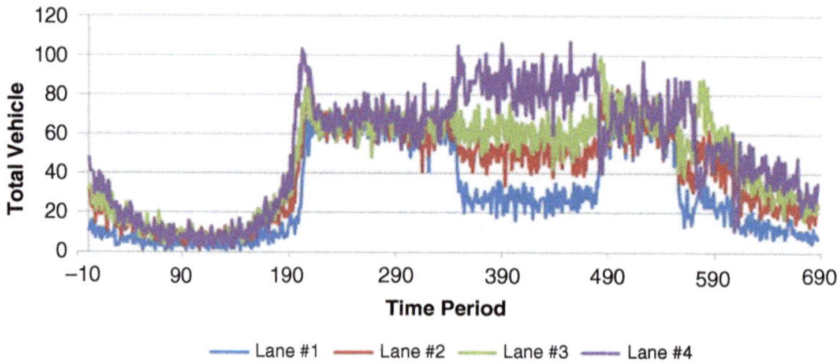

Fig. 4 Total vehicle count measures per lane within 2 min period obtained at upstream RTMS unit during 4th of February 2008

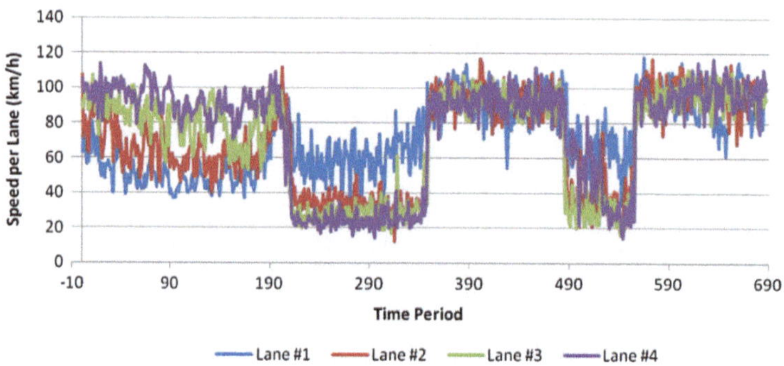

Fig. 5 Speed measures per lane within 2 min period obtained at upstream RTMS unit during 4th of February 2008

Sultan Mehmet bridge, on the European side of Istanbul (see Fig. 3). Measurements on traffic variables, i.e., long vehicle counts, n1(t), the rest vehicle counts, n2(t), speed, s(t), and occupancy, o(t), for the reference time period of 4 downstream lanes labeled consecutively from 1 to 4 where lane number 4 is adjacent to the median.

Each measurement on the variables of flow is obtained by the RTMS unit within a period of 2 min. Total vehicle count and speed measures per lane within 2 min period obtained at upstream RTMS unit during the 4th of February 2008 are respectively given in Figs. 4 and 5.

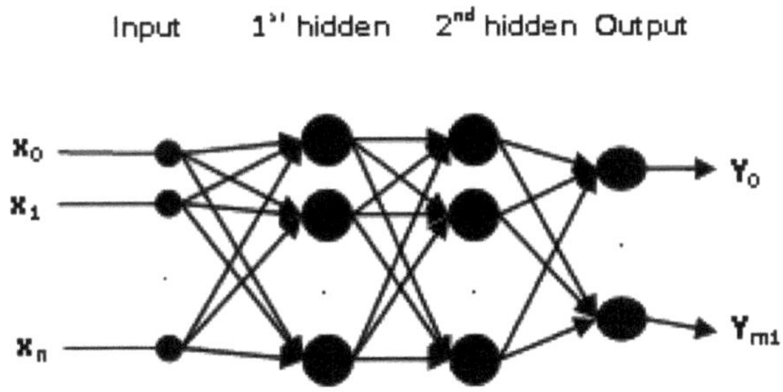

Fig. 6 Configuration of a feed-forward multi layer perceptron

3.2 *Density Derivation Sub-process*

The density derivation sub-process is formulated as a sectional density mapping problem. Solution to this problem is sought by the feed forward back propagation neural network (FFBPNN) method, that is incorporated for neural approximation. In neural network (NN) theory, the neurode (a neural node rather than a biological neuron) is the basic component and a multi layer perceptron structure is a fully interconnected set of layers of neurodes, as shown in Fig. 6. Each neurode (N_{ij}) of a layer is connected to each neurode of the next layer so that only forward transmission through the network is possible, from the input layer to the output layer through the hidden layers.

In this structure, the output Y_i of each neurode of the nth layer is defined by a derivable nonlinear function F as given by

$$Y_i = F \times \left(\sum_j w_i \times y_i \right) \tag{1}$$

where F is the nonlinear activation function, w_{ji} are the weights of the connection between the neurode N_j and N_i, y_j is the output of the neurode of the $(n-1)$th layer. For each input vector presented into the network, the set of connection weights between the different neurodes determines the answer of the network in the output layer. Partial specifications of the problem allow the measurement of output error of the network and the adjustment of its behaviour. An iterative algorithm does the adjustment during the training phase of the neural network. During training phase, a selected set of training patterns is presented to the network. When the error between the network output and the desired output is minimized, the network can be used in a testing phase with test pattern vectors. At this stage, the neural network is described by the optimal weight configuration, which means that theoretically ensures the output error minimization.

As the form of the mapping F a priori is not known, an approximation is sought. The development of NNs offers an alternative to approximate such multivariate functions and had been widely studied beginning from the late 1980s with various results, concerning density or complexity, established for different situations [14–16].

In the following sub-sections the NN methods used for approximation are summarized.

Feed-Forward Back-Propagation Neural Network (FFBPNN) Method

The back-propagation NN, which has only a finite number of parameters, can approximate most bounded functions with arbitrary precision [14] and is used here to approximate F. A multi-layer feed-forward back-propagation neural network FFBPNN($\cdot$) $\in R^m$ is a mapping constructed recursively by vector linear/nonlinear functions, shown with Eq. (2):

$$FFBPNN(\Psi, W, \theta) = \Psi[W_n \Psi(W_{n-1}\Psi(\ldots \Psi(W_1 UI + \Theta_1)) + \theta_{n-1}) + \theta_n] \quad (2)$$

Here; n is the number of layers; W_i is the linear operator of proper dimensions where i $= 1, \ldots,$ n; θ_i is the bias vector, where i $= 1, \ldots,$ n; UI is the input vector of the neural network; ψ is the vector-valued linear/nonlinear mapping (transfer function); W $= [W_1, \ldots, W_n]$; and $\theta = [\theta_{it}, \ldots, \theta_{nt}]$T (T denotes the transpose of vector/matrix). In an FFBPNN, the weight matrices $W_1, W_2, \ldots, W_n$ and bias vectors θ_1, θ_2,..., θ_n are adjusted by a learning rule called back-propagation, to minimize output error with respect to the weights and thresholds [17].

An NN is non-linear if at least some of its transfer functions are non-linear. Typical non-linear transfer functions are utilized in the literature [17].

3.3 *Flow Modelling Sub-process*

The flow modelling component accounts for obtaining the fundamental diagram of flow-density within a macroscopic approach. Macroscopic flow models, having sufficient similarities to make the hydrodynamic theory useful in describing traffic dynamics, trace the collective vehicle dynamics in terms of aggregate variables such as density, flow-rate, and speed and are useful in reproducing network-wide dynamic flow [18].

The fluid dynamic theory of continuous vehicular traffic flow, defined upon the variables of flow-rate q, density k, and speed s, and referred to as the LWR theory, in [19,20], is adopted. This theory assumes that flow is strictly a function of density, $q = f(k)$, or equivalently, speed is strictly a function of density, $s = f(k)$, and enables the fundamental conservation law of fluid dynamics to be written in terms of a continuity equation as given by

$$\frac{\delta k}{\delta t} + \frac{\delta(k \times s(k))}{\delta x} \quad (3)$$

or alternatively as

$$\frac{\delta k}{\delta t} + \left(C(k) \times \frac{\delta k}{\delta x}\right) \tag{4}$$

where $C(k) = \delta f(k)/\delta k$. The LWR theory expresses that slight fluctuations in flow are propagated upstream along kinematic waves, where the speed is given by $c = C(k)$, such as the slope of flow-density curve. Given the appropriate boundary conditions, solution to this model can be obtained by determining the function $k(x,t)$, where x and t represent space and time respectively. Different variations of the macroscopic model given by Eq. (2) can be characterized by the speed-density relationship $s = f(k)$. In the study, we adopt the linear relationship between speed and density [21] as given by

$$s = s_f \times \left(1 - \frac{k}{k_{jam}}\right), \tag{5}$$

$$\frac{\delta k}{\delta t} + \left(s_f \times \frac{\delta k}{\delta x}\right) - \left(\frac{s_f}{k_{jam}}\right) \times \frac{\delta k^2}{\delta x} = 0. \tag{6}$$

The free-flow speed and the occupancy ratio measure, k/k_{jam}, obtained by microwave sensors are input to the macroscopic flow model and the fundamental diagram of flow-density is derived considering that $q = f(s,k)$.

3.4 *Flow State Classification Sub-process*

Flow state classification process involves the segmentation of the fundamental flow-density diagram considering a user defined classification number. n flow states require $n + 1$ flow-density pairs to derive boundaries between flow states, where each state is defined in terms of a density range. This classification approach is analogous to the level of service concept defined over the flow-speed diagram proposed in HCM (2000) [22]. In each time step, dynamic segmentation on the fundamental curve is updated considering the critical value of maximum density where the overall range of density is partitioned to n equal classes to form flow states. Then, density measure mapped by the density derivation sub-process is input and classified within flow states.

4 Problem Formulation and Solution Analysis

The sectional density mapping problem is transformed into a minimum norm problem: the search for the appropriate NN configurations with each NN method to map the density variable specific to each lane existing on the freeway section bounded with the succeeding sensors that will minimize the Euclidean distance, $\| P(UI_k) - NN(UI_k) \|$, where UI_k and Y_k are the actual values vector series (obtained by the two succeeding sensors, downstream and upstream), $UI = [n1^{u1}(t), n1^{u2}(t), n1^{u3}(t), n1^{u4}(t), n2^{u1}(t), n2^{u2}(t), n2^{u3}(t), n2^{u4}(t), s^{u1}(t), s^{u2}(t), s^{u3}(t), s^{u4}(t), n1^{u1}(t), n1^{d2}(t), n1^{d3}(t), n1^{d4}(t), n2^{d1}(t), n2^{d2}(t), n2^{d3}(t), n2^{d4}(t), s^{d1}(t), s^{d2}(t), s^{d3}(t), s^{d4}(t)]$ is the input vector of NN for all four lanes, and $Y_k + 1 = P(UI_k)$.

The solution to the minimum norm problem involves a number of steps. The first one is the choice of the model inputs, and the second step is the attainment of parameters that minimize the norm given above. In order to obtain an approximation to time-varying density variable of each lane, the input variables are selected on purpose to accurately represent the flow state variation within the section bounded with succeeding sensors. Each of the long vehicle counts $n1^{ij}(t)$, the rest vehicle counts $n2^{ij}(t)$, and speed measurements $s^{ij}(t)$ is selected as an input node, where i is the sensor label, $i = u, d$ and j is the counter for lane number, $j = 1, 2, 3, 4$. Therefore, the input layer of the neural network configurations consist of 24 nodes and the 4 output layer nodes represent the density specific to each lane bounded by downstream and upstream sensors.

Considering that the success of neural approximation depends heavily on the availability of a good subset of training data, data partitioning for the NN approximations is carried out considering explicitly the error term computations in all available partitions. The iterative structure of the training process needs a threshold value to stop learning; performance criteria for varying neural network configurations require convergence to some selected error term targets.During the training stage the first half of the data set, out of 687 values were analyzed, the last 344 were then used to examine the performance of the testing phase. The optimum number of training pairs has been selected considering the minima existing after the plot of mean squared error (MSE) terms obtained by scaled training pairs. Following the training period, the networks are applied to the testing data and NN methods' performances are evaluated with the selected statistical criteria.

4.1 Neural Network Component Configuration

Each utilized NN architecture had different configurations for diverging links of both the sample node and the nodes of the sample network. The most appropriate configurations are selected by evaluating the convergence to error term target.

The Levenberg-Marquardt algorithm is utilized to minimize output error with respect to the weights and thresholds in simulations with FFBPNN integration. Back-propagation neural networks having varying numbers of hidden layers and hidden layer units with the hyperbolic tangent functions as their nonlinear transfer functions, $\Psi(x)$, are selected. The FFBP neural mappings existed as in the form given with Eq. (2). In the training process, the optimum iterations are chosen with respect to a fixed mean squared error (MSE) terms on scaled values congestion values. Optimal characteristics of FFBPNN configuration are given in Table 1.

Table 1 FFBPNN configuration charecteristics

Characteristic	Selected measure
Hidden layer number	1
Hidden layer node number	48
Iteration number for training convergence	619
Hidden layer transfer functions	Hyperbolic tangent
Output layer transfer functions	Linear

Table 2 Performance criteria obtained specific to each lane

Criteria	Lane #1	Lane #2	Lane #3	Lane #4
MARE (vkpm-lane)	0.0046	0.0034	0.0037	0.0194
MAE (vkpm-lane)	0.1684	0.0779	0.0939	0.0591
MSE (vkpm-lane)	1.2749	0.0386	0.0262	0.0221
RMSE	1.1291	0.1967	0.1619	0.1487
R^2	0.9942	0.9999	0.9999	0.9999
RMSPE (%)	1.9282	0.4632	0.4948	19.2114

4.2 Simulation Results

To represent the deviation of NN methods' mappings on lane based densities from lane based density data generated by the flow conservation principle, k(t) = (| $(n1^{uj}(t) + n2^{uj}(t)) - (n1^{dj}(t) + n2^{dj}(t))$ |)/(distance between sensors), predictions and conservation principle generated values have been used to calculate the root mean squared error (RMSE) and the coefficient of determination (R^2), shown in Table 2.

To clearly show the deviations of results from observations, the RBFNN predicted lane densities are plotted with the corresponding conservation rule generated values in Figs. 7–10, respectively for lane#1–lane#4.

5 Conclusion

In this study, an approach to flow state classification is proposed as an overall process. Classification of traffic flow states and matching the corresponding real-time flow state is obtained dynamically inputting raw flow measures simultaneously to neural density mapping and traffic flow modeling processes.

Simulation results obtained from density derivation sub-process highlight the fact that, NN mapping method is able to provide accurate density variable derivation. Within the analyses carried out, it is seen that approximating with feed-forward back propagation trained neural networks led to significantly considerable predictions which is due to neuron processing mechanism's flexibility to adapt to non-linear traffic flow relationships. NNs have a distributed processing structure in which

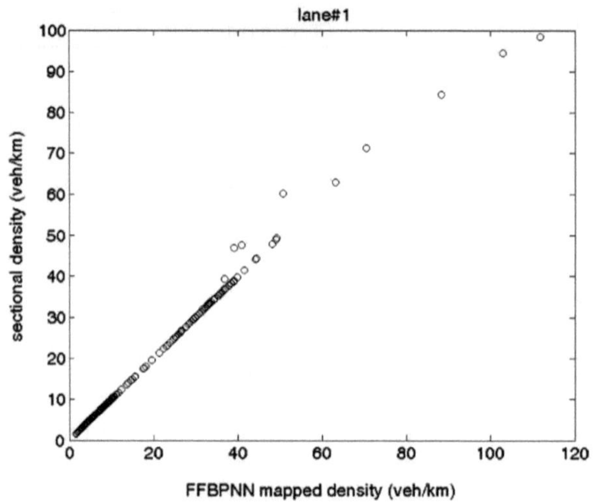

Fig. 7 FFBPNNas predicted lane densities corresponding to conservation rule generated values for lane#1

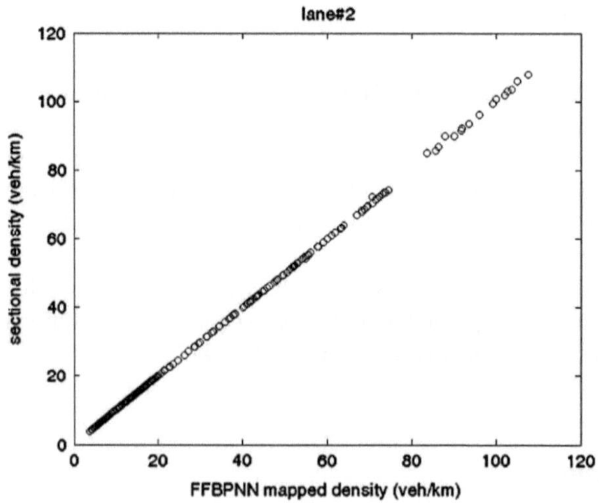

Fig. 8 FFBPNN predicted lane densities corresponding to conservation rule generated values for lane#2

each individual processing unit or the weighted connection between two units is responsible for one small part of the input-output mapping system. Therefore, each component has no more than a marginal influence with respect to the complete solution.

As a result, the neural mechanism functions and generates reasonable lane based sectional density mappings only from speed and vehicle count data. The approach

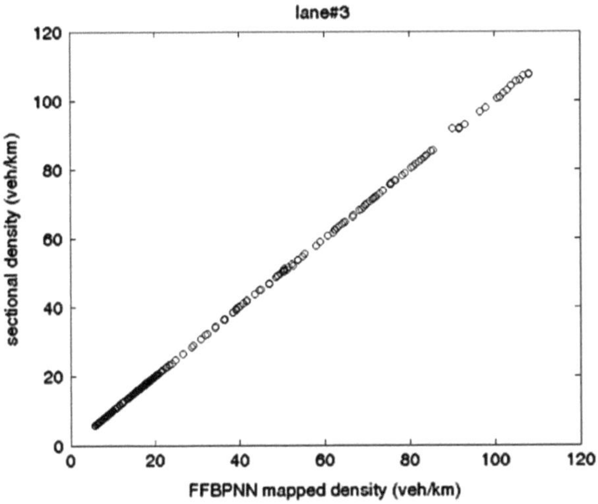

Fig. 9 FFBPNN predicted lane densities corresponding to conservation rule generated values for lane#3

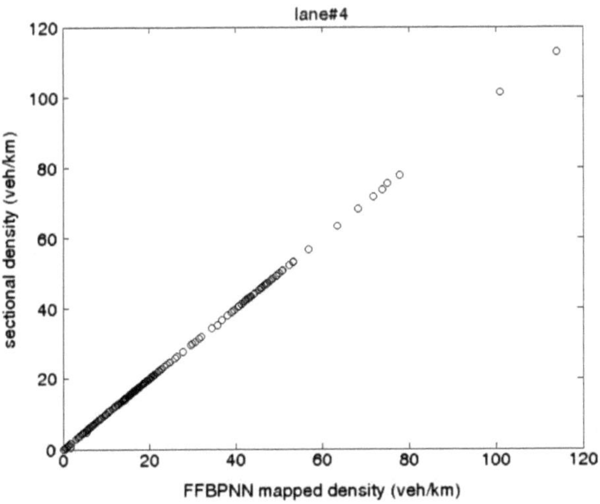

Fig. 10 FFBPNN predicted lane densities corresponding to conservation rule generated values for lane#4

is promising in capturing instantaneous changes on flow states and is open to be utilized within a series of intelligent management strategies including especially incident detection and control, mainstream flow detection for ramp control, and queue propagation analysis.

References

1. M. Zhong, P. Lingras, and S. Sharma, *Estimation of missing traffic counts using factor, genetic, neural, and regression techniques,* Transp. Res. Part C: Emerging Technologies, vol. 12, pp. 139–166, 2004.
2. L. Qu, L. Li, Y. Zhang, and J. Hu, *PPCA-based missing data imputation for traffic flow volume: a systematical approach,* IEEE Transactions on Intelligent Transportation Systems, vol. 10, pp. 512–522, 2009.
3. D.J. Dailey, *A statistical algorithm for estimating speed from single loop volume and occupancy measurements,* Transportation Research Part B: Methodological, vol. 33, no. 5, pp. 313–322, 1999.
4. J. Zhanfeng, C. Chao, C., B. Coifman, and P. Varaiya, *The PeMS algorithms for accurate, real-time estimates of g-factors and speeds from single-loop detectors,* in Proceedings of the 2001 IEEE Intelligent Transportation Systems Conference, pp. 536–541, 2001.
5. B. Coifman, *Improved velocity estimation using single loop detectors,* Transp. Res. Part A: Policy and Practice, vol. 35, pp. 863–880, 2001.
6. D. Yao, X. Gong, and Y. Zhang, *A hybrid model for speed estimation based on single-loop data,* in Proceedings of the 7th International IEEE Conference on Intelligent Transportation Systems, Washington, DC, Oct. 3–6, 2004, pp. 205–209, 2004.
7. M.L. Hazelton, *Estimating vehicle speed from traffic count and occupancy data,* Journal of Data Science, vol. 2, pp. 231–244, 2004.
8. L. Muoz, X. Sun, R. Horowitz, and L. Alvarez, *Traffic density estimation with the Cell Transmission Model,* in Proceedings of the 2003 American Control Conference, Denver, Colorado, vol. 5, pp. 3750–3755, 2003.
9. X. Sun, L. Mufioz, and R. Horowitz, *Highway traffic state estimation using improved mixture Kalman filters for effective ramp metering control,* in Proceedings of the 42nd IEEE Conference on Decision and Control, Maul, Hawaii USA, vol. 6, pp. 6333–6338, 2003.
10. B. Coifman, *Estimating density and lane inflow on a freeway segment,* Transp. Res. Part A: Policy and Practice, vol. 37, pp. 689–701, 2003.
11. US Department of Transportation, Federal Highway Administration, Traffic Detector Handbook, 3rd edition, vol. I, publication no. FHWA-HRT-06-108, 2006.
12. L.A. Klein, Sensor Technologies and Data Requirements for ITS, Artech House, Norwood, MA, 2001.
13. G. Cybenko, *Approximation by superpositions of a sigmoidal function,* Math. Control Signals Syst., vol. 2, pp. 303–314, 1989.
14. K. Hornik, M. Stinchcombe, H. White, *Multilayer feedforward networks are universal approximators,* Neural Netw., vol. 2, pp. 359–366, 1989.
15. R. C. Williamson and U. Hemke, *Existence and uniqueness results for neural network approximations,* IEEE Trans. Neural Netw., vol. 6, pp. 2–13, 1995.
16. D.E. Rumelhart, G.E. Hinton, J.E. Williams, *Learning internal representations by error propagation,* in Parallel Distrubuted Processing-Explorations in the Microstructure of Cognition, vol. 1, D.E. Rumelhart and J.L. McClelland Eds. Cambridge, Mass.: MIT Press, 1986.
17. T. Poggio and F. Girosi, *Networks for approximation and learning,* in Proc. IEEE vol. 9, pp. 1481–1497, 1990.
18. D. F. Specht, *A general regression neural network,* IEEE trans. neural netw., vol. 2, pp. 568–576, Nov. 1991.
19. H. B. Celikoglu, *A dynamic network loading model for traffic dynamics modeling,* IEEE Transactions on Intelligent Transportation Systems, vol. 8, pp. 575–583, 2007.
20. M. J. Lighthill, G. B. Whitham, *On kinematic waves II. A theory of traffic flow on long crowded roads,* in Proc. Roy. Soc. London, part A, vol. 229, pp. 317–345, 1955.
21. P. I. Richards, *Shock waves on the highway,* Operations Research, vol. 4, pp. 42–51, 1956.
22. Transportation Research Board, *Highway Capacity Manual 2000*, Washington DC, 1997.

MIX
Papier aus verantwortungsvollen Quellen
Paper from responsible sources
FSC® C105338

If you have any concerns about our products,
you can contact us on
ProductSafety@springernature.com

In case Publisher is established outside the EU,
the EU authorized representative is:
Springer Nature Customer Service Center GmbH
Europaplatz 3, 69115 Heidelberg, Germany

Printed by Libri Plureos GmbH
in Hamburg, Germany